Advanced Control Technology of Photovoltaic Power Generation Systems

Chenghui Zhang

Advanced Control Technology of Photovoltaic Power Generation Systems

For Safety, Efficiency, Reliability, and Adaptability

Chenghui Zhang
Shandong University
Jinan, China

ISBN 978-981-96-7744-3   ISBN 978-981-96-7745-0 (eBook)
https://doi.org/10.1007/978-981-96-7745-0

This work was supported by National Natural Science Foundation of China.

© The Editor(s) (if applicable) and The Author(s) 2025. This book is an open access publication.

Open Access This book is licensed under the terms of the Creative Commons Attribution 4.0 International License (http://creativecommons.org/licenses/by/4.0/), which permits use, sharing, adaptation, distribution and reproduction in any medium or format, as long as you give appropriate credit to the original author(s) and the source, provide a link to the Creative Commons license and indicate if changes were made.

The images or other third party material in this book are included in the book's Creative Commons license, unless indicated otherwise in a credit line to the material. If material is not included in the book's Creative Commons license and your intended use is not permitted by statutory regulation or exceeds the permitted use, you will need to obtain permission directly from the copyright holder.

The use of general descriptive names, registered names, trademarks, service marks, etc. in this publication does not imply, even in the absence of a specific statement, that such names are exempt from the relevant protective laws and regulations and therefore free for general use.

The publisher, the authors and the editors are safe to assume that the advice and information in this book are believed to be true and accurate at the date of publication. Neither the publisher nor the authors or the editors give a warranty, expressed or implied, with respect to the material contained herein or for any errors or omissions that may have been made. The publisher remains neutral with regard to jurisdictional claims in published maps and institutional affiliations.

This Springer imprint is published by the registered company Springer Nature Singapore Pte Ltd.
The registered company address is: 152 Beach Road, #21-01/04 Gateway East, Singapore 189721, Singapore

If disposing of this product, please recycle the paper.

Preface

Driven by the increasing global energy demand and environmental requirements, the share of renewable energy in modern power systems is steadily increasing. The photovoltaic (PV) power generation has gained significant importance in the global energy landscape due to its potential to provide clean, sustainable, and renewable energy. Both the power generation capacity and installed capacity of PV power generation system have gained rapid increase. According to the statistics from International Renewable Energy Agency, from 2016 to 2023, the total installed capacity of solar PV increased dramatically, reaching 1411 GW in 2023, with an annual addition of 347 GW.

The PV inverter serves as the interface between the PV panels and the power grid and realizes the power conversion, which is the core equipment of the PV power generation system. With the development of PV industry, the requirements of functions or performances for PV inverters are also gradually proposed in practical applications, which consist of safety, generation efficiency, transmitted power quality, robustness to multiple disturbances, grid-friendly, continuity of power supply, and system reliability. The traditional hardware-based methods not only fail to meet these requirements but also incur high costs.

Therefore, the author and his research group have carried out long-term teaching and research, combined with practical experience, and summed up and put forward a series of software-based advanced control technologies. Through these control technologies, the PV power generation system has gradually become a system with high safety, high reliability, high efficiency, and strong adaptability, which serves as a core support in modern power system. To facilitate the understanding, the operating principle, model derivation, and control schemes for PV inverters are presented in this book. It provides step-by-step model derivation and controller design for PV inverter, so that readers can learn and understand quickly. Meanwhile, it provides a comprehensive simulation and experimental results to offer readers a deep insight into the control process of PV inverters. This book serves as a guide for electrical engineers and researchers involved in the development of PV power generation.

This book contains 8 chapters. The contents of each chapter are briefly introduced as follows.

Chapter 1 contains the introduction to this book. It first gives an overview of the development of PV power generation industry. The basics of the PV power generation system are briefly reviewed. The classification of the PV inverters as well as its requirements of functions and performances are introduced. The key factors that influence the functions and performances of inverters are analyzed, which include PV panels, main circuits of PV inverters, nonideal power grids and nonlinear load. Finally, the contents of this book are presented.

Chapter 2 gives the fundamentals of PV inverters, which contains mathematical models, control framework, and modulation methods. First, the mathematical model of the PV inverter is established, which is the basis for designing the system controller. Then, the control framework is illustrated in detail, which consists of phase locked loop (PLL), outer loop for power control, middle loop for DC voltage control, inner loop for AC current control, and auxiliary loop for other objectives. At last, two typical pulse width modulation (PWM) schemes are presented, which serves as the interface between the controller and the main circuit.

To satisfy the requirements of functions and performances of PV inverter, a series of advanced control technologies are presented in Chaps. 3–8, which are elaborated based on the control framework in Chap. 2.

Chapter 3 is concerned with the control technology of PV generation systems for leakage current suppression. At first, the modeling is conducted to reveal the generation mechanism of leakage current. Based on these theoretical analyses, the leakage current suppression technology is elaborated, which includes common-mode circuit (CM) modification methods and exciting source suppression methods. In the former type of methods, the closed-loop control scheme is proposed to further reduce the leakage current the resonance current, which integrated into auxiliary loop, while the latter type of methods is integrated into PWM modulator.

Chapter 4 focuses on the control technology of PV generation systems for maximizing power generation. The conventional MPPT technology and separate MPPT technology are elaborated, respectively. First, the conventional MPPT methods are reviewed for generalized PV applications. Then, several separate MPPT methods are presented to improve total efficiency of MPPTs under the partial shading and panel mismatch conditions in the large-scale PV applications. The essence of MPPT technology is to match the equivalent impedance of PV inverter by adjusting the output impedance of the PV array, which is integrated into outer loop for power control.

Chapter 5 is related to the advanced control technology of PV inverters for high robustness and performance. First, to improve the anti-interference capability of the PV inverter, the nonlinear control technology is presented and integrated into middle loop and inner loop, so as to regulate the DC-link voltage and grid currents, respectively. Then, to improve the power quality and system safety, the control technologies for AC-side current harmonic and ripple suppression associated with DC-side neutral-point voltage (NPV) oscillation suppression are proposed and integrated into auxiliary loop and PWM modulator.

Chapter 6 discusses the advanced control technology of PV inverters under grid faults. At first, the characteristics of grid voltages and grid standard requirements for PV inverter are analyzed and illustrated. Then, a series of advanced control technologies are subsequently presented to ensure the stable operation of the PV inverter and satisfy the grid standard requirements. To maintain the good quality of transmitted power by the PV inverter, the advanced control technology for DC voltage and grid current control is given, which is integrated into middle loop and inner loop. In order to realize the safe and stable operation of three-level inverter, the advanced control technology for NPV balance is proposed, which is integrated into PWM modulator and auxiliary loop.

Chapter 7 gives the control technology of PV inverters for multi-functional operation. At first, the multi-functional modes are reviewed, which include power quality control mode and islanded mode. Then, the control schemes for power quality control mode and islanded mode are presented, respectively. The control scheme for the power quality control mode includes harmonics and reactive current detection, reference current synthesis, and reference current tracking control. The nonlinear control methods for islanded mode operation are presented and embedded into the single-loop voltage control scheme. When the PV inverter operates in the power quality control mode at night and on cloudy days, it has the capability of eliminating the current harmonic and compensating the reactive power. When the PV inverter operates in islanded mode, the continuous power supply can be maintained during grid disturbances and faults.

Chapter 8 focuses on the fault-tolerant control technology of PV inverter for reliability improvement. First, the open-circuit fault diagnosis methods are introduced, which is the base of the fault-tolerant control. Then, three fault-tolerant control strategies with switching loss reduction and NPV oscillation suppression are elaborated, which are integrated into the auxiliary loop and the PWM modulator. The essence of the fault-tolerant control is to avoid using the current paths affected by the faulty power semiconductor switches, so as to ensure the correct AC output current. With the proposed technology in this chapter, the reliability of PV inverters is improved.

Jinan, China   Chenghui Zhang

Acknowledgements This research monograph comprehensively elaborates and summarizes the research on the advanced control techniques for photovoltaic (PV) power generation systems, that has been developed through long-term teaching and research with practical experience performed by the author and his research group. The author hopes that it could be helpful for people involved in the relevant research.

For the completion of this monograph, the author would like to acknowledge the supports of the Innovation Research Group of Natural Science Foundation of China (NSFC) (No. 61821004), Key Program of the NSFC (No. 62333013), National Key Research and Development Program Project (No. 2022YFF0712700), Key Program of Automobile Joint Fund of NSFC (No. U1964207), the National and Local Joint Engineering Research Center of "Renewable Energy and Efficient Energy Conversion Technology", and the Ministry of Education Engineering Research Center of "Power Electronics, Energy Saving Technology and Equipment". It is gratifying that the relevant research works are received awards of the Second Prize of the National Science and Technology Progress Award of the PRC in 2016, the "Key Control Technology and Industrialization of High-Performance Photovoltaic Power Generation System", one of the highest awards given by the State Council of the PRC to recognize citizens and organizations who have made remarkable contributions to scientific and technological progress. This award is the first one of the National Science and Technology Progress Award in the field of PV power generation.

It takes the coordinated efforts of many people to get the final edition off the ground. Special thanks are due to my students, Prof. Xiaoyan Li, Prof. Changwei Qin, Zicheng Zhang, Mengmeng Jing, and Zuohang Hu, who made contributions to extensive proofreading, typesetting, and text finishing work, Prof. Xiaoyan Li, Prof. Changwei Qin, Dr. Tong Liu, Dr. Cheng Fu, Dr. Jie Chen, and Dr. Ying Jiang, who provide me the research efficient materials used in this book, as well as Rui Zhang, Chang Liu, and Zhaowen Zhong, who draw the necessary and creative figures of this book. Meanwhile, I'd also like to express my gratitude to my colleagues Prof. Alian Chen and Prof. Xiangyang Xing in my research team for their helpful materials adopted in this book. The support of the team has been instrumental in making this book what it is. In particular, the author would like to express the sincere appreciation to Prof. Jingyang Fang for his professionalism and wisdom in shaping the structure of this book, which makes the philosophy and approach of the book clearer and easier for the reader to understand.

The author would like to express the deep gratitude to the scholars in this field, in particular, Prof. An Luo at Hunan University, Prof. Jinjun Liu at Xi'an Jiaotong University, Prof. Xing Zhang at Hefei University of Technology, and Prof. Guojun Tan at China University of Mining and Technology, et al. The author is thankful to the cooperative enterprises, which include Shandong Aotai Electric Co. Ltd., Sungrow Power Supply Co. Ltd., China Energy Conservation and Environmental Protection Group (CECEP) Solar Energy Co. Ltd., Tebian Electric Apparatus (TBEA) Co. Ltd., WindSun Science and Technology Co. Ltd., Shandong Taikai Power Electronic Co. Ltd., et al. These enterprises provide the industrial and practical basics for the relevant technology in this book.

The author would like to sincerely thank all the colleagues from Springer, Science Press, China, for their invitation, encouragement, recommendations, and comments. The support and help of Mr. Wayne Hu (the project editor) are greatly appreciated.

Competing Interests The author has no competing interests to declare that are relevant to the content of this manuscript.

Contents

1	**Introduction**	1
	1.1 Photovoltaic Power Generation Industry	1
	1.2 Photovoltaic Power Generation Systems	10
	1.3 Photovoltaic Inverters	14
	1.3.1 Classification of Photovoltaic Inverters	14
	1.3.2 Requirements of Functions and Performances for Inverters	20
	1.4 Key Factors of Functions and Performances for Inverters	25
	1.4.1 Photovoltaic Panels	25
	1.4.2 Main Circuit of Photovoltaic Inverters	27
	1.4.3 Nonideal Power Grids	29
	1.4.4 Nonlinear Loads	29
	1.5 Research Content of This Book	30
	References	31
2	**Fundamentals of Photovoltaic Inverters**	33
	2.1 Working Principles and Mathematical Models	34
	2.1.1 Inverter Structure and Operating Principle	34
	2.1.2 AC-Side Modeling	38
	2.1.3 DC-Side Modeling	43
	2.2 Control Framework	45
	2.2.1 Phase Locked Loop	48
	2.2.2 Outer Loop for Power Control	50
	2.2.3 Middle Loop for DC Voltage Control	52
	2.2.4 Inner Loop for AC Current Control	53
	2.2.5 Auxiliary Loop for Other Objectives	56
	2.3 Pulse Width Modulation Schemes	58
	2.3.1 Continuous Pulse Width Modulation	59
	2.3.2 Discontinuous Pulse Width Modulation	61
	2.4 Summary	64
	References	64

3 Control Technology of Photovoltaic Power Generation Systems for Leakage Current Suppression ... 67
- 3.1 Mechanism Analysis and Modeling ... 69
- 3.2 Leakage Current Suppression Technology ... 71
  - 3.2.1 Common-Mode Circuit Modification Methods ... 71
  - 3.2.2 Exciting Source Suppression Methods ... 90
- 3.3 Experimental Verifications ... 119
  - 3.3.1 Experimental Verifications for Common-Mode Circuit Modification Methods ... 119
  - 3.3.2 Experimental Verifications for Exciting Source Suppression Methods ... 127
- 3.4 Summary ... 142
- References ... 143

4 Control Technology of Photovoltaic Generation Systems for Maximizing Power Generation ... 145
- 4.1 Conventional Maximum Power Point Tracking (MPPT) Technology ... 147
  - 4.1.1 Perturb and Observe Algorithm ... 150
  - 4.1.2 Incremental Conductance Algorithm ... 151
  - 4.1.3 Fractional Open-Circuit Voltage Method ... 153
  - 4.1.4 Fractional Short-Circuit Current Method ... 154
  - 4.1.5 Advanced MPPT Methods ... 154
- 4.2 Separate Maximum Power Point Tracking Technology ... 155
  - 4.2.1 Based on Continuous Modulation Methods ... 161
  - 4.2.2 Based on Discontinuous Modulation Method ... 181
  - 4.2.3 Based on Fault-Tolerant Control Method ... 183
- 4.3 Simulation and Experimental Validations ... 194
  - 4.3.1 Simulation Results and Discussions ... 194
  - 4.3.2 Experimental Results and Discussions ... 199
- 4.4 Summary ... 212
- References ... 213

5 Advanced Technology of Photovoltaic Inverters for High Robustness and Performance ... 215
- 5.1 Nonlinear Control Technology for High Robustness ... 217
  - 5.1.1 Improved Finite-Time Control Method ... 218
  - 5.1.2 Disturbance Observer-Based Finite-Time Control Method ... 224
- 5.2 Control Technology for AC-Side Current Quality Improvement ... 230
  - 5.2.1 AC-Side Current Harmonic Elimination Method ... 231
  - 5.2.2 AC-Side Current Ripple Suppression Methods ... 238

5.3	\multicolumn{2}{l	}{Control Technology for DC-Side Neutral-Point Voltage Oscillation Suppression}	265

5.3 Control Technology for DC-Side Neutral-Point Voltage
    Oscillation Suppression 265
    5.3.1 The Generation Mechanism of Neutral-Point Voltage
          Oscillation 269
    5.3.2 Neutral-Point Voltage Oscillation Suppression
          Methods 270
5.4 Summary 300
References 300

6 Advanced Control Technology of Photovoltaic Inverters Under Grid Faults ... 303
6.1 Overview of Grid Faults 305
6.2 Advanced Control Technology for DC Voltage and Grid
    Current of PV Inverters Under Grid Faults 309
    6.2.1 Dual-Loop Voltage and Current Control Schemes 309
    6.2.2 Dual-Loop Voltage and Power Control Schemes 337
6.3 Advanced Control Technology for Neutral-Point Voltage
    Balance Under Grid Faults 354
    6.3.1 Based on Model Predictive Control 355
    6.3.2 Based on Selective Harmonics Elimination Pulse
          Width Modulation 360
    6.3.3 Based on Common-Mode Voltage Reduction
          Modulation 370
6.4 Summary 389
References 389

7 Control Technology of Photovoltaic Inverters for Multi-functional Operation ... 393
7.1 Overview of Multi-functional Modes 393
7.2 Photovoltaic Inverters in the Power Quality Control Mode .. 396
    7.2.1 Principle of Power Quality Control 396
    7.2.2 Control Schemes for Power Quality Control 399
7.3 Photovoltaic Inverters in the Islanded Mode 436
    7.3.1 Principle of the Islanded Mode 436
    7.3.2 Nonlinear Control Technology in the Islanded Mode ... 438
    7.3.3 Experimental Validations for the Islanded Mode
          Operation 451
7.4 Summary 459
References 460

8 Fault-Tolerant Control Technology of Photovoltaic Inverter for Reliability Improvement ... 463
8.1 Fault Diagnosis Methods 465
    8.1.1 Theory Analysis and Implementation Process 466

		8.1.2 Diagnosis Results for Different Faults of Power Semiconductor Switches	468

8.2 Fault-Tolerant Control Technology 472
    8.2.1 Based on Space Vector Pulse Width Modulation 472
    8.2.2 Based on Continuous Carrier-Based Pulse Width Modulation 488
    8.2.3 Based on Discontinues Carrier-Based Pulse Width Modulation 507
8.3 Summary 526
References 527

About the Author

Chenghui Zhang received his B.S. and M.S. in automation from Shandong University of Technology, Jinan, China, in 1985 and 1988, respectively. He received Ph.D. in operational research and cybernetics from Shandong University, Jinan, China, in 2001.

In 1988, he joined the Shandong University of Technology, where he became a professor in 1998. From July 2000, he joined the Shandong University as a professor, and he is currently a deputy director of the academic committee of Shandong University and also a chair professor and the dean of the School of Control Science and Engineering of Shandong University. He is the founder and the director of the National and Local Joint Engineering Research Center of "Renewable Energy and Efficient Energy Conversion Technology" and the Ministry of Education Engineering Research Center of "Power Electronics, Energy Saving Technology and Equipment".

Professor Zhang is an elected IEEE fellow for his outstanding contributions to the field of renewable energy system control in 2022. He is a fellow of Chinese Association of Automation (CAA) and a Changjiang Scholar distinguished professor awarded by Ministry of Education. Changjiang Scholars Programme is jointly launched and implemented by the Ministry of Education of China and Li Ka Shing Foundation in 1998, which is awarded to the leading talents of disciplines with international influence. Besides, he is the chair of the Technical Committee on the Control of Renewable Energy and Energy Storage Systems within the Asian Control Association, an executive director of CAA, the chairman of the professional committee on control of new energy and energy storage systems of CAA, the deputy chair of Electrical Automation Professional Committee of CAA, the deputy chair of China Power Electronics Society, an executive director of China Electrotechnical Society, and a member of Teaching Committee of Higher Education of Ministry of Education. Professor Zhang has co-authored more than 200 peer-reviewed publications and has a Google Scholar h-index of 71 and i10-index of 323. He was also ranked in the Top 2% career-long and single-year impact scientists worldwide in the Stanford University's Most-Cited Scientists in 2020–2024.

Professor Zhang was awarded the Second Prize of National Science and Technology Progress Award of PRC in 2016, 2020, and 2023, respectively, for his significant contributions in advanced control technology of high-performance photovoltaic power generation system, electric power purification technology in renewable energy power system, and digital-intelligent testing and control technology for large-capacity battery energy storage systems. The National Science and Technology Progress Award is one of the highest honors conferred by the State Council of PRC to recognize citizens and organizations who have made remarkable contributions to scientific and technological progress. He was appointed as the leader of Innovation Research Group of Natural Science Foundation of China (NSFC) in 2018, which is currently the most academically influential and competitive talent team programmer in China and positioned to the top leading innovative research team in a specific frontier area. He won the Ho Leung Ho Lee Science and Technology Progress Award in 2021, which is the highest social reward and awarded by Ho Leung Ho Lee Fund Committee for long-standing eminence in science and the practice of engineering. He also won the Guanghua Engineering Technology Award, the Highest Science and Technology Award of Shandong Province, the National Award for Excellence in Innovation, and the National Advanced Worker, for the outstanding contributions in the field of new energy system control.

Chapter 1
Introduction

Driven by the increasing global energy demand and environmental requirements, the share of renewable energies in modern power systems is steadily increasing. Among various types of renewable energy generation, photovoltaic (PV) power generation possesses the advantages of high flexibility and broad applicability, making it widely adopted worldwide. It is one of the fastest-growing and most promising forms of renewable energy generation. In this chapter, the development of the PV power generation industry is first introduced. Then, the composition of PV systems and typical classifications of PV inverters are elaborated. The requirements of functions and performances for PV inverters are also discussed from the perspective of users' demands and grid standard requests. Thereafter, key factors that influence the performances and functions of PV inverter are systematically summarized. At the end of this chapter, the research contents of this book are given according to the above analyses.

1.1 Photovoltaic Power Generation Industry

Currently, energy shortages and environmental pollution have become major challenges that threaten human society. The gradual depletion of fossil fuels has posed severe challenges to energy supply and constrained sustainable economic development. The widespread use of fossil fuels has led to a continuous increase in carbon dioxide emissions, triggering a series of crises such as climate change and ecological degradation. To address these challenges, many countries have gradually reached a consensus that energy systems must transition toward low-carbon and high-efficiency solutions. In December 2015, 178 Parties worldwide signed the Paris Agreement, which established a global framework for climate action and proposed controlling global temperature rise by reducing greenhouse gas emissions [1]. Subsequently, many countries have enacted relevant laws and policies, leveraging technological innovation and structural optimization to tackle energy and environmental issues. In response to these challenges, the Chinese government further proposed the goals of carbon emission peaking and carbon neutrality ("dual carbon goals") in September 2020, aiming to achieve carbon peaking by 2030 and carbon neutrality by 2060 [2].

Renewable energy generation is regarded as a potential solution to addressing energy and environmental challenges. Renewable energy refers to naturally replenished energy sources that do not diminish with human consumption. Generally, it includes solar energy, hydropower, wind energy, geothermal energy, ocean energy, biomass energy, and tidal energy. Thanks to their inexhaustible and sustainable nature, these energy sources have become vital supports for the global energy transition. Figure 1.1 shows the configuration of the modern power system. As depicted, the entire system is divided into three parts: power generation, transmission, and consumption. In recent years, the installed capacity of renewable energy generation has grown rapidly. As a result, the proportion of renewable energy generation has been increasing year by year, reducing dependence on traditional fossil fuels and alleviating resource and environmental pressures.

According to the International Renewable Energy Agency (IRENA), the global proportion of installed capacity by energy type in 2012 and 2023 is shown in Fig. 1.2 [3]. In 2012, fossil fuel power generation dominated the global energy mix. Renewable energy generation was primarily reliant on hydropower, with wind and PV power accounting for only a small fraction. In recent years, driven by technological advancements and strong policy support, the global development and utilization of renewable energy have expanded significantly, accompanied by a rapid decline in application costs. By 2023, the total installed capacity of renewable energy generation worldwide had surpassed 3800 GW, making up nearly 50% of the global total installed power generation capacity. Notably, wind and PV power have experienced remarkable growth, with their shares of installed capacity rising from 4.8% and 1.9%

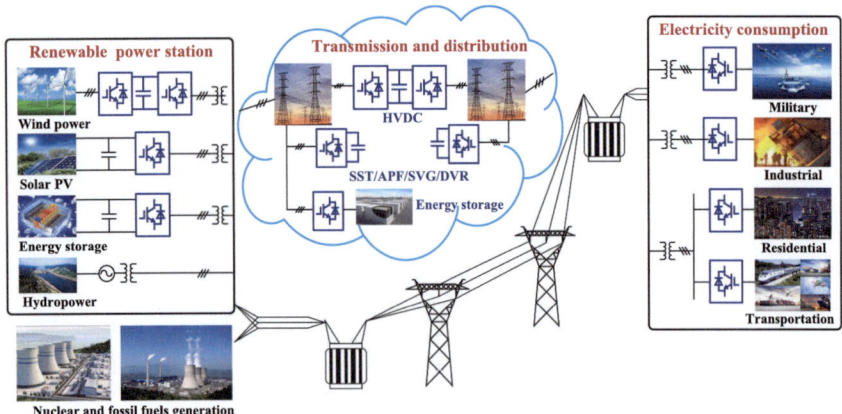

Fig. 1.1 Configuration of the modern power system

1.1 Photovoltaic Power Generation Industry

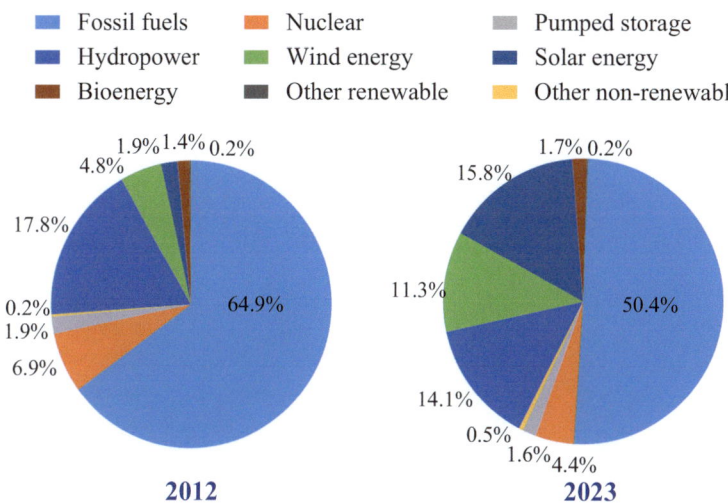

Fig. 1.2 Global proportion of installed capacity by energy type in 2012 and 2023. Reprinted from Ref. [3]

in 2012 to 11.3% and 15.8% in 2023, respectively, earning them the title of the "twin stars" of renewable energy generation.

In recent years, China has made remarkable achievements in the development of renewable energy, becoming a key player in the global transition to clean energy. Owing to abundant natural resources and strong government support, China has significantly expanded its installed renewable energy capacity, particularly in the fields of solar and wind energy. To promote renewable energy development, China enacted the Renewable Energy Law in 2006, providing a legal foundation for the sector's growth. In its 14th Five-Year Plan, the country set a clear target: by 2025, non-fossil energy should account for approximately 20% of primary energy consumption. Additionally, the "carbon peak" and "carbon neutrality" goals proposed by the Chinese government in 2020 have offered a clear direction and strong momentum for renewable energy development.

Figure 1.3 shows the proportion of installed capacity of China by energy type in 2012 and 2023. In 2012, fossil fuels dominated the energy mix, accounting for 70.8% of the total capacity, followed by hydropower at 19.9%. Wind energy and solar power contributed relatively small shares of 5.3% and 0.6%, respectively. By 2023, the energy mix shifted significantly. Fossil fuels saw a major decrease in their share to 46.1%, while hydropower also declined slightly to 15.1%. Wind energy and solar energy experienced notable growth, rising to 20.9% and 12.7%, respectively. This comparison highlights a significant transition in China's energy strategy, with a substantial move toward renewable energy sources, particularly wind and solar energy, while reducing reliance on fossil fuels. In addition, according to IRENA, China's total installed renewable energy capacity had exceeded 1400 GW by 2023,

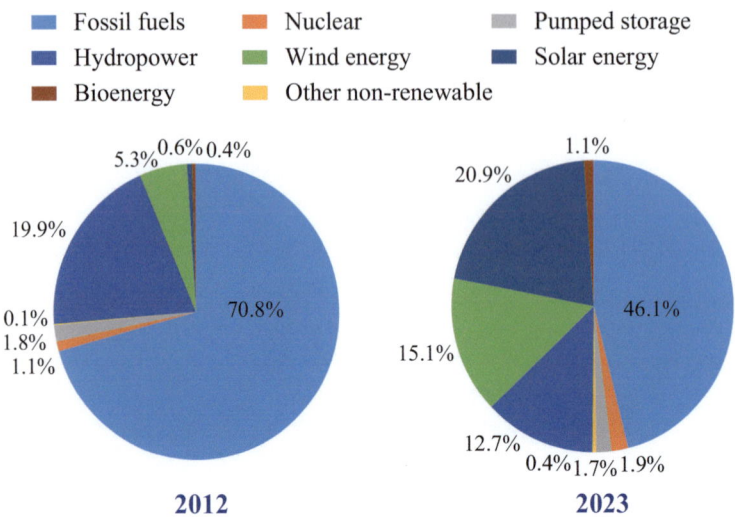

Fig. 1.3 Proportion of installed capacity by energy type in China in 2012 and 2023. Reprinted from Ref. [3]

with solar and wind energy collectively surpassing 800 GW, making it the largest contributor to global renewable energy capacity.

As previously mentioned, PV power generation and wind power generation have developed rapidly in recent years. In the field of renewable energy generation, PV power and wind power have become two critical pillar technologies and occupy a central position in the global energy transition. Wind power generation, as a mature renewable energy technology, can be categorized into onshore and offshore wind power. Onshore wind power was developed earlier, with mature technology and lower costs, while offshore wind power has shown rapid growth in recent years due to its abundant resources and stable wind conditions. Compared to wind power, PV power generation offers greater flexibility and applicability. Benefiting from advancements in related technologies, the efficiency of PV cells has significantly improved, while the installation costs of PV systems have rapidly decreased. These advantages have contributed to its remarkable growth rate. PV power generation has become another "star" in the field of renewable energy.

Figure 1.4 shows the global capacities and annual additions of solar PV and wind energy from 2016 to 2023. As seen, both energy sources have experienced significant growth, reflecting their critical roles in the transition to renewable energy. Notably, PV power generation has shown a faster development rate compared to wind power generation. From 2016 to 2023, the total installed capacity of solar PV increased dramatically, reaching 1411 GW in 2023, with an annual addition of 347 GW. In contrast, wind energy reached a total capacity of 1017 GW in 2023, with an annual addition of 115 GW. While both energy sources are expanding, solar PV

1.1 Photovoltaic Power Generation Industry

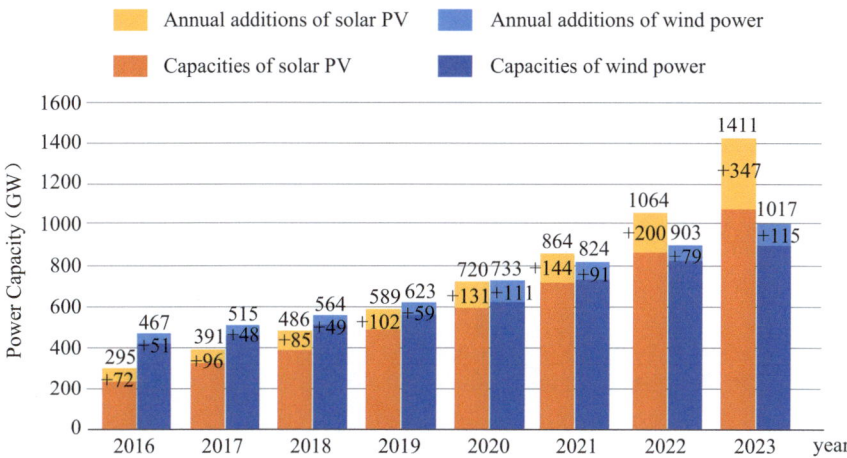

Fig. 1.4 Global capacities and annual additions of solar PV and wind energy from 2016 to 2023. Reprinted from Ref. [3]

has consistently outpaced wind energy in terms of both total capacity and annual growth, particularly since 2020.

Figure 1.5 shows the capacities and annual additions of solar PV and wind energy in China from 2016 to 2023. Similar to the global development of PV and wind power, wind power generation had an earlier start, with a larger capacity of 148 GW in 2016 compared to 78 GW for solar PV. However, between 2016 and 2019, solar PV experienced faster growth in China. During this period, solar PV's annual additions consistently outpaced wind power. By 2019, solar PV capacity had reached 205 GW, narrowing the gap with wind power's 209 GW. In 2020, wind power saw a surge in growth, adding 72 GW compared to solar PV's 49 GW, temporarily surpassing solar in annual additions. However, starting in 2021, solar PV regained its momentum, with steadily increasing annual additions. By 2023, solar PV entered a period of explosive growth, adding a record-breaking 217 GW in a single year, pushing its total capacity to 609 GW. In contrast, wind power added 76 GW in 2023, bringing its total capacity to 441 GW. This trend demonstrates the rapid acceleration of solar PV as a key driver of China's energy transition, fueled by its decreasing costs, technological advancements, and widespread applicability. While wind power remains an essential part of China's renewable energy mix, its growth has been steadier compared to the dramatic rise of solar PV, particularly in recent years. The explosive growth of solar PV in 2023 underscores its increasing dominance in China's renewable energy landscape.

In conclusion, although both PV power generation and wind power generation have experienced rapid growth over the past decade, PV power generation has gradually become one of the core driving forces of energy transition due to its widespread availability and strong flexibility. PV cells generate electricity through the PV effect. When sunlight hits the semiconductor material (usually silicon), it excites electrons,

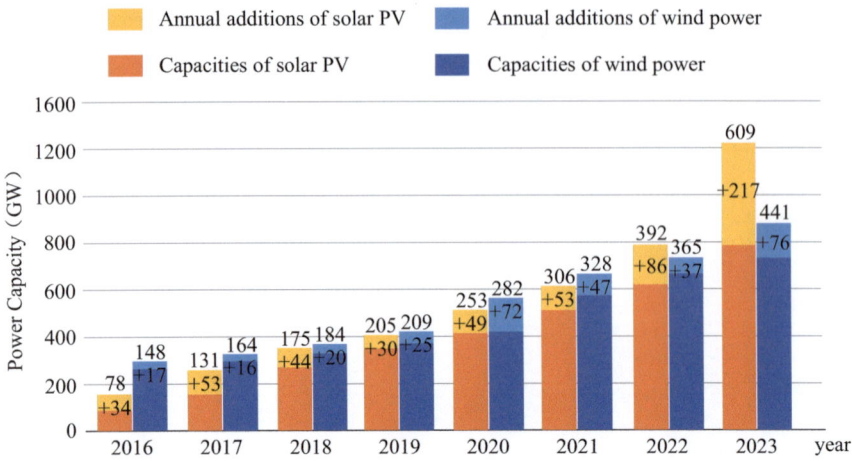

Fig. 1.5 Capacities and annual additions of solar PV and wind energy in China from 2016 to 2023. Reprinted from Ref. [3]

freeing them to create an electric current. With continuous breakthroughs in the efficiency of monocrystalline silicon cells, the advantages of solar PV are steadily strengthening. China, the United States, India, and Europe are currently the major markets for solar PV power generation, with China emerging as the global leader in the solar PV industry.

The rapid development of China's PV industry is benefited from the policy support of the state. Figure 1.6 lists the main policies of PV industry, which are illustrated as follows.

1. Since the enactment of the Renewable Energy Law in 2006, China has introduced a series of policies to promote the development of PV power generation and facilitate the transition of its energy structure [4].
2. In 2009, the Golden Sun Demonstration Project was launched, providing substantial subsidies to eligible PV projects. This initiative aimed to cultivate the domestic market and promote the application of distributed PV systems.

Fig. 1.6 The main policy of China for the PV industry

1.1 Photovoltaic Power Generation Industry

3. In 2011, the Renewable Energy Electricity Price Subsidy Policy implemented the feed-in tariff mechanism, offering fixed electricity price subsidies for grid-connected PV projects. This effectively addressed the high initial costs of PV power generation and attracted significant private investment into the industry.
4. In 2016, the PV Front-runner Program was initiated, focusing on technological advancements and efficiency improvements. This program drives the industry toward high-quality growth and lays the foundation for reducing PV power generation costs. In parallel with the development of the PV industry, China also integrated PV power with poverty alleviation efforts. In March 2016, five ministries jointly issued the Opinions on Implementing PV Power Generation Poverty Alleviation Work, which aimed to fully scale up PV poverty alleviation as a key measure across the country. This initiative utilized PV installations to provide sustainable income sources for impoverished regions.
5. To avoid the disorderly expansion of the PV industry, the "531 Policy" (Notice on Relevant Matters of PV Power Generation in 2018) was introduced. Although it caused a short-term shock to the industry, it also spurred enterprises to innovate and reduce costs.
6. In 2019, the introduction of the Grid Parity Policy marked that China's PV power generation officially entered the "post-subsidy era". The policy required new PV projects to achieve grid parity or operate below the cost of coal-fired power, driving the marketization of PV power generation.
7. In 2021, the State Council of China issued the "Action Plan for Carbon Peaking Before 2030" to further advance the carbon peaking goal. The plan emphasized the necessity to comprehensively promote large-scale development and high-quality growth of solar power generation. It advocated for a dual approach of centralized and distributed solar power deployment, while accelerating the construction of PV power generation bases.
8. In 2022, the County-wide Promotion policy was fully implemented to promote the construction of rural rooftop PV and the development of distributed PV.
9. In 2023, the Promoting High-Quality New Energy Development Plan was outlined to foster the high-quality development of new energy, ensuring sustainable growth and energy security.

Supported by those policies, China has become a leader in the PV power generation field. Figure 1.7 shows the capacities and China's percentage of solar PV from 2016 to 2023. As shown, most listed countries, such as the USA, Germany, Japan, and India, show relatively steady growth in PV installations. In contrast, China's capacity has grown at an exceptionally rapid pace. In 2016, China's share of global PV capacity was 26%. At least by this year, China had already become the country with the largest PV installed capacity in the world. By 2023, China's PV installed capacity accounted for 43%, nearly half of the global installed capacity.

The generalized structure of PV power generation system is shown in Fig. 1.8, which contains three parts: direct current (DC) input side, power conversion unit, and alternating current (AC) output side. In DC side, the PV modules are the main component, which are responsible for converting sunlight into DC electricity through

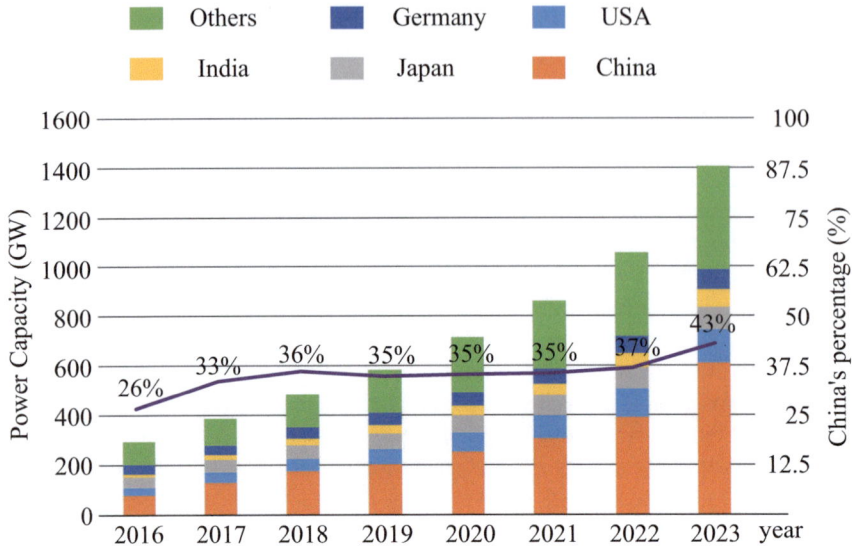

Fig. 1.7 Capacities and China's percentage of solar PV from 2016 to 2023. Reprinted from Ref. [3]

the PV effect. In the power conversion unit, the PV inverter is the core, which is mainly composed of passive components of capacitors and inductors, and power semiconductor switches. The passive components form the filter, and the power semiconductor switches form the main circuit. The capacitors are used to filter the voltage harmonics and obtain the stable voltage. The inductors are used to filter the current harmonics to improve current quality. The power semiconductor switches constitute the main circuit to convert power energy form and control power energy flowing direction by reasonable topology and control algorithm. Specifically, the PV inverters serve as the interfaces between PV modules and the AC power grid or loads. They convert the DC electricity into AC electricity to feed into the power grid or supply power to loads. The advanced control technologies are embedded into the PV inverter controller to meet the requirements of functions and performances. PV inverters also incorporate advanced features such as running status monitoring, energy management and scheduling, which are accomplished in the monitoring or supervisory system. The AC output side includes the power grid or AC loads.

In the PV generation system, the two most critical components are PV modules and PV inverters, both of which are indispensable for solar power generation. As for now, China has been the global leader in both the PV module and PV inverter industries.

In the field of PV modules, China dominates the global market. Chinese PV manufacturers, such as JinkoSolar, Trina Solar, and LONGi, consistently rank among the top suppliers globally. According to the company's annual report, JinkoSolar's PV module shipments reached 75 GW in 2023, ranking first in the world for the year.

1.1 Photovoltaic Power Generation Industry

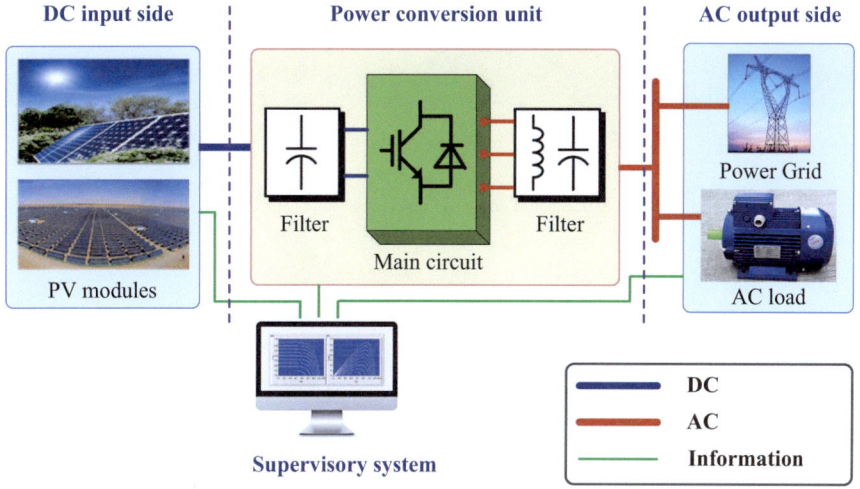

Fig. 1.8 Generalized structure of PV power generation system

According to statistics from Statista, the global PV module production exceeded 600 GW in 2023, with Chinese companies accounting for an astonishing 84.6% of the total production. The share of PV module production by country is shown in Fig. 1.9, with the relevant data sourced from Statista's report [5].

Similarly, China holds a dominant position in the global PV inverter market, with its manufacturers capturing a substantial share of both domestic and international markets. Chinese inverter companies, such as Huawei, Sungrow, and Growatt, are recognized for their cutting-edge technologies, including high-efficiency inverters, smart grid integration, and advanced energy management systems. According to statistics from Wood Mackenzie and Statista's report, the global PV inverter market share ranking by shipment in 2022 is plotted in Fig. 1.10 [6, 7]. As shown in

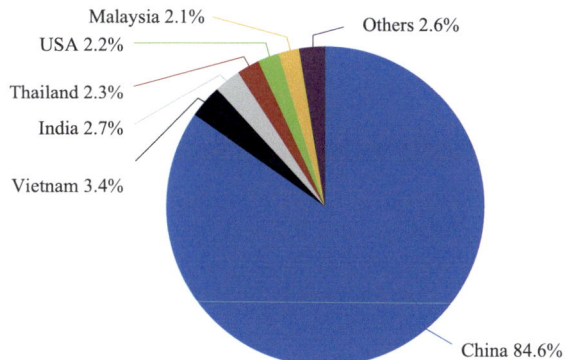

Fig. 1.9 Distribution of solar PV module production worldwide in 2023. Reprinted from Ref. [5]

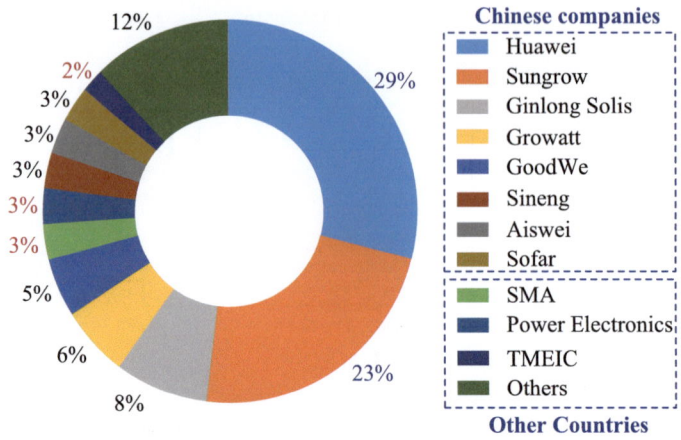

Fig. 1.10 The global PV inverter market share ranking by shipment in 2022. Reprinted from Ref. [7]

the chart, the majority of the companies listed are Chinese enterprises, including Huawei, Sungrow, Ginlong Solis, Growatt, GoodWe, Sineng, Aiswei, and Sofar. The combined shipment volume of these companies accounts for an astonishing 80% of the total market share. Among these, Huawei, with a market share of 29%, and Sungrow, with a market share of 23%, lead the rankings. Other companies, such as SMA from Germany, Power Electronics from Spain, and TMEIC from Japan, share markets of 3%, 3%, and 2%, respectively. In conclusion, China's PV inverter industry has become a key driver in the global transition to renewable energy, further solidifying the country's leadership in the PV sector.

In conclusion, the PV power generation dominates renewable energy generation, which has been developed rapidly in recent years. A series of supporting policies have been proposed around the world to promote the development of the PV industry, which facilitates the transition of its energy structure. China is becoming a global leader in the solar PV industry, especially in the proportions of PV capacities, PV modules, and PV inverters.

1.2 Photovoltaic Power Generation Systems

In this section, the basic structures of PV power generation systems are introduced. The features of each structure are successively described. To begin with, different configurations of solar PV cells and panels are summarized in Fig. 1.11. As shown, the fundamental building block of a PV system is called the PV cell [8]. It directly converts sunlight into electrical energy via the PV effect. A single PV cell produces only a small amount of voltage and current, which limits its practicality for real-world applications. To overcome this limitation, multiple PV cells are electrically

1.2 Photovoltaic Power Generation Systems

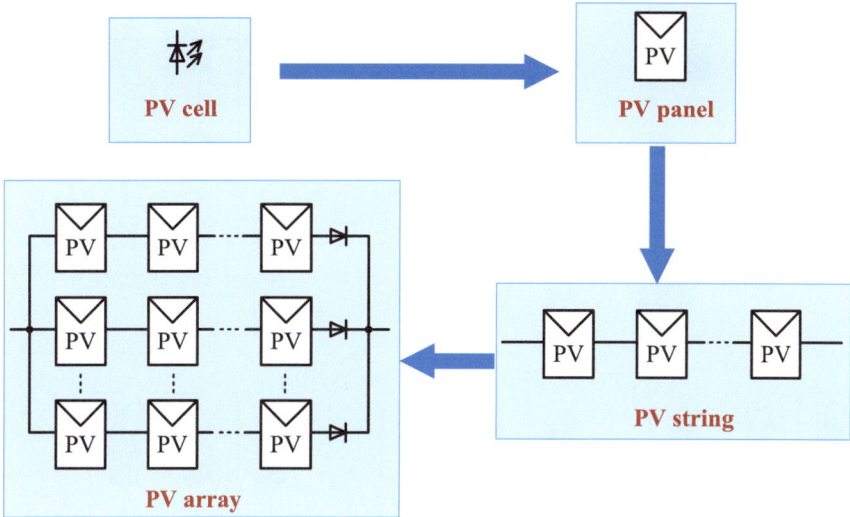

Fig. 1.11 Different configurations of solar PV cells and panels

interconnected in series and parallel to form a PV panel (also referred to as a PV module). A PV panel is a self-contained unit capable of generating a usable amount of electricity, typically encased in a durable frame and protected by a glass cover for outdoor environments. To further increase the system's output voltage and power, multiple PV panels can be connected in series to form a PV string. Additionally, the parallel connection of PV strings is referred to as a PV array. A PV array represents the complete assembly of interconnected solar panels designed to meet the energy generation requirements of a PV system.

Generally, the structure of a PV power generation system directly determines its power output and efficiency. According to the distribution of PV arrays (or strings, panels) and their power rating, PV systems can be classified into the following six types: centralized structure, AC module structure, string structure, multi-string structure, master-slave modular structure, and DC module structure [9–12]. The configurations of these PV systems are depicted in Fig. 1.12.

The centralized structure is shown in Fig. 1.12a. In this structure, the central inverters with large power capacities are employed and connected to the PV array. Usually, the DC combiner boxes are also adopted to aggregate the power of PV strings before being fed into central inverters [13]. The centralized structure is one of the earliest forms of PV systems and is widely used in large-scale ground-mounted power plants. It features high economic efficiency as large-capacity inverters have lower unit costs. Additionally, the centralized layout simplifies system management and maintenance. However, the centralized structure has certain limitations. For instance, when PV panels are connected in series, any performance degradation (e.g., due to shading or aging) of a single panel can impact the entire system's output power.

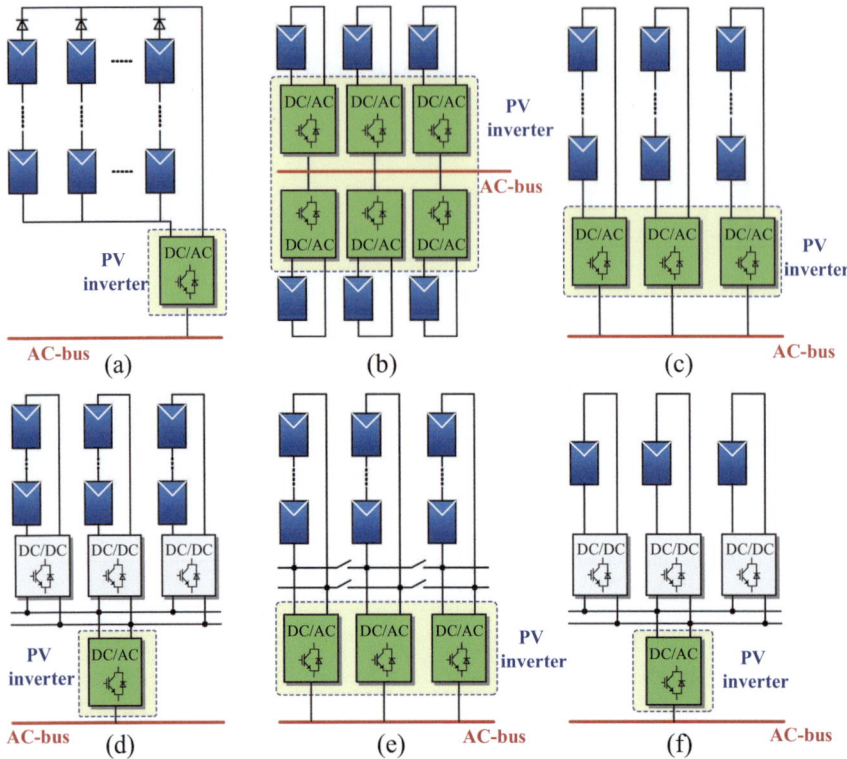

Fig. 1.12 Structures of the PV generation systems **a** centralized structure, **b** AC module structure, **c**, string structure, **d** multi-string structure, **e** master-slave modular structure, **f** DC module structure

Therefore, the maximum power point (MPP) will be decreased if shading of certain PV panels occurs.

The AC module structure is shown in Fig. 1.12b. In this configuration, each PV panel is equipped with a small-capacity micro-inverter. Subsequently, the maximizing power output through maximum power point tracking (MPPT) of each PV panel can be realized individually, eliminating the mismatch problem between different PV panels. As a result, the MPPT efficiency of the entire PV system is significantly improved. The AC module structure is mainly employed in small-scale PV systems, such as residential or small commercial rooftop installations. Since each PV panel is equipped with an individual inverter, this structure provides excellent reliability and plug-and-play capability. In addition, this structure offers flexibility in installation and adaptability to complex terrains or varying installation conditions. However, the cost of this structure is remarkably increased compared with the centralized structure, which limits its widespread application.

The string structure is depicted in Fig. 1.12c. It is commonly used in distributed PV systems. In this structure, a string of PV modules is connected to an independent inverter. A notable feature of the string structure is that each string operates

1.2 Photovoltaic Power Generation Systems

independently, ensuring no interference between strings and significantly enhancing system reliability. Compared to the centralized structure, the string structure also offers a higher efficiency of MPPT. As a trade-off, when this structure is applied to high-power PV generation systems, more inverters are required, leading to an increase in system costs. The string structure is well-suited for small- to medium-scale distributed PV systems, such as industrial and commercial rooftop installations and medium-sized ground-mounted plants.

Figure 1.12d shows the configuration of multi-string structure. In this structure, multiple PV strings are connected to a single inverter with multiple MPPT inputs. Individual DC-DC converters are employed, allowing independent MPPT control of each string. As a result, this structure exhibits the same MPPT performances as the string structure. However, the inverter cost is effectively reduced. In addition, individual control of each string increases the reliability of PV systems. Due to these advantages, the multi-string structure has been widely applied.

The master-slave modular structure is depicted in Fig. 1.12e [14]. As shown, the PV strings are connected by a novel architecture equipped with collaborative switches. Based on the output power of PV arrays, it controls the on-off states of collaborative switches to dynamically adjust the system configuration, thereby achieving the highest energy conversion efficiency. Under low irradiation conditions, the collaborative switches are closed to connect all PV arrays to a single inverter, avoiding the low efficiency caused by multiple inverters operating simultaneously under light load conditions.

The DC modular structure is depicted in Fig. 1.12f. This structure consists of DC/DC modules and a central inverter. Each PV panel is equipped with a DC/DC module to achieve individual MPPT. Therefore, the MPPT efficiency of this structure is identical to the AC module structure and is the highest among all structures. The DC output power of DC/DC modules is aggregated and fed into the central inverter. The central inverter converts DC power to AC power and transmits it into the power grid. As observed, the DC modular structure is complex. The cost of the overall system is remarkably increased compared to the centralized structure.

In conclusion, the selection of a PV system structure depends on multiple factors, such as the capacity of the system, geographical conditions, economic considerations, and operational and maintenance requirements. The centralized structure is suitable for large-scale ground-mounted power plants, while the AC and DC module structures are better suited for small-scale distributed systems. String and multi-string structures perform well in small- to medium-scale PV projects, offering flexibility and high efficiency. The master-slave modular structure stands out in large-scale systems for its ability to optimize efficiency and enhance reliability. As PV technologies continue to advance, these structures will become increasingly refined, providing more flexible and efficient solutions for various application scenarios.

1.3 Photovoltaic Inverters

PV inverters play important roles in the operation of PV systems. In this section, the classifications of PV inverters are introduced, and the basic topologies of PV inverters are successively illustrated. Among various inverter topologies, the three-level topology benefits from high-efficiency and low-output harmonics, and is widely used in PV systems. To ensure the safe operation of PV systems, relevant standards have been specified. The requirements of functions and performances for PV inverter in these standards are also explained.

1.3.1 Classification of Photovoltaic Inverters

The primary function of inverters is to convert the DC electricity into AC electricity, which is compatible with the power grid. This conversion enables the electricity to be seamlessly utilized in both residential and industrial applications. In addition to energy conversion, the efficiency of PV systems heavily depends on the performance of inverters. By ensuring the proper operation of inverters, the MPPT of PV strings or arrays can be achieved, thereby optimizing the power generation efficiency of PV systems. Furthermore, PV inverters are essential for enhancing the safety and reliability of PV systems. They are typically equipped with advanced monitoring and protection features, such as fault detection, over-voltage and over-current protection, and leakage current mitigation. These features ensure stable operation, safeguard the system against potential risks, and contribute to the long-term reliability of PV installations. Consequently, PV inverters are indispensable components of PV systems. The overall performance of PV inverters largely determines the performance of the entire PV system.

There are various classification methods for PV inverters. Based on whether a transformer is used, inverters can be categorized into transformer-based inverters and transformerless inverters [15, 16]. Transformer-based inverters can be further divided into power-frequency transformer-based inverters and high-frequency transformer-based inverters. As the names suggest, power-frequency transformer-based inverters use power-frequency transformers for electrical isolation, while high-frequency transformer-based inverters utilize high-frequency transformers and corresponding high-frequency inverter topologies to achieve electrical isolation. In terms of the number of phases connected to the grid, PV inverters can be classified into single-phase inverters, three-phase inverters, and three-phase four-leg inverters. Based on the number of stages, inverters can be divided into single-stage inverters, two-stage inverters, and multi-stage inverters. Additionally, according to the inverter topology, they can be categorized as two-level inverters, three-level inverters, and multi-level inverters.

Figure 1.13 shows the structures of transformer-based and transformerless inverters. The structure of a power-frequency transformer-based PV inverter is shown

1.3 Photovoltaic Inverters

in Fig. 1.13a. In this topology, the DC power is converted to power-frequency AC power by the inverter and integrated into the power grid by a power-frequency transformer. The transformer enables voltage step-up and provides electrical isolation, thereby enhancing system safety. However, the employment of power-frequency transformers increases system costs. In addition, the large size and weight of the line-frequency transformer reduce the power density of the system.

The structures of high-frequency transformer-based PV inverters are shown in Fig. 1.13b, c. The employment of high-frequency transformers can significantly reduce system size and weight. Figure 1.13b shows the DC/DC conversion-based

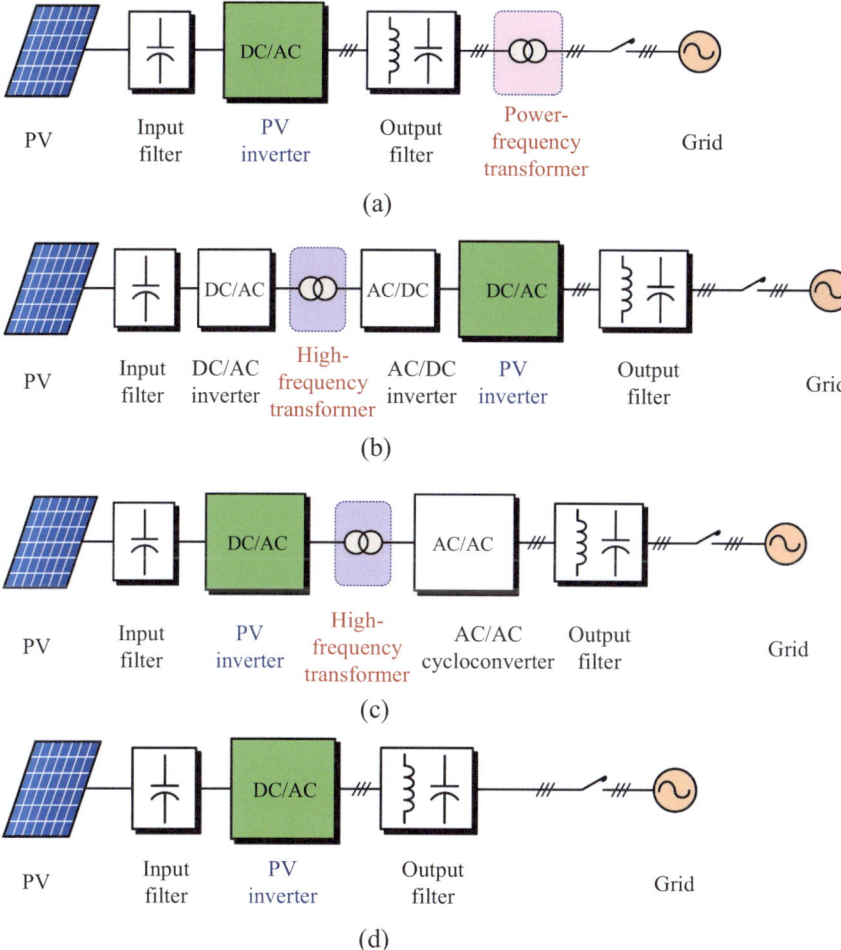

Fig. 1.13 Structures of transformer-based and transformerless inverters **a** power-frequency transformer-based structure, **b** DC/DC conversion-based high-frequency transformer-based structure, **c** cycloconverter-based high-frequency transformer-based structure, **d** transformerless structure

high-frequency structure. In this structure, the DC/AC inverter, high-frequency transformer, and AC/DC converter together form the DC/DC conversion stage. Subsequently, another DC/AC inverter is employed to convert DC power to power-frequency AC power. For the DC/DC conversion stage, the DC/AC inverter converts DC power to high-frequency AC power, and the AC/DC converter provides the opposite function. The high-frequency transformer is used for voltage step-up and provides electrical isolation. However, the multiple power conversion stages result in increased system losses and reduced efficiency. To address this, the cycloconverter-based high-frequency structure, as shown in Fig. 1.13c, has been proposed [17]. In this structure, the AC/DC converter and the following DC/AC inverter are replaced by an AC/AC cycloconverter. Therefore, one power stage is removed and the efficiency is improved.

Nevertheless, the introduction of either power-frequency or high-frequency transformers still adds additional losses to the system. Thanks to advancements in switching devices and power conversion technologies, the efficiency of inverters has been significantly improved. As a result, the inclusion of transformers has become a critical factor limiting further efficiency improvements. In addition, the employment of transformers also increases system weight and cost. In this scenario, the employment of transformerless structures becomes highly attractive. The transformerless structure is shown in Fig. 1.13d, where the transformer is removed and the inverter directly connects to the power grid.

As observed from Fig. 1.13, high-frequency transformer-based structures usually feature two-stage or multi-stage characteristics since high-frequency voltage is required for the application of high-frequency transformers. In comparison, power-frequency transformer-based structures and transformerless structures can adopt single-state, two-stage, and multi-stage inverter systems. However, multi-stage inverter systems are usually avoided to reduce additional power losses. Examples of single-stage and two-stage inverter systems for single-phase and three-phase systems are shown in Fig. 1.14, where the DC/AC inverters adopt the two-level topology for illustration purposes.

For single-stage inverter system, DC/AC inverters need to perform the MPPT algorithm, DC-link voltage control, AC current control, and some auxiliary functions. To ensure successful power inversion and achieve MPPT, the voltage at the MPP of PV arrays should be higher than the peak value of the AC line voltage. For a two-stage inverter system, the MPPT algorithm is implemented by a DC/DC converter. Other functions are implemented by the DC/AC inverter. Usually, the boost converter is employed for DC/DC power conversion. As a result, the voltage at the MPP of PV arrays can be lower than the peak value of the AC line voltage. In this book, we primarily focus on the analysis and control of the DC/AC inverter. The single-stage three-phase topologies are selected as the research objective to cover all control functions in the PV system.

For PV systems, high efficiency, high reliability, and low cost of inverters are crucial for their stable and reliable operation. As shown in Fig. 1.14, two-level DC/AC inverters are used for demonstration purposes. However, two-level topologies suffer from drawbacks such as high output waveform distortion and large filter size.

1.3 Photovoltaic Inverters

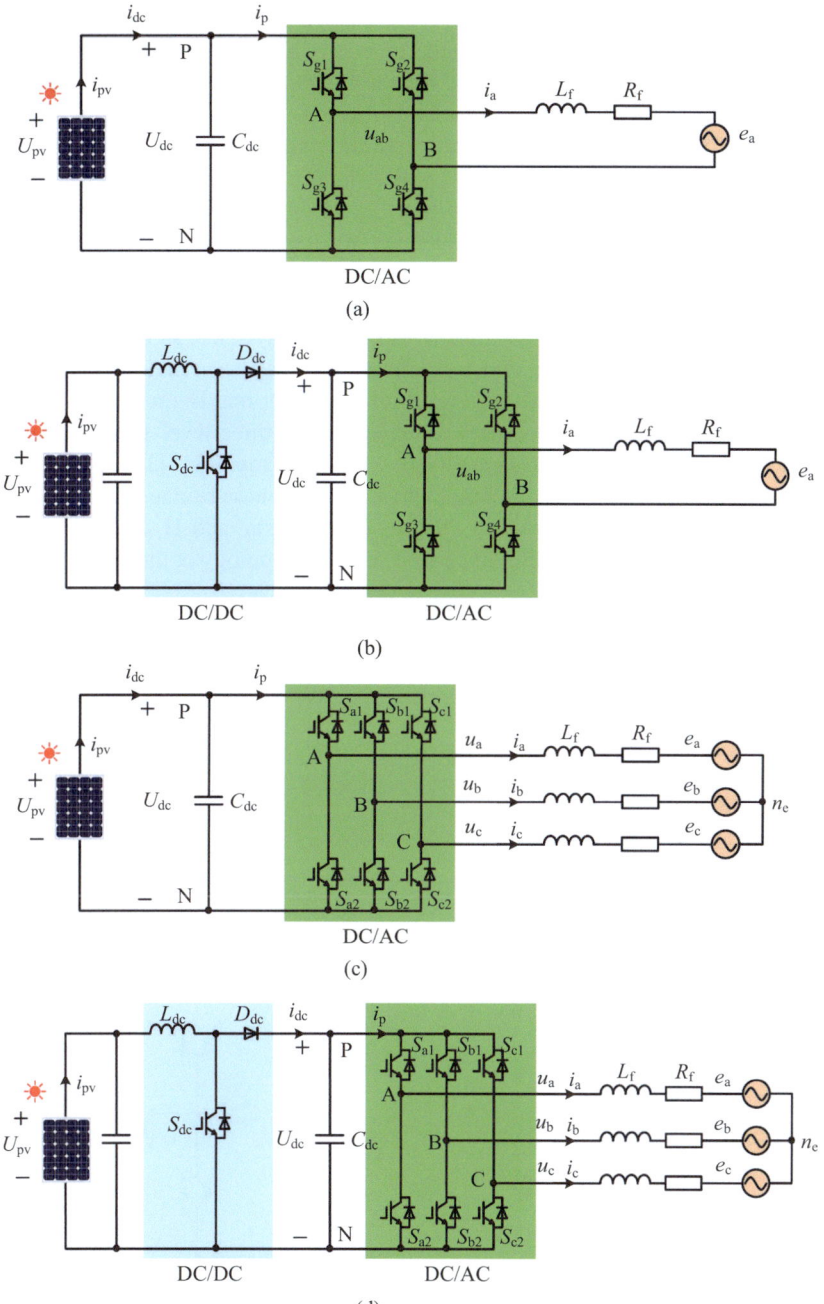

Fig. 1.14 Example illustration of single-stage and two-stage inverters for transformerless structures **a** single-stage single-phase topology, **b** two-stage single-phase topology, **c** single-stage three-phase topology, **d** two-stage three-phase topology

Consequently, the utilization of two-level topologies results in low system efficiency and reliability, which fail to meet the requirements of high-performance PV systems. As a solution, multilevel inverters have gained significant attention from researchers worldwide due to their advantages of lower voltage stress on power devices, reduced output voltage harmonics, and smaller output filter sizes. Among these, three-level inverters strike a balance between system performance and complexity. Compared to conventional two-level inverters, three-level inverters offer lower harmonic content and higher efficiency. Compared to multilevel inverters with more than three levels, they feature simpler control algorithms and lower costs. As a result, three-level inverters act as an ideal solution for achieving high efficiency, high reliability, low harmonic distortion, and low cost in non-isolated PV systems. In the PV industry, three-level inverters have been widely applied.

At present, commonly used three-level topologies primarily include four types: diode-clamped three-level topology, flying capacitor three-level topology, active clamped three-level topology, and T-type three-level topology [18]. Figure 1.15 shows the structures of the four topologies of a single-phase bridge.

In 1980, Japanese researchers A. Nabae, I. Takahashi, and H. Akagi proposed the neutral-point clamped (NPC) three-level inverter topology at the IEEE Industry

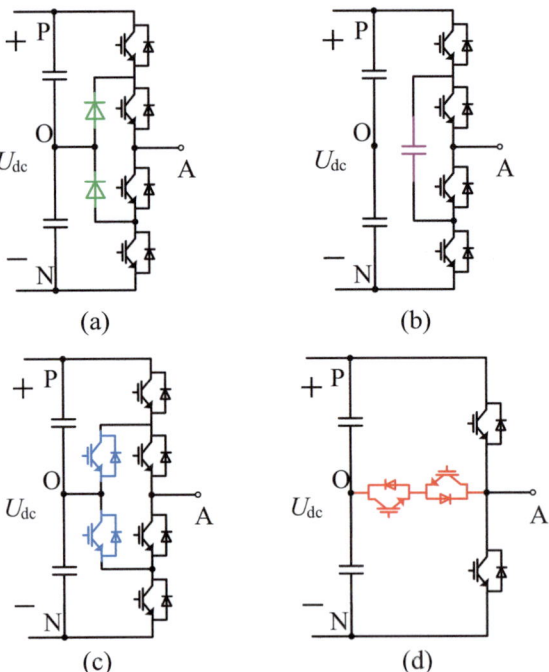

Fig. 1.15 Structures of the four topologies of a single-phase bridge **a** diode-clamped three-level topology, **b** flying capacitor three-level topology, **c** active clamped three-level topology, **d** T-type three-level topology

1.3 Photovoltaic Inverters

Applications Society (IAS) Annual Meeting [19]. This three-level topology is also known as diode neutral-point clamped (or simply as diode-clamped) three-level topology [18]. This topology is shown in Fig. 1.15a. This topology uses four power semiconductor switches for each phase, and two diodes are adopted to clamp the output terminal to the neutral point of the DC side. As a result, three-level voltage outputs can be provided. This topology benefits from low voltage stress on power devices and reduced harmonic content in the output voltage. It has been widely utilized in PV systems, wind power generation, power quality improvement in distribution networks, and motor control. However, due to capacitor capacity deviation and inconsistent switching characteristics of power devices, the NPC topology inherently suffers from neutral-point voltage imbalance problems. Neutral-point balancing control strategies must be designed during system operation. Additionally, uneven current distribution among power devices leads to uneven power losses, which can result in excessive junction temperatures in certain power devices, thereby limiting the system's power level.

To simplify NPC three-level topology, French scholar T. A. Meynard proposed the flying capacitor (FC) three-level inverter topology in 1992, which is depicted in Fig. 1.15b. This topology replaces the clamping diodes in the NPC topology with flying capacitors, resulting in a more balanced distribution of power losses among devices, albeit at the cost of increased system expenses.

To address the issue of uneven power loss in NPC three-level inverters, German scholar T. Bruckner proposed the active neutral-point clamped (ANPC) (or simply active clamped) three-level inverter topology in 2001 [20]. The structure of ANPC topology is shown in Fig. 1.15c. In this topology, clamping diodes in NPC topology are replaced with fully controlled power devices. By generating redundant zero states and increasing current paths, the ANPC topology achieves more balanced power loss distribution among devices and enhances the fault-tolerant operation capability of the system. However, the increased number of power devices leads to higher system costs.

The T-type three-level topology is a novel three-level topology, as shown in Fig. 1.15d. This topology employs two bidirectional switches, implemented using IGBTs connected in a common-emitter configuration, to clamp the output terminal to the neutral point of the DC side [21, 22]. During operation, each IGBT in the bidirectional switches only withstands half of the DC bus voltage, significantly reducing switching losses. This topology combines the advantages of traditional two-level topologies (such as low conduction losses and fewer devices) with those of three-level topologies (such as high output voltage waveform quality and low switching losses). When the switching frequency is between 4 and 30 kHz, the system achieves maximum efficiency, and power loss distribution among devices is relatively balanced. As a result, this topology offers superior overall performance and has become the mainstream solution for PV inverters. In this book, the three-phase three-level T-type inverters ($3LT^2Is$) are selected as the primary research objectives.

1.3.2 Requirements of Functions and Performances for Inverters

The PV inverter serves as an interface between the PV array and the power grid or the load, which implements the safe, efficient, and grid-friendly power conversion. It is important to point out that the standard requirements, regarding the connection of PV systems to the grid, are actually becoming more and more strict with the development of PV generation industry [23]. Similar to traditional thermal and hydroelectric power generation, the requirements can be achieved by the control technology. Besides, the reduction of system costs, minimizing the impact on the power grid, and improvement of PV energy harvesting can also be realized by the corresponding control technology.

The requirements are proposed by the PV station holder, PV energy consumer, and national power grid focusing on aspects of security, efficiency, power quality, and so on. In general, the requirements can be classified into functions and performances.

Requirements of Inverter Functions

From the perspective of PV station holder, PV energy consumer, and national power grid, four main requirements of inverter functions are introduced in this section.

1. **Maximum power point tracking (MPPT)**

    The first functional requirement is concerned with the MPPT, which is the main indicator for PV power station holders. The MPPT technology influences the output power capacity of PV generation system, which directly determines the rate of return of the PV station holder.

    The PV array is a collection of solar panels that convert sunlight into electricity. In practical applications, the utilization of PV array is determined by its internal characteristics and the external environment factors, such as light irradiance, load condition, and environmental temperature. Under different external conditions, the PV array is operated at different but the single maximum power point (MPP). Hence, for the PV power generation system, it is desired to seek the optimal working point to convert the light energy into electrical energy as much as possible.

2. **Low and high voltage ride-through (LVRT/HVRT)**

    Due to the increasing penetration of the PV system, the grid fault happens occasionally. The grid fault will lead to the PV inverter disconnected from the power grid if the control scheme is designed based on normal grid conditions. In return, if lots of PV inverters are disconnected from the power grid, the grid fault will be exacerbated. This vicious circle will bring cascading failure to the power grid and PV power station power, which reduces the stability of the whole power system.

    Thus, the low voltage ride-through (LVRT) and high voltage ride-through (HVRT) functions are crucial for the reliability and robustness of the PV inverters and required by the IEEE 1547-2018 standard [24]. These functions require the PV inverters to remain connected and support the power grid during a grid fault,

1.3 Photovoltaic Inverters

rather than disconnecting. To be specific, PV inverter should remain tied to the power grid and deliver a certain amount of reactive current into the power grid based on the percentage of the system rated current and positive-sequence grid voltage reduction or increment to provide support to the power grid [25]. In addition, the active current must be maintained during LVRT operation, but a reduction of it within the design specification of the PV inverter. In conclusion, during an LVRT/HVRT event, the PV inverter must have the capabilities of staying connected to the power grid, provide reactive power to support the power grid, prevent damage to other electrical equipment, and deliver a certain amount of active power to maintain the power supply. It can be seen that LVRT and HVRT are essential for the reliable integration of PV energy sources into the power grid, ensuring that PV energy sources contribute to grid stability rather than exacerbating disturbances. Thus, the LVRT and HVRT functions of the inverter are the important indicators for PV power station holders and national power grid.

3. **Power quality control**

    The power quality control function refers to the ability of the PV inverters to eliminate the current harmonics and compensate the reactive power.

    On one hand, the output power of the PV generation system is highly dependent on irradiance. Due to the alternation of day and night and the change of weather conditions, PV generation system has the problems of output power intermittence and low utilization rate of inverter capacity, usually less than 30%. Especially, the PV inverters shut down at night. On the other hand, with a large number of power electronic equipment and nonlinear loads connected to the power grid, a large number of current harmonics and reactive power are injected into the power grid, which cause power grid voltage distortion, fluctuations, and other electrical equipment failures. Moreover, in a PV station, the reactive power is also needed due to the presence of reactive power equipment such as transformers, capacitors, and inductors. If the reactive power is transmitted in the power line, the power loss will be inevitably increased.

    Traditionally, the *LC* filters, thyristor-controlled reactor (TCR), thyristor-controlled reactor (TSR), static Var compensator (SVC), static Var generator (SVG), and active power filters (APF) are chosen and applied in PV station to eliminate the harmonic currents and compensate reactive power. However, the additional power quality control equipment will greatly increase the investment cost.

    Since the main circuit structure of PV inverter is consistent with that of APF and SVG, the problem of current harmonics and reactive power can be solved by PV inverters with appropriate control scheme. The function of current harmonics elimination and reactive power compensation for PV inverter is called power quality control function. Thus, the PV inverter operates in power quality control mode at night and on cloudy days, while it operates in power generation mode when the irradiance is sufficient, which achieves green energy conversion and transmission as much as possible. Besides, with this mode, the capacity of PV inverter is fully utilized, and the investment cost of PV power station is reduced.

For the PV energy consumer, the power quality control function can reduce the contents of the harmonic and reactive current, which is beneficial to the safe and efficient use of other electrical equipment. For the PV station holder, the power quality control function makes the PV inverter capacity fully utilized, which reduces the investment cost and increases the rate of return. For the national power grid, the power quality control function can reduce the power grid voltage distortion, fluctuations, which enhance the stability of the power grid.

4. **Anti-islanding protection and islanded operation mode**
On the one hand, the PV inverter commonly operates in grid-connected mode when the power grid is in normal condition. In this mode, the PV inverter connects to the power grid and synchronizes its output with the power grid voltage and frequency. This mode also allows the PV inverter to feed excess power back to the power grid. However, if the PV inverter continues to supply power to the power grid when the power grid is faulty, it will be dangerous for utility workers who might assume the power grid is deenergized during maintenance or repairs. Thus, the anti-islanding protection is required since it is a critical safety feature for PV energy consumers and utility workers. The anti-islanding protection means that the PV inverter detects grid failures and immediately disconnects the inverter from the power grid to ensure safety. Anti-islanding protection is mandated by various international standards to ensure safety and power grid reliability. The international standards include IEEE 1547, UL 1741, and IEC 62116.

On the other hand, with the diversification of electricity applications and the trend of weakening power grid inertia, PV inverters having an islanded operation mode is required. The PV inverter operating in islanded mode refers to a situation where the PV inverter continues to supply power to a local electrical grid or load even when it is disconnected from the main utility grid. In this case, the PV inverter provides voltage and frequency support to electric equipment. PV inverters in islanded mode play a vital role in maintaining power continuity for critical loads and ensuring power system stability during grid disturbances or faults. Besides, islanded mode also provides an effective solution for the areas without power grid coverage or the occasion that requires independent power supply. Additionally, as the main power supply unit of microgrids and backup power system, PV inverter is required to operate in islanded mode to provide stable voltage and frequency support. It should be noted that, in islanded mode, the PV inverter, acting as the sole source for the local power system, requires special control technology to maintain voltage and frequency stability. For PV energy consumers, islanded mode operation guarantees continuity of power, especially for critical loads. For PV station holder, the continuous PV power output can be realized with the islanded mode, which increases the rate of return.

In conclusion, for grid-connected mode of PV inverter, anti-islanding protection is essential to prevent unintended operation during power grid outages. Islanded mode of PV inverters is a critical function for certain applications, but it must be managed carefully to ensure safety and stability.

1.3 Photovoltaic Inverters

Requirements of Inverter Performances

The performances of PV inverter, such as current quality enhancement, reduction of capacitor voltage ripples, leakage current suppression, and fault-tolerant operation ability, have become the focus of researchers in the years to come. These performance requirements are explained as follows.

1. **Current quality enhancement**
   The PWM inverter is the most commonly used topology in the PV power generation system. The main purpose of the inverter is to transfer a high-quality current with a total harmonic distortion (THD) less than 5% into the utility grid. In order to meet the requirements of current quality, the inverters should have good performance in harmonic rejection, which highly depends on the adopted current control strategy [26], and modulation methods.
2. **Reduction of capacitor voltage ripples**
   In a three-level PV inverter, two capacitors are configured in series connection at the dc-side to generate multi-level waveforms at the ac-side. One inherent problem of the three-level PV inverter is neutral-point voltage (NPV) balance, which affects system reliability, good quality of output waveforms, and secure operation of power semiconductor switches. In LVRT operation, the three-level PV inverter faces the challenge of NPV unbalance [27]. Due to the unbalanced grid voltage conditions and low power factor, the NPV will significantly fluctuate to unacceptable levels, which causes power semiconductor switches and dc-link capacitances to operate unsafely. Meanwhile, the output voltage will be distorted remarkably and the output current will contain many harmonic components.
3. **Leakage current suppression**
   The transformerless PV system offers several benefits over transformer-based systems, including their small size and low cost. However, the leakage current is induced by high-frequency common mode voltage (CMV), which can potentially destroy the PV system. The leakage current emerges because of the lack of galvanic isolation in the PV system. Leakage current will cause a series of negative effects on PV generation system, which mainly includes the following aspects.

   (a) Threaten personal safety: Leakage current can pose electric shock hazards to personnel, especially in systems that are not well grounded.
   (b) Lower current quality: Leakage currents lead to grid current distortion and increase the harmonics contents.
   (c) Lower system efficiency: Leakage currents will lose the electrical energy, reducing the overall efficiency of the PV system.
   (d) Damaged electrical equipment: Consistently higher leakage current shortens lifespan of power devices, which will damage PV inverters or other electrical equipment.
   (e) Shutdown PV inverters: High leakage currents can cause grid disturbances or trip residual current devices, leading to PV inverter shutdowns.

(f) High electromagnetic interference: Leakage currents can cause electromagnetic interference, which affects or even damages other electronic devices nearby.

To reduce the negative effects of leakage current as much as possible, lots of requirements are proposed in standard codes.

The standards of IEC 62109-2 and German DIN VDE 0126-1-1 require the PV inverter to be turned off and send out fault signals in less than 0.3 s if the leakage current is higher than the limited values [28, 29]. The limited values are 300 mA for rated capacity lower than 30 kVA, and 10 mA/kVA for the rated capacity higher than 30 kVA. Otherwise, in the case of an instantaneous leakage current, Table 1.1 should be followed. The leakage current is the most important safety specification index for PV inverter users since it is directly related to personnel safety.

4. **Fault-tolerant operation ability**

In the PV inverter, power semiconductor switches are the core components, which realize the conversion of electrical form and determine the energy flowing direction. However, the power semiconductor switches are the most fragile components due to the limitations of the material and the manufacturing process. Statistics indicate that about 34% of the power converter system's failures are caused by the power semiconductor switches and soldering failures. Therefore, it is important to develop fault-tolerant control technology to improve PV inverter's reliability.

When power semiconductor switches fail, the PV inverter should have the capability to maintain operation and minimize disruptions, which is essential for ensuring reliability, safety, and efficiency in electrical systems. Basically, the fault-tolerant operation is classified into hardware-based and software-based solutions. The former solution utilizes multiple inverters or components within an inverter. If one component fails, others can take over the load, maintaining system functionality. The software-based solution runs multiple algorithms and verifies their outputs, which can ensure continued operation even if one algorithm provides a faulty result. It is noted that the system cost will not be increased by the latter solution, which is preferred in practical applications.

It can be concluded that the fault-tolerant operation of PV inverter is a key performance for PV station holder, PV energy consumer, and national power grid, since it is beneficial to the continuous output of electrical energy and the reliability of power system.

Table 1.1 Response time limits for sudden changes in leakage current

RMS value of the leakage current (mA)	Turn-off time (s)
30	0.3
60	0.15
150	0.04

1.4 Key Factors of Functions and Performances for Inverters

As above analyses, the generalized structure diagram of PV power generation system can be divided into three parts: DC input side, power conversion unit, and alternating current (AC) output side. Each part includes different passive and active power components with different characteristics, which have different effects on the performance of the PV power generation system.

This section will analyze the key factors in each part that influence the performance of the PV power generation system, including PV panels, power components of PV inverters, and nonideal power grids and load.

1.4.1 Photovoltaic Panels

1. **Input power**

    The PV panels are a crucial component in solar energy systems, converting sunlight into electricity through the PV effect. The utilization rate of PV panels determines how much of the sunlight can be converted into usable electricity. Typically, for commercial PV panels products, the utilization rate ranges from 15 to 22%. Higher utilization rate of PV panels can generate more electricity from the same surface area, making them advantageous for space-constrained installations.

    It is well known that the utilization of PV array is not only determined by its internal characteristics, but also by the external environment factors. The internal characteristics determined by the materials, which mainly affect the PV effect. The external environment factors mainly include light irradiance, load condition, and environmental temperature. Figure 1.16 shows the power-voltage (P-U) curves of PV panel with different irradiance and different temperature. It can be seen that the output voltage and power of PV panel exhibit typical nonlinear characteristics.

    Usually, there is one MPP in each P-U curve. If the PV panel is always operating in the MPP, the maximum output power of the PV power generation system can be realized. The MPPT control technology of PV inverter is used to search and track the MPP. However, the control accuracy and speed are difficult to achieve at the same time due to the nonlinear characteristics.

    Besides, the PV array is composed of the PV panels connected in series and parallel. If partial shading and panels mismatch conditions occur, the I-V curves become different as shown in Fig. 1.17b. Multiple local peaks may occur in and the P-U curve as shown in Fig. 1.17b, which brings significant challenges for MPPT.

2. **Leakage current**

   The PV systems without transformers are the suitable option to minimize the cost of the total system. However, the leakage current is generated by the parasitic capacitor. The parasitic capacitor is between the PV panels and the PV frame connected to the ground. If there is no voltage difference between the PV panel and the PV frame, the leakage current cannot be induced. However, the positive voltage and negative voltage of the PV panel with respect to the grid ground have different values according to the inverter switching states. The generated positive and negative voltages are called CMV, which causes the leakage current.

   The leakage current path is shown in Fig. 1.18, which shows that the parasitic capacitor is the essential reason for the closed current path. The CMV (u_{cm}) is the exciting source generated by the PV inverter. The leakage current will lead to a series of negative effects, such as threatening personal safety, reducing current quality and system efficiency, damaging electrical equipment, causing high electromagnetic interference, and so on. Especially in a system without good grounding, when someone touches a PV panel or PV inverter by mistake, the leakage current will flow into the human body, resulting in electric shock.

   As in the above presentation, the leakage current has lots of negative effects on the PV inverter system. Thus, many countries restrict the leakage current to a low value, and the rule about the leakage current should be satisfied for the accreditation of inverter equipment.

To summarize, PV panels are responsible for converting sunlight into DC electricity, which determines the input power of the PV generation system. The leakage current is an inevitable problem that leads to a series of negative issues, especially for safety, and must be solved.

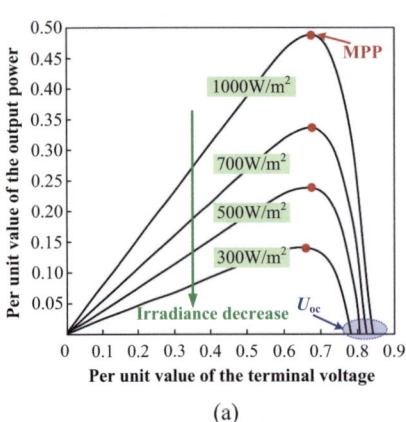 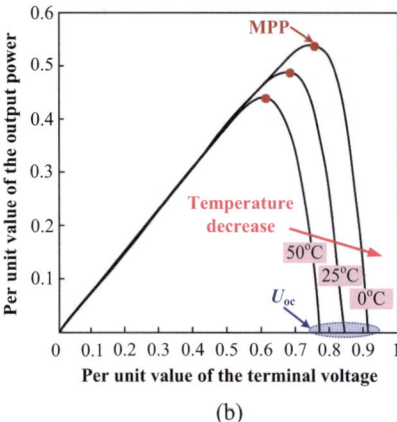

Fig. 1.16 The *P-U* curves of PV panel with the different environment conditions **a** same temperature but different irradiance conditions, **b** same irradiance but different temperature conditions

1.4 Key Factors of Functions and Performances for Inverters

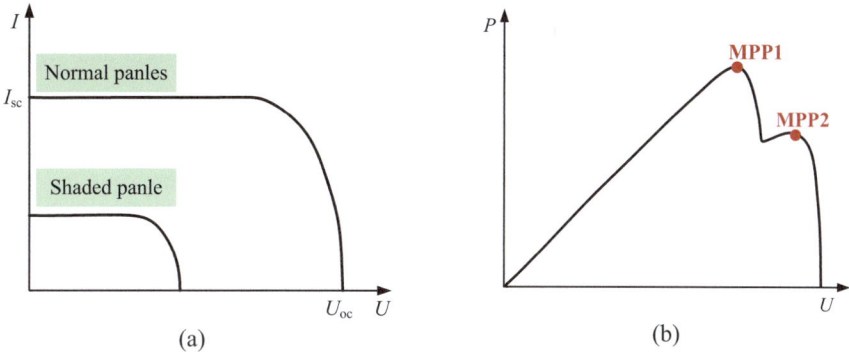

Fig. 1.17 The PV array characteristics when one PV panel is shaded **a** *I-V* curves of PV panels, **b** *P-U* curve of PV array

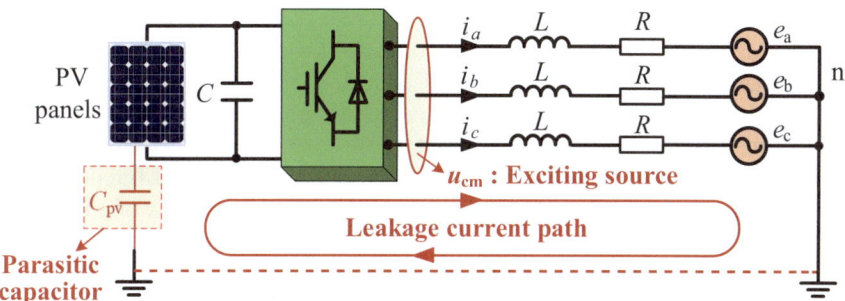

Negative effects:
- **Threatening personal safety**
- **Reducing current quality and system efficiency**
- **Damaging electrical equipment**
- **Causing high electromagnetic interference** ···

Fig. 1.18 The flow path of leakage current and its negative effects

1.4.2 Main Circuit of Photovoltaic Inverters

1. **AC-side filter**

   To attenuate the high-frequency current harmonics caused by the pulse width modulation (PWM), the AC-side filter is usually used as an interface between the output side of the PV inverter and the power grid. Two typical types of AC-side filters consist of *L* filter and *LCL* filter. Compared with *L* filter, *LCL* filter has stronger suppression capability for high-frequency harmonic. However, *LCL* filter is a third-order system with a resonance peak, which requires a more complex current controller to maintain system stability. In conclusion, the AC-side filter directly affects the harmonic contents of output current and system stability.

2. **DC-side capacitors**

   As mentioned above, the three-level PV inverter faces one inherent problem of NPV, which degrades the quality of output waveforms, threatens the safe operation of power semiconductor switches, and reduces the system reliability. To suppress the unbalanced NPV, large-volume electrolytic capacitors are usually adopted at the dc link, which decreases the power density and reliability of the three-level PV inverter. In steady state, with conventional modulation strategies, the NPV has a fundamental frequency which is three times the output fundamental frequency. This requires a large dc-link capacitor value for constant NPV. In addition, the NPV control is required to avoid the output-voltage distortion and overvoltage stress on semiconductors. That is to say, the dc-link capacitor affects the stability of the input voltage of PV inverter.

3. **Power semiconductor switches**

   The power semiconductor switch is one of the vulnerable components in PV inverters. As shown in Fig. 1.19, semiconductor and soldering failures in device modules account for 34% of failures in the power converter system. According to a survey based on over 200 products of 80 companies, power semiconductor switches have been selected by 31% of responders as the most fragile components. Temperature stressor has the most impact on the reliability of power electronic components and systems. Other factors such as humidity and vibration are very closely related to the failure of power semiconductor switches. We can conclude that the reliability of power semiconductor switches is closely related to the reliability of the overall power electronics system. Therefore, power semiconductor switches faults and their handling methods in the PV inverters must be investigated for improving the reliability of PV power generation systems.

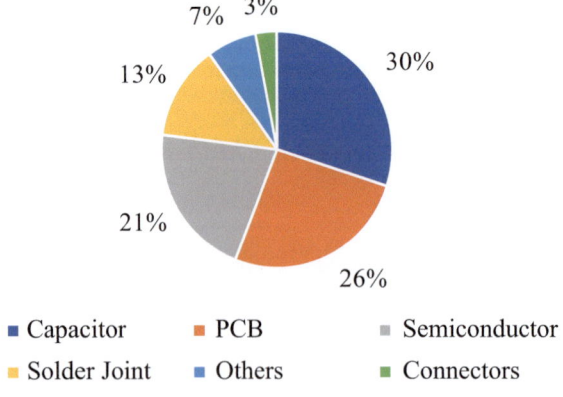

Fig. 1.19 Distributions of faults in power converter

All in all, the PV inverter consists of DC- and AC-side filters and power semiconductor switches, which affect the input and output waveforms quality and the system reliability.

1.4.3 Nonideal Power Grids

The influence of a nonideal power grid on PV systems is significant, which affects both the performance and reliability of PV generation systems. Here are some of the key impacts. First, the nonideal power grids may experience fluctuations in voltage levels due to varying loads and poor grid management. PV systems are designed to operate within a specific voltage range, and excessive deviations can lead to inverter tripping, reduced efficiency, or even damage. Second, the PV inverters are typically designed to operate at a specific frequency (usually 50 or 60 Hz). Variations can affect their ability to synchronize with the grid, which can influence power quality and the safe operation of the system. In addition, nonideal grids may have harmonic distortions caused by other electrical equipment. Harmonics can impact the performance of PV systems, potentially leading to overheating in inverters and reduced lifespan. To mitigate the effects of a nonideal grid, advanced control technology has been proposed to ensure high-quality continuous power conversion and transmission, and to maintain the stable operation of the PV inverter. The advanced control technology will be elaborated in Chap. 6.

1.4.4 Nonlinear Loads

Along with grid voltage distortion, the presence of nonlinear loads in the local load also causes a negative impact on the PV system [30]. The current of nonlinear loads does not vary in proportion to the voltage applied. Instead, they create a nonlinear relationship between voltage and current, resulting in harmonic distortion. Common examples of nonlinear loads include power electronics (like variable frequency drives), LED lighting, computer power supplies and servers, and switching power supplies. Nonlinear loads introduce harmonics into the electrical system. Harmonics will distort the waveform of current and voltage. The presence of harmonic currents leads to increased losses in both the PV system and the distribution grid. Harmonics can affect the efficiency of power inverters connected to PV arrays, potentially leading to overheating and degrading the system performance. The total harmonic distortion (THD) caused by nonlinear loads can lead to poor power quality. Poor power quality can damage sensitive electronic equipment and also result in inefficient operation of PV systems. Nonlinear loads may cause unbalanced current conditions in three-phase systems, leading to uneven loading on the phases. If not managed properly, this can result in inefficiencies and even damage to PV system components.

1.5 Research Content of This Book

The worldwide energy situation is first introduced in this chapter. As one of the fastest-growing and most promising forms of renewable energy generation, the PV system has attracted a great deal of attention due to its salient advantages of high flexibility and broad applicability. The PV power generation serves as a promising solution for solving the global problems of energy crisis and environmental pollution.

As described above, the PV inverter, as the interface between the PV panels and the power grid, is responsible for the conversion and transmission of electrical energy. With the development of PV industry, the power generation capacity and installed capacity are constantly increasing. The requirements of functions or performances for PV inverters are also gradually proposed in practical applications. The main requirements include safety, generation efficiency, transmitted power quality, robustness to multiple disturbances, grid-friendly, continuity of power supply and system reliability.

Traditional hardware-dependent methods not only fail to meet the aforementioned requirements but also incur high costs. Thus, the advanced control technologies are urgently needed, which are embedded into the PV inverter controller. Therefore, the author and his research group have carried out long-term teaching and research with practical experience, and summed up and put forward a series of advanced control technologies. Through these control technologies, PV power generation system has gradually become a high safety, high reliability, high efficiency, and strong adaptability system, which becomes a core support in modern power system.

The content of this book is organized according to the gradually proposed requirements of PV inverters. The requirements and corresponding control technology in each chapter are briefly introduced as follows.

Chapter 2 introduces the fundamentals of PV inverter, which includes models, control framework, and pulse width modulation technology of PV inverter. These fundamentals ensure PV inverter realizes basic normal operation. To satisfy the requirement of safety, the generation mechanism of leakage current is given and suppression technology is proposed subsequently in Chap. 3. To improve the power generation efficiency of PV system as much as possible, the conventional MPPT technology and separate MPPT technology are elaborated in Chap. 4. To improve the anti-interference capability and quality of AC-side current and DC-side voltage, the nonlinear control technology for robustness improvement, the control technology for AC-side current harmonic and ripple suppression and DC-side NPV oscillation suppression are presented in Chap. 5. In Chap. 6, under the grid fault condition, the advanced control technology for DC voltage and grid current control is given, which maintains the high-quality power energy conversion and transmission and satisfies the standard codes requirements. Besides, advanced control technology for NPV balance is also introduced, which ensures PV inverter in safe and stable operation. Chapter 7 illustrates the control technology of PV inverter for multi-functional operation, including power quality control mode and islanded mode. When a PV inverter operates in the power quality control mode at night and on cloudy days, it has the ability of current harmonic elimination and reactive power compensation. When PV inverters operate in islanded mode, the continuous power supply can be maintained during grid disturbances or faults. To improve the reliability and ensure the normal operation of PV inverter when power semiconductor switches fault, the fault-tolerant control technology is proposed in Chap. 8.

We hope that the contents of this book will contribute to PV power generation and become a reference book for engineers involved in the development and application of renewable power generation.

References

1. J. Rockström, O. Gaffney, J. Rogelj, M. Meinshausen, N. Nakicenovic, H.J. Schellnhuber, A roadmap for rapid decarbonization. Science **355**, 1269–1271 (2017)
2. Z. Liu, Z. Deng, G. He, et al., Challenges and opportunities for carbon neutrality in China. Nat. Rev. Earth Environ. **3**, 141–155 (2022)
3. IRENA, *Renewable Capacity Statistics* (IRENA, 2024)
4. B. Bai, Z. Wang, J. Chen, Shaping the solar future: an analysis of policy evolution, prospects and implications in China's photovoltaic industry. Energ. Strat. Rev. **54**, 101474 (2024)
5. https://www.statista.com/statistics/668749/regional-distribution-of-solar-pv-module-manufacturing/
6. Wood Mackenzie, *Global Solar PV Inverter and Module-Level Power Electronics (MLPE) Market Share* (Wood Mackenzie, 2023)
7. https://www.statista.com/statistics/1003705/global-pv-inverter-market-share-shipments/
8. M.G. Villalva, J.R. Gazoli, E.R. Filho, Comprehensive approach to modeling and simulation of photovoltaic arrays. IEEE Trans. Power Electron. **24**(5), 1198–1208 (2009)
9. S. Kouro, J.I. Leon, D. Vinnikov, L.G. Franquelo, Grid-connected photovoltaic systems: an overview of recent research and emerging PV converter technology. IEEE Ind. Electron. Mag. **9**(1), 47–61 (2015)
10. K. Zeb, et al., A comprehensive review on inverter topologies and control strategies for grid connected photovoltaic system. Renew. Sust. Energ. Rev. **94**, 1120–1141 (2018)
11. W. Li, Y. Gu, H. Luo, W. Cui, X. He, C. Xia, Topology review and derivation methodology of single-phase transformerless photovoltaic inverters for leakage current suppression. IEEE Trans. Ind. Electron. **62**(7), 4537–4551 (2015)
12. S.B. Kjaer, J.K. Pedersen, F. Blaabjerg, A review of single-phase grid-connected inverters for photovoltaic modules. IEEE Trans. Ind. Appl. **41**(5), 1292–1306 (2005)
13. T. Xu, F. Gao, T. Hao, X. Meng, Z. Ma, C. Zhang, H. Chen, Two-layer global synchronous pulse width modulation method for attenuating circulating leakage current in PV station. IEEE Trans. Ind. Electron. **65**(10), 8005–8017 (2018)
14. F. Gao, D. Niu, H. Tian, C. Jia, N. Li, Y. Zhao, Control of parallel-connected modular multilevel converters. IEEE Trans. Power Electron. **30**(1), 372–386 (2015)
15. C. Hou, C. Shih, P. Cheng, A.M. Hava, Common-mode voltage reduction pulse width modulation techniques for three-phase grid-connected converters. IEEE Trans. Power Electron. **28**(4), 1971–1979 (2013)
16. W. Li, Y. Wang, J. Hu, H. Yang, C. Li, X. He, Common-mode current suppression of transformerless nested five-level converter with zero common-mode vectors. IEEE Trans. Power Electron. **34**(5), 4249–4258 (2019)
17. R.K. Surapaneni, D.B. Yelaverthi, A.K. Rathore, Cycloconverter-based double-ended microinverter topologies for solar photovoltaic AC module. IEEE J. Emerg. Sel. Topics Power Electron. **4**(4), 1354–1361 (2016)

18. J. Rodríguez, L.G. Franquelo, S. Kouro, et al., Multilevel converters: an enabling technology for high-power applications, in *Proceedings of the IEEE*, (2009), pp. 1786–1818
19. A. Nabae, I. Takahashi, H. Akagi, A neutral-point clamped PWM inverter. IEEE Trans. Ind. Appl. **17**(5), 518–523 (1981)
20. P. Barbosa, Active neutral-point-clamped multilevel converters, in *IEEE Power Electronics Specialists Conference*, (2005), pp. 2296–2301
21. M. Schweizer, T. Friedli, J.W. Kolar, Comparative evaluation of advanced three-phase three-level inverter/converter topologies against two-level systems. IEEE Trans. Ind. Electron. **60**(12), 5515–5527 (2013)
22. M. Schweizer, J.W. Kolar, Design and implementation of a highly efficient three-level T-type converter for low-voltage applications. IEEE Trans. Power Electron. **28**(2), 899–907 (2013)
23. J.M. Carrasco et al., Power-electronic systems for the grid integration of renewable energy sources: a survey. IEEE Trans. Ind. Electron. **53**(4), 1002–1016 (2006)
24. IEEE Standard for Interconnection and Interoperability of Distributed Energy Resources With Associated Electric Power Systems Interfaces (IEEE Standard, 2018), pp. 1547–2018
25. Z. Shao, X. Zhang, F. Wang, R. Cao, H. Ni, Analysis and control of neutral-point voltage for transformerless three-level PV inverter in LVRT operation. IEEE Trans. Power Electron. **32**(3), 2347–2359 (2017)
26. Q. Zhao, Y. Ye, A PIMR-type repetitive control for a grid-tied inverter: structure, analysis, and design. IEEE Trans. Power Electron. **33**(3), 2730–2739 (2018)
27. Y. Li, X. Yang, W. Chen, T. Liu, F. Zhang, Neutral-point voltage analysis and suppression for NPC three-level photovoltaic converter in LVRT operation under imbalanced grid faults with selective hybrid SVPWM strategy. IEEE Trans. Power Electron. **34**(2), 1334–1355 (2019)
28. K.B. Tawfiq, H. Zeineldin, A. Al-Durra, E.F. El-Saadany, Mitigation of leakage current and current harmonics in PV grid-connected systems using a new H10 three-phase inverter. IEEE J. Emerg. Sel. Topics Power Electron. **13**, 214–229 (2025). https://doi.org/10.1109/JESTPE.2024.3500071
29. C. Wen, X. Xing, M. Jing, B. Sun, C. Liu, F. Blaabjerg, Alleviation of leakage current and neutral point voltage deviation for using an eight-switch inverter. IEEE Trans. Ind. Electron. **71**(8), 8793–8807 (2024)
30. Y. Ye, Y. Xiong, UDE-based current control strategy for LCCL-type grid-tied inverters. IEEE Trans. Ind. Electron. **65**(5), 4061–4069 (2018)

Open Access This chapter is licensed under the terms of the Creative Commons Attribution 4.0 International License (http://creativecommons.org/licenses/by/4.0/), which permits use, sharing, adaptation, distribution and reproduction in any medium or format, as long as you give appropriate credit to the original author(s) and the source, provide a link to the Creative Commons license and indicate if changes were made.

The images or other third party material in this chapter are included in the chapter's Creative Commons license, unless indicated otherwise in a credit line to the material. If material is not included in the chapter's Creative Commons license and your intended use is not permitted by statutory regulation or exceeds the permitted use, you will need to obtain permission directly from the copyright holder.

Chapter 2
Fundamentals of Photovoltaic Inverters

As introduced in Chap. 1, the photovoltaic (PV) inverters are the key link responsible for converting solar energy into electricity. The topology and control technology directly determine the investment costs, conversion efficiency, and output performance of the PV generation system. There are many topologies for connecting PV modules to the grid. Among these, the three-level T-type inverter ($3LT^2I$) is preferred in PV applications for several exclusive features in operation, performance, and structural issues. Especially, $3LT^2I$ is an efficiency-effective inverter in low-voltage applications. Thus, in this chapter, the $3LT^2I$ is taken as the typical topology to introduce the operation principle, modeling, control framework, and modulation schemes of PV inverters.

To begin with, the configuration and schematics of $3LT^2I$ are demonstrated, followed by the mathematical models of the AC side and DC side, respectively. The mathematical models act as the foundations for the control of $3LT^2I$. Then, the basic control framework for $3LT^2I$ is illustrated, including the phase locked loop (PLL), the outer loop for power control, the middle loop for DC voltage control, the inner loop for AC current control, and the auxiliary loop for other objectives. The other objectives are responsible for special requirements of $3LT^2I$ in PV applications. The leakage current suppression and the neutral-point voltage (NPV) control are taken as the typical application cases to elaborate the control principles and implementation process of the auxiliary loop. After those contents, the pulse width modulation (PWM) technologies are elaborated and the continuous PWM (CPWM) and discontinuous PWM (DPWM) are successively introduced.

2.1 Working Principles and Mathematical Models

It should be pointed out that the 3LT^{2}I can be used in three-phase and single-phase circuits. As the three-phase 3LT^{2}I is the mainstream topology for a vast number of applications, the 3LT^{2}I for the three-phase circuit is the topic in this whole chapter. Without special instructions, 3LT^{2}I refers to three-phase 3LT^{2}I.

The mathematical model plays an important role in the analysis and controller design of 3LT^{2}I. In this section, we aim to develop the mathematical models of 3LT^{2}I, including the AC-side and DC-side models. For convenience, we first introduce the structure and operating principle of 3LT^{2}I. The output states of the power circuit are described and the switching function is defined. Additionally, we introduce fundamental concepts of modulation techniques, which aid in the modeling process. A detailed description of modulation techniques will be elaborated in Sect. 2.3. On the basis of these introductions, we successively establish the AC and DC-side mathematical models of 3LT^{2}I.

2.1.1 Inverter Structure and Operating Principle

Figure 2.1 describes the basic structure of 3LT^{2}I. As shown, each phase consists of four power semiconductor switches, which are denoted as S_{x1}, S_{x2}, S_{x3}, and S_{x4} (where $x =$ a, b, c, representing three phases), respectively. The power semiconductor switches employ the insulated gate bipolar transistors (IGBTs), and each switch is equipped with an anti-parallel diode. The DC side of 3LT^{2}I connects to the PV array, where P and N denote the positive and negative DC bus, respectively. The capacitors C_1 and C_2 are adopted to filter DC voltage ripples and reduce voltage variations. In addition, the neutral point, represented by O, is provided by the two capacitors. The power modules of each phase connect to P, O, and N, thereby allowing a three-level output. U_{dc} and i_{dc} denote the dc-link voltage and current, respectively. U_{pv} and i_{pv} denote the PV voltage and current, respectively. As seen, for single-stage PV inverter, $U_{dc} = U_{pv}$ and $i_{dc} = i_{pv}$. U_p and U_n are the voltages of C_1 and C_2, respectively. i_p, i_n, and i_o represent the positive, negative, and NP currents of the DC bus. i_{c1} and i_{c2} are the currents flowing through C_1 and C_2, respectively.

As for the AC side, the power modules connect to the power grid through an L filter, where L_f denotes the filter inductance. R_f represents the equivalent series resistance of the L filter and the transmission line. u_a, u_b, and u_c denote the output voltages of the inverter. i_a, i_b, and i_c are the output currents. The power grid is modeled by the ideal three-phase voltage sources e_a, e_b, and e_c. The point n_e denotes the common point of the power grid.

Taking point O as the reference, the basic switch combinations, the output states, the output voltages, and the current flowing passes are summarized in Table 2.1 [1]. Here, the turn-on and turn-off states of power semiconductor switches are represented by 1 and 0, respectively, and the names of output states correspond to points P, O,

2.1 Working Principles and Mathematical Models

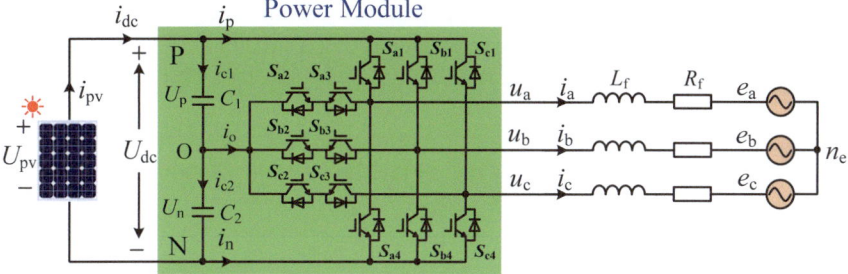

Fig. 2.1 The structure of 3LT^{2}I

and N of the DC side. For example, if S_{x1} and S_{x2} switch on while S_{x3} and S_{x4} switch off, the output of phase x connects to point P. Then, the output state of phase x is denoted as [P]. Similarly, the output state is denoted as [N] when S_{x1} and S_{x2} switch off while S_{x3} and S_{x4} switch on. In this case, the output of phase x connects to point N. When S_{x2} and S_{x3} switch on while S_{x1} and S_{x4} switch off, the output of phase x connects to point O and the output state is defined as [O].

For the convenience of the following analysis, the switching function S_x is introduced, and its value is calculated according to the switch combinations as

$$S_x = \begin{cases} 1 & (S_{x1}, S_{x2}, S_{x3}, S_{x4}) = (1, 1, 0, 0) \\ 0 & (S_{x1}, S_{x2}, S_{x3}, S_{x4}) = (0, 1, 1, 0) \\ -1 & (S_{x1}, S_{x2}, S_{x3}, S_{x4}) = (0, 0, 1, 1) \end{cases} \quad (2.1)$$

When the NPV is balanced (i.e., $U_p = U_n = U_{dc}/2$), the output voltages u_x of each phase can be uniformly expressed as

$$u_x = \frac{U_{dc}}{2} S_x \quad (2.2)$$

During the operation of 3LT^{2}I, the output states of each phase transition continuously among states [P], [O], and [N]. The transition instants are decided by the PWM process. Figure 2.2 shows a typical implementation of the three-level PWM process, where u_{ca1} and u_{ca2} are the periodic triangular carrier waves, and m_x (x = a, b, or c) denotes the modulation wave. T_s represents the period of u_{ca1} and u_{ca2}, which

Table 2.1 Relationship among switch combinations, output states, output voltages, and current flowing path of 3LT^{2}I

Switch combinations ($S_{x1}, S_{x2}, S_{x3}, S_{x4}$)	Output state	Output voltage u_x	Current flowing path
(1, 1, 0, 0)	[P]	U_p	S_{x1}
(0, 1, 1, 0)	[O]	0	S_{x2}, S_{x3}
(0, 0, 1, 1)	[N]	$-U_n$	S_{x4}

denotes the switching period. The reciprocal of T_s, denoted as f_s, is the switching frequency. Typically, the switching frequency is much smaller than the frequency of the modulation signals. Therefore, m_x is considered a constant within one switching period.

Figure 2.2a shows the waveform of output state when $m_x \geq 0$. As shown, m_x intersects only with u_{ca1}, where t_1 and t_2 are the two intersection instants. When

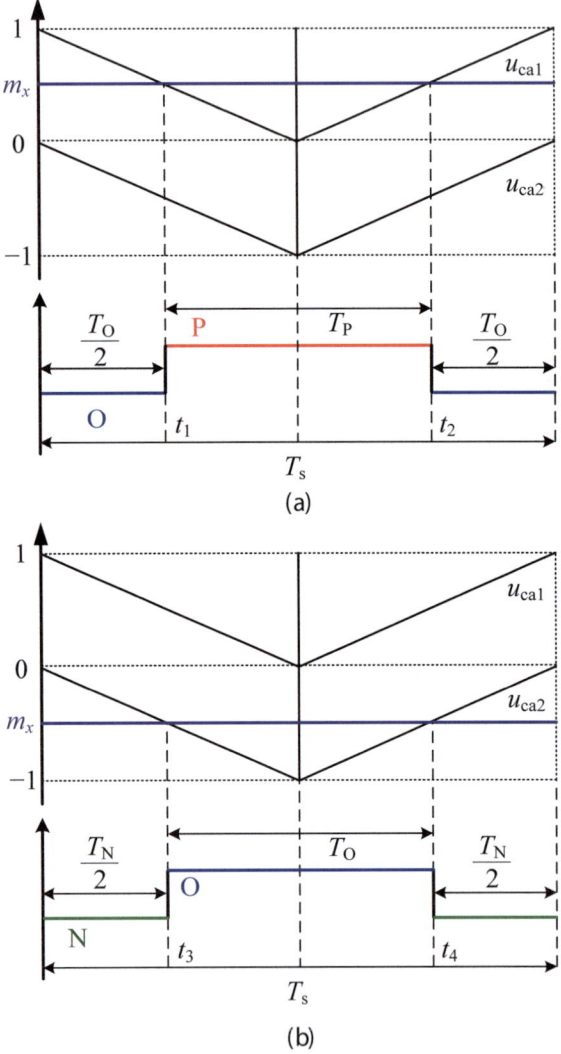

Fig. 2.2 Typical implementation of the three-level PWM process **a** $m_x \geq 0$, **b** $m_x < 0$

2.1 Working Principles and Mathematical Models

$m_x \geq u_{ca1}$, the phase x outputs [P] state. When $m_x < u_{ca1}$, the phase x outputs [O] state. T_P and T_O denote the dwell time of the states [P] and [O], respectively. According to the principle of similar triangles, T_P can be calculated as

$$T_P = T_s m_x \quad (2.3)$$

Figure 2.2b shows the waveform of the output state when $m_x < 0$. As shown, m_x intersects only with u_{ca2}, where t_3 and t_4 are the two intersection instants. When $m_x \geq u_{ca2}$, the phase x outputs [O] state. When $m_x < u_{ca2}$, the phase x output N state. T_N denotes the dwell time of the state [N]. Similarly, T_N can be calculated as

$$T_N = T_s m_x \quad (2.4)$$

According to the above analyses, the averaged value of S_x over one switching period, for both $m_x \geq 0$ and $m_x < 0$, equals m_x. As a result, the averaged voltage of u_x over one switching period, for both $m_x \geq 0$ and $m_x < 0$, is derived as

$$u_x = \frac{U_{dc}}{2} m_x \quad (2.5)$$

It should be noted that Eq. (2.2) provides an exact representation of u_x, which includes the on-state and off-state behaviors of the power devices. This equation captures both the low-frequency and high-frequency components of the output voltage. In contrast, Eq. (2.5) presents an averaged representation of u_x, incorporating only the low-frequency components of the output voltage.

As described previously, each of the three phases has three output states. As a result, there are total $3^3 = 27$ output state combinations for 3LT^{2}I. In the three-phase system, the concept of space vector is commonly used for analysis. Taking the inverter output voltage as an example, the space vector of u_a, u_b, and u_c, denoted as U_{abc}, is defined as

$$U_{abc} = \frac{2}{3}\left(u_a + e^{j\frac{2\pi}{3}} u_b + e^{-j\frac{2\pi}{3}} u_c\right) \quad (2.6)$$

where e is the base of the natural logarithm and j is the imaginary unit.

Similarly, the space vectors of inverter currents and grid voltages, denoted respectively as I_{abc} and E_{abc}, are defined as

$$I_{abc} = \frac{2}{3}\left(i_a + e^{j\frac{2\pi}{3}} i_b + e^{-j\frac{2\pi}{3}} i_c\right) \quad (2.7)$$

$$E_{abc} = \frac{2}{3}\left(e_a + e^{j\frac{2\pi}{3}} e_b + e^{-j\frac{2\pi}{3}} e_c\right) \quad (2.8)$$

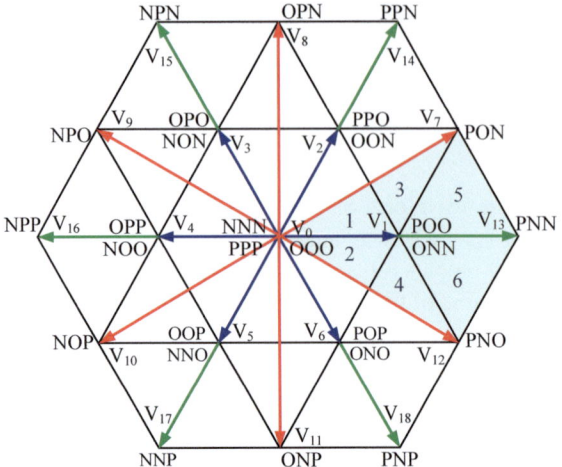

Fig. 2.3 Space voltage vector diagram of 3LT^{2}I

Based on the definition of the space vector, we can calculate the 27 output voltage space vectors that correspond to the 27 output state combinations. The diagrams of these space vectors are illustrated in Fig. 2.3 [2, 3].

For convenience, the voltage space vectors are classified into four types according to their amplitudes. The four types consist of zero, small, medium, and large voltage vectors, respectively. Table 2.2 shows the detailed classification. As demonstrated, there are 3 zero voltage vectors, 12 small voltage vectors, 6 medium voltage vectors, and 6 large voltage vectors. Among the four types of voltage vectors, the small voltage vectors are further classified as the P-type small voltage vectors (only consisting of [P] and [O] states) and the N-type small voltage vectors (only consisting of [N] and [O] states). In addition, each P-type small voltage vector corresponds to an N-type small voltage vector, both located at the same position. Therefore, there are six pairs of small voltage vectors. We refer to each pair of small voltage vectors as mutually redundant, as they are located at the same position.

2.1.2 AC-Side Modeling

In this part, the AC-side models of 3LT^{2}I are constructed. It should be noted that L and LCL filters are usually adopted on the side for connecting the power grid. For different filters, the AC-side models of 3LT^{2}I are different.

2.1 Working Principles and Mathematical Models

Table 2.2 The classification of voltage space vectors

Voltage vector	Magnitude	Output state	
Zero vector	0	[PPP] [OOO] [NNN]	
Small vector	$\frac{1}{3}U_{dc}$	P-type	N-type
		[POO]	[ONN]
		[PPO]	[OON]
		[OPO]	[NON]
		[OPP]	[NOO]
		[POP]	[ONO]
		[OOP]	[NNO]
Medium vector	$\frac{\sqrt{3}}{3}U_{dc}$	[PON] [OPN] [NPO]	
		[NOP] [ONP] [PNO]	
Large vector	$\frac{2}{3}U_{dc}$	[PNN] [PPN] [NPN]	
		[NPP] [NNP] [PNP]	

Based on L Filter

When L is selected as the AC-side filter, the structure $3LT^2I$ is shown in Fig. 2.1. Selecting point O as the reference point, the AC-side mathematical model of $3LT^2I$ in the *abc* frame is derived as [1, 4]

$$L_f \frac{d}{dt}\begin{bmatrix} i_a \\ i_b \\ i_c \end{bmatrix} = -R_f \begin{bmatrix} i_a \\ i_b \\ i_c \end{bmatrix} + \begin{bmatrix} u_a \\ u_b \\ u_c \end{bmatrix} - \begin{bmatrix} e_a \\ e_b \\ e_c \end{bmatrix} - u_{nO}\begin{bmatrix} 1 \\ 1 \\ 1 \end{bmatrix} \quad (2.9)$$

where u_{nO} denotes the voltage between point O and point n_e.

By combining (2.9) with (2.2), an exact mathematical model of the AC side, defined as the switching model, can be derived as

$$L_f \frac{d}{dt}\begin{bmatrix} i_a \\ i_b \\ i_c \end{bmatrix} = -R_f \begin{bmatrix} i_a \\ i_b \\ i_c \end{bmatrix} + \frac{U_{dc}}{2}\begin{bmatrix} S_a \\ S_b \\ S_c \end{bmatrix} - \begin{bmatrix} e_a \\ e_b \\ e_c \end{bmatrix} - u_{nO}\begin{bmatrix} 1 \\ 1 \\ 1 \end{bmatrix} \quad (2.10)$$

Similarly, by combining (2.9) with (2.5), the averaged model of the AC side can be derived as

$$L_f \frac{d}{dt}\begin{bmatrix} i_a \\ i_b \\ i_c \end{bmatrix} = -R_f \begin{bmatrix} i_a \\ i_b \\ i_c \end{bmatrix} + \frac{U_{dc}}{2}\begin{bmatrix} m_a \\ m_b \\ m_c \end{bmatrix} - \begin{bmatrix} e_a \\ e_b \\ e_c \end{bmatrix} - u_{nO}\begin{bmatrix} 1 \\ 1 \\ 1 \end{bmatrix} \quad (2.11)$$

As indicated by (2.1), the switching function S_x is discontinuous. As a result, the switching model is discontinuous and is not suitable for controller design. As an alternative, the averaged model is continuous and can be utilized for controller design. However, the voltage u_{nO} in the averaged model cannot be easily measured, which brings additional difficulties to the controller design. To address this issue, the coordinate transformations are usually introduced. Generally, there are two commonly used coordinate transformations. The first one is the transformation from the abc frame to the $\alpha\beta$ frame, and the second one is the transformation from the $\alpha\beta$ frame to the dq frame.

Taking the transformation of inverter currents as an example, the relationships of currents in different frames are demonstrated in Fig. 2.4. Figure 2.4a shows the relationship between abc frame and $\alpha\beta$ frame, where i_α and i_β denote the currents of the α axis and the β axis, respectively. θ_i is the phase angle of the current space vector $\boldsymbol{I}_{abc}$. As shown, the α axis coincides with a coordinate axis, and the β axis leads the α axis by 90°. Denoting $\boldsymbol{T}_{abc/\alpha\beta}$ as the transformation matrix from abc frame to $\alpha\beta$ frame, $\boldsymbol{T}_{abc/\alpha\beta}$ can be derived as

$$\boldsymbol{T}_{abc/\alpha\beta} = \frac{2}{3}\begin{bmatrix} 1 & -\frac{1}{2} & -\frac{1}{2} \\ 0 & \frac{\sqrt{3}}{2} & -\frac{\sqrt{3}}{2} \end{bmatrix} \tag{2.12}$$

As shown in Fig. 2.4a, the amplitude of $\boldsymbol{I}_{abc}$ remains unchanged after the transformation. By multiplying $\boldsymbol{T}_{abc/\alpha\beta}$ with the three-phase inverter current i_a, i_b, and i_c, i_α and i_β can be calculated as

$$\begin{bmatrix} i_\alpha \\ i_\beta \end{bmatrix} = \frac{2}{3}\begin{bmatrix} 1 & -\frac{1}{2} & -\frac{1}{2} \\ 0 & \frac{\sqrt{3}}{2} & -\frac{\sqrt{3}}{2} \end{bmatrix}\begin{bmatrix} i_a \\ i_b \\ i_c \end{bmatrix} \tag{2.13}$$

Conversely, the variables in the abc frame can also be obtained from the variables in the $\alpha\beta$ frame. The corresponding transformation matrix $\boldsymbol{T}_{\alpha\beta/abc}$ and transformation

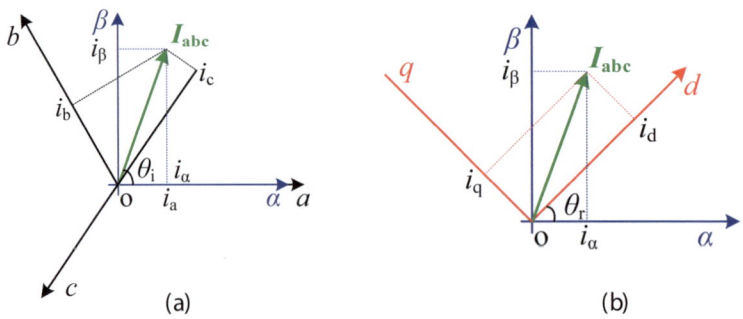

Fig. 2.4 The relationship of the abc, $\alpha\beta$, and dq frames **a** relationship between abc frame and $\alpha\beta$ frame, **b** relationship between $\alpha\beta$ frame and dq frame

2.1 Working Principles and Mathematical Models

equation are respectively expressed as

$$T_{\alpha\beta/abc} = \begin{bmatrix} 1 & 0 \\ -\frac{1}{2} & \frac{\sqrt{3}}{2} \\ -\frac{1}{2} & -\frac{\sqrt{3}}{2} \end{bmatrix} \qquad (2.14)$$

$$\begin{bmatrix} i_a \\ i_b \\ i_c \end{bmatrix} = \begin{bmatrix} 1 & 0 \\ -\frac{1}{2} & \frac{\sqrt{3}}{2} \\ -\frac{1}{2} & -\frac{\sqrt{3}}{2} \end{bmatrix} \begin{bmatrix} i_\alpha \\ i_\beta \end{bmatrix} \qquad (2.15)$$

As observed, the multiplication of $T_{abc/\alpha\beta}$ and $T_{\alpha\beta/abc}$ equals a two-dimensional identity matrix. By multiplying $T_{\alpha\beta/abc}$ with the three-phase averaged model in (2.11), the averaged model of the AC side in the $\alpha\beta$ frame is derived as

$$L_f \frac{d}{dt}\begin{bmatrix} i_\alpha \\ i_\beta \end{bmatrix} = -R_f \begin{bmatrix} i_\alpha \\ i_\beta \end{bmatrix} + \frac{U_{dc}}{2}\begin{bmatrix} m_\alpha \\ m_\beta \end{bmatrix} - \begin{bmatrix} e_\alpha \\ e_\beta \end{bmatrix} \qquad (2.16)$$

where the subscripts α and β denote variables in the α axis and the β axis, respectively. The grid voltages e_a, e_b, and e_c are assumed to be balanced.

In (2.16), the voltage u_{nO} has been removed in the mathematical model. Therefore, this model can be used for controller design. The variables in $\alpha\beta$ frame are sinusoidal signals. When aiming to track the reference signals of the inverter output currents, the proportional-integral (PI) regulator cannot achieve zero steady-state control errors. Instead, the proportional resonance (PR) controller is usually adopted when the model of (2.16) is selected.

Figure 2.4b shows the relationship between $\alpha\beta$ frame and dq frame, where i_d and i_q denote the currents of the d axis and the q axis, respectively. The dq frame rotates counterclockwise at the fundamental angular frequency ω_1, and θ_r denotes the phase angle between the d axis and the α axis. In addition, the q axis leads the d axis by 90°. It should also be noted that the derivative of θ_r equals ω_1, since both the abc frame and the $\alpha\beta$ frame are stationary frames. Denoting $T_{\alpha\beta/dq}$ as the transformation matrix from the $\alpha\beta$ frame to the dq frame, $T_{\alpha\beta/dq}$ can be derived as

$$T_{\alpha\beta/dq} = \begin{bmatrix} \cos\theta_r & \sin\theta_r \\ -\sin\theta_r & \cos\theta_r \end{bmatrix} \qquad (2.17)$$

By multiplying $T_{\alpha\beta/dq}$ with the inverter current i_α and i_β in the $\alpha\beta$ frame, i_d and i_q can be calculated as

$$\begin{bmatrix} i_d \\ i_q \end{bmatrix} = \begin{bmatrix} \cos\theta_r & \sin\theta_r \\ -\sin\theta_r & \cos\theta_r \end{bmatrix}\begin{bmatrix} i_\alpha \\ i_\beta \end{bmatrix} \qquad (2.18)$$

Conversely, the variables in the $\alpha\beta$ frame can also be obtained from the variables in the dq frame. The corresponding transformation matrix $T_{dq/\alpha\beta}$ and transformation equation are respectively expressed as

$$\boldsymbol{T}_{dq/\alpha\beta} = \boldsymbol{T}_{\alpha\beta/dq}^{-1} = \begin{bmatrix} \cos\theta_r & -\sin\theta_r \\ \sin\theta_r & \cos\theta_r \end{bmatrix} \quad (2.19)$$

$$\begin{bmatrix} i_\alpha \\ i_\beta \end{bmatrix} = \begin{bmatrix} \cos\theta_r & -\sin\theta_r \\ \sin\theta_r & \cos\theta_r \end{bmatrix} \begin{bmatrix} i_d \\ i_q \end{bmatrix} \quad (2.20)$$

As observed, $\boldsymbol{T}_{dq/\alpha\beta}$ is the inverse of $\boldsymbol{T}_{\alpha\beta/dq}$. By multiplying $\boldsymbol{T}_{\alpha\beta/abc}$ with the three-phase averaged model in (2.16), the averaged model of the AC side in the dq frame is derived as

$$L_f \frac{d}{dt}\begin{bmatrix} i_d \\ i_q \end{bmatrix} = -\begin{bmatrix} R_f & -\omega_1 L_f \\ \omega_1 L_f & R_f \end{bmatrix}\begin{bmatrix} i_d \\ i_q \end{bmatrix} + \frac{U_{dc}}{2}\begin{bmatrix} m_d \\ m_q \end{bmatrix} - \begin{bmatrix} e_d \\ e_q \end{bmatrix} \quad (2.21)$$

where the subscripts d and q denote variables in d axis and q axis, respectively.

After $\alpha\beta/dq$ transformation, the fundamental frequency sinusoidal variables are converted into DC variables. As a result, zero steady-state control errors of inverter currents can be realized by the PI regulator. Nevertheless, the state variables i_d and i_q are coupled with each other, which brings extra difficulties in the analysis and controller design of the inverter system.

Based on *LCL* Filter

To achieve better attenuation of high-frequency harmonics, an *LCL* filter can be employed. Figure 2.5 describes the topology of 3LT²I with *LCL* filter. As shown, the *LCL* filter is composed of L_1, L_2, and C_f. L_g represents the inductance of the transmission line. The 3LT²I connects to the power grid through the point of common coupling (PCC). i_{a1}, i_{b1}, and i_{b1} are the converter currents. i_{a2}, i_{b2}, and i_{b2} denote the grid currents. i_{fa}, i_{fb}, and i_{fc} represent the AC capacitor currents. v_{fa}, v_{fb}, and v_{fc} are the AC capacitor voltages. v_{ga}, v_{gb}, and v_{gc} denote the voltages of PCC. n_f is the common node of AC capacitors.

Based on Kirchhoff's voltage and current laws, the mathematical model of 3LT²I with *LCL* filter is derived as [5, 6]

$$\begin{cases} L_1 \frac{di_{x1}}{dt} = -v_{fx} + u_x \\ C_f \frac{dv_{fx}}{dt} = i_{x1} - i_{x2} \\ L_{gs} \frac{di_{x2}}{dt} = v_{fx} - e_x - u_{nf} \end{cases} \quad (2.22)$$

where $L_{gs} = L_2 + L_g$. u_{nf} denotes the voltage between point n_e and point n_f.

By applying $abc/\alpha\beta$ transformation, the models of 3LT²I with *LCL* filter in the $\alpha\beta$ frame are expressed as

2.1 Working Principles and Mathematical Models

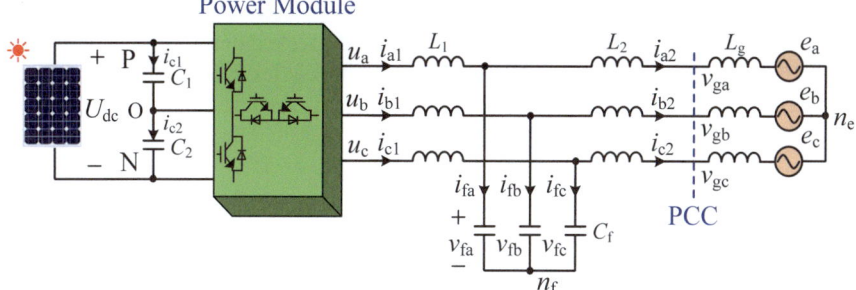

Fig. 2.5 The topology of 3LT^{2}I with *LCL* filter

$$\begin{cases} L_1 \frac{d}{dt} \begin{bmatrix} i_{\alpha 1} \\ i_{\beta 1} \end{bmatrix} = -\begin{bmatrix} v_{f\alpha} \\ v_{f\beta} \end{bmatrix} + \begin{bmatrix} u_\alpha \\ u_\beta \end{bmatrix} \\ C_f \frac{d}{dt} \begin{bmatrix} v_{f\alpha} \\ v_{f\beta} \end{bmatrix} = \begin{bmatrix} i_{\alpha 1} \\ i_{\beta 1} \end{bmatrix} - \begin{bmatrix} i_{\alpha 2} \\ i_{\beta 2} \end{bmatrix} \\ L_{gs} \frac{d}{dt} \begin{bmatrix} i_{\alpha 2} \\ i_{\beta 2} \end{bmatrix} = \begin{bmatrix} v_{f\alpha} \\ v_{f\beta} \end{bmatrix} - \begin{bmatrix} e_\alpha \\ e_\beta \end{bmatrix} \end{cases} \quad (2.23)$$

Similarly, the models of 3LT^{2}I with *LCL* filter in the *dq* frame are derived as

$$\begin{cases} L_1 \frac{d}{dt} \begin{bmatrix} i_{d1} \\ i_{q1} \end{bmatrix} = \begin{bmatrix} 0 & \omega_1 L_1 \\ -\omega_1 L_1 & 0 \end{bmatrix} \begin{bmatrix} i_{d1} \\ i_{q1} \end{bmatrix} - \begin{bmatrix} v_{fd} \\ v_{fq} \end{bmatrix} + \begin{bmatrix} u_d \\ u_q \end{bmatrix} \\ C_f \frac{d}{dt} \begin{bmatrix} v_{fd} \\ v_{fq} \end{bmatrix} = \begin{bmatrix} 0 & \omega_1 C_f \\ -\omega_1 C_f & 0 \end{bmatrix} \begin{bmatrix} v_{fd} \\ v_{fq} \end{bmatrix} + \begin{bmatrix} i_{d1} \\ i_{q1} \end{bmatrix} - \begin{bmatrix} i_{d2} \\ i_{q2} \end{bmatrix} \\ L_{gs} \frac{d}{dt} \begin{bmatrix} i_{d2} \\ i_{q2} \end{bmatrix} = \begin{bmatrix} 0 & \omega_1 L_{gs} \\ -\omega_1 L_{gs} & 0 \end{bmatrix} \begin{bmatrix} i_{d2} \\ i_{q2} \end{bmatrix} + \begin{bmatrix} v_{fd} \\ v_{fq} \end{bmatrix} - \begin{bmatrix} e_d \\ e_q \end{bmatrix} \end{cases} \quad (2.24)$$

Notably, the corresponding switching models and averaged models of 3LT^{2}I with *LCL* filter in three frames can be obtained by substituting (2.2) and (2.5) into the above models, respectively. These models are not detailed anymore in this book.

2.1.3 DC-Side Modeling

Generally, the DC-side capacitors C_1 and C_2 are identical, and their capacitance is denoted as C_{dc}. According to circuit theory, the differential equations of the DC-side capacitors can be derived as

$$\begin{cases} \frac{d}{dt} U_p = \frac{i_{c1}}{C_{dc}} \\ \frac{d}{dt} U_n = \frac{i_{c2}}{C_{dc}} \end{cases} \quad (2.25)$$

As shown in Fig. 2.1, U_{dc} equals the sum of U_p and U_n. In addition, we denote $\Delta U_{pn} = U_p - U_n$ as the difference between U_p and U_n. To derive a more useful mathematical model for the DC side, we select U_{dc} and ΔU_{pn} as the states. Considering $i_{c1} = i_{dc} - i_p$, $i_{c2} = i_{dc} + i_n$, the derivatives of U_{dc} and ΔU_{pn} are respectively expressed as

$$\frac{d}{dt}U_{dc} = \frac{2i_{dc} - i_p + i_n}{C_{dc}} \tag{2.26}$$

$$\frac{d}{dt}\Delta U_{pn} = \frac{i_o}{C_{dc}} \tag{2.27}$$

According to Table 2.1, the values of i_p, i_n, and i_o are closely related to the output states of each phase. When phase x (where $x = $ a, b, or c) is in [P] state, the corresponding output current i_x contributes to i_p. Similarly, when phase x is in [O] or [N] states, the output current i_x contributes to i_o or i_n, respectively. As a result, $-i_p + i_n$ and i_o can be respectively represented in terms of i_x and S_x as

$$-i_p + i_n = -S_a i_a - S_b i_b - S_c i_c \tag{2.28}$$

$$i_o = -[(1 - |S_a|)i_a + (1 - |S_b|)i_b + (1 - |S_c|)i_c] \tag{2.29}$$

Subsequently, the switching models of the DC-side are derived as

$$\frac{d}{dt}U_{dc} = \frac{2i_{dc} - S_a i_a - S_b i_b - S_c i_c}{C_{dc}} \tag{2.30}$$

$$\frac{d}{dt}\Delta U_{pn} = -\frac{(1 - |S_a|)i_a + (1 - |S_b|)i_b + (1 - |S_c|)i_c}{C_{dc}} \tag{2.31}$$

As described previously, the averaged value of S_x over one switching period equals the corresponding modulation wave m_x. Assuming that the ripples of the inverter currents are sufficiently small, the average models of the DC side can be further derived as

$$\frac{d}{dt}U_{dc} = \frac{2i_{dc} - m_a i_a - m_b i_b - m_c i_c}{C_{dc}} \tag{2.32}$$

$$\frac{d}{dt}\Delta U_{pn} = -\frac{(1 - |m_a|)i_a + (1 - |m_b|)i_b + (1 - |m_c|)i_c}{C_{dc}} \tag{2.33}$$

Moreover, by applying the *abc/αβ* transformation and the *αβ/dq* transformation, the following equation can be obtained as

$$m_a i_a + m_b i_b + m_c i_c = 1.5(m_d i_d + m_q i_q) \tag{2.34}$$

As a result, the model of (2.32) can be rewritten as

2.2 Control Framework

$$\frac{d}{dt}U_{dc} = \frac{2i_{dc} - 1.5m_d i_d - 1.5m_q i_q}{C_{dc}} \quad (2.35)$$

It is observed that the model of (2.35) is nonlinear because of the multiplication terms $m_d i_d$ and $m_d i_q$. In addition, it is difficult to find a suitable control variables from (2.35) to regulate the dc-link voltage, since there are too many variables to affect this voltage. Based on the law of power conservation, we can derive another form of the DC-side voltage model [7]. First, the instantaneous power that flows into the power module is calculated by the following equation

$$p_{in} = U_{dc} i_{dc} + \frac{1}{2} C_{dc} U_{dc} \dot{U}_{dc} \quad (2.36)$$

In the above equation, the contribution of i_o to the capacitor power is neglected as the averaged value of i_o is zero for a balanced NPV. As for the AC side, the active and reactive power injected into the power grid can be calculated according to the instantaneous power theory as

$$\begin{cases} p_g = \frac{3}{2}(e_d i_d + e_q i_q) \\ q_g = \frac{3}{2}(e_q i_d - e_d i_q) \end{cases} \quad (2.37)$$

Considering the equivalent series resistance R_f is relatively small, the power dissipation on R_f is neglected. Based on the law of power conservation, the dynamic model of the U_{dc} is expressed as

$$\frac{1}{2} C_{dc} U_{dc} \frac{d}{dt} U_{dc} = U_{dc} i_{dc} - \frac{3}{2}(e_d i_d + e_q i_q) \quad (2.38)$$

By employing the PLL, the phase angle of the power grid can be obtained and is typically adopted as the coordinate transformation angle. Therefore, the q-axis variable of grid voltage e_q is approximately equal to zero. Then, the d-axis variable of the grid current i_d can be selected as the control variable of the dc-link voltage.

2.2 Control Framework

In this section, the control framework for PV inverter with generalized topology is introduced, which is depicted in Fig. 2.6. The control framework includes five stages for different control objectives. In stage 1, the outer loop implements the algorithm of maximum power point tracking (MPPT) to track the maximum power of PV arrays [8]. The MPPT algorithm yields the dc-link voltage reference. In stage 2, the middle loop regulates the dc-link voltage to its reference value [9]. This is achieved by regulating the active current (i.e., the d-axis current i_d) since the dc-link voltage is directly affected by active power. Naturally, the output of the middle loop is the

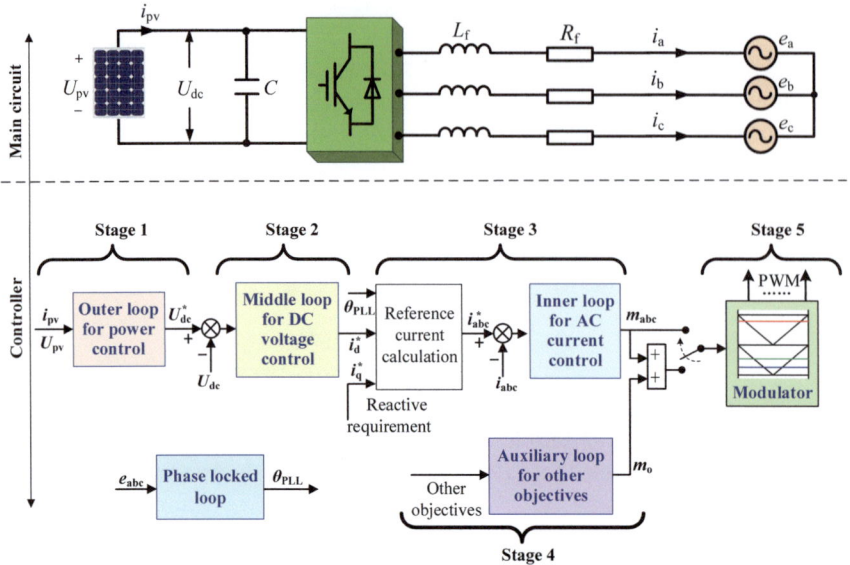

Fig. 2.6 The control framework for PV inverter with generalized topology

active current reference or the *d*-axis current reference i_d^*. In stage 3, the inner loop controls the output current to track their reference values. It should be noted that the reference currents (i_{abc}^*) are calculated according to the active current reference (i.e., i_d^*), the reactive current reference (i.e., the *q*-axis current reference i_q^*), and the phase of grid voltage (θ_{PLL}). The reactive current reference is manually set to meet the requirement of reactive power control [1, 10]. The PLL is employed to obtain phase information of the power grid [11], which provides the synchronization signal of the grid voltage to the current controller. The outputs of the current inner loop are the three-phase sinusoidal modulation signals m_{abc}.

Stage 1, stage 2, and stage 3 form the typically cascaded control structure. It should be noted that the power, DC voltage, and AC current are the basic control objectives for the commonly used PV inverter, which merely satisfy the basic normal operation requirements. However, to meet the requirements of commercialization of PV power generation, some special performance indexes have to be satisfied.

In stage 4, the other control objectives or goals are achieved by the auxiliary loop. The other control objectives include the leakage current suppression for improving the security of users, switching loss reduction for high efficiency, current harmonic elimination for high power quality, and so on. Moreover, for specific topologies of PV inverter, other control objectives are also realized in the auxiliary loop. These objectives cover the controllability of NPV in $3LT^2I$, suppression of circulating current in the modular multilevel converter (MMC), equalization of capacitor voltage of each modular in the cascaded H-bridge converter (CHB), etc. The output of auxiliary loop is the zero-sequence modulation waveform.

2.2 Control Framework

In addition to the above control modules, the modulator is also one of the most important modules, which is located in stage 5. The modulator acts as an actuator in PV inverter system and connects the output of controllers to the power module of PV inverter. The modulator determines the actions of power semiconductor switches. In other words, it converts the continuous output signals from the control loops into the discrete pulse signals to control ON and OFF states of the power semiconductor switches. The modulator also impacts system performances. For instance, the power losses and the output waveform quality highly depend on the modulation methods.

For clarity, the commonly used topology $3LT^2I$ is taken as the example to explain the specific control framework as shown in Fig. 2.7. The MPPT algorithm is integrated into the outer loop to realize maximum power generation. The input variables are i_{pv} and U_{pv} of the PV array, and the output is the dc-link voltage reference U^*_{dc}. The voltage controller is integrated into the middle loop, which tracks the desired dc-link voltage U^*_{dc} obtained by the MPPT algorithm. The output variable of the middle loop is the active reference current i_d^*. For the inner current loop, the three-phase grid voltage and current e_{abc} and i_{abc} in abc frame are transformed into DC variables e_{dq} and i_{dq} in the dq frame and input the current controller. The current controller is used to track the reference active current i_d^* and reactive current i_q^*, and the output variables are the modulation signals in dq frame m_{dq}. Then dq/abc transformation is applied to transmit m_{dq} into three-phase sinusoidal modulation signals m_{as}, m_{bs}, and m_{cs}.

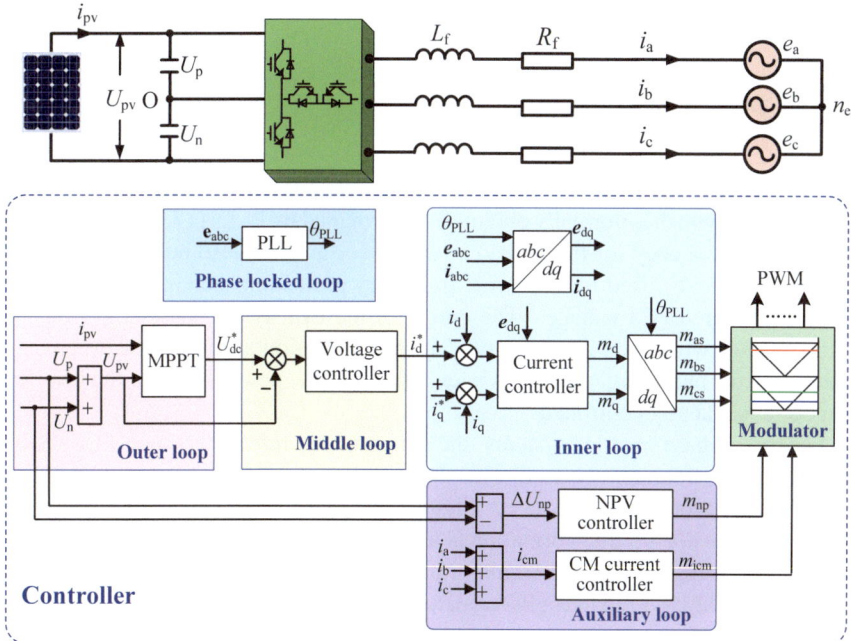

Fig. 2.7 The control framework for $3LT^2I$ of PV inverter

Besides the objectives of PV power, dc-link voltage and grid current, leakage current and NPV are another two objectives for the particularity of PV application and $3LT^2I$ topology [12, 13]. The reasons are listed as follows. In the special application of PV power generation, leakage current i_{cm} is a critical issue, because it distorts the output currents quality and threatens people safety. For the inherent characteristic of $3LT^2I$ topology, NPV directly affects the output performance and determines the safe and stable operation of the inverter. Thus, for the $3LT^2I$ topology applied in a PV power generation system, leakage current and NPV are another two control objectives. The NPV controller is used to adjust the NPV in its reference value, and its output is m_{np}. and the common-mode (CM) controller is used to suppress the leakage current or CM resonance current, and its output is m_{icm}. These two controllers are integrated into the auxiliary loop.

It should be noted that although the above two objectives can be achieved by adding hardware to modify the inverter topology, we categorize them into the auxiliary loop in this book. The reason is that adding hardware to modify the inverter topology will cause other problems that need to be solved by appropriate control methods in the auxiliary loop. For example, in the engineer application, a modified *LCL* (*MLCL*) filter is usually employed in $3LT^2I$ to reduce leakage currents [14–16]. However, the common-mode resonant current (CMRC) is induced by the *MLCL* filter, which can be suppressed by the CM current controller.

The output variables of the current inner loop and auxiliary loop are fed into the modulator to produce the gating signals to drive power semiconductor switches ON or OFF. The detailed control process is illustrated in the following section.

2.2.1 Phase Locked Loop

The grid synchronization of $3LT^2I$ necessitates the phase angle information of the grid voltages, which is normally obtained and provided by PLLs [17, 18]. In addition, the angular frequency (or frequency) of the power grid can also be obtained by the PLL, which can be used for abnormal frequency protection. The inputs of the PLL are the measured grid voltages. The output of the PLL is the phase angle θ_{PLL} of grid voltages, which is utilized for *abc/dq* transformation as mentioned previously. Therefore, the phase angle of the inverter current (or the power factor of the inverter) can be subsequently controlled.

For three-phase inverter systems, the synchronous reference frame PLL (SRF-PLL) method is commonly used. This method features the advantages of a simple structure and a fast response speed. Figure 2.8 describes the block diagram of the SRF-PLL. The *abc/dq* transformation is first applied to measured grid voltages. The rotating phase angle used for the transformation adopts θ_{PLL}. Then, the *q*-axis variable of the grid voltages is fed into a PI regulator. The output of the PI regulator is denoted as $\Delta\omega_{PLL}$, which can be recognized as the variation of the angular frequency. The sum of $\Delta\omega_{PLL}$ and ω_1, denoted as ω_{PLL} is the angular frequency. The output phase angle is further obtained by integrating ω_{PLL}, thereby forming the closed loop.

2.2 Control Framework

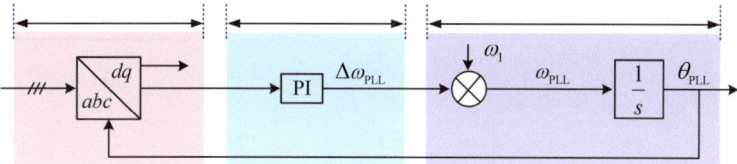

Fig. 2.8 The schemes of SRF-PLL

For balanced three-phase sinusoidal grid voltages, the corresponding dq-axis voltages are DC values. When θ_{PLL} equals the grid-phase angle, the d-axis voltage e_q equals zero. Conversely, $e_q = 0$ implies the θ_{PLL} equals the grid-phase angle. Therefore, the objective of the PLL is realized.

However, this method assumes that the grid voltages are balanced and contain low levels of harmonics. When the grid voltages are unbalanced, the double-frequency ripples will be introduced in e_q. When the grid voltages contain high-order harmonics, the high-frequency ripples will also be introduced in e_q. As a result, the phase information is inaccurate. To address these issues, the dual second-order generalized integrator-based PLL (DSOGI-PLL) is proposed [19, 20]. As indicated by the name of the improved PLL, it contains two second-order generalized integrators (SOGIs), which are used to generate the virtual orthogonal signal.

The implementation of the SOGI is demonstrated in Fig. 2.9, where the α-axis grid voltage e_α is taken as an example [21]. As shown, there are two outputs of the SOGI, namely $e_{\alpha I}$ and $qe_{\alpha I}$. When e_α is a pure sinusoidal signal, e_α and qe_α are in phase, and $qe_{\alpha I}$ lags e_α by 90°. Therefore, $e_{\alpha I}$ and $qe_{\alpha I}$ are orthogonal signals. The parameter k_{sogi} is the damping coefficient. According to Fig. 2.9, the transfer functions from e_α to $e_{\alpha I}$ and $qe_{\alpha I}$ can be derived respectively as

$$D_e(s) = \frac{e_{\alpha I}(s)}{e_\alpha(s)} = \frac{k_{sogi}s}{s^2 + k_{sogi}s + \omega_1^2} \qquad (2.39)$$

$$D_{qe}(s) = \frac{qe_{\alpha I}(s)}{e_\alpha(s)} = \frac{k_{sogi}\omega_1^2}{s^2 + k_{sogi}s + \omega_1^2} \qquad (2.40)$$

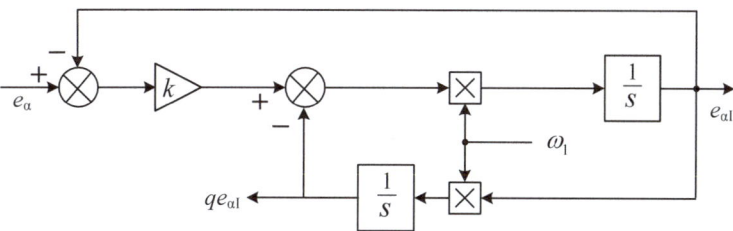

Fig. 2.9 The diagram of SOGI

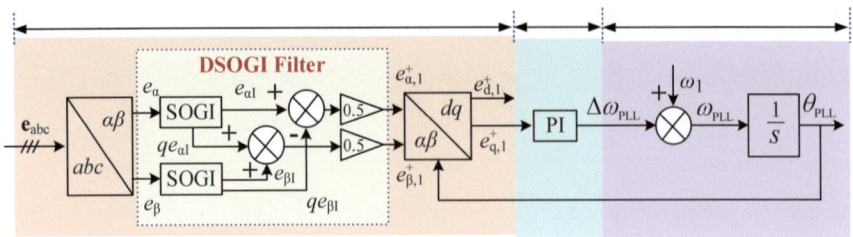

Fig. 2.10 The diagram of DSOGI-PLL

In addition to generating orthogonal signals, the SOGI also benefits from the filtering capabilities of harmonics according to the above transfer functions. When the power angular frequency changes, ω_1 in (2.39) and (2.40) can be replaced by ω_{PLL} with a low-pass filter, forming an adapted version of the SOGI. In Fig. 2.10, the implementation of the DSOGI-PLL is illustrated. Compared with the SRF-PLL method, a DSOGI-based filter is adopted before $\alpha\beta/dq$ transformation. The DSOGI-PLL adopts two SOGI modules for α and β axes, respectively. Then, the positive sequence fundamental components of the grid voltages are calculated as

$$\begin{cases} e_{\alpha,1}^+ = \frac{1}{2}(e_{\alpha I} - qe_{\beta I}) \\ e_{\beta,1}^+ = \frac{1}{2}(qe_{\alpha I} + e_{\beta I}) \end{cases} \quad (2.41)$$

2.2.2 Outer Loop for Power Control

The outer loop employs the MPPT algorithm to track the MPP of PV arrays. For renewable energy generation, it is important to maximize power output as much as possible. Figure 2.11 shows the characteristics of PV arrays under the same irradiance but different temperature conditions, where Fig. 2.11a, b describe the corresponding voltage-current (*U-I*) and power-voltage (*P-U*) characteristics, respectively. Figure 2.12 shows the characteristics of PV arrays under the same temperature but different irradiance conditions, where Fig. 2.12a, b describe the corresponding voltage-current (*U-I*) and power-voltage (*P-U*) characteristics, respectively. As observed from these curves, PV arrays exhibit typical nonlinear characteristics due to the influence of irradiance and temperature. The output power reaches its maximum when the inverter operates at a specific voltage point. However, when part of the PV array is shaded or a fault occurs, these characteristic curves become more complex. On these conditions, multiple local voltage peaks may occur, which brings significant challenges for MPPT.

Generally, the performance of an MPPT algorithm can be evaluated by the MPPT efficiency. Denoting η_{mppt} as MPPT efficiency, its definition is described as

2.2 Control Framework

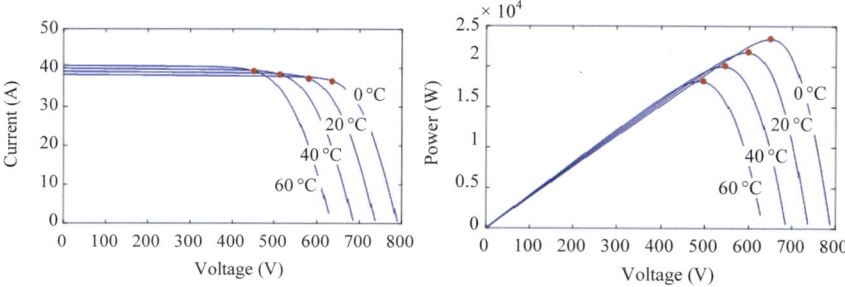

Fig. 2.11 Characteristics of PV arrays under the same irradiance but different temperature conditions **a** *U-I* curve, **b** *P-U* curve

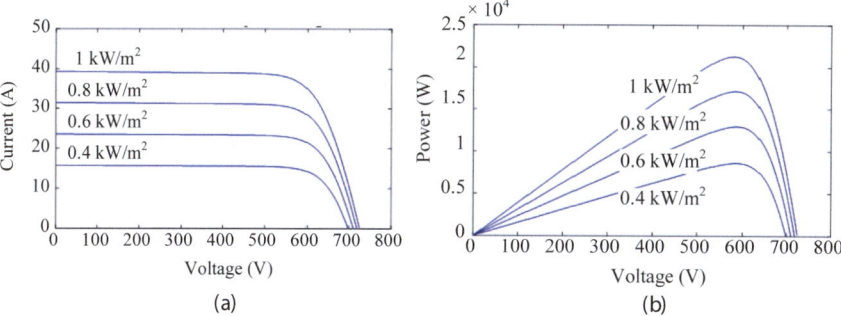

Fig. 2.12 Characteristics of PV arrays under the same temperature but different irradiance conditions **a** *U-I* curve, **b** *P-U* curve

$$\eta_{\text{mppt}} = \frac{\text{Actual output power}}{\text{Maximum available power}} \times 100\% \qquad (2.42)$$

According to the definition, the higher the MPPT efficiency, the more output power can be obtained. Commonly used MPPT algorithms include: perturb and observe (P&O) method, incremental conductance (INC) method, fuzzy logic control (FLC) method, particle swarm optimization (PSO) method, neural network (NN) based method, and others [22–24]. Among those methods, the P&O method is the most commonly used in engineering applications. However, it usually suffers from steady-state oscillations. To overcome this drawback. Improved versions, such as the hysteresis comparison-based P&O method and variable step size-based P&O method are proposed. The basic operating principles of the above methods are presented in Chap. 4.

2.2.3 Middle Loop for DC Voltage Control

For 3LT²I, the stability of the middle loop for DC voltage control is important. On one hand, this control loop serves as the actuator of the outer loop. Accurate tracking of dc-link voltage reference guarantees the MPPT. On the other hand, a stable middle loop ensures the safe operation of 3LT²I. As observed from (2.21), the dc-link voltage directly affects the output voltage of the power module. If this loop is out of control, it will be difficult to regulate the output currents.

Typically, the PI regulator is employed for DC voltage control. To ensure a successful connection to the power grid, the dc-link voltage should be controlled higher than the peak value of the AC line voltage. As described, the dc-link voltage is regulated by adjusting the d-axis current reference. For convenience, we rewrite the DC-side mathematical model of (2.38) as follows

$$\frac{1}{2}C_{dc}U_{dc}\frac{d}{dt}U_{dc} = U_{dc}i_{dc} - \frac{3}{2}(e_d i_d + e_q i_q) \quad (2.43)$$

Obviously, the above model features a nonlinear characteristic. As observed, $U_{dc}i_{dc}$ represents the output power of PV arrays, and $3(e_d i_d + e_q i_q)/2$ is the AC-side active power. The equality of the two terms indicates that the derivative of U_{dc} equals zero. As a result, U_{dc} remains unchanged. Since the rotating angle for $\alpha\beta/dq$ transformation employs the phase angle of PLL, e_q is relatively small. Consequently, the DC-side model can be simplified as

$$\frac{1}{2}C_{dc}U_{dc}\frac{d}{dt}U_{dc} = U_{dc}i_{dc} - \frac{3}{2}e_d i_d \quad (2.44)$$

The above model is still nonlinear, which is not convenient for control parameter tuning. By applying small-signal linearization on (2.44), the corresponding small-signal model can be derived as

$$\frac{1}{2}C_{dc}U_{dc0}\frac{d}{dt}\Delta U_{dc} = I_{dc0}\Delta U_{dc} + U_{dc0}\Delta i_{dc} - \frac{3}{2}e_d \Delta i_d \quad (2.45)$$

where prefix Δ represents the small-signal perturbations. e_d is considered a constant without small-signal variations. U_{dc0} and I_{dc0} are the steady-state values of the U_{dc} and i_{dc}, respectively.

For cascaded multi-loop control, the bandwidths of outer and inner loops are typically designed to be separate. Specifically, the bandwidth of the outer loop should be much lower than that of the inner loop. As a result, the dynamics of the inner loop can be neglected. For the middle loop for DC voltage control and inner loop for AC current control, the bandwidth of inner loop should be larger than that of the middle loop. On this assumption, the inner loop can be considered as a unit gain when designing the middle loop. The simplified block diagram of the middle loop

2.2 Control Framework

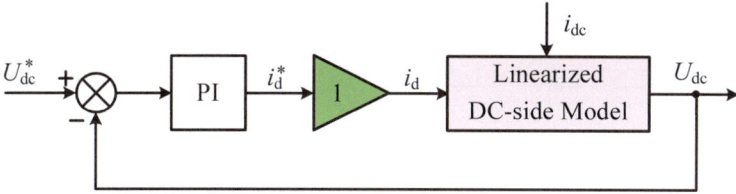

Fig. 2.13 Simplified block diagram of the middle loop

can be formulated as Fig. 2.13. Then, the middle loop controller can be conveniently designed.

2.2.4 Inner Loop for AC Current Control

The fundamental objective of inner loop lies in regulating d- and q-axis current to their reference values. As for 3LT²I with *LCL* filter, additional damping (passive or active) should be added into the current inner loop to suppress the resonant currents introduced by the *LCL* filter. Here, we only provide some basic control concepts of current control for 3LT²I with *L* filter.

As described in the above section, the bandwidth of the inner loop should be larger than that of the middle loop. As a result, the dc-link voltage can be considered as a constant when designing the current controller. In Fig. 2.7, the d-axis reference $i_d{}^*$ comes from the middle loop and the q-axis reference $i_q{}^*$ is set manually. Since the rotating angle for *abc*/*dq* transformation adopts the grid phase angle, $i_d{}^*$ and $i_q{}^*$ are recognized as the active and reactive currents, respectively. The decoupling control of active and reactive power is achieved by this method. Generally, the $i_q{}^*$ is set to zero corresponding to the requirement of unit power factor operation. However, it is set to a nonzero value if the PV inverter receives instructions from the state grid or the converter operates in high- or low-voltage ride-through conditions.

Figure 2.14 illustrates the block diagram of the AC current controller in dq frame. The current controller entirely contains three terms. First, the grid voltage feedforward is adopted to mitigate its impact on the output current. Then, the current feedforward is introduced for the decoupling of d and q axes. Through the above two parts, the d and q-axis models are transferred into the independent first-order inertia transfer functions, which will be easy to tune. Finally, two PI regulators are adopted as the current feedback term to eliminate the current control errors. The overall current controller is described as follows

$$\begin{cases} m_d = \frac{2}{U_{dc}}\left[\left(K_{iP} + \frac{K_{iI}}{s}\right)\left(i_d^* - i_d\right) - \omega_1 L_f i_q + e_d\right] \\ m_q = \frac{2}{U_{dc}}\left[\left(K_{iP} + \frac{K_{iI}}{s}\right)\left(i_q^* - i_q\right) + \omega_1 L_f i_d + e_q\right] \end{cases} \quad (2.46)$$

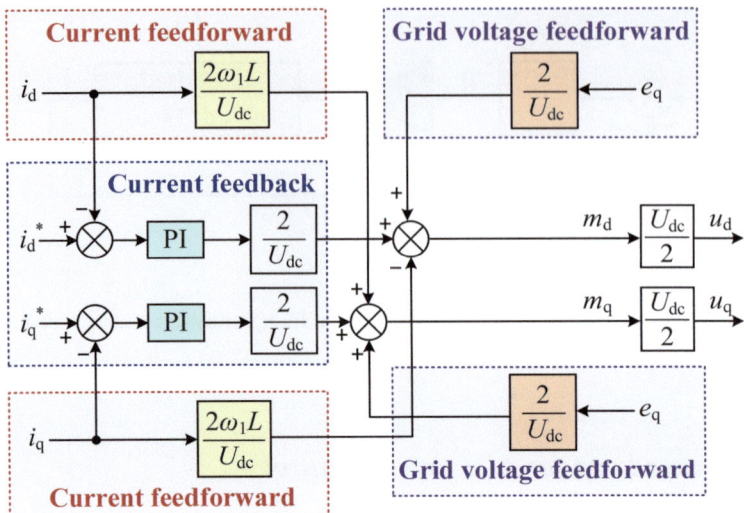

Fig. 2.14 The block diagram of the AC current controller in dq frame

where K_{iP} and K_{iI} are the proportional and integral coefficients of PI regulators, respectively.

Substitute the current controller model into the averaged converter model, we can obtain

$$\frac{d}{dt}\begin{bmatrix} i_d \\ i_q \end{bmatrix} = \begin{bmatrix} \frac{K_{iP}+\frac{K_{iI}}{s}-R_f}{L_f} & 0 \\ 0 & \frac{K_{iP}+\frac{K_{iI}}{s}-R_f}{L_f} \end{bmatrix} \begin{bmatrix} i_d \\ i_q \end{bmatrix} + \frac{1}{L_f}\left(K_{iP}+\frac{K_{iI}}{s}\right)\begin{bmatrix} i_d^* \\ i_q^* \end{bmatrix} \quad (2.47)$$

In addition to the above type of current controller, we can also design the current controller in $\alpha\beta$ frame. Differently, PR regulators are employed instead of PI regulators. The corresponding control block diagram is demonstrated in Fig. 2.15. As shown, the current feedforward terms are removed, since there are no coupling terms between α- and β-axes. The grid voltage feedforward terms are preserved and the PI regulators are replaced by the PR regulators.

The ideal transfer function of the PR regulator is expressed as

$$K_{PR}(s) = K_{iP} + \frac{K_{iR}s}{s^2 + \omega_1^2} \quad (2.48)$$

where K_{iR} is the resonant coefficient.

According to (2.48), the PR regulator has an infinite gain at ω_1 rad/s. In comparison, the PI regulator has an infinite gain at 0 rad/s. To further explain the control effects of the two controllers, we make the following definitions: $L_{icd}(s)$ and $G_{icd}(s)$ denote the open-loop and closed-loop transfer functions of the current control loop

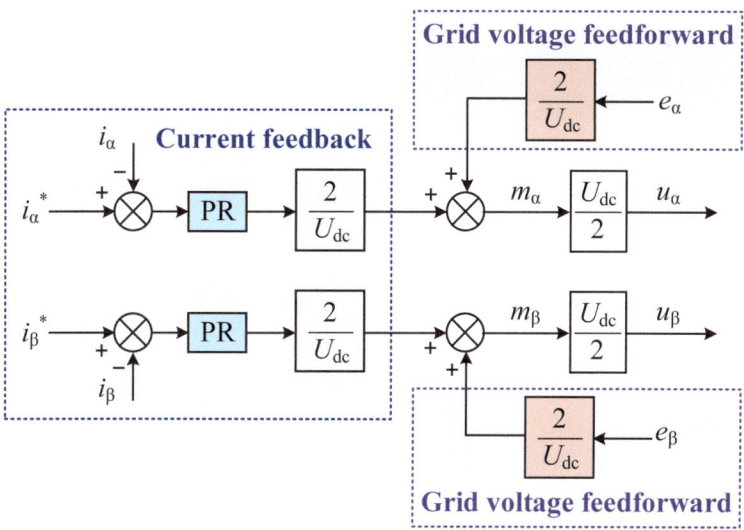

Fig. 2.15 The block diagram of the AC current controller in $\alpha\beta$ frame

for the d axis. $L_{ic\alpha}(s)$ and $G_{ic\alpha}(s)$ denote the open-loop and closed-loop transfer functions of the current control loop for the α-axis. We obtain $G_{icd}(s) = L_{icd}(s)/[1 + L_{icd}(s)]$ and $G_{ic\alpha}(s) = L_{ic\alpha}(s)/[1 + L_{ic\alpha}(s)]$. Based on the above analysis, we know that $L_{icd}(j0)$ and $L_{ic\alpha}(j\omega_1)$ equal infinity. As a result, $G_{icd}(j0)$ and $G_{ic\alpha}(j\omega_1)$ equal one. Therefore, the PI regulator can track constant reference with zero steady-state error, while the PR regulator can track sinusoidal reference at ω_1 with zero steady-state error. Due to this reason, the resonant controller is also recognized as the generalized integrator.

However, the PR controller is not suitable for engineering applications, since the magnitudes of other frequencies drop rapidly. The performance of PR regulator deteriorates significantly even with a small variation of ω_1. As an alternative, we employ the quasi-proportional resonant (QPR) controller, and its transfer function is described as

$$K_{\text{QPR}}(s) = K_{\text{iP}} + \frac{2K_{\text{iR}}\omega_{\text{c\_res}}s}{s^2 + 2K_{\text{iR}}\omega_{\text{c\_res}}s + \omega_1^2} \tag{2.49}$$

where $\omega_{\text{c\_res}}$ denotes the damping coefficient. For comparison, the Bode diagrams of PI and QPR regulator are plotted in Fig. 2.16.

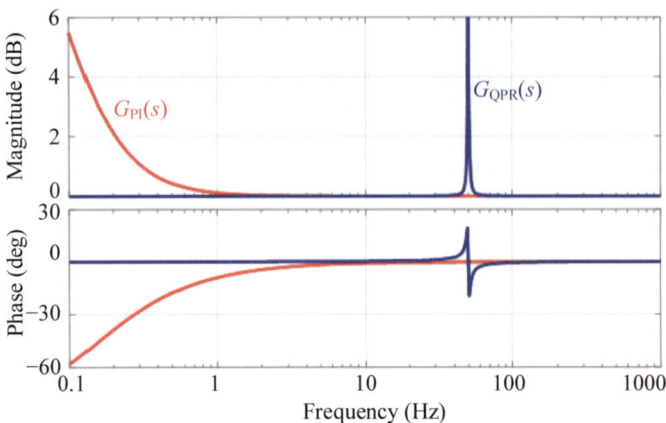

Fig. 2.16 Bode diagrams of PI and QPR regulators

2.2.5 Auxiliary Loop for Other Objectives

The auxiliary loop of 3LT²I mainly includes leakage current suppression and NPV balancing. The leakage currents are introduced by parasitic capacitance between the PV array and the ground, while the NPV unbalance is an inherent problem of three-level inverters. In this subsection, mechanism and control methods for the above two problems in the auxiliary loop are briefly introduced.

Leakage Current Suppression

In PV systems, there are parasitic capacitance between PV array and the ground [25, 26]. Figure 2.17 presents an illustration of the parasitic capacitance, denoted as C_{pv}. As shown, parasitic capacitances exist between both the positive and negative terminals of the PV array and the ground. Additionally, the neutral line of the grid is typically connected to the ground. Therefore, a common-mode (or leakage) current path is formed and the leakage currents are inevitably generated. Large leakage currents pose a threat to personal safety, and therefore must be suppressed to a safe range. According to VDE 0126-1-1, the root mean square (RMS) value of leakage currents should be restricted lower than 300 mA when the PV inverter connects to the power grid.

In the early stages of PV inverter development, isolated transformers were commonly used to realize galvanic isolation and prevent the leakage currents. However, the isolated transformer increases the PV inverter system size, weight, and cost, which decreases the system efficiency. The transformerless PV inverter is preferred in engineering applications. Much of the literature has investigated methods to suppress the leakage currents. For single-phase PV inverters, the leakage currents are mainly suppressed by topology improvement. For example, the H5, H6, and Heric

2.2 Control Framework

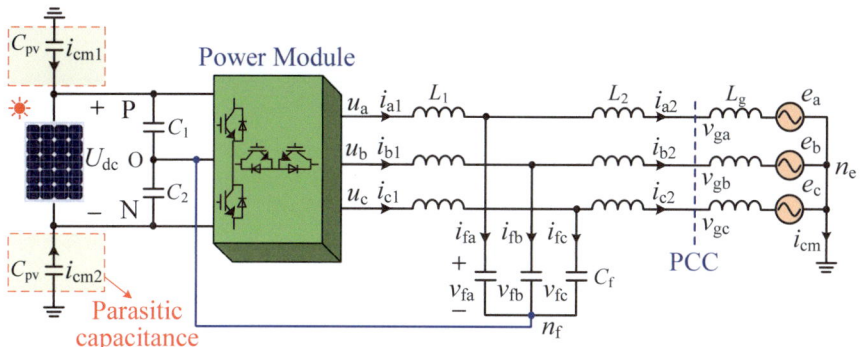

Fig. 2.17 Illustration of parasitic capacitance and the topology of 3LT^{2}I with *MLCL* filter

topologies have been proposed to realize constant common-mode voltage (CMV) and low leakage currents. For three-phase PV inverters, improvement of topology usually needs more auxiliary diodes or switches, which increases the system complexity and decreases the system efficiency. Instead, many modified modulation strategies with reduced common-mode voltage (CMV) behavior were proposed for leakage current suppression, as the CMV acts as the main exciting source of the leakage current. However, these methods are usually at the price of increased harmonics of output currents and magnetic core losses of the filter inductance.

In addition to these solutions, a *MLCL* has also been proposed by connecting the common node of AC capacitors and the neutral point O, as illustrated by the red line shown in Fig. 2.17. The *MLCL* provides a low-impedance common-mode pass that can bypass the leakage current. Therefore, the leakage current can be effectively suppressed. Compared with other solutions, this method benefits from low complexity, low cost, and high efficiency. Moreover, the high-quality output current can be guaranteed as the modulation process is not changed. One major disadvantage of this approach lies in the common-mode resonant risk, which may introduce common-mode resonant current (CMRC) and threaten the stability of the system. Fortunately, the CMRC can be suppressed by the control algorithm without any hardware cost. This will be introduced in detail in Chap. 3.

Neutral-Point Voltage Balance

The NPV should be controlled to remain balanced to prevent the over-voltage damage of the DC capacitors and power devices. In addition, unbalanced NPV will introduce increased harmonic pollution in output currents. As described, the NPV unbalance is an inherent problem of the three-level topology. This problem can be explained by the DC-side mathematical model. For convenience, we rewrite Eq. (2.27) as follows

$$\frac{d}{dt}\Delta U_{pn} = \frac{i_o}{C_{dc}} \tag{2.50}$$

Table 2.3 The impact of output voltage space vectors on NPV

Voltage vector	Switching states	Current i_o	Switching states	Current i_o
Zero vector	[PPP]	0	[NNN]	0
	[OOO]	0		
Small vector	[POO]	$-i_a$	[ONN]	i_a
	[PPO]	i_c	[OON]	$-i_c$
	[OPO]	$-i_b$	[NON]	i_b
	[OPP]	i_a	[NOO]	$-i_a$
	[POP]	i_b	[ONO]	$-i_b$
	[OOP]	$-i_c$	[NNO]	i_c
Medium vector	[PON]	i_b	[OPN]	i_a
	[NPO]	i_c	[NOP]	i_b
	[ONP]	i_a	[PNO]	i_c
Large vector	[PNN]	0	[PPN]	0
	[NPN]	0	[NPP]	0
	[NNP]	0	[PNP]	0

As observed, ΔU_{pn} relates only to the NP current i_o. When the averaged NP current is not zero, the imbalance of the NPV will occur. In the same way, if the NPV has been unbalanced, we can drive it to a balanced condition by regulating i_o. This principle set the foundation of NPV control.

To further investigate the control of NPV, we summarize the impact of output voltage space vectors on NPV in Table 2.3. As shown, zero vectors and large vectors do not affect the NPV, while medium vectors and small vectors can affect the NPV. In addition, the redundant small voltage vectors have opposite effects on NPV. Thus, the redundant small voltage vectors are preferred for NPV control. In other words, the NPV can be controlled by changing the dwell time of P-type and N-type small vectors. Conveniently, the change in dwell time can be realized by injecting the zero-sequence voltage (ZSV) into the three-phase modulation signals. The control diagram of NPV control is described in Fig. 2.7.

2.3 Pulse Width Modulation Schemes

The PWM technology is essential in the safe operation of the $3LT^2I$. In Sect. 2.1.1, we have introduced the basic principles of PWM for each phase and defined the concept of the voltage space vector. In this section, we will provide a more comprehensive explanation of the PWM technology, delving deeper into its principles and applications. Generally, PWM methods for the $3LT^2I$ can be classified into continuous PWM (CPWM) methods and discontinuous PWM (DPWM) methods. Both classes have been widely applied in engineering practice. In addition, there are also other

2.3 Pulse Width Modulation Schemes

PWM methods. For example, the selective harmonic elimination pulse width modulation (SHEPWM) method is commonly used in applications requiring extremely low switching frequencies. In this book, we primarily focus on CPWM and DPWM methods due to their widespread applicability.

Notably, there are two kinds of implementation methods for CPWM and DPWM strategies. The first one is the space vector-based realization and the second one is the carrier-based realization. The space vector-based realization utilizes the principle of vector synthesis. The carrier-based realization injects certain zero-sequence voltages into three-phase modulation signals. Essentially, the two kinds of strategies are equivalent. Compared to space vector-based realization, carrier-based realization is simpler and requires less implementation code. The PWM methods using carrier-based realization are further classified as carrier-based PWM (CBPWM) [27, 28].

2.3.1 Continuous Pulse Width Modulation

When applying the CPWM methods, the modulation signals of the three phases update every switching period. As a result, the power semiconductor switches of each phase switch in each switching period. The commonly used CPWM methods include sinusoidal pulse width modulation (SPWM), space vector pulse width modulation (SVPWM), saddle pulse width modulation (SAPWM), and third harmonic injection pulse width modulation (THIPWM). In the following, we successively introduce the carrier-based realization of the four strategies.

SPWM

In the SPWM method, the three-phase sinusoidal signals are compared with two carrier waves as shown in Fig. 2.2 to obtain the gate signals of power semiconductor switches. The three-phase sinusoidal signals are derived from inner current controllers. The output states or voltages of each phase follow the principle illustrated in Fig. 2.2. The main disadvantage of the SPWM method lies in the low DC voltage utilization. In PV applications, this disadvantage reduces the overall power generation of PV arrays.

SVPWM

Originally, the idea of SVPWM came from the space vector-based realization. It uses adjacent vectors to synthesize reference voltage vectors within a switching period [29, 30]. First, the dwell times of each adjacent vector are calculated. Then the voltage vectors are sorted to generate the switching sequence. For three-level inverters, the commonly used switching sequence is the symmetrical seven-segment. Compared with SPWM, SVPWM features the advantages of reduced harmonics and improved DC-voltage utilization. However, the space vector-based realization of SVPWM is quite complex, especially for three-level inverters. As an alternative, the carrier-based realization can be performed. The corresponding ZSV can be derived as

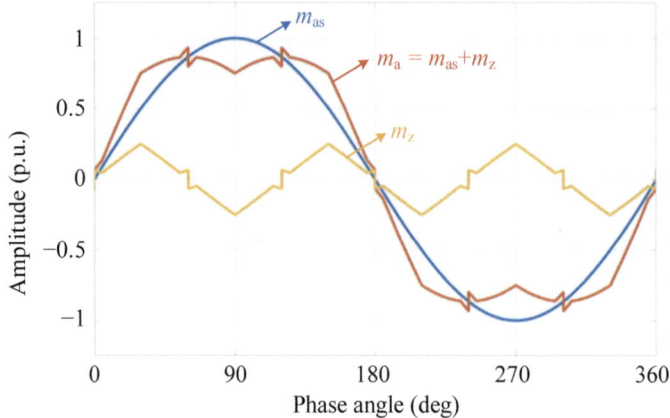

Fig. 2.18 Modulation signals of phase A and ZSV for SVPWM

$$m_z = \frac{1}{2}[1 - \max\{m_{xs} + k_x\} - \min\{m_{xs} + k_x\}] \quad (2.51)$$

where m_z denotes the ZSV. k_x (x = a, b, c) is an intermediate variable. It equals zero when $m_{xs} \geq 0$ and equals one when $m_{xs} < 0$.

Figure 2.18 shows the modulation signals of phase A and ZSV for SVPWM, where m_x denotes the modulation signal after ZSV injection. As observed, the shape of m_x looks like a claw.

SAPWM

The ZSV to realize SAPWM can be derived as [3]

$$m_z = -\frac{1}{2}[\max\{m_{as}, m_{bs}, m_{cs}\} + \min\{m_{as}, m_{bs}, m_{cs}\}] \quad (2.52)$$

Essentially, the distance between $\max\{m_{as}, m_{bs}, m_{cs}\}$ and 1 is always equal to the distance between $\min\{m_{as}, m_{bs}, m_{cs}\}$ and -1 for SAPWM. The modulation signals of phase A and ZSV for SAPWM are shown in Fig. 2.19. As observed, the shape of m_x looks like a saddle. In addition, the SAPWM is equivalent to the SVPWM of two-level inverters.

THIPWM

In the THIPWM method, a pure sinusoidal third-order harmonic is selected as the ZSV [31–33]. Assume $m_{as} = A_{amp}\cos(\omega_1 t + \theta_1)$, where A_{amp} and θ_1 are the amplitude and initial phase angle of m_{as}. Then, the corresponding ZSV is calculated as

$$m_z = -\lambda_{thi}A_{amp}\cos(3\omega_1 t + 3\theta_1) \quad (2.53)$$

where λ_{thi} is usually set to 1/6 to achieve maximum utilization of dc-link voltage.

2.3 Pulse Width Modulation Schemes

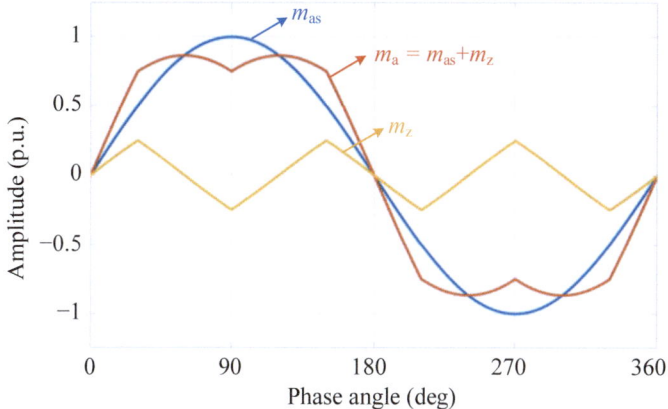

Fig. 2.19 Modulation signals of phase A and ZSV for SAPWM

In practical applications, m_{as}, m_{bs}, and m_{cs} are obtained through *dq/abc* transformation of m_d and m_q. To be specific, m_{as} is calculated by

$$m_{as} = m_d \cos\theta_{PLL} - m_q \sin\theta_{PLL} \tag{2.54}$$

According to the trigonometric formula, m_{as} can be further represented as

$$m_{as} = A_{amp} \cos(\theta_{PLL} + \theta_1) \tag{2.55}$$

where

$$\begin{cases} A_{amp} = \sqrt{m_d^2 + m_q^2} \\ \theta_1 = \arctan(m_q/m_d) \end{cases} \tag{2.56}$$

Then, the ZSV for THIPWM can be calculated by m_d, m_q, and θ_{PLL} as

$$m_z = -\lambda_{thi}\sqrt{m_d^2 + m_q^2} \cos\left[3\theta_{PLL} + 3\arctan(m_q/m_d)\right] \tag{2.57}$$

The modulation signals of phase A and ZSV for THIPWM are shown in Fig. 2.20.

2.3.2 Discontinuous Pulse Width Modulation

When applying the DPWM methods, only two of the three-phase modulation signals update every switching period. As a result, the power semiconductor switches of one phase remain on or off for a certain time. The phase with an unchanged modulation signal is referred to the clamping phase. Then, the switching losses can be

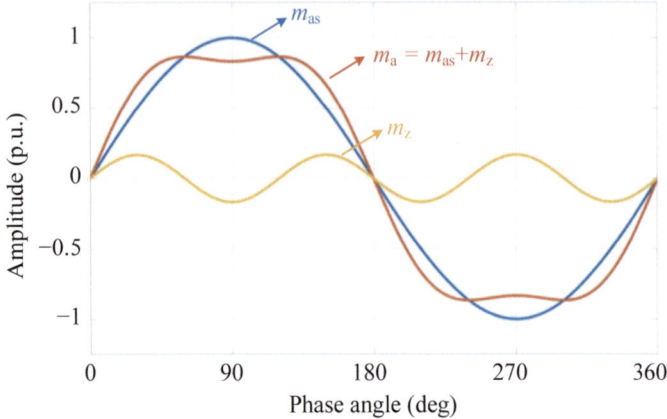

Fig. 2.20 Modulation signals of phase A and ZSV for THIPWM

reduced. There are many types of DPWM methods, including DPWM0, DPWM1, DPWM2, DPWM3, DPWMA, DPWMMIN, DPWMMAX, and others [34, 35]. The DPWMMIN and DPWMMAX methods lead to asymmetric switching losses in power devices and are therefore rarely used. In the following, the realizations of the first five methods are successively introduced.

The clamping phases of DPWM0, DPWM1, DPWM2, DPWM3, and DPWMA for 0°–60° are illustrated in Fig. 2.21. The ZSVs injected to three-phase sinusoidal signals for each small region are marked in the figure. According to the symmetric principle of three phases, clamping phases within 60°–360° can be correspondingly derived. As observed, the main difference among various DPWM methods lies in the clamped phase in each region.

Among those DPWM methods, DPWMA benefits from low common-mode voltage and low harmonics, making it widely used. The ZSV for realization of DPWMA can be expressed as

$$m_z = \begin{cases} \min(m_{hx}) & \text{if } \min(|m_{hx}|) \leq \min(|m_{lx}|) \\ \max(m_{lx}) & \text{if } \min(|m_{hx}|) > \min(|m_{lx}|) \end{cases} \quad (2.58)$$

where

$$m_{hx} = \begin{cases} 1 - m_{xs}, & \text{if } m_{xs} \geq 0 \\ - m_{xs}, & \text{if } m_{xs} < 0 \end{cases}$$
$$m_{lx} = \begin{cases} -m_{xs}, & \text{if } m_{xs} \geq 0 \\ -1 - m_{xs}, & \text{if } m_{xs} < 0 \end{cases} \quad (2.59)$$

The modulation signals of phase A and ZSV for DPWMA are shown in Fig. 2.22.

2.3 Pulse Width Modulation Schemes

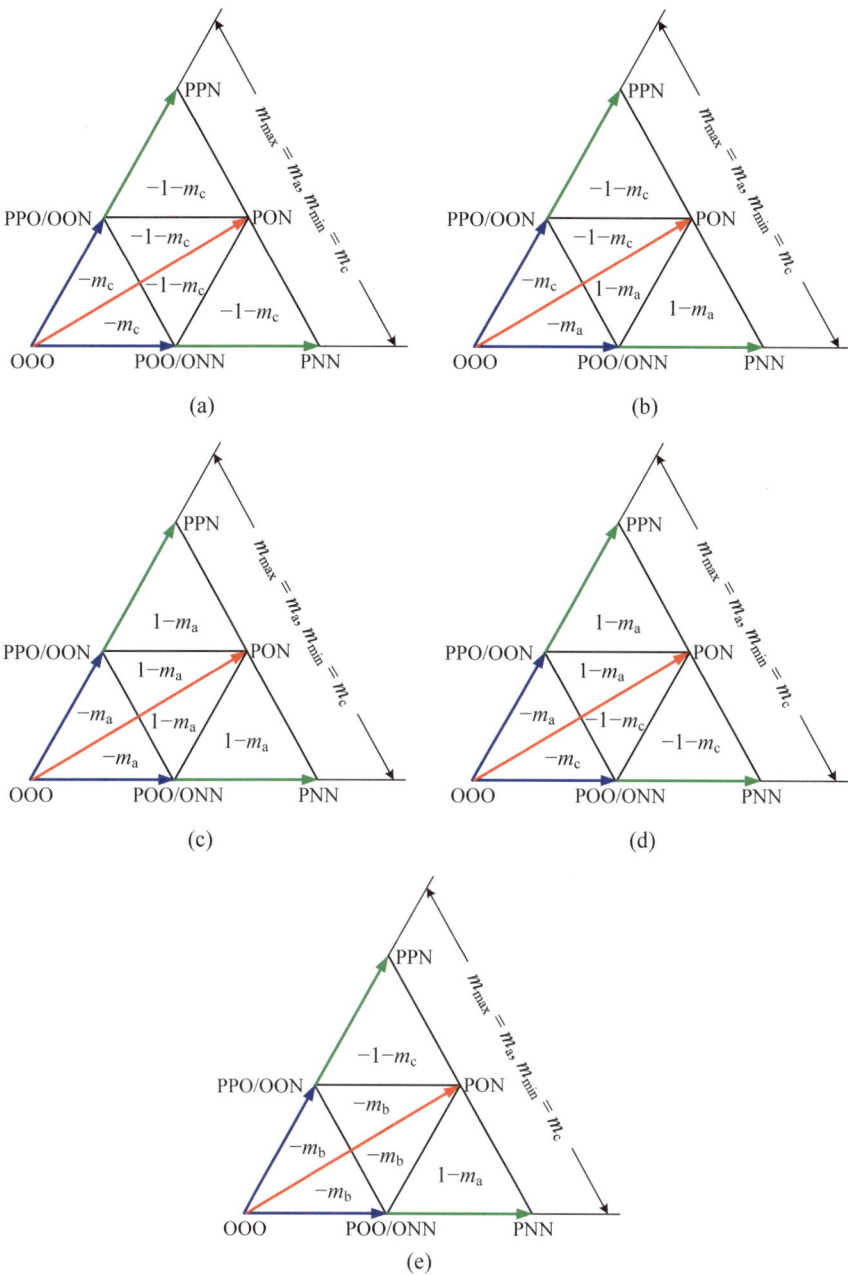

Fig. 2.21 Clamping phases of DPWM0, DPWM1, DPWM2, DPWM3, and DPWMA for 0°–60°
a DPWM0, **b** DPWM1, **c** DPWM2, **d** DPWM3, **e** DPWMA

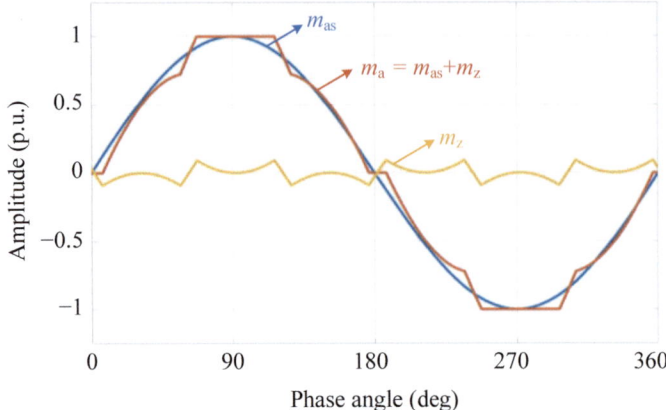

Fig. 2.22 Modulation signals of phase A and ZSV for DPWMA

2.4 Summary

In this chapter, we first introduce the fundamental operating principles of $3LT^2I$. On this basis, the mathematical models of the AC-side and DC-side are derived, including the switching models and the averaged models. Then, a typical control framework is described. The PLL, the outer loop for power control, the middle loop for DC voltage control, and the inner loop for AC current control are introduced, successively. In addition, the auxiliary loop, which consist of the suppression of the leakage current and the balance control of the NPV, are described. Finally, the PWM schemes are elaborated and the continuous PWM and discontinuous PWM are respectively introduced.

References

1. Z. Zhang, *Research on Control Strategy of Parallel T-Type Three-Level Inverter System* (Shandong University, Master, 2017)
2. S. Busquets-Monge et al., The nearest three virtual space vector PWM—a modulation for the comprehensive neutral-point balancing in the three-level NPC inverter. IEEE Power Electron. Lett. **2**(1), 11–15 (2004)
3. D. Holmes, T. Lipo, *Pulse Width Modulation for Power Converters: Principles and Practice* (Wiley, New York, 2003)
4. A. Chen, Z. Zhang, X. Xing, K. Li, C. Du, C. Zhang, Modeling and suppression of circulating currents for multi-paralleled three-level t-type inverters. IEEE Trans. Ind. Appl. **55**(4), 3978–3988 (2019)
5. Y. Jiao, F.C. Lee, LCL filter design and inductor current ripple analysis for a three-level NPC grid interface converter. IEEE Trans. Power Electron. **30**(9), 4659–4668 (2015)
6. E. Twining, D.G. Holmes, Grid current regulation of a three-phase voltage source inverter with an LCL input filter. IEEE Trans. Power Electron. **18**(3), 888–895 (2003)

References

7. C. Fu, C. Zhang, G. Zhang, J. Song, C. Zhang, B. Duan, Disturbance observer-based finite-time control for three-phase AC–DC converter. IEEE Trans. Ind. Electron. **69**(6), 5637–5647 (2022)
8. H.A. Sher, A.F. Murtaza, A. Noman, K.E. Addoweesh, K. Al-Haddad, M. Chiaberge, A new sensorless hybrid MPPT algorithm based on fractional short circuit current measurement and P&O MPPT. IEEE Trans. Sustain. Energy **6**(4), 1426–1434 (2015)
9. T.S. Win, Y. Hisada, T. Tanaka et al., Novel simple reactive power control strategy with dc capacitor voltage control for active load balancer in three-phase four-wire distribution systems. IEEE Trans. Ind. Appl. **51**(5), 4091–4099 (2015)
10. C. Bao, X. Ruan, X. Wang, W. Li, D. Pan, K. Weng, Step-by-step controller design for LCL-type grid-connected inverter with capacitor–current-feedback active-damping. IEEE Trans. Power Electron. **29**(3), 1239–1253 (2014)
11. M. Ciobotaru, R. Teodorescu, F. Blaabjerg, A new single-phase PLL structure based on second order generalized integrator, in *2006 37th IEEE Power Electronics Specialists Conference* (2006), pp. 1–6
12. J.S. Lee, K.B. Lee, New modulation techniques for a leakage current reduction and a neutral-point voltage balance in transformerless photovoltaic systems using a three-level inverter. IEEE Trans. Power Electron. **29**(4), 1720–1732 (2014)
13. Z. Liang, X. Lin, X. Qiao et al., A coordinated strategy providing zero-sequence circulating current suppression and neutral-point potential balancing in two parallel three-level converters. IEEE J. Emerg. Sel. Topics Power Electron. **6**(1), 363–376 (2018)
14. H. Akagi, S. Tamura, A passive EMI filter for eliminating both bearing current and ground leakage current from an inverter-driven motor. IEEE Trans. Power Electron. **21**(5), 1459–1469 (2006)
15. J.C. Giacomini, L. Michels, H. Pinheiro, C. Rech, Active damping scheme for leakage current reduction in transformerless three-phase grid-connected PV inverters. IEEE Trans. Power Electron. **33**(5), 3988–3999 (2018)
16. X. Li, X. Xing, C. Zhang, A. Chen, C. Qin, G. Zhang, Simultaneous common-mode resonance circulating current and leakage current suppression for transformerless three-level T-type PV inverter system. IEEE Trans. Ind. Electron. **66**(6), 4457–4467 (2019)
17. S. Golestan, J.M. Guerrero, J.C. Vasquez, Three-phase PLLs: A review of recent advances. IEEE Trans. Power Electron. **32**(3), 1894–1907 (2017)
18. Y. Han, M. Luo, X. Zhao, J.M. Guerrero, L. Xu, Comparative performance evaluation of orthogonal-signal-generators-based single-phase PLL algorithms—a survey. IEEE Trans. Power Electron. **31**(5), 3932–3944 (2016)
19. A.A. Nazib, D.G. Holmes, B.P. McGrath, Decoupled DSOGI-PLL for improved three phase grid synchronization, in *2018 International Power Electronics Conference* (2018), pp. 3670–3677
20. A. Ranjan, S. Kewat, B. Singh, DSOGI-PLL with in-loop filter based solar grid interfaced system for alleviating power quality problems. IEEE Trans. Ind. Appl. **57**(1), 730–740 (2021)
21. F. Xiao, L. Dong, L. Li, X. Liao, A frequency-fixed SOGI-based PLL for single-phase grid-connected converters. IEEE Trans. Power Electron. **32**(3), 1713–1719 (2017)
22. M. Eltawil, Z. Zhao, MPPT techniques for photovoltaic applications. Renew. Sustain. Energy Rev. **25**, 793–813 (2013)
23. M. Brito, L. Galotto, L. Sampaio, G. Melo, C. Canesin, Evaluation of the main MPPT techniques for photovoltaic applications. IEEE Trans. Ind. Electron. **60**, 1156–1167 (2013)
24. N. Femia, G. Petrone, G. Spagnuolo, M. Vitelli, Optimization of perturb and observe maximum power point tracking method. IEEE Trans. Power Electron. **20**(4), 963–973 (2005)
25. M.C. Cavalcanti, K.C. de Oliveira, A.M. de Farias, F.A.S. Neves, G.M.S. Azevedo, F.C. Camboim, Modulation techniques to eliminate leakage currents in transformerless three-phase photovoltaic systems. IEEE Trans. Ind. Electron. **57**(4), 1360–1368 (2010)
26. C. Morris, D. Han, B. Sarlioglu, Reduction of common mode voltage and conducted EMI through three-phase inverter topology. IEEE Trans. Power Electron. **32**(3), 1720–1724 (2017)
27. L.M. Tolbert, T.G. Habetler, Novel multilevel inverter carrier-based PWM method. IEEE Trans. Ind. Appl. **35**(5), 1098–1107 (1999)

28. P. Josep, Z. Jordi, C. Salvador et al., A carrier-based PWM strategy with zero-sequence voltage injection for a three-level neutral-point-clamped converter. IEEE Trans. Power Electron. **27**(2), 642–651 (2012)
29. F. Rojas, R. Cárdenas, R. Kennel et al., A simplified space-vector modulation algorithm for four-leg NPC converters. IEEE Trans. Power Electron. **32**(11), 8371–8380 (2017)
30. F. Rojas, R. Kennel, R. Cardenas et al., A new space-vector-modulation algorithm for a three-level four-leg NPC inverter. IEEE Trans. Energy Convers. **32**(1), 23–35 (2017)
31. M.A.A. Younis, N.A. Rahim, S. Mekhilef, Simulation of grid connected THIPWM-three-phase inverter using SIMULINK, in *Proceedings of the 2011 IEEE Symposium on Industrial Electronics and Applications* (2011), pp. 133–137
32. J.-H. Park, J.-S. Lee, K.-B. Lee, Sinusoidal harmonic voltage injection PWM method for Vienna rectifier with an LCL filter. IEEE Trans. Power Electron. **36**(3), 2875–2888 (2019)
33. S. Albatran, A.R.A. Khalaileh, A.S. Allabadi, Minimizing total harmonic distortion of a two-level voltage source inverter using optimal third harmonic injection. IEEE Trans. Power Electron. **35**(3), 3287–3297 (2020)
34. W. Zhu, C. Chen, S. Duan, T. Wang, P. Liu, A carrier-based discontinuous PWM method with varying clamped area for Vienna rectifier. IEEE Trans. Ind. Electron. **66**(9), 7177–7188 (2019)
35. L. Dalessandro, S.D. Round, U. Drofenik, J.W. Kolar, Discontinuous space-vector modulation for three-level PWM rectifiers. IEEE Trans. Power Electron. **23**(2), 530–542 (2008)

Open Access This chapter is licensed under the terms of the Creative Commons Attribution 4.0 International License (http://creativecommons.org/licenses/by/4.0/), which permits use, sharing, adaptation, distribution and reproduction in any medium or format, as long as you give appropriate credit to the original author(s) and the source, provide a link to the Creative Commons license and indicate if changes were made.

The images or other third party material in this chapter are included in the chapter's Creative Commons license, unless indicated otherwise in a credit line to the material. If material is not included in the chapter's Creative Commons license and your intended use is not permitted by statutory regulation or exceeds the permitted use, you will need to obtain permission directly from the copyright holder.

Chapter 3
Control Technology of Photovoltaic Power Generation Systems for Leakage Current Suppression

In photovoltaic (PV) power generation systems, the parasitic capacitor is introduced between PV panels and the grounds, which inevitably leads to the leakage current. The leakage current causes electromagnetic interference, distorts the output currents quality and threatens people safety. Thus, the leakage current suppression is the primary issue in the design of PV inverter, and attracts wide attention. Different countries have developed standard requirements for leakage current. For example, the VDE 0126-1-1 restricts the root mean square (RMS) value of leakage current under 300 mA when the PV inverter power level below 30 kVA [1].

Conventionally, the transformer was used to realize galvanic isolation and prevent the leakage current [2]. However, the isolated transformer increases the PV inverter systems size, weight, and cost, which decreases the system efficiency and stability. Therefore, the transformerless PV inverter is widely used. Nevertheless, the high leakage current will be unavoidably produced. In order to effectively attenuate the leakage current, different inverter topologies have been proposed. In low-power applications, the single-phase inverters including the Heric, H6 and neutral-point-clamped topology are proposed to realize constant common-mode voltage (CMV) and low leakage [3]. In the high-power application, the three-phase H7 and H8 [4], topologies were proposed to reduce CMV amplitude, and thus attenuate leakage current. However, all the above modified inverter topologies need auxiliary diodes or power semiconductor switches, which increases the system complexity and decreases the system efficiency.

How to effectively suppress leakage current without increasing system cost and reducing system efficiency is an important problem. Driven by the above issues, the author studies the generation mechanism of leakage current, and found that the necessary and sufficient conditions of leakage current generation are variable exciting source and closed common-mode (CM) circuit. Based on the mechanism analysis, two types leakage current suppression methods are proposed. The first type method is proposed from the perspective of CM circuit, which is called CM circuit modification method. In this method, a capacitive branch is introduced and paralleled

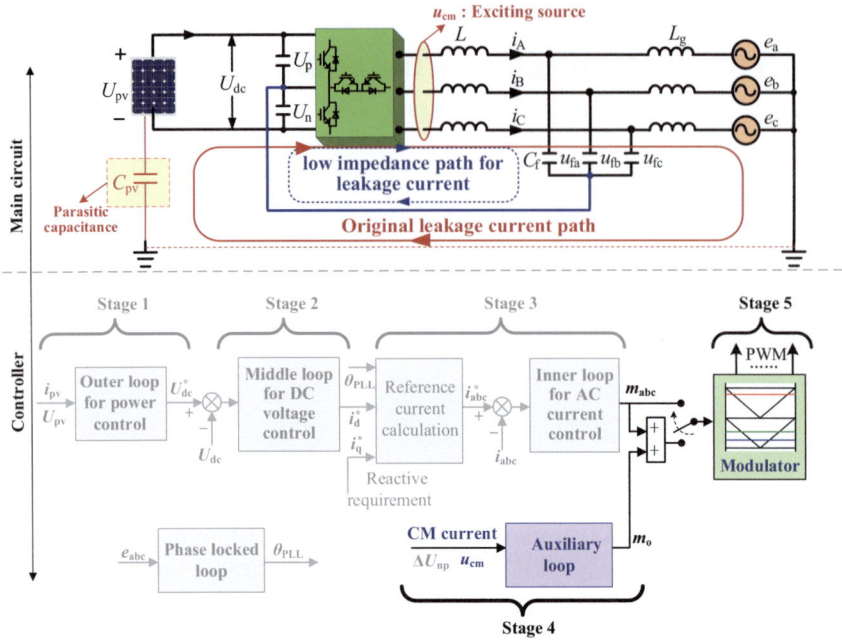

Fig. 3.1 The modules associated with leakage current suppression in overall control diagram

into the original CM circuit to offer a low impedance path for leakage current, so as to suppress the leakage current. The CM circuit modification is realized by adding a conductor wire inside a PV inverter. Besides, the closed-loop control scheme for CM circuit is proposed for the first time to further reduce the leakage current and eliminate the resonance current caused by the introduced capacitive branch. The second type method is proposed from the perspective of exciting source. In this method, the leakage current is reduced by suppressing the exciting source. Using these methods, no additional auxiliary diodes or switches are required, which has minimal impact on system cost and efficiency. As shown in Fig. 3.1, the additional conductor wire is represented by a blue line, which connects the common point of the filter capacitors directly to the dc-link neutral-point. The closed-loop control scheme for CM circuit is integrated into auxiliary loop. The method for exciting source suppression is integrated into PWM modulator. The related modules have been highlighted in the overall control diagram.

For the convenience of readers, this chapter is organized as follows. Firstly, the generation mechanism of leakage current is analyzed and corresponding models are established. Secondly, according to these theoretical knowledges, the two types of leakage current suppression methods are elaborated. Then, the experimental results are given to demonstrate the effectiveness of the presented theoretical analyses and proposed methods. Finally, the summary is given.

3.1 Mechanism Analysis and Modeling

As we all know the three-level T-type inverter (3LT^{2}I) is widely used in PV power generation systems for its outstanding advantage in efficiency in 4–30 kHz [5]. Thus, in this chapter, 3LT^{2}I is selected as the research object, and the relevant theories and technologies are given.

The configuration of transformless PV inverter is shown in Fig. 3.2. On the DC side, the PV panel is the dc source, which supplies the power to the inverter. C_{pv} is the equivalent parasitic capacitor between the PV panel and the ground. i_{cm1} and i_{cm2} are the leakage currents flowing into C_{pv}. u_{pv} is the voltage of C_{pv}. P and N represent the positive dc-bus and negative dc-bus, respectively. O is the neutral-point of dc-bus. C_1 and C_2 are the dc-link capacitors, and their corresponding voltages are U_p and U_n, respectively. On the AC side, the LCL filter is used for getting better performance of the output AC current. L, L_g, and C_f are inverter-side inductor, grid-side inductor, and filter capacitor, respectively. n_1 is the common point of C_f. n is the common point of the power grid.

Choosing the negative dc-bus N as the reference point. The average model of the three-phase inverter in a stationary frame is obtained based on Kirchhoff's voltage law, which is expressed as

$$\begin{cases} u_{AN} = L\frac{di_A}{dt} + L_g\frac{di_a}{dt} + e_a + u_{nN} \\ u_{BN} = L\frac{di_B}{dt} + L_g\frac{di_b}{dt} + e_b + u_{nN} \\ u_{CN} = L\frac{di_C}{dt} + L_g\frac{di_c}{dt} + e_c + u_{nN} \end{cases} \quad (3.1)$$

where u_{AN}, u_{BN}, and u_{CN} are the inverter-side output voltages; i_A, i_B, and i_C are inverter-side currents; i_a, i_b, and i_c are grid currents; e_a, e_b, and e_c are grid voltages;

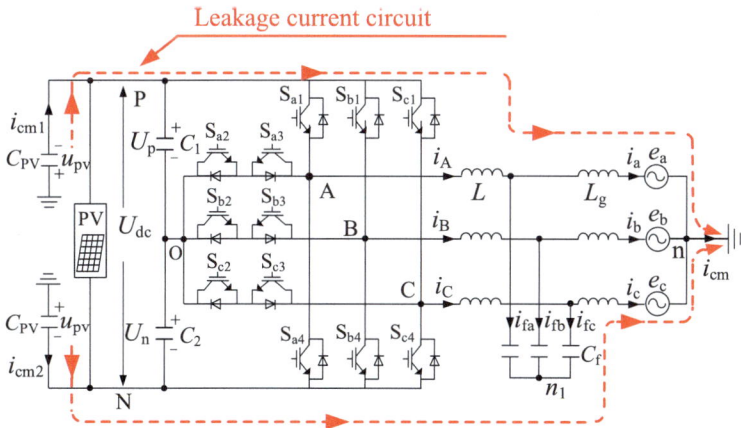

Fig. 3.2 The configuration of transformless PV 3LT2I system with LCL filter

and u_{nN} is the voltage between the grid common point n and the negative side of dc-link N.

Considering the balanced three-phase grid system, $e_a + e_b + e_c = 0$, the sum of average model of the three-phase inverter can be expressed as

$$\sum_{J=A,B,C} u_{JN} = \sum_{J=A,B,C} L\frac{di_J}{dt} + \sum_{j=a,b,c} L_g\frac{di_j}{dt} + 3u_{nN} \tag{3.2}$$

As shown in transformless PV power generation system with 3LT²I, the voltage u_{nN} can be expressed as

$$\begin{cases} u_{nN} = \frac{1}{C_{pv}} \int i_{cm1} dt + U_{dc} \\ u_{nN} = \frac{1}{C_{pv}} \int i_{cm2} dt \end{cases} \tag{3.3}$$

Based on the above equation, the voltage u_{nN} is rewritten as

$$u_{nN} = \frac{1}{2C_{pv}} \int (i_{cm2} + i_{cm1}) dt + \frac{1}{2} U_{dc} \tag{3.4}$$

The leakage current i_{cm} can be expressed in two ways

$$\begin{cases} i_{cm} = \sum_{J=A,B,C} i_J = \sum_{j=a,b,c} i_j \\ i_{cm} = i_{cm1} + i_{cm2} \end{cases} \tag{3.5}$$

The CMV is represented by u_{cm}, which is defined as

$$u_{cm} = \frac{1}{3}(u_{AN} + u_{BN} + u_{CN}) \tag{3.6}$$

Substituting the expressions of u_{nN}, i_{cm} and u_{cm} into the sum of average model of the three-phase inverter, the CM mathematical model with *LCL* filter can be obtained as

$$3u_{cm} = (L + L_g)\frac{di_{cm}}{dt} + \frac{3}{2C_{pv}} \int i_{cm} dt + \frac{3}{2} U_{dc} \tag{3.7}$$

The above equation presents the relationship between CMV u_{cm} and the leakage current i_{cm}. According to this relationship, the CM equivalent circuit of the inverter with an *LCL* filter can be derived, which is depicted in Fig. 3.3.

In the CM equivalent circuit, the circulation path is closed, which consists of the impedance components of the inductors of L and L_g, and the parasitic capacitor C_{pv}. The inductors and the capacitor are connected in series. The power sources include u_{cm} and $3U_{dc}/2$. Consequently, the leakage current is induced by the power sources in the closed circulation path.

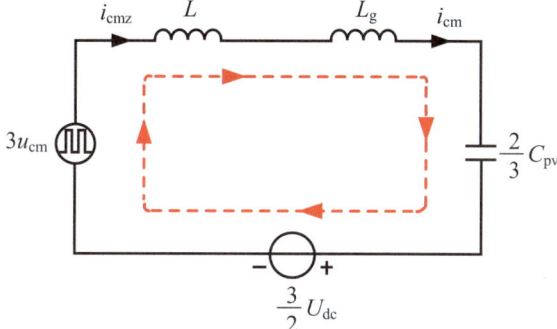

Fig. 3.3 The CM equivalent circuit of transformless PV 3LT²I system with *LCL* filter

It should be noted that only the inductors and capacitor are included in the CM circuit, and the power source of $3U_{dc}/2$ is a constant value. Therefore, the influence of the power source of $3U_{dc}/2$ on the leakage current can be ignored. While the power source of u_{cm} is a variable, which is composed of high-frequency and low-frequency components. Thus, in this CM circuit, the exciting source of leakage current is merely the u_{cm}.

3.2 Leakage Current Suppression Technology

According to the above analyses, it can be found that the necessary and sufficient conditions of leakage current generation include two key points. The first point is the exciting source, which should be a variable quantity. The second point is the closed CM circuit, which is composed of inductors and capacitors.

Inspired by these two points, two types leakage current suppression methods are proposed. The first type method is proposed from the perspective of CM circuit. In this method, the capacitive branch is added into the CM circuit to offer a low impedance path for leakage current, so as to suppress the leakage current. The second type method is proposed from the perspective of exciting source. In this method, the leakage current is reduced by suppressing the exciting source. In these two methods, no additional auxiliary diodes or switches are required, which have few impacts on system cost and efficiency. The detailed analyses of these two types methods are given in the following section.

3.2.1 Common-Mode Circuit Modification Methods

In 2006, H. Akagi and S. Tamura proposed a passive EMI filter to eliminate both bearing current and ground leakage current in inverter-driven motor systems. Driven by this idea, an improved *LCL* (*ILCL*) filter is proposed in PV inverter systems. The

ILCL can be used to reduce the leakage current by changing the behavior of the CM circuit. The high-frequency leakage current to the ground is drastically reduced without additional auxiliary switches.

It should be pointed out that in the first type method of leakage current suppression, three key points are necessary. They are the *ILCL* filter, the closed-loop controller for CM circuit and the pulse width modulation (PWM) module. The *ILCL* filter suppresses the high-frequency component of leakage current by dividing the CM current. The closed-loop controller of CM circuit suppresses the low-frequency component of leakage current and resonance CM current by closed-loop regulation. The PWM module is used to control the power semiconductor switches ON or OFF, which is the actuator for the system controllers to implement the control strategies.

Presently, the space voltage pulse width modulation (SVPWM) is widely applied for higher dc-link voltage utilization and is easily realized by a digital controller. Usually, the SVPWM can be divided into the continuous modulation method and the discontinuous modulation method. The continuous modulation method has superior performance of output voltage and output current. The discontinuous modulation method has distinct advantage in system power losses reduction.

The leakage current suppression method based on CM circuit modification method is further classified into the continuous modulation method [6] and the discontinuous modulation method [7, 8]. These two methods are elaborated in the following sections.

Based on Continuous Modulation

In order to suppress high-frequency leakage current, the *ILCL* filter is adopted. The configuration of transformless PV $3LT^2I$ system with *ILCL* is shown in Fig. 3.4. In this configuration, the common point of the filter capacitors n_1 is directly connected to the neutral-point O.

In this case, the average model of transformless PV $3LT^2I$ with *ILCL* filter is also satisfy the following expression

$$\begin{cases} u_{AN} = L\frac{di_A}{dt} + \frac{1}{C_f} \int i_{fa} dt + U_n \\ u_{BN} = L\frac{di_B}{dt} + \frac{1}{C_f} \int i_{fb} dt + U_n \\ u_{CN} = L\frac{di_C}{dt} + \frac{1}{C_f} \int i_{fc} dt + U_n \end{cases} \quad (3.8)$$

where i_{fa}, i_{fb}, and i_{fc} are the currents on the filter capacitor; and U_n is the voltage across the bottom capacitor C_2. Similarly, the sum of above equation can be expressed as

$$\sum_{J=A,B,C} u_{JN} = \sum_{J=A,B,C} L\frac{di_J}{dt} + \sum_{j=a,b,c} \frac{1}{C_f} \int i_{fjd} t + 3U_n \quad (3.9)$$

3.2 Leakage Current Suppression Technology

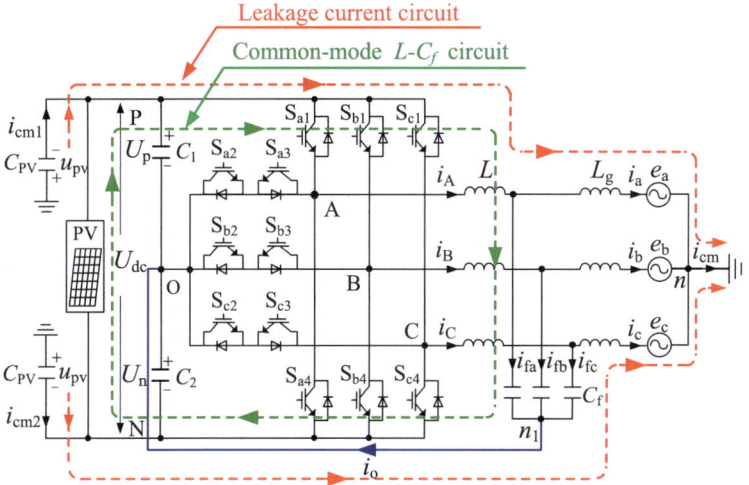

Fig. 3.4 The configuration of transformless PV 3LT²I system with *ILCL* filter

For ease of understanding, the common mode circulating current (CMCC) is presented by i_{cmz} and is defined as

$$i_{cmz} = \sum_{J=A,B,C} i_J = \sum_{j=a,b,c} i_j + \sum_{j=a,b,c} i_{fj} \quad (3.10)$$

Since the directly connection of filter capacitors n_1 and the neutral-point O, the leakage current flowing path is changed. In this configuration, according to the impact of leakage current on safety and its original definition, the i_{cm} is rederived as

$$i_{cm} = \sum_{j=a,b,c} i_j \quad (3.11)$$

According to the expression of i_{cm}, i_{cmz} and u_{cm}, the summary equation of the average model of transformless PV 3LT²I with *ILCL* filter can be derived as

$$3u_{cm} = L\frac{di_{cmz}}{dt} + \frac{1}{C_f}\int i_o dt + 3U_n \quad (3.12)$$

where i_o is the sum of i_{fa}, i_{fb}, and i_{fc}, which satisfy $i_o = i_{fa} + i_{fb} + i_{fc}$.

It should be noted that, when the common point of the filter capacitors n_1 is directly connected to the neutral-point O, the average model of CM equivalent circuit of the inverter with *ILCL* filter should be modified as

$$3u_{cm} = L\frac{di_{cmz}}{dt} + L_g\frac{di_{cm}}{dt} + \frac{3}{2C_{pv}}\int i_{cm}dt + \frac{3}{2}U_{dc} \quad (3.13)$$

Fig. 3.5 The CM equivalent circuit of transformless PV 3LT^{2}I system with *ILCL* filter

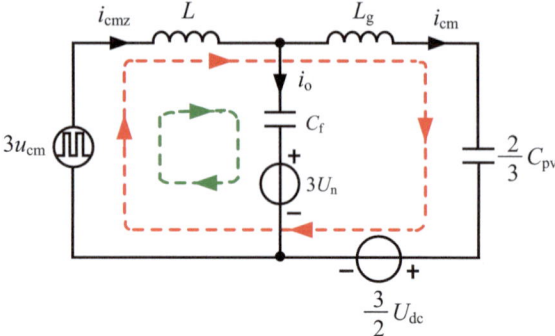

According to the two average models, the CM equivalent circuit of the inverter with *ILCL* filter can be easily derived, which is depicted in Fig. 3.5.

Compared with the CM equivalent circuit of the 3LT^{2}I system with an *LCL* filter, the C_f branch is added, which offers a low impedance path for high-frequency leakage current. So, the high-frequency leakage current is drastically reduced.

For better comparison of the leakage current suppression performance between the *LCL* and *ILCL* filter, the corresponding transfer functions of leakage current are respectively derived as

$$\begin{cases} G_{CM1}(s) = \dfrac{I_{cm}(s)}{3U_{cm}(s)} = \dfrac{2C_{pv}s}{2C_{pv}(L+L_g)s^2+3} \\ G_{CM2}(s) = \dfrac{I_{cm}(s)}{3U_{cm}(s)} = \dfrac{2C_{pv}s}{2C_{pv}C_fLL_gs^4+[3LC_f+2C_{pv}(L+L_g)]s^2+3} \end{cases} \quad (3.14)$$

where, $G_{CM1}(s)$ and $G_{CM1}(s)$ are transfer functions of the CM equivalent circuit with an *LCL* filter and an *ILCL* filter, respectively. It can be seen that the CM circuit with an *ILCL* filter is a fourth-order system, which has high-frequency leakage current attenuation ability. The magnitude response of transfer functions of leakage current is shown in Fig. 3.6. The parameters for the magnitude response are $L = 1$ mH, $L_g = 0.15$ mH, $C_f = 10$ μF, $C_{pv} = 0.15$ μF. It can be seen that the high-frequency leakage current attenuation gain with $G_{CM2}(s)$ is higher than that with $G_{CM1}(s)$. So, the leakage current with an *ILCL* filter is much lower than that with an *LCL* filter.

However, the transfer function of $G_{CM2}(s)$ has two resonance peaks, which result in common mode resonance circulating current (CMRCC) and system instability. The resonance frequencies are approximated as

$$\begin{cases} f_{r1} = \dfrac{1}{2\pi\sqrt{LC_f}} \\ f_{r2} = \dfrac{\sqrt{3}}{2\pi\sqrt{2L_gC_{pv}}} \end{cases} \quad (3.15)$$

As the above equation indicates, the resonance frequency f_{r1} is decided by the parameters of L and C_f. The frequency of f_{r2} depends on the parameters of L_g and C_{pv}, which is difficult to determine in practical applications. Generally, the frequency f_{r1}

3.2 Leakage Current Suppression Technology

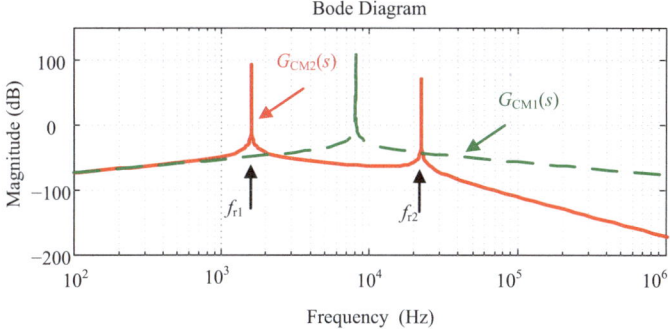

Fig. 3.6 Magnitude response of CM equivalent circuit

is smaller than the switching frequency, whereas the resonance frequency f_{r2} is much higher than the switching frequency. So, the resonance peak in f_{r2} is uncontrollable. In practice, the CMRCC will be excited by the variational $3u_{cm}$ at f_{r1} rather than f_{r2}. Therefore, the resonance frequency f_{r1} is analyzed in detail in this paper.

The CMRCC causes inverter-side current significant oscillation and leakage current increment, which results in current protection maloperation and degrades system safety performance. So, it is urgent to suppress the CMRCC.

The damping resistor can be used to suppress the CMRCC by modifying the impedance of C_f branch. However, the high-frequency leakage current attenuation performance is decreased, and power losses are increased. Therefore, an improved CMCC model and a hybrid control strategy composed of PI and feedforward control are proposed to realize better CMRCC suppression in this chapter.

When an *ILCL* filter is adopted, the CM equivalent circuit shows that the impedance components consist of the inductors L and L_g, the filter capacitor and the parasitic capacitor C_{pv}. Since these impedance components are all inductors and capacitors, the constant power source cannot induce any current in this capacitive-inductive CM equivalent circuit. Thus, the constant power source of $3/2U_{dc}$ and $3U_n$ can be ignored.

Assuming the leakage current is effectively suppressed, that is $i_{cm} = 0$, the CMCC satisfies

$$3u_{cm} = L\frac{di_{cmz}}{dt} + \frac{1}{C_f}\int i_{cmz}dt \tag{3.16}$$

Based on the above expression, the CMCC is influenced by L, C_f, and $3u_{cm}$. L and C_f determine the resonance frequency f_{r1}, which is constant once the filter is selected. So, the $3u_{cm}$ can be used to suppress CMRCC by adjusting the amplitude at the frequency f_{r1}.

The amplitude adjustment of $3u_{cm}$ is realized by the PWM module. The output states of each phase of 3LT^{2}L can be defined as [P], [O], and [N]. The space vector diagram of the three-level inverter is shown in Fig. 3.7. The space vectors are classified into four groups based on the magnitude. There are six large vectors, six medium vectors, and six small vectors with double redundancy and one zero vector with triple redundancy. The redundant small vectors can be used to realize different control objectives, as they will not affect output voltage.

The continuous modulation method has superior performance in output voltage and output current. In order to obtain a superior output waveform, the continuous modulation method based on a seven-segment switching sequence is adopted in this section. The switching sequences are shown in Table 3.1, which starting and ending with N-type small vectors.

Assuming the dc-link voltage is constant and neutral point voltage (NPV) is balanced, the output voltage is expressed as

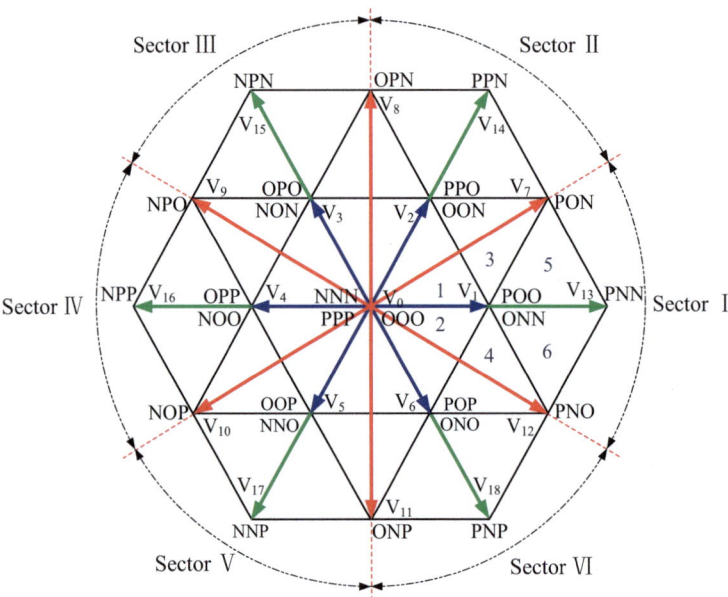

Fig. 3.7 Space vector diagram of the $3LT^2I$

Table 3.1 Corresponding switching sequences in different sector using continuous modulation

Sectors	Switching sequences
I	ONN → ⋯ → POO → ⋯ → ONN
II	OON → ⋯ → PPO → ⋯ → OON
III	NON → ⋯ → OPO → ⋯ → NON
IV	NOO → ⋯ → OPP → ⋯ → NOO
V	NNO → ⋯ → OOP → ⋯ → NNO
VI	ONO → ⋯ → POP → ⋯ → ONO

3.2 Leakage Current Suppression Technology

$$u_{JN}(t) = \begin{cases} \frac{U_{dc}}{2}(\varepsilon(t-t_J) - \varepsilon(t-T_s+t_J)) & [N] \to [O] \to [N] \\ \frac{U_{dc}}{2}(\varepsilon(t-t_J) - \varepsilon(t-T_s+t_J)) + \frac{U_{dc}}{2} & [O] \to [P] \to [O] \end{cases} \quad (3.17)$$

where $\varepsilon(t)$ is the unit step function; T_s is the sample period; t_J is state-switching time; and the time t can change from 0 to T_s. The initial voltage is $U_{dc}/2$ if the switching transition is $[O] \to [P] \to [O]$. Thus, the average value of output phase voltages can be re-written as

$$u'_{JN}(t) = \begin{cases} \frac{U_{dc}}{2} \frac{T_s-2t_J}{T_s} & [N] \to [O] \to [N] \\ \frac{U_{dc}}{2} \frac{T_s-2t_J}{T_s} + \frac{U_{dc}}{2} & [O] \to [P] \to [O] \end{cases} \quad (3.18)$$

Based on this assumption, the CMV is rewritten as

$$3u_{cm} = \sum_{J=A,B,C} u'_{JN}(t) = \frac{U_{dc}}{2}[D_z(t) + n] \quad (3.19)$$

where $D_z(t)$ is defined zero-sequence duty ratio, which can be used to adjust the dwell time of redundant small vectors; and n is the number of switching transition of $O \to P \to O$. $D_z(t)$ and n are expressed as

$$D_z(t) = \sum_{J=A,B,C} D_J(t) = \frac{1}{T_s} \sum_{J=A,B,C} (T_s - 2t_J) \quad (3.20)$$

$$n = \begin{cases} 1 & \text{sectors I, III, V} \\ 2 & \text{sectors II, IV, VI} \end{cases} \quad (3.21)$$

Based on the above deriving process and using the using Laplace transform, the improved CMCC model with respect to zero-sequence duty ratio is obtained as

$$I_{cmz}(s) = \frac{C_f s}{LC_f s^2 + 1} \frac{U_{dc}}{2}[D_z(s) + n] \quad (3.22)$$

It can be observed that the improved CMCC model is a second-order system, and influenced by filter parameters, zero-sequence duty ratio and the number of switching transitions.

In the second-order model, the CMRCC is excited at the frequency f_{r1}. Therefore, the hybrid control strategy composed of PI and feedforward control is proposed to realize CMRCC suppression. The PI controller is used to suppress the CMRCC. Meanwhile, the feedforward controller is used to eliminate the disturbances of $D_z(s)$ and n. The proposed control strategy is shown in Fig. 3.8.

In order to mitigate the CMRCC to zero, the reference value of I^*_{cmz} is set to zero. The error between CMCC and its reference value is fed into the hybrid controller to generate the control variable d_z, which is expressed as

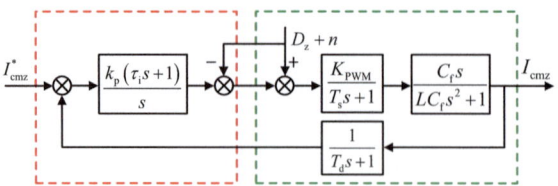

Fig. 3.8 PI and feed-forward control schema for CM circuit control

$$d_z = \left(I^*_{cmz} - I_{cmz}\right)\left[\frac{k_p(\tau_i s + 1)}{s}\right] - (D_z + n) \quad (3.23)$$

where k_p and τ_i are parameters of PI controller. Then, the model in (3.21) is rewritten as

$$I_{cmz}(s) = \frac{C_f s}{LC_f s^2 + 1} \frac{U_{dc}}{2}[D_z(s) - 6d_z + n] \quad (3.24)$$

It can be seen that the variable d_z is used to modify the zero-sequence duty ratio $D_z(s)$ to realize CMRCC suppression. The modified state-switching times of each phase can be expressed as

$$\begin{cases} t_{A(1)} = t_A + d_z \cdot T_s \\ t_{B(1)} = t_B + d_z \cdot T_s \\ t_{C(1)} = t_C + d_z \cdot T_s \end{cases} \quad (3.25)$$

In order to maintain system stable, the control variable d_z should satisfy

$$-\frac{\min(t_A, t_B, t_C)}{T_s} \le d_z \le \frac{\min(t_A, t_B, t_C)}{T_s} \quad (3.26)$$

Assuming that the reference voltage vector is located in the region 5 of Sector I, the modified switching sequence is shown in Fig. 3.9.

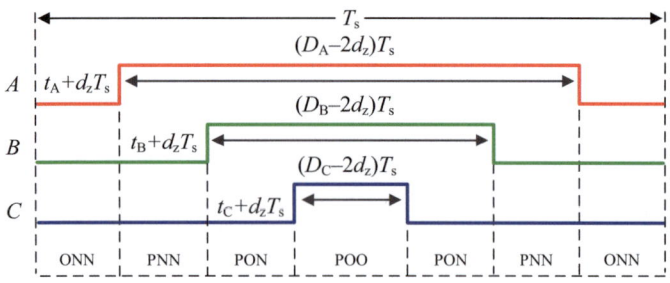

Fig. 3.9 Modified switching sequence in region 5 of Sector I using continuous modulation

3.2 Leakage Current Suppression Technology

The open-loop transfer function of the CMCC loop can be obtained as

$$G_{op}(s) = G_c(s)G_{PWM}(s)G_{plant}(s)H(s)$$
$$= \frac{k_p(\tau_i s + 1)}{s} \cdot \frac{K_{PWM}}{T_s s + 1} \cdot \frac{C_f s}{LC_f s^2 + 1} \cdot \frac{1}{T_d s + 1} \quad (3.27)$$

where K_{PWM} and T_s are the equivalent gain and time constant of PWM module; and T_d is the time delay of digital processor. Since T_s and T_d are both small enough, the small time constant can be merged as

$$\frac{1}{T_m s + 1} = \frac{1}{T_s s + 1} \cdot \frac{1}{T_d s + 1} \quad (3.28)$$

where $T_m = T_s + T_d$. Thus, the open-loop transfer function can be simplified as

$$G_{op}(s) = \frac{k_p K_{PWM} C_f(\tau_i s + 1)}{(T_m s + 1)(LC_f s^2 + 1)} \quad (3.29)$$

So the characteristic equation of the closed-loop transfer function is expressed as

$$F(s) = 1 + G_c(s)G_{PWM}(s)G_{plant}(s)H(s) \quad (3.30)$$

The characteristic polynomial of characteristic equation is obtained as

$$q(s) = (T_m s + 1)(LC_f s^2 + 1) + k_p K_{PWM} C_f(\tau_i s + 1) \quad (3.31)$$

Using the Routh-Hurwitz criterion, $\tau_i > T_m$ is a necessary and sufficient condition to keep the system stable. Meanwhile, when the switching frequency f_s is 15 kHz, $L = 1$ mH, and $C_f = 10$ μF, the T_m satisfy the following conditions

$$\begin{cases} T_m = T_s + T_d \approx 1.5 T_s \\ \sqrt{LC_f} > T_m \\ T_s = \frac{1}{f_s} \end{cases} \quad (3.32)$$

At low frequencies, the open-loop transfer function is approximately a proportional. When $|LC_f \omega^2| > 1$, the open-loop transfer function is further simplified as

$$G_{op}(s) = \frac{k_p K_{PWM} C_f(\tau_i s + 1)}{LC_f s^2 (T_m s + 1)} \quad (3.33)$$

In order to keep the system stable, the following condition should be satisfied

$$\sqrt{LC_f} > \tau_i > T_m \quad (3.34)$$

Fig. 3.10 Bode diagram of the CMRCC control system

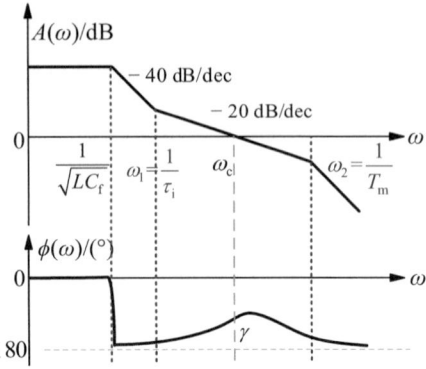

Therefore, the Bode diagram of simplified open-loop transfer function is obtained and drawn in Fig. 3.10.

In the mid-to-high-frequency band, the CMRCC control system can be the typical type-II system. For the sake of analysis, a new variable h is introduced, which is regarded as a

$$h = \frac{\omega_2}{\omega_1} = \frac{\frac{1}{T_m}}{\frac{1}{\tau_i}} = \frac{\tau_i}{T_m} \quad (3.35)$$

As shown in the Bode diagram, h is the width of the mid-frequency band with a slop of -20 dB/dec.

Based on the design rules of control systems engineering, $\omega_1 + \omega_2 = 2\omega_c$ should be satisfied. Meanwhile, the tradeoff between overshoot and tracking speed of the control system should be considered when h is selected. Subsequently, the parameters are obtained as

$$\begin{cases} k_p = \frac{L}{K_{PWM}} \frac{h+1}{2h^2 T_m^2} \\ \tau_i = hT_m \end{cases} \quad (3.36)$$

In order to determine the stability of the system, the transfer function of the steady-state error is expressed as

$$\frac{E(s)}{I^*_{cmz}} = \frac{1}{1 + G_{op}(s)}$$

$$= \frac{1}{1 + G_c(s)G_{PWM}(s)G_{plant}(s)H(s)}$$

$$= \frac{1}{1 + \frac{k_p(\tau_i s + 1)}{s} \cdot \frac{K_{PWM}}{T_{PWM}s + 1} \cdot \frac{C_f s}{LC_f s^2 + 1} \cdot \frac{1}{T_d s + 1}}$$

3.2 Leakage Current Suppression Technology

$$= \frac{(T_m s + 1)(LC_f s^2 + 1)}{(T_m s + 1)(LC_f s^2 + 1) + k_p K_{PWM} C_f (\tau_i s + 1)} \qquad (3.37)$$

With the final-value theorem of Laplace transform and the unit step input function, the steady-state error of CMCC can be obtained as

$$e_{ss} = \lim_{t \to \infty} e(t) = \lim_{s \to 0} sE(s) = \frac{1}{k_p K_{PWM} C_f + 1} \qquad (3.38)$$

It can be seen from steady-state error that the expression of e_{ss} is a constant, which is decreased with k_p.

It is worth noting that, the parameter of the grid-side inductance L_g will affect leakage current, and it is not considered in the mathematical description of the proposed method. For better understanding, the magnitude response of leakage current is depicted in Fig. 3.11 with different grid inductance L_g. It can be observed that f_{r2} tends to reduce with the increase of the grid inductance. The reduced f_{r2} may match with the switching frequency components of CMV, which increases leakage current. Moreover, inverter-side current oscillation will be excited for the reduced frequency of f_{r2}.

As we all know, the damping resistor R can be used to suppress the resonance peak. When the damping resistor is adopted, the CM equivalent circuit is shown in Fig. 3.12.

Assuming the leakage current is effectively suppressed, that is $i_{cm} = 0$, the CMCC i_{cmz} satisfies

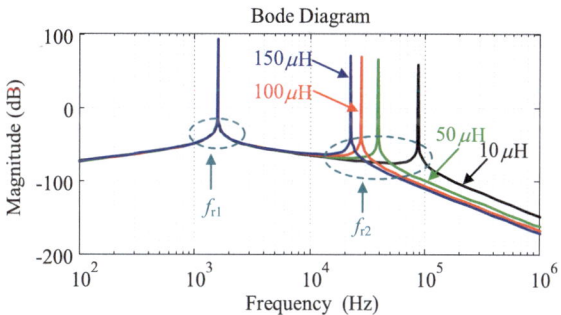

Fig. 3.11 Magnitude response of CM equivalent with different L_g

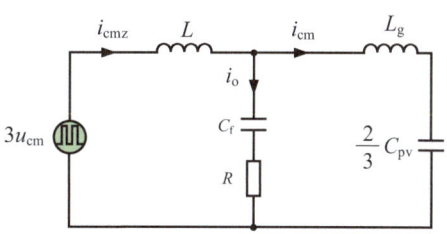

Fig. 3.12 The CM equivalent circuit using different damping resisters

Fig. 3.13 Magnitude response of CM equivalent using different damping resisters

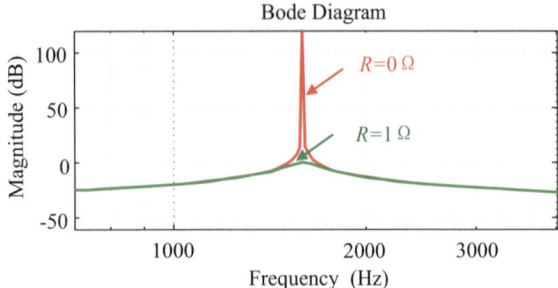

$$3u_{cm} = L\frac{di_{cmz}}{dt} + \frac{1}{C_f}\int i_{cmz}dt + Ri_{cmz} \tag{3.39}$$

Using the Laplace transform, the transfer function is obtained as

$$\frac{I_{cmz}(s)}{3U_{cm}(s)} = \frac{C_f s}{LC_f s^2 + RC_f s + 1} \tag{3.40}$$

The magnitude response is shown in Fig. 3.13 with different damping resistor value.

It can be seen that the CMCC has a high resonance peak when $R = 0\,\Omega$. With the increase of the damping resistor, the resonance peak decreases. When the damping resistor $R = 1\,\Omega$, the resonance peak is 0 dB. However, the power losses increase with the damping resistor, which reduces system efficiency. Moreover, the damping resistor impacts the high-frequency harmonic attenuation ability in differential-mode (DM) circuit.

Since the *ILCL* filter is adopted in the $3LT^2I$, the current resonance in DM circuit is also an important issue. The current resonance in DM circuit degrades system performance and leads to system instability. Thus, for DM circuit, current resonance suppression controller method should be designed.

As shown in $3LT^2I$ system with an *ILCL* filter, the averaged model of DM circuit is expressed as

$$\begin{cases} u_{AN} = L\frac{di_A}{dt} + u_{ca} + U_n \\ u_{BN} = L\frac{di_B}{dt} + u_{cb} + U_n \\ u_{CN} = L\frac{di_C}{dt} + u_{cc} + U_n \end{cases} \tag{3.41}$$

where u_{ca}, u_{cb}, and u_{cc} are the voltages on the filter capacitor. The DM equivalent circuit is shown in Fig. 3.14.

To meet industrial requirements, current sensors are integrated on the inverter side and single-loop inverter-side current control. The transfer function of the inverter side current to the output current in the DM circuit model is expressed as

$$G_{DM}(s) = \frac{I_J(s)}{U_{inv}(s)} = \frac{L_g C_f s^2 + 1}{LL_g C_f s^3 + (L + L_g)s} \tag{3.42}$$

3.2 Leakage Current Suppression Technology

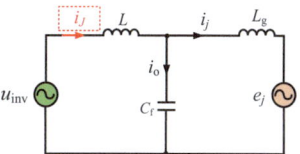

Fig. 3.14 DM equivalent circuit

It can be observed that the transfer function is a third order model. Selecting the same parameters as the CM circuit, the magnitude response of $G_{DM}(s)$ is shown in Fig. 3.15.

It can be seen that the magnitude response has a resonance peak, which results in inverter-side current oscillation and system instability. Therefore, an effective control method is needed to damp the resonance peak. The resonance frequency is calculated as

$$f_r = \frac{1}{2\pi}\sqrt{\frac{L+L_g}{LL_gC_f}} \quad (3.43)$$

In order to ensure system stable and reduce system costs, a single-loop inverter-side current control with capacitor-voltage-feedforward (CVF) is adopted in this chapter. The control diagram with CVF is shown in Fig. 3.16.

$M(s)$ is the feedforward coefficient. So, the inverter side current is derived as

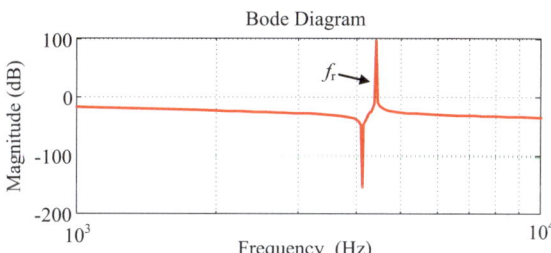

Fig. 3.15 Magnitude response of transfer function in DM equivalent circuit

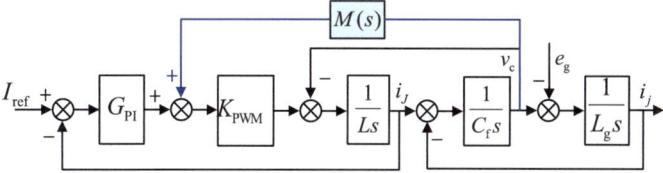

Fig. 3.16 Control diagram for DM circuit with capacitor voltage feedforward

$$I_J(s) = \frac{(L_gC_fs^2+1)G_{PI}K_{PWM}I_{ref} - (1-M(s)K_{PWM})e_g}{LL_gC_fs^3 + G_{PI}K_{PWM}L_gC_fs^2 + (L+L_g - L_gM(s)K_{PWM})s + G_{PI}K_{PWM}} \quad (3.44)$$

The open-loop transfer function is obtained as

$$G(s)H(s) = \frac{(L_gC_fs^2+1)G_{PI}K_{PWM}}{LL_gC_fs^3 + [L+L_g - L_gM(s)K_{PWM}]s} \quad (3.45)$$

Considering unit feedforward is the common choice, so the feedforward coefficient is chosen as

$$M(s) = \frac{1}{K_{PWM}} \quad (3.46)$$

Based on the above equations, the open-loop transfer function simplified as

$$G(s)H(s) = \frac{G_{PI}K_{PWM}}{Ls} \quad (3.47)$$

With the CVF, the three-order LCL filter turns to be a single L filter, and the resonance peak is totally eliminated.

Besides, the PI controller is used to realize the inverter-side current regulating and tracking. The transfer function of the PI controller is

$$G_{PI}(s) = K_p + \frac{K_i}{s} \quad (3.48)$$

So the open-loop transfer function of DM circuit is rewritten as

$$G(s)H(s) = \left(K_p + \frac{K_i}{s}\right)\frac{K_{PWM}}{Ls} \quad (3.49)$$

According to the theoretical analyses in [9], the crossover frequency $f_c = f_s/10$ is selected to attenuate high-frequency noise. The corner frequency of PI controller is $f_L = K_i = (2\pi K_p)$. In order to obtain a good system performance, the condition of $f_L < f_c < f_r$ should be satisfied. Here, the $f_L = 1$ kHz is selected to avoid the decrease of the phase margin resulted from PI controller. The magnitude of $|G_{PI}(s)|$ can be approximated to K_p at the frequency of f_c. Meanwhile the magnitude of $|G(s)H(s)|_{f_c} = 1$. So, the parameter of K_p and K_i are obtained as

$$\begin{cases} K_p = \frac{2\pi f_c L}{K_{PWM}} \\ K_i = 2\pi K_p f \end{cases} \quad (3.50)$$

So, the close-loop transfer function is derived as

3.2 Leakage Current Suppression Technology

$$G_{cl} = \frac{K_p s + K_i}{Ls^2 + K_p s + K_i} \quad (3.51)$$

The bandwidth is the frequency f_B, at which the frequency response has declined 3 dB from the low-frequency range. Based on the parameters in this paper, the bandwidth of the DM control system is 2.38 kHz.

The controller bandwidth of the DM circuit does not affect the circulating current control in CM circuit. While the NPV balancing control will influence the circulating current control. Thus, to reduce the harmonics of the output waveforms, the NPV should be balanced. The proportional (P) controller is adopted to control NPV balance. The output of the P controller is used to adjust the dwell time of small vectors, thereby controlling NPV balance.

The second modified state-switching times can be expressed as

$$\begin{cases} t_{A(2)} = t_{A(1)} - y_z \cdot T_s \\ t_{B(2)} = t_{B(1)} - y_z \cdot T_s \\ t_{C(2)} = t_{C(1)} - y_z \cdot T_s \end{cases} \quad (3.52)$$

where y_z is the output value of the NPV balance controller. Similarly, y_z should satisfy

$$-\frac{\min(t_{A(1)}, t_{B(1)}, t_{C(1)})}{T_s} \leq y_z \leq \frac{\min(t_{A(1)}, t_{B(1)}, t_{C(1)})}{T_s} \quad (3.53)$$

The overall block diagram of the system is shown in Fig. 3.17. The hybrid controller for the CM circuit with the *ILCL* filter realizes CMRCC and leakage current suppression. Meanwhile, the PI and CVF controllers achieve current control and resonance reduction in the DM circuit. Besides, the P controller is used to balance NPV.

Based on Discontinuous Modulation

As mentioned above, while the continuous modulation method has superior performance of output voltage and output current, it also presents some drawbacks. Due to the frequent operation of switching devices in continuous modulation, the system experiences higher switching and conduction losses, leading to reduced overall efficiency. To address this issue, the discontinuous modulation method is introduced. By reducing the turn-on and turn-off times of switching devices, the discontinuous modulation method significantly lowers system power losses and improves efficiency. Despite its advantages in reducing power losses, the discontinuous modulation method introduces new challenges in leakage current suppression based on the CM circuit modification method.

Conventionally, the discontinuous modulation method can be classified into two types: space vector-based discontinuous modulation and carrier-based discontinuous modulation. To simplify both calculation and implementation, the carrier-based

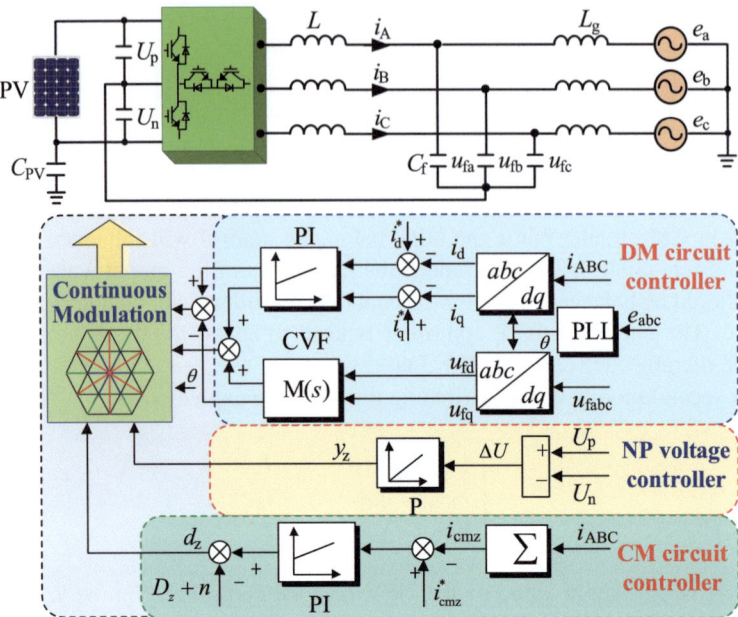

Fig. 3.17 Block diagram of CM circuit modification method with continuous modulation

discontinuous modulation is adopted in this section. Actually, these two methods are mathematically equivalent [10].

According to the analyses in above section, to mitigate the CMRCC, a control variable d_z is introduced to modify the switching sequence. However, the switching sequence in the discontinuous modulation method cannot be modified by an additional control variable to maintain the discontinuity. To solve the issue, in this section, two key points must be considered. The first point is the decoupling of CMRCC mitigation and the discontinuity of the discontinuous modulation method. The second point is the leakage current suppression method based on discontinuous modulation with a small dc-link capacitance value, which aims to further save cost. The two points are elaborated as follows.

1. **The decoupling of CMRCC mitigation and discontinuity of discontinuous modulation method**

    Similar to the continuous modulation, the high-frequency leakage current is suppressed by adopting the *ILCL* filter, which offers a low impedance path for the high-frequency leakage current. Therefore, the control variable d_z is also needed to mitigate the CMRCC to zero. To solve the coupling issue mentioned above, the generating mechanism of the three-phase modulation waveforms in the discontinuous modulation is reconstructed. The three-phase sinusoidal reference voltages can be expressed as

3.2 Leakage Current Suppression Technology

$$\begin{cases} V_a = V_m \sin(\omega t) \\ V_b = V_m \sin\left(\omega t - \frac{2\pi}{3}\right) \\ V_c = V_m \sin\left(\omega t + \frac{2\pi}{3}\right) \end{cases} \quad (3.54)$$

where V_m is the magnitude of the sinusoidal reference voltages.

The discontinuous modulation method is realized by comparing the sinusoidal reference voltages with a square waveform. According to three-phase sinusoidal reference voltages, the square waveform is defined as

$$V_{squ} = k_m \, \text{sgn}(|V_{max}| - |V_{min}|) + d_z \quad (3.55)$$

where k_m is the magnitude of the square waveform, sgn( ) is the sign function, V_{max} and V_{min} are the maximum and minimum values of the sinusoidal reference voltages, d_z is the control variable that achieves the CMRCC mitigation.

The sinusoidal reference voltages and square waveform are shown in Fig. 3.18. It can be seen that, 60° clamping period of the traditional DPWM1 is equally divided into three intervals, e.g., [π/3, 4π/9], [4π/9, 5π/9] and [5π/9, 2π/3], which can be realized by adjusting the magnitude of the square waveform k_m to make the square waveform intersects with the sinusoidal reference voltages. Therefore, regardless of the control variable d_z, to divide the clamping period, the square waveform should intersect with the sinusoidal reference voltage of phase A at 4π/9 and 5π/9 by changing the value of k_m. Thus, k_m is calculated as

$$k_m = V_m \sin\left(\frac{4\pi}{9}\right) = V_m \sin\left(\frac{5\pi}{9}\right) \quad (3.56)$$

After the value of k_m is determined, it is necessary to determine the clamping state for each interval. If the inverter operates at unity power factor, the phase current of the load is almost consistent with the phase voltage. To minimize switching losses, the clamping states should clamp near the peak voltage. Therefore, the zero-sequence voltage of the discontinuous modulation is defined as

$$V_{z,dpwm} = \begin{cases} \frac{V_{dc}}{2} - V_{max} & (V_{max} > V_{squ} > 0 \text{ or } V_{squ} < V_{min} < 0) \\ -\frac{V_{dc}}{2} - V_{min} & (V_{squ} > V_{max} > 0 \text{ or } V_{min} < V_{squ} < 0) \\ 0 & \text{else} \end{cases} \quad (3.57)$$

As shown in Fig. 3.19, the discontinuous modulation reference voltages $V_{j,dpwm}$ (j = a, b, c) are the sum of the sinusoidal reference voltages V_j and the zero-sequence voltage $V_{z,dpwm}$, which is expressed as

$$V_{j,dpwm} = V_j + V_{z,dpwm} \quad (j = a, b, c) \quad (3.58)$$

As the square waveform definition shows that by changing the control variable d_z, the switching sequence can be modified without destroying the discontinuity of the discontinuous modulation. The decoupling of CMRCC mitigation and discontinuity is achieved.

The proposed CMRCC control strategy based discontinuous modulation is shown in Fig. 3.20. In order to mitigate the CMRCC to zero, the reference value of i^*_{cmz} is set to zero. The error between CMCC and its reference value is fed into the PI controller to generate the control variable d_z, which is used to adjust the DPWM reference voltages, thereby changing the three-phase switching sequence, and thus the CMRCC can be suppressed.

2. **The discontinuous modulation with small dc-link capacitance value**

   In this section, the leakage current suppression based on discontinuous modulation method with the small dc-link capacitance value is investigated, which has a distinct advantage in costs reduction.

   As mentioned in above section, an improved discontinuous modulation is presented to mitigate the high-frequency leakage current and the generated CMRCC without destroying the discontinuity. However, undesirable resonance currents will be excited due to the abrupt changes in the reference voltages. To avoid exciting resonance currents, the DPWMA method is more suitable for the *ILCL* filter, which can smooth the transition process of reference voltages [11]. The control variable d_z is also necessary to mitigate CMRCC by this discontinuous modulation method due to the adoption of the *ILCL* filter.

   To maintain the discontinuity of the discontinuous modulation, in the process of injecting the control variable d_z, when one phase is clamped, only the other two phases are injected. Due to the asymmetrical injection, the three-phase reference voltages are asymmetrical and the reference voltages are also asymmetrical between the positive and negative half cycles. Therefore, the inverter-side output voltages u_{JN} (J = A, B, C) contain multiples of 3rd harmonics, especially the low-frequency harmonics such as 3rd, 6th, 9th and 12th harmonics.

   In addition, to save costs, the dc-link capacitance value is typically small, which results in a large NP voltage ripple. Such a large NP voltage ripple will introduce low-frequency harmonics in the grid currents, especially the fifth and seventh harmonics. To suppress these low-frequency harmonics, a PIR controller instead of conventional PI controller is adopted in the current control loop. It should be noted that the fifth and seventh current harmonics are characterized as sixth harmonics in the dq reference frame [12]. Thus, the transfer function of the PIR controller is written as

$$G_{PIR}(s) = \frac{K_p s + K_i}{s} + \frac{2K_r \omega_c s}{s^2 + 2\omega_c s + \omega_0^2} \qquad (3.59)$$

where K_p is the proportion gain, K_i is the integral gain, K_r is the resonant gain, ω_c is the resonance bandwidth, ω_0 is the central angular frequency, which is selected as 600π rad/s to suppress the sixth harmonics in the dq reference frame.

Unfortunately, the *ILCL* filter offers low impedance for high-frequency leakage current attenuation, it also introduces the new low-frequency harmonics

3.2 Leakage Current Suppression Technology

in the grid currents due to the complex coupling between the low-frequency harmonics in output voltages and the high peak gain of the PIR controller. The detailed analyses are as follows.

Since the inverter-side output voltages u_{JN} contain the 3rd, 6th, 9th and 12th harmonics, the CMV u_{cm} will also contain the 3rd, 6th, 9th and 12th harmonics. According to the CM equivalent circuit of the transformless PV 3LT²I system with the *ILCL* filter, the CMV will excite the *ILCL* CM current i_o, the transfer function from u_{cm} to i_o is obtained as

$$G_{io}(s) = \frac{i_o(s)}{u_{cm}(s)} = \frac{3C_f s}{LC_f s^2 + 1} \tag{3.60}$$

Thus, the *ILCL* CM current i_o is expressed as

$$i_o(s) = G_{io}(s) \cdot u_{cm}(s) = \frac{3C_f s}{LC_f s^2 + 1} \cdot u_{cm}(s) \tag{3.61}$$

According to the above expression, the harmonics in u_{cm} will be transferred to i_o through the transfer function $G_{io}(s)$ and the low-frequency harmonics in u_{cm} are still characterized as the 3rd, 6th, 9th and 12th harmonics after passing through the transfer function $G_{io}(s)$. Therefore, the *ILCL* CM current i_o contains the 3rd, 6th, 9th and 12th harmonics.

Since the *ILCL* filter connects the common point of the filter capacitors to the dc-link NP O, the *ILCL* CM current i_o containing the 3rd, 6th, 9th and 12th harmonics is injected into the dc-link NP O. Therefore, the dc-link NP voltage will also contain the 3rd, 6th, 9th and 12th harmonics.

In the *dq* reference frame, the harmonic suppression effect of the PIR controller whose center frequency is six times the fundamental frequency is characterized as the 5th and 7th harmonics in the $\alpha\beta$ reference frame. The 5th and 7th harmonics suppression action of the PIR controller will interact with the 6th, 9th and 12th harmonics in the dc-link, resulting in the presence of new low-frequency harmonics such as 11th, 13th, 14th, 16th, 17th and 19th harmonics in the output currents.

The generating mechanism of the newly appeared harmonics is summarized as Fig. 3.21. It can be seen that the asymmetrical three-phase reference voltages caused by the injection of the active damping strategy lead to the harmonics that are multiples of the 3rd harmonics such as the 6th, 9th and 12th harmonics in the inverter-side output voltages, the CMV, the *ILCL* CM current and the dc-bus. The 6th, 9th and 12th harmonics in the dc-bus will act with the PIR controller to generate the 11th, 13th, 14th, 16th, 17th and 19th harmonics in the output currents.

The newly appeared low-frequency 11th, 13th, 14th, 16th, 17th and 19th harmonics are coupled with the *ILCL* filter and PIR controller. Based on the generating mechanism, suppressing the 6th, 9th and 12th harmonics in the *ILCL* CM current i_o could be a potential solution. Therefore, a notch filter $G_{NF}(s)$ is introduced in the CM circuit control loop, which blocks the feedback path of the 6th, 9th and 12th harmonics from the inverter output side to the dc-bus to suppress the 11th, 13th, 14th, 16th, 17th and 19th harmonics. The expression of $G_{\mathrm{NF}}(s)$ is

$$G_{\mathrm{NF}}(s) = \frac{s^2 + \omega_n^2}{s^2 + s\frac{\omega_n}{Q} + \omega_n^2} \qquad (3.62)$$

where Q is the quality factor, ω_n is the central angular frequency of the notch filter. The purpose of using the notch filter is to suppress the low-frequency harmonics such as 6th, 9th and 12th harmonics in the *ILCL* CM current i_o. Therefore, the notch frequency ω_n is selected as 9 times the fundamental frequency. The range of the stop band is set between 300 and 600 Hz.

The control block diagram including the notch filter is shown in Fig. 3.22, which consists of a current control loop and a CM circuit control loop. In the current control loop, the PIR controller is used to suppress the fifth and seventh harmonics in the output currents caused by the NP voltage ripple. In the CM circuit control loop, the notch filter is introduced to filter out the multiples of the third harmonics in the *ILCL* CM current i_o caused by the unbalanced three-phase reference voltages, thereby eliminating the low-frequency 11th, 13th, 14th, 16th, 17th and 19th harmonics in the output currents.

In summary, the discontinuous modulation methods for leakage current and CMRCC suppression with the *ILCL* filter is presented in detailed. The key points are to achieve decoupling between discontinuity and CMRCC mitigation. For the first point, the generating mechanism of the three-phase modulation waveforms of the discontinuous modulation is reconstructed to apply the control variable d_z. For the second point, based on the small dc-link capacitance value, the DPWMA method with a smooth transition process of reference voltages combined with a PIR controller and a notch filter is presented to suppress the low-frequency harmonics, and the control variable d_z is still needed to mitigate CMRCC.

3.2.2 Exciting Source Suppression Methods

Based on the model of leakage current, the exciting source of leakage current is merely the u_{cm}. Thus, the second type method is proposed from the perspective of exciting source suppression. The exciting source suppression strategy can be classified into the CMV elimination method and the CMV reduction method. Once the topology of a PV inverter is selected, the CMV is directly decided by the modulation methods. Therefore, these two methods will be derived based on modulation strategies.

It should be noted that the traditional three-level inverter merely has a voltage buck capability. That is to say, the ac output voltage cannot exceed the dc input

3.2 Leakage Current Suppression Technology

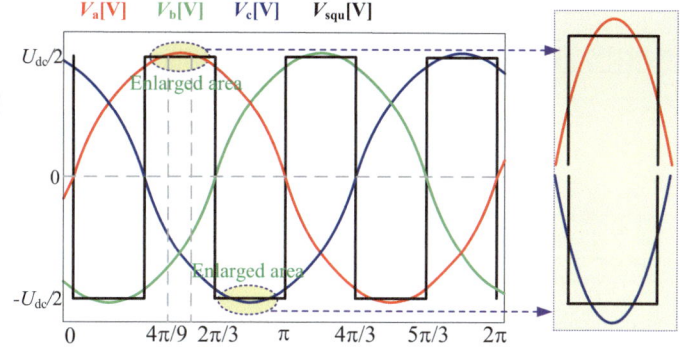

Fig. 3.18 Relationship of three-phase voltages and square waveform for the discontinuous modulation

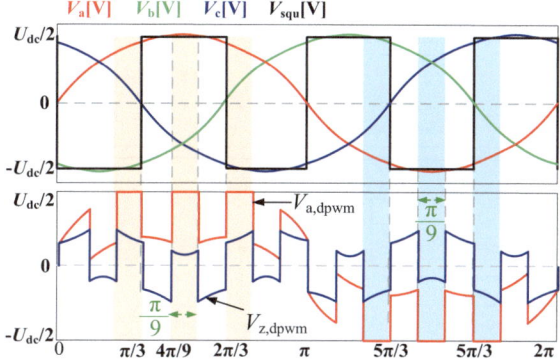

Fig. 3.19 Waveforms of the proposed discontinuous modulation method

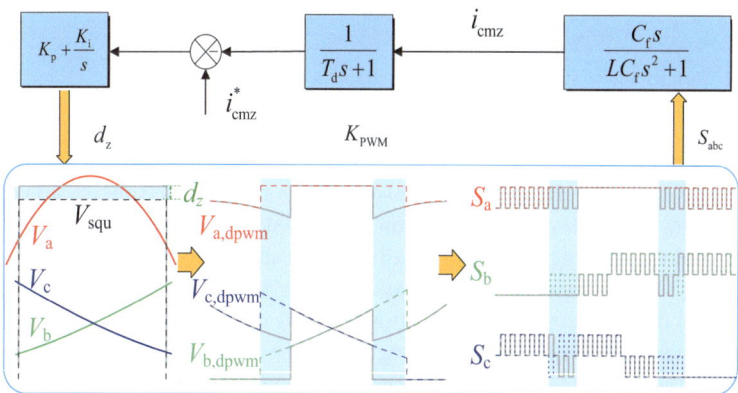

Fig. 3.20 Block diagram of the CMRCC control strategy with the proposed discontinuous modulation method

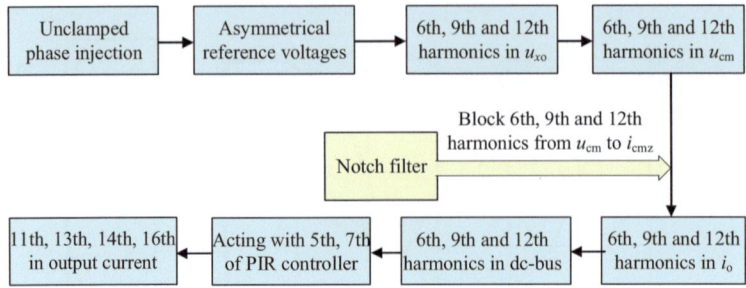

Fig. 3.21 Generating mechanism of the 11th, 13th, 14th and 16th harmonics

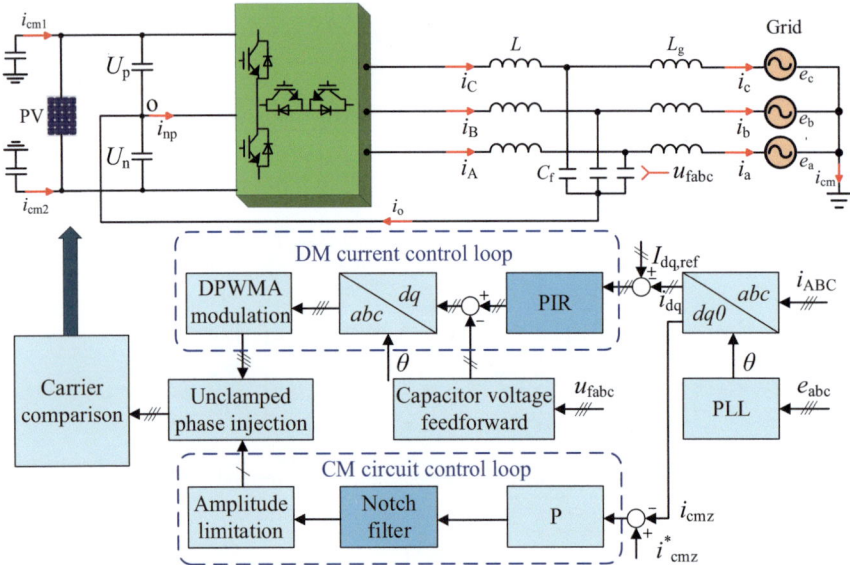

Fig. 3.22 Control block diagram of the proposed method

voltage. When the output voltage of a PV array is reduced due to the reduction of light radiation or temperature, the PV inverter cannot achieve the energy conversion, and the power of the PV power generation system will be reduced. To solve this issue, it is required that the circuit with high voltage gain should be embedded between the PV array and the inverter. The quasi-Z-source three-level inverter has been proposed to surmount the shortcomings of traditional PWM inverters, which can implement buck and boost operations through a single stage power conversion [13]. Moreover, the reliability of this inverters is enhanced due to the avoidance of dead-time. Thus, in this section, the quasi-Z-source three-level T-type inverter (QZS-3LT^{2}I) is used as an example to illustrate the CMV suppression method.

Common-Mode Voltage Elimination

The CMV elimination is realized by only adopting the basic vectors with zero CMV magnitude to synthesize the output voltage [14]. The basic voltage vectors with zero CMV magnitude include six medium vectors, one zero vector and three shoot-through vectors. Specifically, two medium vectors, one zero vector and one shoot-through vector (2M1Z-ST) are used to synthesize the output voltage in one switching period. Thus, this CMV elimination method is called 2M1Z-ST method. With this method, the CMV is maintained as a constant value and does not cause leakage current. For the QZS-3LT^{2}I, the CMV elimination method is elaborated from four steps in the following section.

Step 1: Basic Operating Principle of the Quasi-Z-Source Three-Level T-Type Inverter

The basic topology of the QZS-3LT^{2}I is depicted in Fig. 3.23, which combines a three-level quasi-Z-source (QZS) network with a 3LT^{2}I. U_{dc} represents the output voltage of the PV array, and v_{dc} represents the dc-link voltage, which is the output of the QZS network. It should be pointed out that the inductor L and resistor R are adopted as the load considering the practical industrial applications. The 3LT^{2}I is configured between the QZS network and the load.

It is known that the conventional 3LT^{2}I has three switching states: [P], [O], and [N]. For the QZS-3LT^{2}I, there exists an additional shoot-through state (denoted by the [F] state in this paper), which is achieved by turning on 4 power semiconductor switches configured in one phase leg [15]. It is emphasized that the shoot-through state is essential to implement the voltage boosting capability, which is not permitted for the conventional three-level inverter. One phase leg is considered to describe the operational principle of the QZS-3LT^{2}I. Selecting the common point O of the two quasi-Z-source networks as the reference, the switching states and the corresponding output voltages of the QZS-3LT^{2}I are summarized in Table 3.2, where V_{dc} is the magnitude of the dc-link voltage.

Considering that the quasi-Z-source network is symmetrical ($C_1 = C_4$, $C_2 = C_3$, $L_1 = L_3$, and $L_2 = L_4$), then $v_{C1} = v_{C4}$, $v_{C2} = v_{C3}$, $v_{L1} = v_{L4}$, and $v_{L2} = v_{L3}$. The

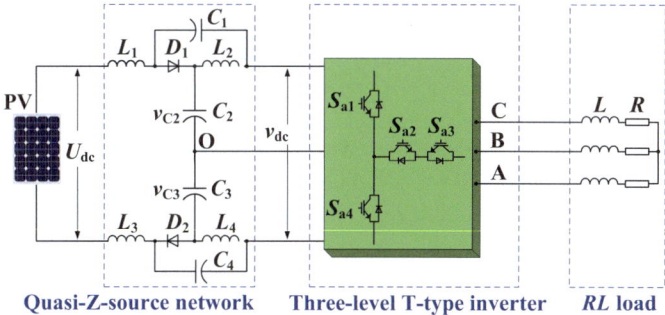

Fig. 3.23 Basic topology of QZS-3LT^{2}I

Table 3.2 Corresponding switching states of quasi-Z-source three-level T-type inverter

Switching states	ON power semiconductor switches	Output voltage
[P]	S_{x1}, S_{x2}	$+V_{dc}/2$
[O]	S_{x2}, S_{x3}	0
[N]	S_{x3}, S_{x4}	$-V_{dc}/2$
[F]	$S_{x1}, S_{x2}, S_{x3}, S_{x4}$	0

boost factor is defined as the ratio of the dc-link voltage magnitude V_{dc} to the dc input voltage, can be expressed as

$$B = \frac{V_{dc}}{U_{dc}} = \frac{1}{1 - 2 \cdot \frac{t_{ST}}{T}} = \frac{1}{1 - 2d_{ST}} \quad (3.63)$$

where t_{ST} and d_{ST} are the dwell time and duty cycle of shoot-through vector.

In steady state, the sum of capacitor voltages v_{C2} and v_{C3} can be expressed as

$$v_{c2} + v_{c3} = \frac{1 - d_{ST}}{1 - 2d_{ST}} U_{dc} \quad (3.64)$$

Since the shoot-though state does not affect the output voltage, the CMV generated by the QZS-3LT²I is similar to that of the traditional 3LT²I. The CMV is defined as one-third of the sum of three output phase voltages, which is expressed as

$$u_{cm} = \frac{1}{3}(u_{AO} + u_{BO} + u_{CO}) \quad (3.65)$$

where u_{AO}, u_{BO}, and u_{CO} are three-phase output voltages when the common point O of the two quasi-Z-source networks as the reference.

The magnitudes of large vectors, medium vectors, and zero vector [OOO] are expressed as

$$\begin{cases} \left|\vec{V_L}\right| = \frac{2}{3} B \cdot U_{dc} \\ \left|\vec{V_L}\right| = \frac{\sqrt{3}}{3} B \cdot U_{dc} \\ \left|\vec{V_L}\right| = 0 \end{cases} \quad (3.66)$$

Since shoot-through states can be inserted in Phase A, Phase B, or Phase C, there exist three shoot-through vectors: [FOO], [OFO], and [OOF]. It is worth mentioning that the shoot-through vectors and the zero vector have the same output voltage across the load.

According to the magnitude of basic voltage vectors and the definition of CMV, the basic voltage vectors and CMVs of the QZS-3LT²I are listed in Table 3.3. Each basic

3.2 Leakage Current Suppression Technology

voltage vector can be categorized into large, medium, small, zero and shoot-through vectors based on its magnitude.

It can be seen that the CMVs of the voltage vectors highlighted in blue are equal to zero. If the modulation scheme uses only these vectors with zero CMV magnitudes, the CMV generated by the inverter can be maintained at zero level. Based on these analyses, the proposed scheme will be presented in the next section.

Step 2: Basic Voltage Vectors Selection for Output Voltage Synthesizing
Figure 3.24 displays the space vector diagram (SVD) of the QZS-3LT^{2}I, which is divided into 12 sectors. In this SVD, six medium vectors (represented by red arrows), a zero vector [OOO] (highlighted in purple) and three shoot-through vectors (highlighted in sky blue) have zero CMV magnitude, which are selected to synthesize the output voltage.

In each sector, two medium vectors, one zero vector and one shoot-through vector (2M1Z-ST) are used to synthesize the output voltage during one switching period. Thus, zero CMV can be guaranteed.

Step 3: Duty Cycle Calculation of Basic Voltage Vectors with Zero CMV Magnitude
As a typical example, Sector 1 of the SVD is considered to illustrate the duty cycle calculation. In Sector 1, voltage vectors $\vec{V}_7$, $\vec{V}_{12}$, $\vec{V}_0$, and $\vec{V}_{FA}$ are adopted to produce the ac output voltage. The voltage-second balance equation is written as

$$\begin{cases} V_{\text{ref}} = \vec{V}_7 \cdot d_7 + \vec{V}_{12} d_{12} + \vec{V}_0 \cdot d_Z + \vec{V}_{FA} \cdot d_{\text{ST-A}} \\ d_7 + d_{12} + d_Z + d_{\text{ST-A}} = 1 \end{cases} \quad (3.67)$$

Table 3.3 Voltage vectors and CMVs of QZS-3LT^{2}I for 2M1Z-ST modulation method

Vectors	States	CMV	States	CMV	States	CMV
Large	PNN	$-V_{dc}/6$	PPN	$+V_{dc}/6$	NPN	$-V_{dc}/6$
	NPP	$+V_{dc}/6$	NNP	$-V_{dc}/6$	PNP	$+V_{dc}/6$
Medium	PON	0	OPN	0	NPO	0
	NOP	0	ONP	0	PNO	0
Small P-type	POO	$+V_{dc}/6$	PPO	$+V_{dc}/3$	OPO	$+V_{dc}/6$
	OPP	$+V_{dc}/3$	OOP	$+V_{dc}/6$	POP	$+V_{dc}/3$
Small N-type	ONN	$-V_{dc}/3$	OON	$-V_{dc}/6$	NON	$-V_{dc}/3$
	NOO	$-V_{dc}/6$	NNO	$-V_{dc}/3$	ONO	$-V_{dc}/6$
Zero	OOO	0	PPP	$+V_{dc}/2$	NNN	$-V_{dc}/2$
Shoot-through	FOO	0	OFO	0	OOF	0

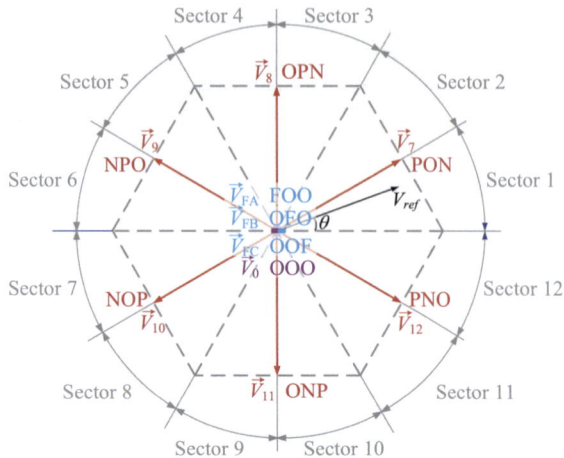

Fig. 3.24 The space vector diagram of QZS-3LT^{2}I using the 2M1Z-ST method: medium vectors (blue), zero vector (purple), and shoot-through vectors (sky blue)

where V_{ref} denotes the reference vector. d_7, d_{12}, d_Z, and d_{ST-A} are the duty cycles of $\vec{V}_7, \vec{V}_{12}, \vec{V}_0$, and $\vec{V}_{FA}$, respectively. As mentioned previously, the shoot-through vector should be inserted within the duration of zero vector, and the duty cycles for the basic vectors are attained as

$$\begin{cases} d_7 = m \cdot \sin\left(\theta + \frac{\pi}{6}\right) \\ d_{12} = m \cdot \sin\left(\frac{\pi}{6} - \theta\right) \\ d_0 = 1 - d_7 - d_{12} - d_{ST-A} \end{cases} \quad (3.68)$$

where m means the modulation index, whose definition is given as

$$m = \frac{\sqrt{3} \cdot V_R}{B \cdot U_{dc}} \quad (3.69)$$

where V_R, B and U_{dc} are the magnitudes of reference vector, boost factor, and dc input voltage, respectively.

To synthesize the reference vector correctly, the maximum magnitude of reference vector is expressed as

$$V_{R,max,2M1Z-ST} = \frac{\sqrt{3}}{3} \cdot B \cdot U_{dc} \cdot \cos\frac{\pi}{6} = \frac{1}{2} \cdot B \cdot U_{dc} \quad (3.70)$$

where $V_{R,max,2M1Z-ST}$ is the maximum value of the magnitude of the reference vector V_R. According to the expression of the modulation index and the maximum magnitude of the reference vector, the maximum modulation index for the 2M1Z-ST method is derived as

$$m_{max-2M1Z-ST} = \frac{\sqrt{3} \cdot V_{R,max,2M1Z-ST}}{B \cdot U_{dc}} = \frac{\sqrt{3}}{2} < 1 \quad (3.71)$$

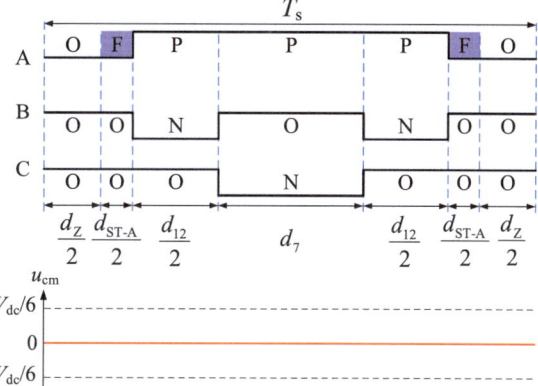

Fig. 3.25 The switching sequences and CMV waveform in Sector 1 using 2M1Z-ST modulation method

It can be seen that the maximum modulation index is lower than that of the conventional SVM scheme. The reason is that the large vectors in the SVD are abandoned.

Step 4: Optimal Switching Sequence Designing
When designing the switching sequence, two principles need to be considered. At first, to reduce the switching losses of power semiconductor switches and improve system efficiency, the number of switching transitions should be restricted. Second, to mitigate the current ripple of the inverter, the symmetric characteristics should be considered. Moreover, the shoot-through state has to be inserted to realize voltage boosting.

In Sector 1, voltage vectors $\vec{V}_7$, $\vec{V}_{12}$, $\vec{V}_0$, and $\vec{V}_{FA}$ are adopted to produce the ac output voltage. Considering the above switching sequence designing principles, the switching sequence is designed as OOO-FOO-PNO-PON-PNO-FOO-OOO, which has seven segments. The shoot-through state in Phase A is inserted within the duration of the zero vector $\vec{V}_0$ to implement voltage boosting. The switching sequence is shown in Fig. 3.25. It can be seen that the CMV is maintained at a constant value of zero during a switching cycle with the 2M1Z-ST modulation method.

Similarly, the switching sequences of the 2M1Z-ST in other sectors can also be obtained in this way, and the switching sequences are summarized in Table 3.4. It can be seen that each phase is selected as the shoot-through phase in one-third of all sectors, so relatively balanced losses of power devices can be guaranteed.

Common-Mode Voltage Reduction

The modulation method of 2M1Z-ST only adopts the medium, zero and shoot-through vectors to eliminate the CMV completely. However, this approach reduces the maximum output voltage. As the above analyses reveal, the maximum modulation

Table 3.4 The switching sequences in all sectors using the 2M1Z-ST modulation method

Sectors	Switching sequences
1	OOO-FOO-PNO-PON-PNO-FOO-OOO
2	OOO-OFO-OPN-PON-OPN-OFO-OOO
3	OOO-OOF-PON-OPN-PON-OOF-OOO
4	OOO-FOO-NPO-OPN-NPO-FOO-OOO
5	OOO-OFO-OPN-NPO-OPN-OFO-OOO
6	OOO-OOF-NOP-NPO-NOP-OOF-OOO
7	OOO-FOO-NPO-NOP-NPO-FOO-OOO
8	OOO-OFO-ONP-NOP-ONP-OFO-OOO
9	OOO-OOF-NOP-ONP-NOP-OOF-OOO
10	OOO-FOO-PNO-ONP-PNO-FOO-OOO
11	OOO-OFO-ONP-PNO-ONP-OFO-OOO
12	OOO-OOF-PON-PNO-PON-OOF-OOO

index is only 86.6% of the traditional method, which reduces the dc voltage utilization. Additionally, since no small vectors are employed, the 2M1Z-ST method does not have NPV recovery capability. To solve this issue, three types of CMV reduction modulation methods are presented in this section. These methods are defined as: (1) large, medium, zero, and shoot-through voltage vectors (LMZ-ST) modulation [16], (2) large, medium, zero, small and shoot-through voltage vectors (LMZS-ST) modulation [17], (3) hybrid space vector modulation (HSVM) [18]. The detailed analyses of each modulation are elaborated as follows.

1. **Large, medium, zero, and shoot-through voltage vectors (LMZ-ST) method**

    LMZS-ST modulation adopts large, medium, zero, and shoot-through voltage vectors to synthesize the reference voltage. By properly selecting the shoot-through phase according to the sector number, shoot-through states are inserted within zero vector [OOO] to realize voltage boosting. The zero vector and shoot-through voltage vectors have the same output voltage across the load, i.e., 0 V, so the normalized volt-second balance is not affected. The dwell times of voltage vectors are calculated through the modified volt-second balance principle. Although the CMV is not completely eliminated, its magnitude can be restricted within one-sixth of the dc-link voltage, and its slew rate is also reduced. Moreover, the modulation index is not affected, and high dc-link voltage utilization can be maintained. Four steps are required to implement LMZS-ST modulation, which are elaborated as follows.

Step 1: Basic Voltage Vectors Selection for CMV Reduction
The basic vectors and their corresponding CMVs are listed in Table 3.5. It can be seen that the CMVs of the vectors highlighted in blue are lower than $V_{dc}/6$. In which, the CMVs of six large vectors are equal to $V_{dc}/6$ or $-V_{dc}/6$. The CMVs of six medium vectors, one zero vector [OOO] and shoot-through vector are equal to 0.

3.2 Leakage Current Suppression Technology

Table 3.5 Voltage vectors and CMVs of QZS-3LT^{2}I for LMZ-ST modulation method

Vectors	States	CMV	States	CMV	States	CMV
Large	PNN	$-V_{dc}/6$	PPN	$+V_{dc}/6$	NPN	$-V_{dc}/6$
	NPP	$+V_{dc}/6$	NNP	$-V_{dc}/6$	PNP	$+V_{dc}/6$
Medium	PON	0	OPN	0	NPO	0
	NOP	0	ONP	0	PNO	0
Small P-type	POO	$+V_{dc}/6$	PPO	$+V_{dc}/3$	OPO	$+V_{dc}/6$
	OPP	$+V_{dc}/3$	OOP	$+V_{dc}/6$	POP	$+V_{dc}/3$
Small N-type	ONN	$-V_{dc}/3$	OON	$-V_{dc}/6$	NON	$-V_{dc}/3$
	NOO	$-V_{dc}/6$	NNO	$-V_{dc}/3$	ONO	$-V_{dc}/6$
Zero	OOO	0	PPP	$+V_{dc}/2$	NNN	$-V_{dc}/2$
Shoot-through	FOO	0	OFO	0	OOF	0

The corresponding SVD is shown in Fig. 3.26. The large vectors are represented by green arrows, the medium vectors are represented by red arrows, the zero vector [OOO] is highlighted in purple and the shoot-through vectors are highlighted in sky blue. Since shoot-through states can be inserted in Phase A, Phase B, or Phase C, there exist three shoot-through vectors: [FOO], [OFO], and [OOF]. To balance the losses of power devices, each phase is selected as the shoot-through phase for one-third of all sectors. Shoot-through phase has been represented with different background colors.

In each sector, if only one large, one medium, one zero vector [OOO], and one corresponding shoot-through vectors are adopted to synthesize the reference voltage in each switching period, the CMV magnitude of the PV inverter can be restricted within $V_{dc}/6$, which is only a half of that with the traditional modulation method.

Step 2: Dwell Time Calculation Without Shoot-Through

Since the shoot-through voltage vectors are inserted within the zero vector [OOO] to realize voltage boosting, the ac output voltage is not affected. So, the dwell times of the large, medium and zero vectors can be calculated by the volt-second balance principle without shoot-through voltage vectors. Then the zero vectors dwell time is modified by the dwell time of shoot-through voltage vectors.

The detailed duty cycle calculation will be presented in this section. Sector 1 and Sector 2 in the proposed SVD will be considered as representative examples for odd and even sectors to explain the proposed scheme.

Assuming the reference voltage vector V_{ref} is located in Sector 1, the large vector $\vec{V}_{13}$, medium vector $\vec{V}_7$, and zero vector $\vec{V}_0$ are adopted to synthesize the reference vector. According to the volt-second balance principle

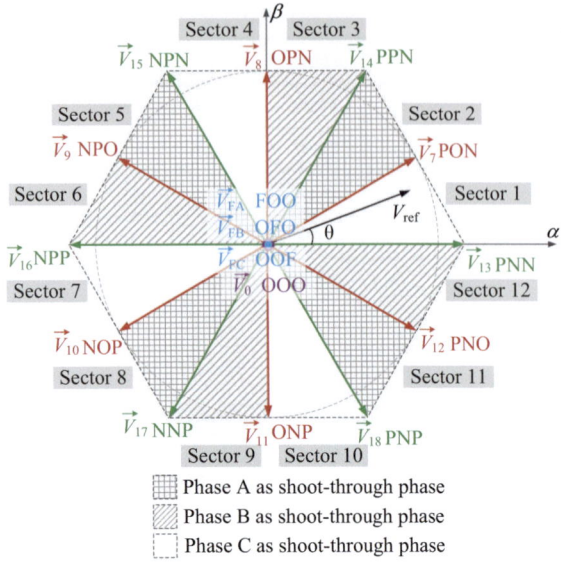

Fig. 3.26 SVD of QZS-3LT²I using the LMZ-ST method: large vectors (green), medium vectors (red), zero vector (purple), and shoot-through vectors (sky green)

$$V_{\text{ref}} = \vec{V}_{13} t_L + \vec{V}_7 \cdot t_M + \vec{V}_0 \cdot t_Z \tag{3.72}$$

where voltage vectors $\vec{V}_{13}$, $\vec{V}_7$, $\vec{V}_0$, and V_{ref} is expressed as

$$\begin{cases} \vec{V}_{13} = \frac{2}{3} B \cdot U_{\text{dc}} \\ \vec{V}_7 = \frac{\sqrt{3}}{3} B \cdot U_{\text{dc}} \cdot e^{j\frac{\pi}{6}} \\ \vec{V}_0 = 0 \\ V_{\text{ref}} = V_R \cdot e^{j\theta} \end{cases} \tag{3.73}$$

The dwell times of large vector $\vec{V}_{13}$, medium vector $\vec{V}_7$, and zero vector $\vec{V}_0$, can be calculated as

$$\begin{cases} t_L = \sqrt{3} m T_s \sin\left(\frac{\pi}{6} - \theta\right) \\ t_M = 2 m T_s \sin\theta \\ t_Z = T_s - t_L - t_M \end{cases} \tag{3.74}$$

where t_L, t_M, and t_Z are dwell times of large vector $\vec{V}_{13}$, medium vector $\vec{V}_7$, and zero vector $\vec{V}_0$, respectively. T_s is the switching period, and $0 \leq \theta < \pi/6$. m is the modulation index. The maximum magnitude of the reference voltage vector is equal to the length of medium vector, and can be found from

$$V_{R,\max,\text{LMZ-ST}} = \frac{\sqrt{3} B \cdot U_{\text{dc}}}{3} \tag{3.75}$$

3.2 Leakage Current Suppression Technology

According to the definition of modulation index, the maximum modulation index with the LMZ-ST modulation can be derived as

$$m_{\text{max-LMZ-ST}} = \frac{\sqrt{3} \cdot V_{R,\text{max,LMZ-ST}}}{B \cdot U_{\text{dc}}} = 1 \tag{3.76}$$

It can be seen that the dc-link voltage utilization of the proposed scheme is the same as that of the conventional SVM method. Although small vectors are not adopted to synthesize the reference voltage vector, the dc-link voltage utilization is not affected.

The selected switching sequence in Sector 1 is OOO-PON-PNN-PON-OOO, and the state transition times for Phase A, B, and C can be calculated as

$$\begin{cases} T_a = \frac{t_Z}{2} \\ T_b = \frac{t_Z + t_M}{2} \\ T_c = \frac{t_Z}{2} \end{cases} \tag{3.77}$$

It can be seen that the maximum positive pulse width exists in Phase A, while the maximum negative pulse width exists in Phase C. The dwell times of these two pulses are identical.

When the reference voltage vector V_{ref} is located in Sector 2, the large vector $\vec{V}_{14}$, medium vector $\vec{V}_7$, and zero vector $\vec{V}_0$ are adopted to synthesize the reference vector. According to the volt-second balance principle

$$V_{\text{ref}} T_s = \vec{V}_{14} t_L + \vec{V}_7 \cdot t_M + \vec{V}_0 \cdot t_Z \tag{3.78}$$

where voltage vectors $\vec{V}_{14}$, $\vec{V}_7$, $\vec{V}_0$, and V_{ref} is expressed as

$$\begin{cases} \vec{V}_{14} = \frac{2}{3} B \cdot U_{\text{dc}} e^{j\frac{\pi}{3}} \\ \vec{V}_7 = \frac{\sqrt{3}}{3} B \cdot U_{\text{dc}} \cdot e^{j\frac{\pi}{6}} \\ \vec{V}_0 = 0 \\ V_{\text{ref}} = V_R \cdot e^{j\theta} \end{cases} \tag{3.79}$$

The dwell times of large vector $\vec{V}_{14}$, medium vector $\vec{V}_7$, and zero vector $\vec{V}_0$, can be calculated as

$$\begin{cases} t_L = \sqrt{3} m T_s \sin\left(\theta - \frac{\pi}{6}\right) \\ t_M = 2 m T_s \sin\left(\frac{\pi}{3} - \theta\right) \\ t_Z = T_s - t_L - t_M \end{cases} \tag{3.80}$$

where $\pi/6 \leqslant \theta < \pi/3$.

The selected switching sequence in Sector 2 is OOO-PON-PPN-PON-OOO, and the state transition times for Phase A, B, and C can be calculated as

$$\begin{cases} T_{\rm a} = \frac{t_{\rm Z}}{2} \\ T_{\rm b} = \frac{t_{\rm Z}+t_{\rm M}}{2} \\ T_{\rm c} = \frac{t_{\rm Z}}{2} \end{cases} \quad (3.81)$$

Step 3: Shoot-Through Insertion and Switching Sequence Designing
When shoot-through states are inserted into the aforementioned switching sequences, the modified volt-second balance principle can be expressed as

$$\begin{cases} V_{\rm ref}T_{\rm s} = \vec{V}_{\rm L}t_{\rm L} + \vec{V}_{\rm M}t_{\rm M} + \vec{V}_0 t_{\rm Z} + \vec{V}_{\rm ST}t_{\rm ST} \\ T_{\rm s} = t_{\rm L} + t_{\rm M} + t_{\rm Z} + t_{\rm ST} \end{cases} \quad (3.82)$$

where $\vec{V}_{\rm ST}$ is the shoot-through voltage vector, $T_{\rm s}$ is the switching period, and $t_{\rm L}$, $t_{\rm M}$, and $t_{\rm Z}$ are the dwell times of the large vector, medium vector, zero vector $\vec{V}_0$, respectively.

Shoot-through state can be inserted in Phase A, Phase B, or Phase C, which has three types of shoot-through vectors ([FOO], [OFO], [OOF]). For the LMZ-ST modulation, shoot-through states are inserted within the time interval of the zero vector. The total shoot-through interval is divided into two identical parts per switching period, with each part inserted between the zero vector and the corresponding medium vector in each sector. The symmetrical characteristic in one switching period can be guaranteed in this way.

When the reference voltage vector $V_{\rm ref}$ is located in Sector 1, for the proposed scheme, the shoot-through interval is inserted in Phase C, and shoot-through vector [OOF] is additionally utilized to synthesize the reference vector. A shoot-through interval $t_{\rm ST\text{-}C}$ is subtracted from the dwell time of the zero vector, and the modified dwell time of the zero vector can be expressed

$$t'_{\rm Z} = t_{\rm Z} - t_{\rm ST\text{-}C} \quad (3.83)$$

The shoot-through interval is equally divided into two parts, and symmetrically inserted into one switching cycle. The state transition times for Phase A and Phase B remain unchanged, while the state transition time for Phase C should be modified as

$$\begin{cases} T'_{\rm a} = T_{\rm a} \\ T'_{\rm b} = T_{\rm b} \\ T'_{\rm c} = T_{\rm c} - \frac{t_{\rm ST\text{-}C}}{2} \end{cases} \quad (3.84)$$

As shown in Fig. 3.27a, the designed switching sequence in Sector 1 is OOO-OOF-PON-PNN-PON-OOF-OOO. During the insertion of the shoot-through state [OOF], the output of Phase C is connected to the neutral-point, which is the same as state O. The line-to-line voltage will not be affected in this way.

3.2 Leakage Current Suppression Technology

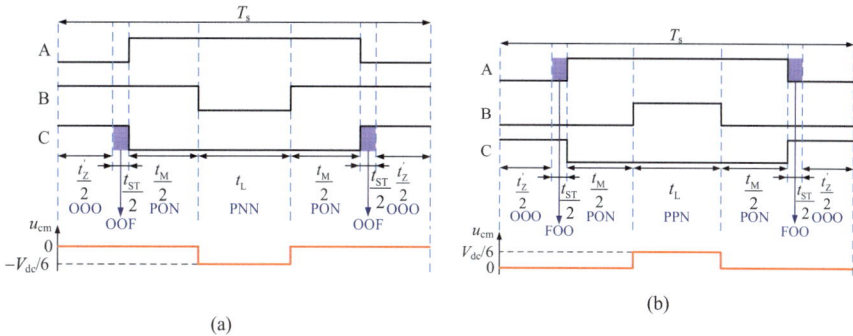

Fig. 3.27 Switching sequences in Sector 1 and Sector 2 with LMZ-ST modulation method **a** Sector 1, **b** Sector 2

When the reference voltage vector V ref is located in Sector 2, the shoot-through interval is inserted in Phase A. Shoot-through vector [FOO] is additionally utilized to synthesize the reference vector. The state transition times for Phase B and Phase C remain unchanged, while the state transition time for Phase A should be modified as

$$\begin{cases} T'_a = T_a - \frac{t_{ST\_A}}{2} \\ T'_b = T_b \\ T'_c = T_c \end{cases} \quad (3.85)$$

As shown in Fig. 3.27b, the designed switching sequence in Sector 2 is OOO-FOO-PON-PPN-PON-FOO-OOO. During the insertion of shoot-through state [FOO], the output of Phase A is connected to the neutral-point, which is the same as state O. The line-to-line voltage will not be affected in this way.

Similarly, the switching sequences of the proposed scheme in other sectors can also be obtained in this way. The switching sequences are summarized in Table 3.6. It can be seen that each phase is selected as the shoot-through phase in one-third of all sectors, so relatively balanced losses of the power devices can be guaranteed.

2. **Large, medium, zero, small and shoot-through voltage vectors (LMZS-ST) modulation**

   The LMZ-ST modulation method uses large, medium, zero, and shoot-through vectors to reduce the CMV. However, this scheme lacks NPV recovery capability, because no small vectors are employed. In practical applications, the NPV shifts due to the effects of non-ideal factors, such as capacitance differences and different characteristics of power semiconductor switches. The NPV imbalance causes current harmonics, reduces output power quality, and even damages power semiconductor switches. Therefore, how to control NPV balance with low CMV magnitude becomes an essential issue.

   To achieve that, this section presents a novel modulation method for QZS-3LT^{2}I to reduce the CMV, control the NPV balance, and realize voltage boosting

Table 3.6 The switching sequences in all sectors using the LMZ-ST modulation method

Sectors	Switching sequences
1	OOO-OOF-PON-PNN-PON-OOF-OOO
2	OOO-FOO-PON-PPN-PON-FOO-OOO
3	OOO-OFO-OPN-PPN-OPN-OFO]-OOO
4	OOO-OOF-OPN-NPN-OPN-OOF-OOO
5	OOO-FOO-NPO-NPN-NPO-FOO-OOO
6	OOO-OFO-NPO-NPP-NPO-OFO-OOO
7	OOO-OOF-NOP-NPP-NOP-OOF-OOO
8	OOO-FOO-NOP-NNP-NOP-FOO-OOO
9	OOO-OFO-ONP-NNP-ONP-OFO-OOO
10	OOO-OOF-ONP-PNP-ONP-OOF-OOO
11	OOO-FOO-PNO-PNP-PNO-FOO-OOO
12	OOO-OFO-PNO-PNN-PNO-OFO-OOO

simultaneously. The novel modulation method adopts five basic vectors to synthesize the reference voltage vector, which includes large vector, medium vector, small vector with low CMV magnitude, zero vector, and shoot-through vector (abbreviated to LMSZ-ST). Besides the small vector in current sector, the small vector in adjacent sector can be also selected, which enhances the NPV balance control capability.

The implementation of the LMZS-ST includes three steps, which will be illustrated as follows.

Step 1: Basic Vectors Selection for CMV Reduction and NPV Balance

The basic voltage vectors and CMVs of the QZS-3LT^{2}I are listed in Table 3.7. It can be seen that the CMVs of vectors highlighted in blue are lower than $V_{dc}/6$. Specifically, the CMVs of six large vectors are equal to $V_{dc}/6$ or $-V_{dc}/6$, and the CMVs of six medium vectors, one zero vector [OOO] and shoot-through vector are equal to 0. Meanwhile, the CMVs of partial small vectors are equal to $V_{dc}/6$ or $-V_{dc}/6$.

The SVD of QZS-3LT^{2}I using LMZS-ST modulation method is shown in Fig. 3.28, which is divided into 12 sectors. In this SVD, large vectors are represented by green arrows, medium vectors are represented by red arrows, small vectors with $V_{dc}/6$ CMV magnitude are represented by blue arrow, zero vector [OOO] is highlighted in purple, and shoot-through vectors are highlighted in green. Similar to LMS-ST, the shoot-through states can be inserted in Phase A, Phase B, or Phase C, and thus three shoot-through vectors: [FOO], [OFO], and [OOF] are included. Meanwhile, the insertion rules of shoot-through states also remain the same as LMZ-ST method.

Being different from the LMZ-ST modulation method, in addition to the large, medium, zero, shoot-through vectors, six small vectors with a $V_{dc}/6$ CMV magnitude

3.2 Leakage Current Suppression Technology

Table 3.7 Voltage vectors and CMVs of QZS-3LT^{2}I for LMZ-ST modulation method

Vectors	States	CMV	States	CMV	States	CMV
Large	PNN	$-V_{dc}/6$	PPN	$+V_{dc}/6$	NPN	$-V_{dc}/6$
	NPP	$+V_{dc}/6$	NNP	$-V_{dc}/6$	PNP	$+V_{dc}/6$
Medium	PON	0	OPN	0	NPO	0
	NOP	0	ONP	0	PNO	0
Small P-type	POO	$+V_{dc}/6$	PPO	$+V_{dc}/3$	OPO	$+V_{dc}/6$
	OPP	$+V_{dc}/3$	OOP	$+V_{dc}/6$	POP	$+V_{dc}/3$
Small N-type	ONN	$-V_{dc}/3$	OON	$-V_{dc}/6$	NON	$-V_{dc}/3$
	NOO	$-V_{dc}/6$	NNO	$-V_{dc}/3$	ONO	$-V_{dc}/6$
Zero	OOO	0	PPP	$+V_{dc}/2$	NNN	$-V_{dc}/2$
Shoot-through	FOO	0	OFO	0	OOF	0

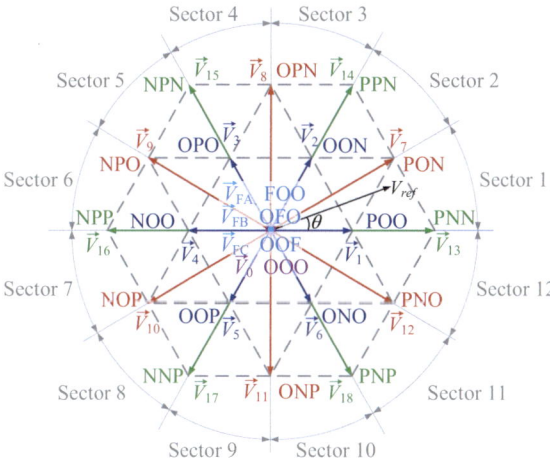

Fig. 3.28 SVD of QZS-3LT^{2}I using the LMZS-ST method: large vectors(green), medium vectors (red), small vector (blue), zero vector (purple), and shoot-through vectors (sky green)

are also adopted in LMZS-ST. Since the CMV magnitude of all vectors adopted in the LMZ-ST modulation method is lower than $V_{dc}/6$, the CMV magnitude of the PV inverter can be restricted within $V_{dc}/6$, which is the same as that with LMZ-ST method.

The maximum magnitude of the reference voltage vector is equal to the length of the medium vector, and can be found from

$$V_{R,\max,\text{LMZ-ST}} = \frac{\sqrt{3}B \cdot U_{dc}}{3} \qquad (3.86)$$

Submitting the maximum magnitude of the reference voltage vector to the definition of modulation index, the maximum modulation index is derived as

$$m_{\max\text{-LMZ-ST}} = \frac{\sqrt{3} \cdot V_{R,\max,\text{LMZ-ST}}}{B \cdot U_{dc}} = 1 \qquad (3.87)$$

It can be seen that the maximum dc-link voltage utilization of the LMZS-ST scheme is the same as the LMZ-ST method. Although small vectors with high CMV magnitudes are not adopted, the maximum dc-link voltage utilization is not reduced.

Step 2: Dwell Times Calculation and Switching Sequence Designing

Sector 1 is considered as a representative example to explain the dwell time calculation and switching sequence design of the LMZS-ST scheme. The voltage difference between the two inner dc-link capacitors C_2 and C_3 is calculated as

$$v_{\text{dif}} = v_{C2} - v_{C3} \qquad (3.88)$$

where v_{C2} and v_{C3} are the voltages across capacitors C_2 and C_3, respectively. According to the sign of the voltage difference between the two inner dc-link capacitors, the following analyses are divided into two cases:

1. $v_{\text{dif}} > 0$: First, consider the case that the voltage difference v_{dif} is greater than zero. There exists one P-type small vector $\vec{V}_1$ [POO] in Sector 1, and its CMV magnitude is equal to one-sixth of the dc-link voltage. Because the P-type small vector is connected between the positive dc-link and the neutral-point, the neutral current flows into the neutral-point, as shown in Fig. 3.29a. This small vector can be used to discharge the upper dc-link capacitor C_2, and the voltage difference v_{dif} is reduced in this way.

    To avoid the switching transitions between switching cycles, the designed switching sequence starts and ends with the zero vector $\vec{V}_0$ [OOO]. The selected P-type small vector $\vec{V}_1$ is placed between zero vector V0 and medium vector $\vec{V}_7$ in order not to generate additional switching losses. After selecting the small vector $\vec{V}_1$ [POO], the state transition time of Phase A becomes the minimum among three phases. Therefore, the shoot-through state is subsequently inserted in Phase A within the interval of zero vector $\vec{V}_0$ [OOO]. In other words, shoot-through vector $\vec{V}_{FA}$ [FOO] is adopted to realize voltage boosting, and the output voltage will not be affected in this way. Considering the switching numbers of power devices, the switching sequence is designed as OOO-FOO-POO-PON-PNN-PON-POO-FOO-OOO, as shown in Fig. 3.30a.

    It can be clearly seen that the CMV magnitude is restricted within one-sixth of the dc-link voltage after the selection of small vector $\vec{V}_1$ and the insertion of shoot-through state. The NPV balance can be controlled in this way. After adopting the small vector $\vec{V}_1$ [POO] and the shoot-through vector $\vec{V}_{FA}$ [FOO], the volt-second balance equation is expressed as

$$\begin{cases} V_{ref}T_s = \vec{V}_{13}t'_{13} + \vec{V}_7 t'_7 + \vec{V}_1 t'_1 + \vec{V}_0 t'_0 + \vec{V}_{FA} t'_{ST\text{-}A} \\ T_s = t'_{13} + t'_7 + t'_1 + t'_Z + t'_{ST\text{-}A} \end{cases} \quad (3.89)$$

where T_s is the switching period. t'_{13}, t'_7, t'_1, t'_Z, and $t'_{ST\text{-}A}$ are the modified dwell times of voltage vectors $\vec{V}_{13}$, $\vec{V}_7$, $\vec{V}_1$, $\vec{V}_0$, and $\vec{V}_{FA}$, respectively. The Voltage vectors $\vec{V}_{13}$, $\vec{V}_7$, $\vec{V}_1$, $\vec{V}_0$, and $\vec{V}_{FA}$ are expressed as

$$\begin{cases} \vec{V}_{13} = \frac{2}{3}BU_{in} \\ \vec{V}_7 = \frac{\sqrt{3}}{3}BU_{in} \cdot e^{j\frac{\pi}{6}} \\ \vec{V}_1 = \frac{1}{3}BU_{dc} \\ \vec{V}_0 = 0 \\ \vec{V}_{FA} = 0 \end{cases} \quad (3.90)$$

Thus, the vol-second balance equation can be re-written that

$$\begin{cases} V_R T_s \cos\theta = \frac{2}{3}BU_{dc}t'_{13} + \frac{1}{2}BU_{dc}t'_7 + \frac{1}{3}BU_{dc}t'_1 \\ V_R T_s \sin\theta = \frac{\sqrt{3}}{6}BU_{dc}t'_7 \end{cases} \quad (3.91)$$

It is important to note that the shoot-through duty cycle should not be modified before and after the selection of P-type small vector $\vec{V}_1$, so as to guarantee the voltage boosting capability. Assume that the dwell time of small vector $\vec{V}_1$ is equal to y_t. Comparing the volt-second balance equations before and after the adoption of small vector $\vec{V}_1$, the dwell times of other vectors are updated as

$$\begin{cases} t'_{13} = t_{13} - \frac{1}{2}y_t \\ t'_7 = t_7 \\ t'_Z = t_Z - \frac{1}{2}y - t'_{ST\text{-}A} \end{cases} \quad (3.92)$$

where t_{13}, t_7, t_Z are the dwell time with LMZ-ST modulation method. In order to realize the designed switching sequence, the dwell times of the selected vectors should be positive, so the variable y_t should satisfy

$$0 < y_t < \min\left\{2t_{13}, 2t_Z - 2t'_{ST\text{-}A}\right\} \quad (3.93)$$

2. $v_{dif} < 0$: Next, consider the case that the voltage difference v_{dif} is less than zero. It is indicated that an N-type small vector should be selected to reduce the voltage

across the lower capacitor. However, N-type small vector with low CMV magnitude does not exist in Sector 1. To overcome this limitation, an alternative N-type small vector with low CMV magnitude in adjacent sectors should be used. The N-type small vector $\vec{V}_2$ [OON] in Sector 2 is adopted in the proposed scheme. Since the three phases are connected between the neutral-point and the negative dc-link, as shown in Fig. 3.29b, the N-type small vector reduces the voltage across the lower capacitor C_3. In addition, the shoot-through state should be properly inserted in order to realize voltage boosting.

After selecting the small vector $\vec{V}_2$ [OON], the state transition time of Phase C becomes the minimum among the three phases. Therefore, the shoot-through state is inserted in Phase C within the interval of zero vector $\vec{V}_0$ [OOO], that is to say, the shoot-through vector $\vec{V}_{FC}$ [OOF] is adopted to realize voltage boosting. The ac output voltage is not affected by this means. The switching sequence is designed as OOO-OOF-OON-PON-PNN-PON-OON-OOF-OOO, as shown in Fig. 3.30b. It is clear that the CMV magnitude is also restricted within one-sixth of the dc-link voltage after the selection of small vector $\vec{V}_2$ [OON] and the insertion of shoot-through state. The NPV balance can be controlled in this way.

After the adoption of small vector $\vec{V}_2$ [OON] and shoot through vector $\vec{V}_{FC}$ [OOF], the volt-second balance equation is expressed as

$$\begin{cases} V_{ref}T_s = \vec{V}_{13}t'_{13} + \vec{V}_7 t'_7 + \vec{V}_2 t'_2 + \vec{V}_0 t'_0 + \vec{V}_{FC} t'_{ST\text{-}C} \\ T_s = t'_{13} + t'_7 + t'_2 + t'_Z + t'_{ST\text{-}C} \end{cases} \quad (3.94)$$

where t_2' and $t'_{ST\text{-}C}$, are the modified dwell times of voltage vectors $\vec{V}_2$, and $\vec{V}_{FC}$, respectively. The small vector $\vec{V}_2$ [OON] is expressed as

$$\vec{V}_2 = \frac{1}{3}BU_{dc} \cdot e^{j\frac{\pi}{3}} \quad (3.95)$$

Thus, the volt-second balance equation can be expressed as

$$\begin{cases} V_R T_s \cos\theta = \frac{2}{3}BU_{dc}t'_{13} + \frac{1}{2}BU_{dc}t'_7 + \frac{1}{6}BU_{dc}t'_2 \\ V_R T_s \sin\theta = \frac{\sqrt{3}}{6}BU_{dc}t'_7 + \frac{\sqrt{3}}{6}BU_{dc}t'_2 \end{cases} \quad (3.96)$$

The shoot-through duty cycle should not be modified before and after the selection of N-type small vector $\vec{V}_2$, so the voltage boosting capability is not affected. Assume that the dwell time of small vector $\vec{V}_2$ is equal to y_t. Comparing the volt-second balance equations before and after the adoption of small vector $\vec{V}_1$, the dwell times of other vectors are updated as

3.2 Leakage Current Suppression Technology

$$\begin{cases} t'_{13} = t_{13} + \frac{1}{2}y_t \\ t'_7 = t_7 - y_t \\ t'_Z = t_Z - \frac{1}{2}y_t - t'_{ST\text{-}C} \end{cases} \tag{3.97}$$

The dwell times of the selected vectors should be positive, so the variable y_t should be limited as

$$0 < y_t < \min\left\{t_7, 2t_Z - 2t'_{ST\text{-}C}\right\} \tag{3.98}$$

To sum up, the basic voltage vector selection, shoot-through state insertion, dwell times calculation, and switching sequence design of the proposed scheme in Sector 1 have been explained in detail. According to symmetry of the space vector diagram, the cases in other sectors can be similarly analyzed. The selected small vectors, shoot-through phases, switching sequences and constraints of the dwell time of the selected small vectors in all sectors are summarized in Table 3.8.

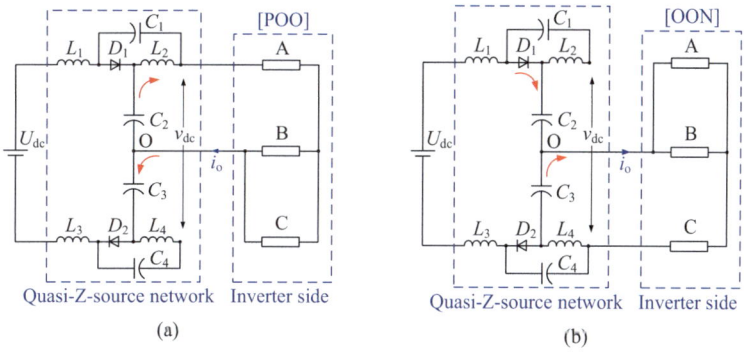

Fig. 3.29 Effect of small vectors on the NPV balance **a** P-type small vector [POO], **b** N-type small vector [OON]

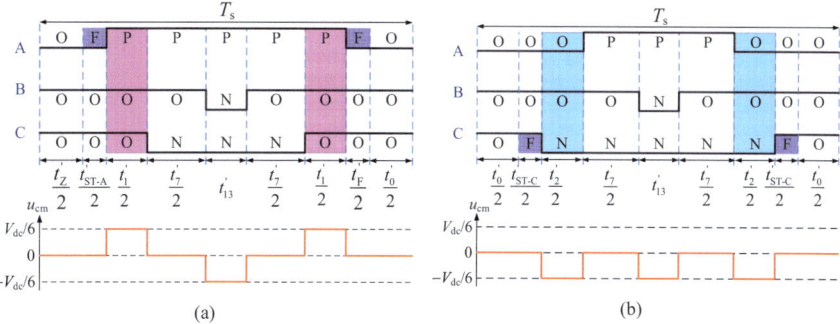

Fig. 3.30 Switching sequences in Sector 1 for LMZS-ST method **a** $v_{\text{dif}} > 0$, **b** $v_{\text{dif}} < 0$

Table 3.8 Small vectors, shoot-through phases, switching sequences and dwell time constraints

Sectors	v_{dif}	Small vector	Shoot-through phase	Switching sequences	Constraints of variable y_t
1	>0	POO	A	OOO-FOO-POO-PON-PNN-⋯	$0 < y_t < \min\{2t_{13}, 2t_Z - 2t_{ST}\}$
	<0	OON	C	OOO-OOF-OON-PON-PNN-⋯	$0 < y_t < \min\{t_7, 2t_Z - 2t_{ST}\}$
2	>0	POO	A	OOO-FOO-POO-PON-PPN-⋯	$0 < y_t < \min\{t_7, 2t_Z - 2t_{ST}\}$
	<0	OON	C	OOO-OOF-OON-PON-PPN-⋯	$0 < y_t < \min\{2t_{14}, 2t_Z - 2t_{ST}\}$
3	>0	OPO	B	OOO-OFO-OPO-OPN-PPN-⋯	$0 < y_t < \min\{t_8, 2t_Z - 2t_{ST}\}$
	<0	OON	C	OOO-OOF-OON-OPN-PPN-⋯	$0 < y_t < \min\{2t_{14}, 2t_Z - 2t_{ST}\}$
4	>0	OPO	B	OOO-OFO-OPO-OPN-NPN-⋯	$0 < y_t < \min\{2t_{15}, 2t_Z - 2t_{ST}\}$
	<0	OON	C	OOO-OOF-OON-OPN-NPN-⋯	$0 < y_t < \min\{t_8, 2t_Z - 2t_{ST}\}$
5	>0	OPO	B	OOO-OFO-OPO-NPO-NPN-⋯	$0 < y_t < \min\{2t_{15}, 2t_Z - 2t_{ST}\}$
	<0	NOO	A	OOO-FOO-NOO-NPO-NPN-⋯	$0 < y_t < \min\{t_9, 2t_Z - 2t_{ST}\}$
6	>0	OPO	B	OOO-OFO-OPO-NPO-NPP-⋯	$0 < y_t < \min\{t_9, 2t_Z - 2t_{ST}\}$
	<0	NOO	A	OOO-OFO-OPO-NPO-NPP-⋯	$0 < y_t < \min\{2t_{16}, 2t_Z - 2t_{ST}\}$
7	>0	OOP	C	OOO-OOF-OOP-NOP-NPP-⋯	$0 < y_t < \min\{t_{10}, 2t_Z - 2t_{ST}\}$
	<0	NOO	A	OOO-FOO-NOO-NOP-NPP-⋯	$0 < y_t < \min\{2t_{16}, 2t_Z - 2t_{ST}\}$

(continued)

3.2 Leakage Current Suppression Technology   111

Table 3.8 (continued)

Sectors	v_{dif}	Small vector	Shoot-through phase	Switching sequences	Constraints of variable y_t
8	>0	OOP	C	OOO-OOF-OOP-NOP-NNP-⋯	$0 < y_t < \min \{2t_{17}, 2t_Z - 2t_{ST}\}$
	<0	NOO	A	OOO-FOO-NOO-NOP-NNP-⋯	$0 < y_t < \min \{t_{10}, 2t_Z - 2t_{ST}\}$
9	>0	OOP	C	OOO-OOF-OOP-ONP-NNP-⋯	$0 < y_t < \min \{2t_{17}, 2t_Z - 2t_{ST}\}$
	<0	ONO	B	OOO-OFO-ONO-ONP-NNP-⋯	$0 < y_t < \min \{t_{11}, 2t_Z - 2t_{ST}\}$
10	>0	OOP	C	OOO-OOF-OOP-ONP-PNP-⋯	$0 < y_t < \min \{t_{11}, 2t_Z - 2t_{ST}\}$
	<0	ONO	B	OOO-OFO-ONO-ONP-PNP-⋯	$0 < y_t < \min \{2t_{18}, 2t_Z - 2t_{ST}\}$
11	>0	POO	A	OOO-FOO-POO-PNO-PNP-⋯	$0 < y_t < \min \{t_{12}, 2t_Z - 2t_{ST}\}$
	<0	ONO	B	OOO-OFO-ONO-PNO-PNP-⋯	$0 < y_t < \min \{2t_{18}, 2t_Z - 2t_{ST}\}$
12	>0	POO	A	OOO-FOO-POO-PNO-PNN-⋯	$0 < y_t < \min \{2t_{13}, 2t_Z - 2t_{ST}\}$
	<0	ONO	B	OOO-OFO-ONO-PNO-PNN-⋯	$0 < y_t < \min \{t_{12}, 2t_Z - 2t_{ST}\}$

Step 3: Coordinate Control Strategy

As is well known that the sum of modulation index (m) and shoot-through duty cycle (d_{ST}) should not exceed 1 to maintain the normal operation of QZS-3LT^{2}I. To realize the NPV balance control and guarantee the normal voltage boosting capability, a coordinate control strategy between the NPV balance control and the voltage boosting is presented in this section, which is depicted in Fig. 3.31.

The dc-link voltage control is realized by controlling the shoot-through duty cycle, and an indirect control method is used here. The voltages across the inner dc-link capacitors are sampled. The actual magnitude of the dc-link voltage (V_{dc}) is subsequently calculated. Then the error between the reference value and the actual value of the dc-link voltage going through a PI controller produces the shoot-through duty cycle, which is expressed as

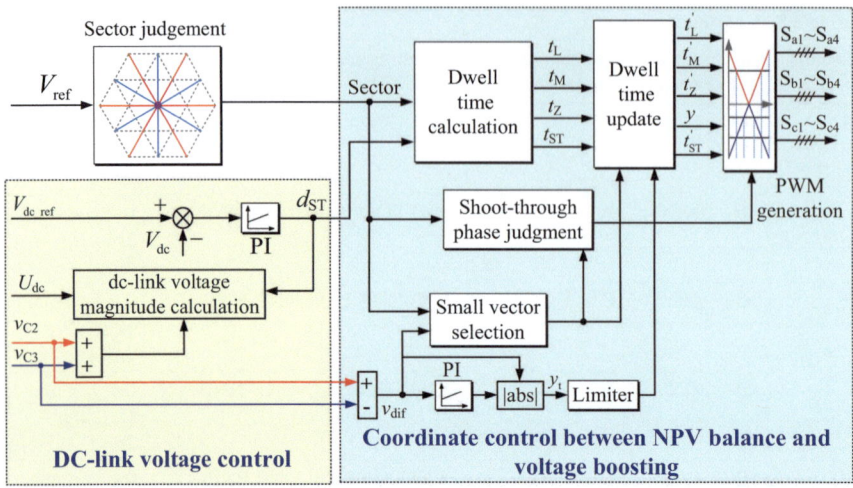

Fig. 3.31 Coordinate control strategy between the NPV balance control and the voltage boosting using LMZS-ST method

$$d_{ST} = \left(k_{p,F} + \frac{k_{i,F}}{s}\right) \cdot \left(V_{dc\_ref} - V_{dc}\right) \quad (3.99)$$

where V_{dc} and V_{dc_ref} are the magnitudes of the dc-link voltage and its reference value, respectively. $k_{p,F}$ and $k_{i,F}$ are the proportional and integral coefficients of the PI controller for dc-link voltage control. When the dc input voltage falls, the shoot-through duty cycle can be increased by the closed-loop controller, so as to make the magnitude of the dc-link voltage track its reference value.

The small vector with low CMV magnitude is selected based on the sector number and the sign of NPV difference. A PI controller is adopted, and the dwell time of the selected P-type or N-type small vector is obtained as

$$y_t = \left|\left(k_{p,NP} + \frac{k_{i,NP}}{s}\right) \cdot v_{dif}\right| \cdot T_s \quad (3.100)$$

After selecting the small vector, the dwell times of basic vectors are accordingly updated. It is important to note that the output variable (y_t) of the NPV controller should be limited by considering the shoot-through duty cycle. By using this coordinated control strategy, the voltage boosting capability is not affected by the NPV balance control. If the power factor is low, the neutral current from the medium vector is increased, which results in larger voltage ripples of the dc-link capacitors. In this case, the dwell time of small vector can be increased through closed-loop regulation, so as to control NPV balance. To realized voltage boosting, the shoot-through phase is selected, and the shoot-through states are inserted within zero vector

3.2 Leakage Current Suppression Technology

to realize voltage boosting. Finally, the updated dwell times of the basic vectors are used to generate PWM drive signals for the power semiconductor switches in the QZS-3LT²I.

3. **Hybrid space vector modulation (HSVM)**

The LMZS-ST method can reduce CMV magnitude and balance NPV by adding the small vectors with low CMV magnitude. However, the number of transitions in CMV is increased significantly, leading to CM EMI and large leakage current. That is to say, minimizing both the CMV magnitude and frequency is an important issue.

Thus, a hybrid space vector modulation (HSVM) scheme for QZS-3LT²I to reduce the CMV magnitude and frequency, control the NPV balance, and boost the dc input voltage simultaneously is proposed.

Different from the LMZS-ST method, the HSVM scheme excludes six large vectors but adopts six medium vectors, six small vectors with low CMV magnitudes, one zero vector, and three shoot-through vectors to generate the ac output voltage. The SVD of QZS-3LT²I using HSVM modulation method is shown in Fig. 3.32, which is divided into 12 sectors. In this SVD, and medium vectors are represented by red arrow, and small vectors with $V_{dc}/6$ CMV magnitude represented by blue arrow, and zero vector [OOO] is highlighted in purple color, and shoot-through vectors are highlighted in green color. The shoot-through states include [FOO], [OFO], and [OOF], and the insertion rules same as the above modulation methods.

The HSVM scheme is derived from 2M1Z-ST modulation method. Based on the 2M1Z-ST method, small vectors with low CMV magnitude are selected according to the v_{dif} condition and boundary conditions, and then the proper switching sequence is determined. By doing so, the NPV balance is actively controlled. Sector 1 of the SVD is considered to illustrate the presented HSVM scheme.

When small vectors are not adopted, vectors $\vec{V}_7$, $\vec{V}_{12}$, $\vec{V}_0$, and $\vec{V}_{FA}$ are used to synthesize reference voltage. The volt-second balance equation is the same as that in the 2M1Z-ST method, which is re-depicted as

$$\begin{cases} V_{ref} = \vec{V}_7 \cdot d_7 + \vec{V}_{12} d_{12} + \vec{V}_0 \cdot d_Z + \vec{V}_{FA} \cdot d_{ST\text{-}A} \\ d_7 + d_{12} + d_Z + d_{ST\text{-}A} = 1 \end{cases} \quad (3.101)$$

The duty cycles for basic vectors are also re-written as

$$\begin{cases} d_7 = m \cdot \sin\left(\theta + \frac{\pi}{6}\right) \\ d_{12} = m \cdot \sin\left(\frac{\pi}{6} - \theta\right) \\ d_Z = 1 - d_7 - d_{12} - d_{ST\text{-}A} \end{cases} \quad (3.102)$$

If only voltage vectors $\vec{V}_7$, $\vec{V}_{12}$, $\vec{V}_0$, and $\vec{V}_{FA}$ are employed, the CMV can be maintained at zero. Nevertheless, active NPV balance cannot be guaranteed.

To reduce the CMV magnitude and frequency, control the NPV balance, and boost the dc input voltage simultaneously, an appropriate small vector with low CMV magnitude should be included, and the duty cycles of the basic vectors in 2M1Z-ST method should be updated.

The predefined NPV difference band is denoted by $v_{\text{dif\_band}}$. By considering the value of the v_{dif}, the HSVM scheme is divided into three cases.

Case 1: $-v_{\text{dif\_band}} \leq v_{\text{dif}} \leq v_{\text{dif\_band}}$

When the value of v_{dif} does not exceed the predefined voltage difference band, two medium vectors, a zero vector, and a shoot-through vector are adopted to synthesize the reference vector to reduce the number of commutations. The detailed calculations of duty cycles have been provided in the above section, and the switching sequence is defined as Type-I. It should be pointed out that the duty cycles and switching sequence are identical to those used in the 2M1Z-ST method. Thus, the detailed analyses are not illustrated any more.

Case 2: $v_{\text{dif}} > 0$, $v_{\text{dif}} > v_{\text{dif\_band}}$

In this case, the P-type small vector should be chosen to decrease the voltage across capacitor C_2 and increase the voltage across capacitor C_3. Thus, the small vector $\vec{V}_1$ [POO] can be utilized when reference voltage is in Sector 1. The volt-second balance equation can be further modified as follows:

$$\begin{cases} V_{\text{ref}} = \vec{V}_7 d'_7 + \vec{V}_{12} d'_{12} + \vec{V}_0 d'_z + \vec{V}_{\text{FA}} d'_{\text{ST-A}} + \vec{V}_1 y_d \\ d'_7 + d'_{12} + d'_z + d'_{\text{ST-A}} + y_d = 1 \end{cases} \quad (3.103)$$

where y_d is the duty cycle of the selected small vector $\vec{V}_1$ [POO]. Since five basic vectors participate in synthesizing the reference vector, the duty cycles of these five

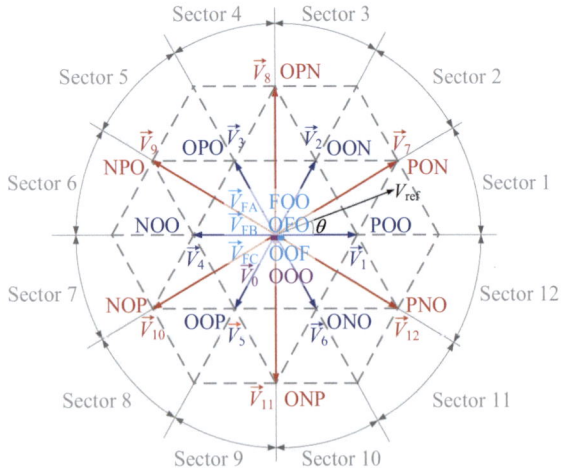

Fig. 3.32 SVD of QZS-3LT^{2}I using the HSVM method: medium vectors (red), small vector (blue), zero vector (purple), and shoot-through vectors (sky green)

3.2 Leakage Current Suppression Technology

basic vectors cannot be directly obtained by solving the volt-second balance equation. To cope with this difficulty, a PI controller is introduced to regulate the NPV in a closed-loop form, and the duty cycle of the selected small vector is obtained as

$$y_d = \left| \left(k_{p,NP} + \frac{k_{i,NP}}{s} \right) \cdot v_{dif} \right| \tag{3.104}$$

Then we can use the duty cycles of the small vector and shoot-through vector to express the duty cycles of other vectors. According to the equivalence of the volt-second balance, the duty cycles of the medium vectors and the zero vector are updated as follows:

$$\begin{cases} d'_7 = d_7 - \frac{1}{3} y_d \\ d'_{12} = d_{12} - \frac{1}{3} y_d \\ d'_Z = d_Z - \frac{1}{3} y_d \end{cases} \tag{3.105}$$

It is obvious that the duty cycles should not be less than 0 to guarantee the correct generation of the switching sequence. By taking this fact into consideration, the boundary condition for the duty cycle of the small vector $\vec{V}_1$ [POO] is obtained as

$$0 \leqslant y \leqslant \min\{3d_7, 3d_{12}, 3d_Z\} \tag{3.106}$$

The input of the PI controller is the voltage difference between the two dc-link capacitors, and the output of the PI controller is further used to obtain the duty cycle of the selected small vector. In essence, this strategy belongs to the closed-loop control method, because the voltage difference between the two dc-link capacitors is utilized as feedback. Therefore, the appropriate duty cycle of the small vector can be adaptively obtained. To be specific, when the voltage difference is not very large, the PI controller generates an appropriate duty cycle for the small vector, which will be smaller than the maximum value set by the boundary conditions. In this way, the over-regulation of the voltage difference to the opposite direction can be avoided. When a large voltage difference occurs, the duty cycle generated by the PI controller accordingly becomes large, and its maximum value is determined by the boundary conditions. By doing so, the NPV unbalance can be mitigated with fast speed. To sum up, the adoption of the PI controller is suitable for different NPV conditions.

To satisfy the requirements of reduced switching numbers and lower harmonic distortions, the symmetric characteristics of the switching sequence are considered. As given in Fig. 3.33a, the switching sequence for this case is designed as OOO-FOO-PNO-PON-POO-PON-PNO-FOO-OOO, which has 9 segments. The small vector $\vec{V}_1$ is configured in the middle of the switching sequence, while the shoot-through state in Phase A is inserted within the duration of the zero vector $\vec{V}_0$ to implement the voltage boosting.

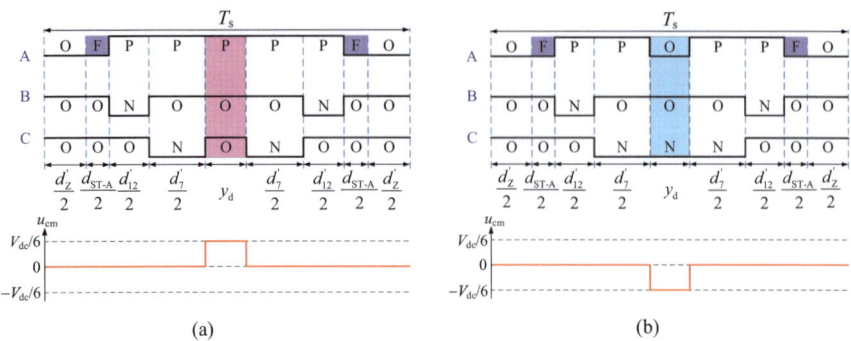

Fig. 3.33 The switching sequences in Sector 1 for HSVM. **a** The switching sequence after adopting the P-type small vector (Type-II). **b** The switching sequence after adopting the N-type small vector (Type-III)

Case 3: $v_{dif} < 0$, $v_{dif} < -v_{dif\_band}$

When the negative value of NPV occurs, the N-type small vector should be selected accordingly to decrease the voltage difference between capacitors C_2 and C_3. However, there does not exist N-type small vector in Sector 1. To cope with this difficulty, the N-type small vector $\vec{V}_2$ [OON] is selected, which is located in the adjacent sector. In this case, the volt-second balance equation can be revised as follows:

$$\begin{cases} V_{ref} = \vec{V}_7 d_7' + \vec{V}_{12} d_{12}' + \vec{V}_0 d_Z' + \vec{V}_{FA} d_{ST-A}' + \vec{V}_2 y_d \\ d_7' + d_{12}' + d_Z' + d_{ST-A}' + y_d = 1 \end{cases} \quad (3.107)$$

where y_d is the duty cycle of the selected small vector $\vec{V}_2$ [OON]. Similar to the case 1 condition, the duty cycle of the N-type small vector is also acquired through a PI controller. As analyzed in the previous case, the equivalence of the volt-second balance should be considered as well to produce the ac output voltage correctly. The updated duty cycles of the fundamental vectors are expressed as

$$\begin{cases} d_7' = d_7 - \frac{2}{3} y_d \\ d_{12}' = d_{12} + \frac{1}{3} y_d \\ d_Z' = d_Z - \frac{2}{3} y_d \end{cases} \quad (3.108)$$

Under this condition, the boundary condition for the duty cycle of the N-type small vector $\vec{V}_2$ [OON] is expressed as

$$0 \leqslant y \leqslant \min\left\{\frac{3}{2} d_7, \frac{3}{2} d_Z\right\} \quad (3.109)$$

3.2 Leakage Current Suppression Technology

The switching sequence is arranged as OOO-FOO-PNO-PON-OON-PON-PNO-FOO-OOO, as depicted in Fig. 3.33b. It is clear that the N-type small vector $\vec{V}_2$ is placed in the middle of the switching sequence.

It is noted that Type-II and Type-III switching sequences can be generated when the boundary conditions are satisfied. When the duty cycle of the small vector is restricted to zero, the switching sequence will become Type-I switching sequence.

The switching transitions for different phases are analyzed as follows. In Type-I switching sequence, Phase B has two switching transitions, while the other phases have one switching transition in one switching period. In Type-II and Type-III switching sequences, an additional phase has two switching transitions in one switching period, which increases the switching losses of power devices to some extent. It is noted that the Type-II switching sequence is selected only when the capacitor voltage difference exceeds the predefined voltage difference band and the corresponding boundary condition for the P-type small vector is satisfied. A similar principle is obtained for Type-III switching sequence. It is evident that the continuous utilizations of Type-II and Type-III switching sequences are avoided, which is beneficial to reduce the number of commutations.

For other sectors of the SVD, the corresponding theoretical analyses can be conducted by using a similar methodology of Sector 1. Figure 3.34 depicts the control diagram of the HSVM scheme, and the detailed procedures are summarized as follows.

1. By judging the sector where the reference vector is located, the duty cycles of the zero vector, medium vector, and shoot-through vector are calculated.
2. The values of the v_{dif} and v_{dif_band} are compared by a comparator module to generate a flag indicating NPV conditions. Specifically, when $-v_{dif\_band} \leq v_{dif} \leq v_{dif\_band}$, the generated flag is equal to 1. When $v_{dif} > 0$,

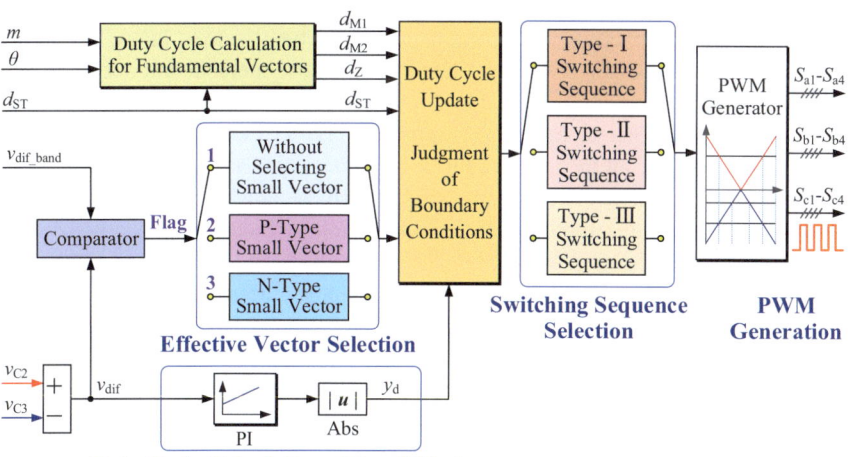

Fig. 3.34 Control diagram of the HSVM scheme

$v_{dif} > v_{dif\_band}$, the generated flag is equal to 2. When $v_{dif} < 0$, $v_{dif} < -v_{dif\_band}$, the generated flag is equal to 3. When the flag is equal to 2 or 3, a small vector is adopted to balance the NPV, whose duty cycle is obtained by regulating the NPV through a PI controller, as mentioned previously.
3. The duty cycles of all basic vectors are updated by considering the equivalence of the voltage-second balance, and the boundary conditions for the duty cycle of the small vector are subsequently obtained.
4. According to the flag and the corresponding boundary conditions, the type of switching sequence is determined, and the PWM drive signals for power semiconductor switches are generated. When the P-type or N-type boundary condition is satisfied, a small vector with low CMV magnitude is used to realize NPV balance, and the shoot-through state is inserted within the zero vector [OOO] to implement the voltage boosting. In both cases, the switching sequences with 9 segments are generated, which has symmetrical characteristics. Otherwise, four fundamental vectors participate in synthesizing the reference vector, and the CMV is equal to 0.

Table 3.9 summarizes the switching sequences in all sectors for the HSVM scheme. It should be noted that the Type-I switching sequence is the same as that used in the 2MLZ-ST method, which has been given in previous section. Thus, the Type-I switching sequence is not listed in this section. Since the switching sequences are distributed symmetrically on both sides, the former half parts of the switching sequences are provided for simplicity. By considering the flag and the boundary conditions, the type of switching sequence can be determined. If the flag is equal to 2, and the boundary condition of the P-type small vector is satisfied, the Type-II switching sequence is used. If the flag is equal to 3, and the boundary condition of the N-type small vector is satisfied, the Type-III switching sequence is used. Otherwise, the Type-I switching sequence is applied to avoid the generation of additional commutations.

In summary, the exciting source suppression can be classified into CMV elimination methods and CMV reduction methods. All these methods are derived based on modulation strategies. In the CMV elimination method, two medium vectors, one zero vector and one shoot-through vector (2M1Z-ST) are used to synthesize the output voltage. Since all the magnitudes of voltage vectors adopted in 2M1Z-ST are zero, the CMV of the inverter is maintained at zero. Meanwhile, three CMV modulation methods are presented in this section, which includes LMZ-ST, LMZS-ST, and HSVM. Using these methods, the CMV magnitude of the inverter is restricted within $V_{dc}/6$. All these methods are benefit to leakage current suppression.

Table 3.9 The switching sequences in all sectors using the HSVM modulation method

Sectors	Adopting P-type small vector (Type-II)	Adopting N-type small vector (Type-III)
1	OOO-FOO-PNO-PON-POO-···	OOO-FOO-PNO-PON-OON-···
2	OOO-OFO-OPN-PON-POO-···	OOO-OFO-OPN-PON-OON-···
3	OOO-OOF-PON-OPN-OPO-···	OOO-OOF-PON-OPN-OON-···
4	OOO-FOO-NPO-OPN-OPO-···	OOO-FOO-NPO-OPN-OON-···
5	OOO-OFO-OPN-NPO-OPO-···	OOO-OFO-OPN-NPO-NOO-···
6	OOO-OOF-NOP-NPO-OPO-···	OOO-OOF-NOP-NPO-NOO-···
7	OOO-FOO-NPO-NOP-OOP-···	OOO-FOO-NPO-NOP-NOO-···
8	OOO-OFO-ONP-NOP-OOP-···	OOO-OFO-ONP-NOP-NOO-···
9	OOO-OOF-NOP-ONP-OOP-···	OOO-OOF-NOP-ONP-ONO-···
10	OOO-FOO-PNO-ONP-OOP-···	OOO-FOO-PNO-ONP-ONO-···
11	OOO-OFO-ONP-PNO-POO-···	OOO-OFO-ONP-PNO-ONO-···
12	OOO-OOF-PON-PNO-POO-···	OOO-OOF-PON-PNO-ONO-···

3.3 Experimental Verifications

To validate the effectiveness and correctness of the presented models and control methods, comprehensive experiments are conducted. The experimental results are obtained based on the CM circuit modification method and the exciting source reduction method.

3.3.1 Experimental Verifications for Common-Mode Circuit Modification Methods

Based on Continuous Modulation

To verify the presented method, some experiments are conducted. The controller of the experimental prototype is the TMS320F28335 DSP and the switching device for the insulated gate bipolar translator is the 10-FZ12NMA080SH01-M260F from Vincotech. The parasitic capacitance is connected to the negative side of dc-link. The parameters used in the experiment are shown in Table 3.10.

Four methods are compared in simulations and experiments. Method-1 is *LCL* filter and CVF for DM circuit. Method-2 is *ILCL* filter and CVF for DM circuit. Method-3 is *ILCL* filter and damping resistor for CM and DM circuits. Method-4 is the proposed method, in which the *ILCL* filter, hybrid control for CM circuit, and CVF for DM circuit are used.

Figure 3.35 shows experimental results of grid currents, leakage currents, and leakage current spectrum using the four methods when $i_d^* = 20$ A and $i_q^* = 0$ A. In Fig. 3.35a, the peak value of leakage current is 2 A with Method-1, which

Table 3.10 Experiment parameters under ideal power grid conditions

Parameters	Values
Output power	10 kW
Grid voltage/frequency	110 V/50 Hz
DC voltage U_{dc}	340 V
DC-link capacitance C_1, C_2	940 μF
Switching frequency f_s	15 kHz
Sampling frequency f_c	15 kHz
Inverter-side inductance L	1 mH
Grid-side inductance L_g	0.15 mH
AC-side filter capacitance C_f	10 μF
Parasitic capacitance C_{pv}	0.5 μF
Damping resister in Method-3	1 Ω

fails to comply with the standard level of 300 mA. The high-frequency components around f_s are the main reason of high leakage current as indicated in the leakage current spectrum. When Method-2 is adopted, the high-frequency components are greatly reduced. Thus, the leakage current is well suppressed, and the magnitude is about 50 mA, as shown in Fig. 3.35b. However, the leakage current increases at f_{r1} caused by CMRCC. From Fig. 3.35c, it can be seen that the leakage current at f_{r1} is well reduced, while the high-frequency components slightly increase with Method-3. Meanwhile, power losses are increased. Consequently, the proposed Method-4 is used and the experimental waveforms are shown in Fig. 3.35d. The leakage current is effectively suppressed, and the amplitude is lower than 20 mA. It represents a significant reduction of 90% in comparison with Method-1.

Compared to Method-4, the power losses of Method-3 are increased because of the damping resistor. Figure 3.36 shows the experimental power losses in the damping resistor. It is observed that the power losses caused by the damping resistor increase with the output power. Additionally, in lower power conditions, the damping power losses represent a large portion of the output power.

Aside from that, the CMRCC directly results in inverter-side current oscillation. The inverter-side currents and their frequency spectrum are shown in Fig. 3.37. With Method-2, the inverter-side currents are oscillated greatly at f_{r1}, as shown in Fig. 3.37a. The peak value and the THD of the inverter-side currents are 32 A and 55.03%, respectively. When Methods 3 and 4 are applied, the oscillations of the inverter-side currents are significantly suppressed, and better performances are achieved as shown in Fig. 3.37b, c, respectively.

Moreover, with Method-4, the THD of the grid currents is lower than that with Method-3, as shown in Table 3.11. The reason is that the filter high-frequency attenuation ability is degraded by the damping resistor. Furthermore, total power losses are reduced since the damping resistor is not needed. The peak value of leakage current and THDs of grid current and inverter-side current are also summarized in Table 3.11.

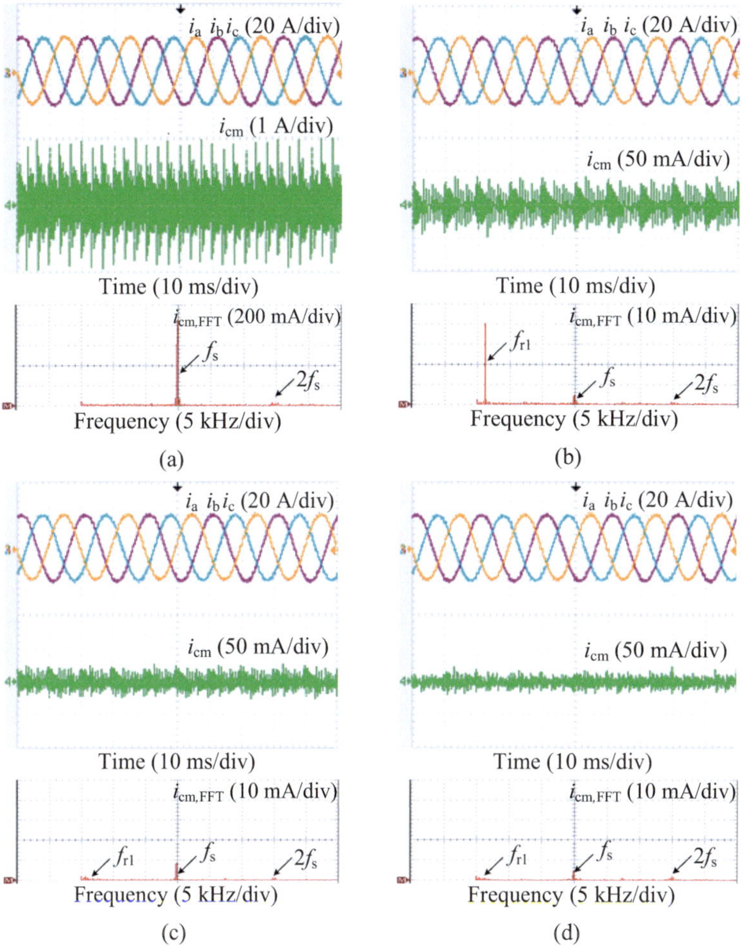

3.3 Experimental Verifications 121

Fig. 3.35 Experimental results of grid currents, leakage currents, and leakage current spectrums **a** Method-1, **b** Method-2, **c** Method-3, **d** Method-4

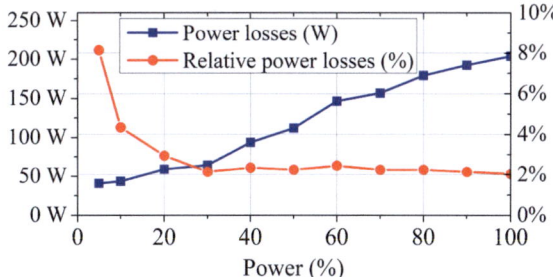

Fig. 3.36 Experimental power losses in the damping resistor with Method-3

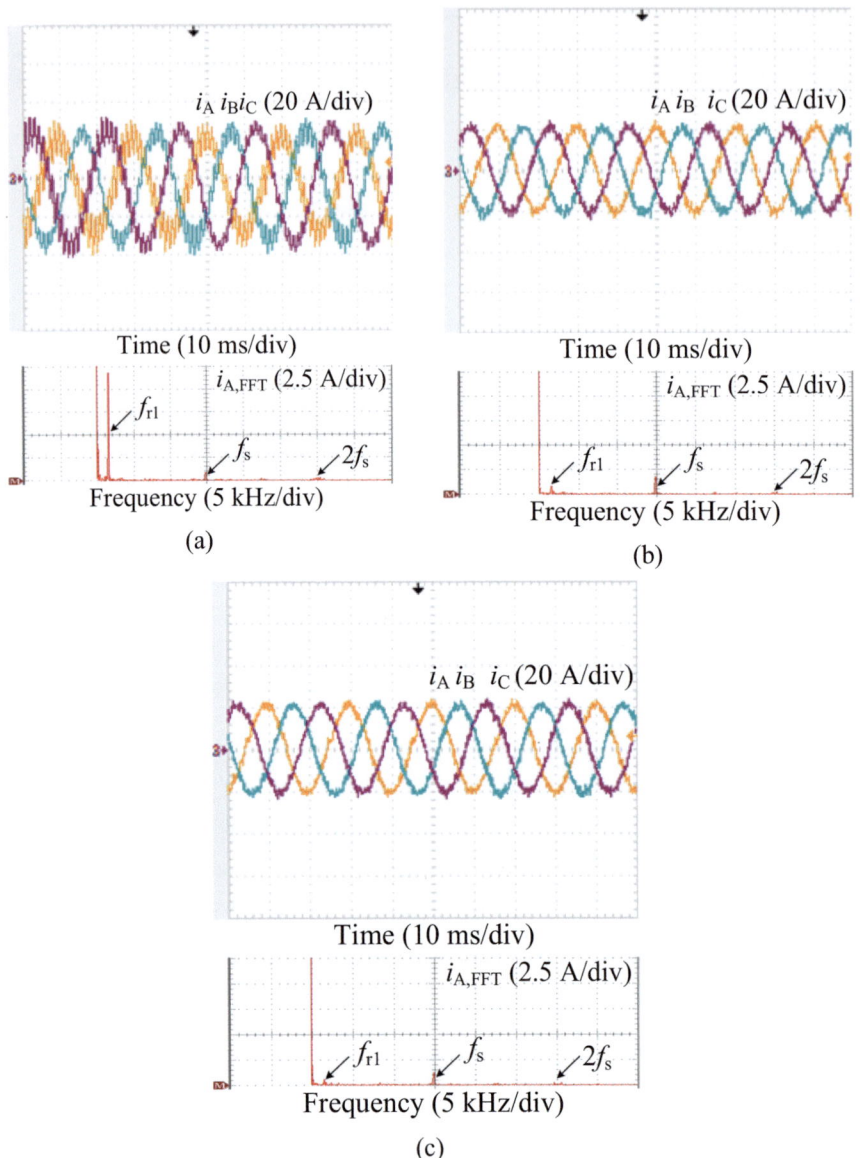

Fig. 3.37 Experimental results of inverter-side currents and inverter-side currents spectrums a Method-2, b Method-3, c Method-4

3.3 Experimental Verifications

Table 3.11 The leakage currents and current THDs comparison with different methods

Parameters	Method-1	Method-2	Method-3	Method-4
i_{cm}	2 A	50 mA	30 mA	20 mA
i_A THD	14.54%	56.03%	14.53%	14.49%
i_a THD	6.03%	4.01%	4.32%	3. 94%

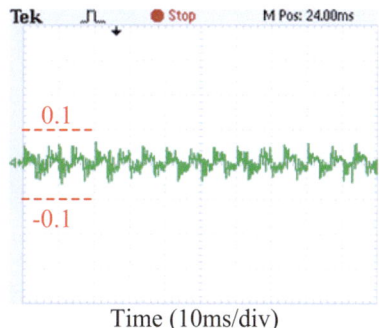

Fig. 3.38 Experimental result of output control variable d_z

The output of the CM circuit controller d_z realizes CMRCC suppression by adjusting the zero-sequence duty ratio D_z. The experimental waveform of d_z is shown in Fig. 3.38.

The dynamic experimental results are shown in Fig. 3.39. Along with good steady-state performance, the proposed Method-4 also offers fast-dynamic performance when the reference current steps from 16 to 24 A. As shown in Fig. 3.39a, the leakage current and inverter-side currents oscillation is well suppressed. The peak value of the leakage current is about 25 mA during the step transition, which is far below the standard limitation. The inverter-side currents are still kept in good performance. Figure 3.39b shows a comparative result with Methods 2 and 4. It can be seen that the inverter-side current oscillates greatly, and the peak value is about 38 A using Method-2. To solve this problem, the proposed Method-4 is applied. The distortions of inverter-side currents are well-eliminated. The performance of inverter-side currents is improved. Moreover, the leakage current with Method-4 is lower than that with Method-2.

The efficiency of the experimental prototype has been measured by the WT3000 power analyser. The efficiency curves with four methods are shown in Fig. 3.40. With Method-1, the highest leakage current decreases the system efficiency for the increased electromagnetic loss. To suppress the high leakage current, an *ILCL* filter is adopted in Method-2. However, the resonance of inverter-side currents is excited, which greatly increases current ripples. The high current ripples increase magnetic losses of the inverter system. Thus, a damping resistor is used in Method-3 to reduce the resonance current. Nevertheless, the damping resistor inserts additional power

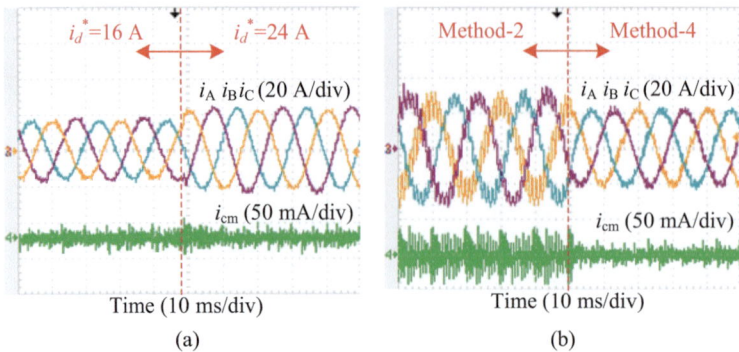

Fig. 3.39 Dynamic experimental result of inverter-side currents and leakage current. **a** Step response with Method-4. **b** Control methods changing from Method-2 to Method-4

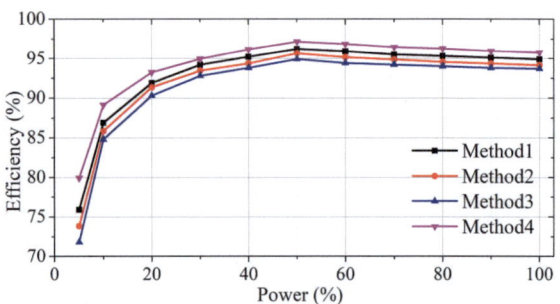

Fig. 3.40 Experimental efficiency of the transformerless PV inverter

losses, increases leakage current, and impacts the capability of attenuating high-frequency harmonics. These weaknesses lead to the lowest system efficiency among all methods. In order to address these issues, a hybrid controller is proposed in Method-4. The proposed method avoids using damping resistor and realizes CMRCC and leakage current suppression simultaneously, which greatly decreases power losses and increases the system efficiency.

Based on Discontinuous Modulation

To verify the proposed method, experiments are carried out in a 5 kW three-phase T-type inverter with an *ILCL* filter. Two key points as the theory indicated are considered. The parameters used in experiments are shown in Table 3.12.

1. **The decoupling of CMRCC mitigation and discontinuity of discontinuous modulation method**

    The proposed discontinuous modulation method is compared with the conventional SVPWM and DPWM1 method to prove the validity. All the three methods use the *ILCL* filter. In the conventional SVPWM method, an active damping

3.3 Experimental Verifications

Table 3.12 Experiment parameters under ideal power grid conditions

Parameters	Values
Output power	5 kW
Grid voltage/frequency	110 V/50 Hz
DC voltage U_{dc}	300 V
DC-link capacitance C_1, C_2	2820 µF/220 µF
Switching frequency f_s	10 kHz
Sampling frequency f_c	10 kHz
Inverter-side inductance L	1 mH
Grid-side inductance L_g	0.5 mH
AC-side filter capacitance C_f	4.7 µF
Parasitic capacitance C_{pv}	0.5 µF

strategy is used to suppress the CMRCC. However, to maintain the discontinuous modulation, the active damping strategy cannot be used in the conventional DPWM1 method. Therefore, a damping resistor is introduced between the common point of the filter capacitors and the dc-link NP O for conventional DPWM1 method to suppress the CM resonance current.

Figure 3.41 shows the experimental waveforms of the inverter-side currents, line-to-line voltage, phase voltage and the spectrum of the inverter-side currents in the three methods. According to Fig. 3.41a, the conventional CBPWM method shows the best inverter-side current quality since the reference voltages are continuous and an active damping strategy is used. However, the reference voltages of this method are changed continuously, leading to higher switching losses. The DPWM1 method reduces the switching losses by allowing certain switches to remain inactive during a time interval as shown in Fig. 3.41b. To maintain the discontinuity of reference voltages, a passive damping strategy is used in DPWM1 method to suppress the CM resonance current, which has limited harmonic suppression capability and increases system losses due to the damping resistor. Compared with the DPWM1 method, the proposed DPWM method shortens the clamping period as shown in Fig. 3.41c. Moreover, the harmonics currents are lower than those in the DPWM1 method since the active damping strategy is used with the proposed DPWM method.

Figure 3.42 shows the upper dc-link capacitor voltage and the leakage current of the three methods. The *ILCL* filter introduces an additional low impedance CM loop for leakage current. Therefore, no matter what modulation strategy is used, the leakage currents are effectively attenuated. When the active damping strategies of SVPWM and the proposed discontinuous modulation are used, the leakage current is further reduced in Fig. 3.42a, c. In addition, the proposed discontinuous modulation can also reduce the NPV ripple compared to DPWM1 since the clamping period is reduced.

Figure 3.43 shows the dynamic experimental waveforms of the CM resonance current suppression strategy of the proposed DPWM method. It can be seen that the performance of the inverter-side currents is improved after the proposed CM resonance suppression strategy is activated.

Figure 3.44 shows the power losses of the three methods at different modulation indices. It is seen that the power losses of the three methods decrease as the modulation index increases. The conduction losses of the three methods have no significant difference, especially when the modulation index is large. Compared with the conventional CBPWM method and proposed DPWM method, the conventional DPWM1 method has the lowest switching losses since the switches will not commutate in a large clamping period. The switching losses of the proposed DPWM method are close to those of the DPWM1 method and lower than those of the CBPWM method.

The efficiency of the three methods is given in Fig. 3.45. It can be seen that the DPWM1 method exhibits the highest efficiency due to its lower power losses compared to the other two methods. In the proposed DPWM method, the efficiency is slightly lower than that of the DPWM1 method. However, the proposed DPWM method has better performance in NP voltage ripple reduction and CM resonance suppression than that of the DPWM1 method, and the efficiency is higher than that of the CBPWM method.

2. **The discontinuous modulation with small dc-link capacitance value**

   There are also three methods compared in the experiments. Method I uses a PI controller in the DM current control loop. Method II uses a PIR controller in the DM current control loop. Method III is the proposed method, in which a PIR controller and a notch filter are used in the current control loop and the CM circuit control loop, respectively.

   Figure 3.46 shows the experimental results of grid currents, CMCC and grid currents FFT diagrams of the three methods. As shown in Fig. 3.46a, when the PI controller is adopted, the inverter-side currents is significantly distorted. Meanwhile, it can be seen from the FFT diagram that the grid currents contain a large number of fifth and seventh harmonics because of the large NP voltage ripple. When the PIR controller is applied in Fig. 3.46b, the distortion of the grid currents is reduced and the fifth and seventh harmonics are effectively suppressed, which proves the effectiveness of the PIR controller. Nevertheless, the CMCC is increased and the resonance is introduced at the peak of the grid currents. Moreover, the low-frequency 11th, 13th, 17th and 19th harmonics in the grid currents are increased caused by the use of the PIR controller. However, due to the symmetrical nature of the three-phase power system, even harmonics are eliminated, leaving only odd harmonics. After the proposed scheme is used, the distortion of the grid currents and the low-frequency harmonics are suppressed as shown in Fig. 3.46c. Thus, the grid currents of the inverter can be significantly improved even in the presence of large NP voltage ripple.

   Figure 3.47 shows the transient switching process from Method I to Method III. When Method I is applied, the fifth and seventh harmonics cause significant distortion in the grid currents. After Method III is adopted, the distortion and resonance in the grid currents are suppressed.

3.3.2 Experimental Verifications for Exciting Source Suppression Methods

In this section, simulation and experimental results are given to illustrate the effects and characteristics of different CMV suppression methods. The detailed parameters are given in Table 3.13. It is worth mentioning that the passive elements in the QZS network are designed on the basis of the guidelines in [19] with the following equations.

$$\begin{cases} L_1 = L_2 = L_3 = L_4 \geqslant \frac{U_{dc}^2 \cdot (1-d_{ST}) \cdot d_{ST}}{2P_{out} \cdot f_s \cdot k_L \cdot (1-2d_{ST})} \\ C_1 = C_4 \geqslant \frac{2P_{out}}{V_{dc}^2 \cdot k_{C1,4} \cdot f_s} \cdot (1 - 2d_{ST}) \\ C_2 = C_3 \geqslant \frac{2P_{out}}{V_{dc}^2 \cdot k_{C2,3} \cdot f_s} \cdot \frac{1-2d_{ST}}{1-d_{ST}} \cdot d_{ST} \end{cases} \quad (3.110)$$

where U_{dc}, d_{ST}, P_{out}, f_s, and k_L are the input dc voltage, shoot-through duty cycle, output power of the inverter system, switching frequency, and the current ripple factor of inductors, respectively. $k_{C1,4}$ is the voltage ripple factor of capacitors C_1 and C_4, while $k_{C2,3}$ is the voltage ripple factor of capacitors C_2 and C_3.

Four methods are included, which are the conventional SVM method, LMZ-ST method, LMZS-ST method, and HSVM method. It should be noted that the 2M1Z-ST method has a lower maximum modulation index, and leads to the reduction of dc-link voltage utilization rate. Moreover, the 2M1Z-ST method cannot actively control the NPV because all small vectors are abandoned. Thus, the results using 2M1Z-ST method are not given in this section.

1. **Simulation results**
   Simulation results based on MATLAB/Simulink platform are provided in this section. Moreover, for the QZS-3LT²I, there exists coupling between the modulation index and the shoot-through duty cycle. To be specific, the sum of modulation index and shoot-through duty cycle cannot exceed 1, and this principle is obeyed in simulated and experimental tests.

   Three cases including non-boost operation mode, boost operation mode, and NPV balance control ability for HSVM are conducted in simulation results. To demonstrate the ability of CMV reduction for different modulation methods, a zoom-in waveform of the CMV is displayed at the bottom of each figure.

Case 1: Non-boost Operation Mode
The obtained results in the non-boost operation are displayed in Fig. 3.48. In the non-boost operation, the dc input voltage is fixed at 180 V, and the modulation index and shoot-through duty cycle are given as 0.95 and 0, respectively. It is evident that the output voltage of the QZS network is equal to the dc input voltage. The line-to-line voltages of four methods are five-level waveforms, and three-phase sinusoidal current waveforms are obtained, which validate the normal operation of the QZS-3LT²I system. Since conventional SVM method adopts all basic vectors to produce the ac

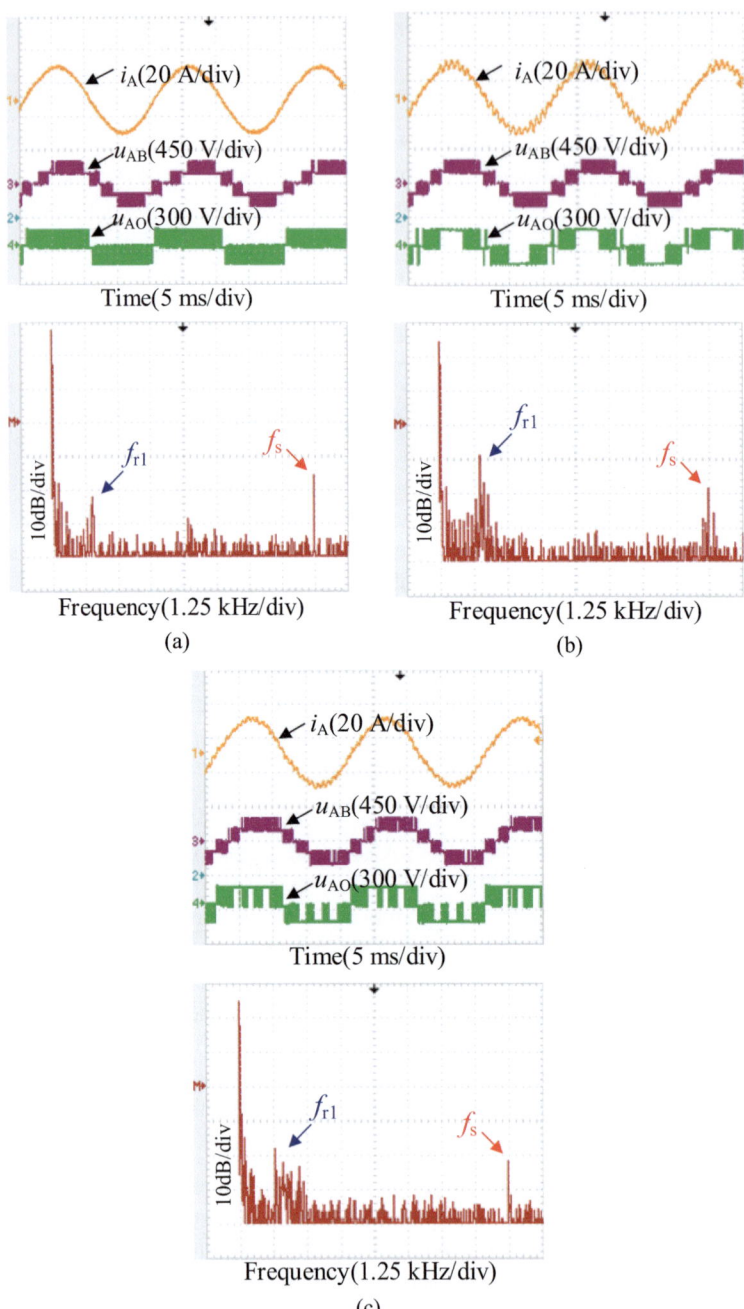

Fig. 3.41 Experimental results of the inverter-side currents, line-to-line voltage, phase voltage and the spectrum of the inverter-side currents. **a** SVPWM method. **b** DPWM1 method. **c** Proposed DPWM method

3.3 Experimental Verifications

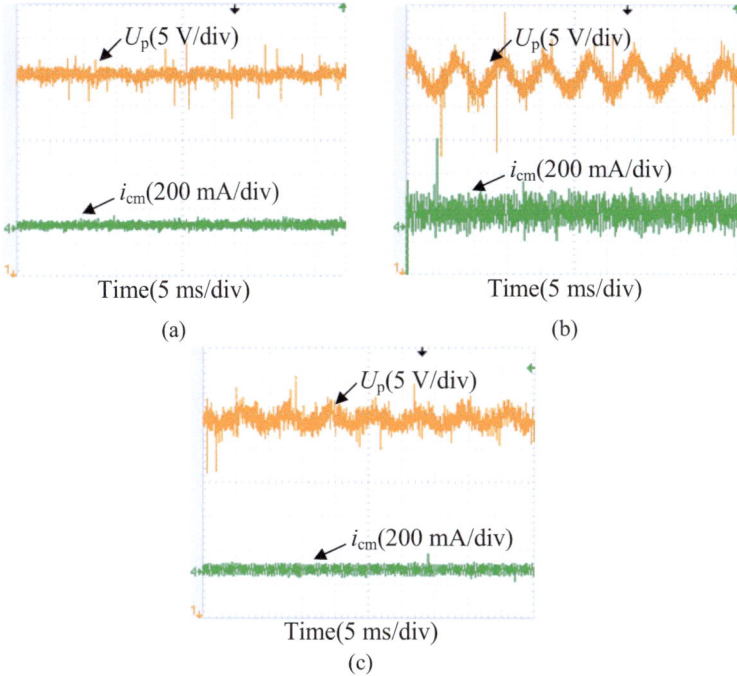

Fig. 3.42 Experimental results of the upper dc-link voltage and the leakage current. **a** SVPWM method. **b** DPWM1 method. **c** Proposed DPWM method

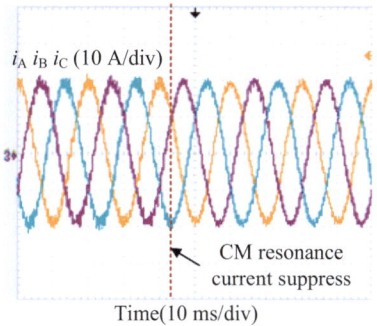

Fig. 3.43 Dynamic experimental result of inverter-side currents

output voltage, the CMV magnitude is about 60 V, which is one-third of the dc-link voltage. The other three methods can reduce the CMV magnitude from $V_{dc}/3$ to $V_{dc}/6$. The LMZ-ST method avoids utilizing all small vectors to suppress the CMV slew rate and magnitude, so the NPV balance cannot be actively controlled. The LMZS-ST method adopts six small vectors with low CMV magnitudes to balance the NPV actively, which however increases the CMV frequency, as depicted in Fig. 3.48c. For

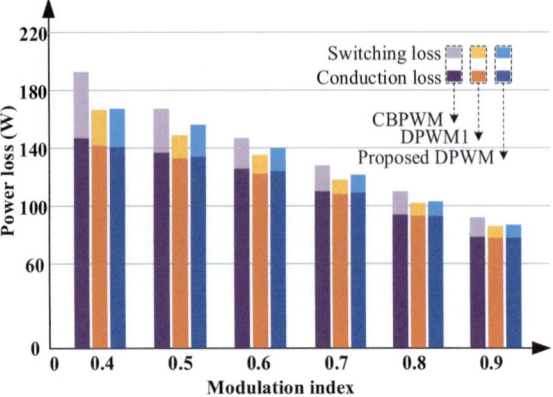

Fig. 3.44 Power losses in the three methods under different modulation indices

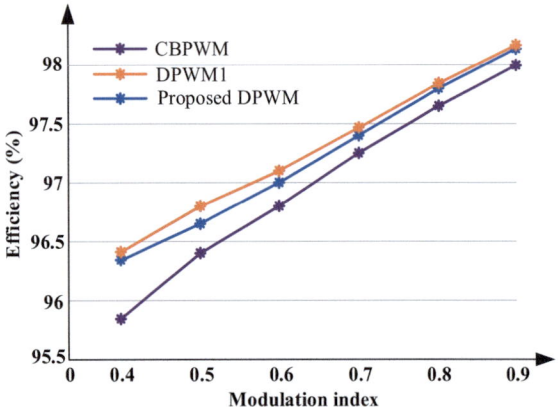

Fig. 3.45 System efficiency in the three methods under different modulation indices

the HSVM method, there exists at most one change in the CMV waveform during one switching period, which can suppress the CMV frequency, as displayed in Fig. 3.48d.

Case 2: Boost Operation Mode

To validate the boost operation of QZS-3LT^{2}I applying different methods, the dc input voltage is reduced to 135 V, and the modulation index and shoot-through duty cycle are given as 0.84 and 0.125, respectively. The obtained simulated waveforms are depicted in Fig. 3.49. It is clearly observed that the dc-link voltages of all methods are higher than the input voltages, which prove that the normal boost function is achieved by incorporating shoot-through states. Since the conventional SVM method utilizes the nearest three vectors to generate the ac output voltage, the quality of output voltage and current is the best among four methods. However, the CMV magnitude produced by the conventional SVM method is twice as high as that of other three methods since all the small vectors are adopted. When the HSVM method is applied,

3.3 Experimental Verifications

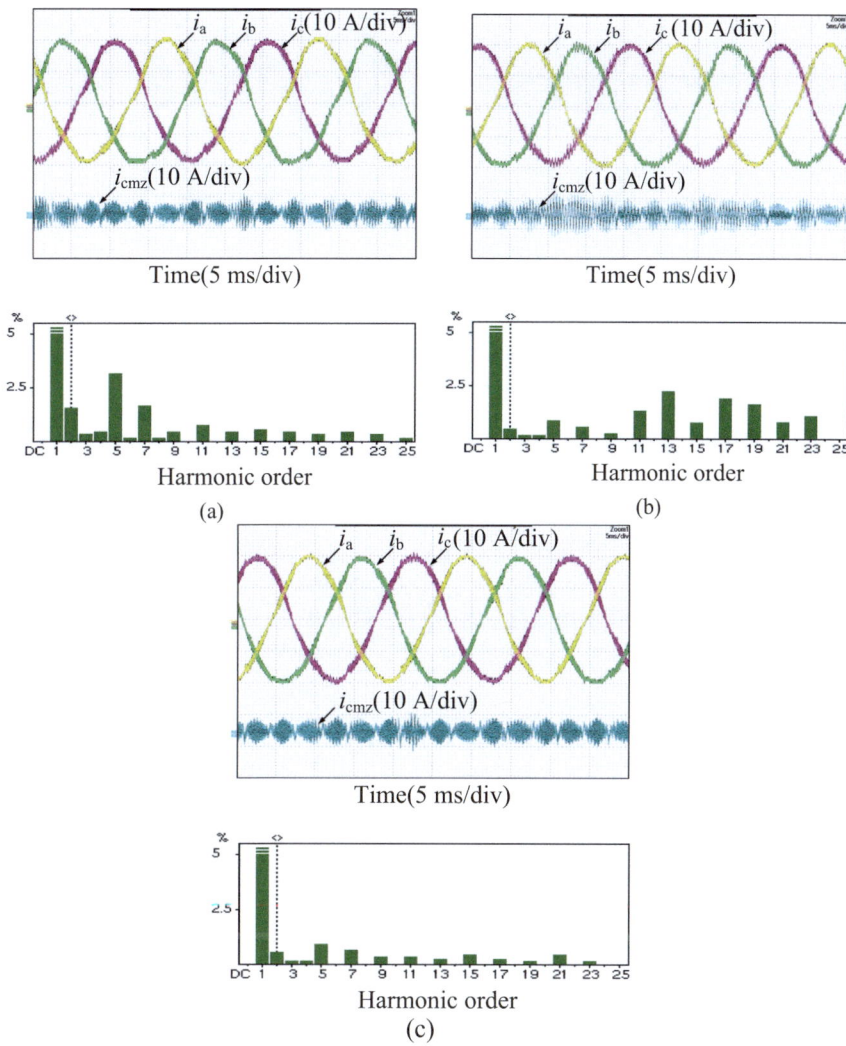

Fig. 3.46 Experimental results of the grid currents, the CMCC and the grid currents FFT diagrams **a** Method I, **b** Method II, **c** Method III

the CMV magnitude is not equal to zero only when the small vector is selected. Otherwise, the zero CMV can be maintained in one switching period.

Case 3: NPV Balance Control Ability of HSVM

To obtain the duty cycle of the small vector of HSVM, two approaches are compared. To facilitate the following description, the NPV balance method based on the PI controller is denoted as NPV-Method-1, while the NPV balance method by setting the duty cycle of the small vector to the maximum value under boundary conditions

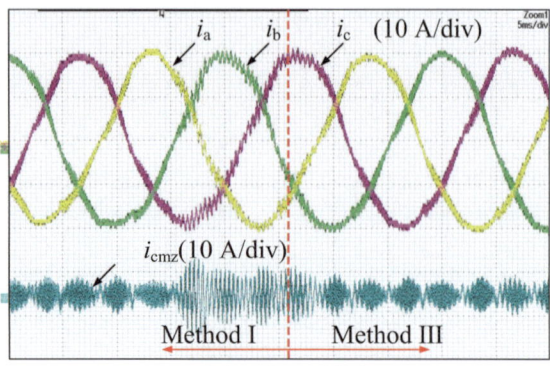

Fig. 3.47 Dynamic experimental result of grid currents and CMCC when switching Method I to Method III

Table 3.13 Experiment parameters under ideal power grid conditions

Parameters	Values
Fundamental frequency of ac output f_1	50 Hz
DC input voltage U_{dc}	90–180 V
Switching frequency f_s	12.5 kHz
Sampling time T_s	80 μs
Inductance L_1, L_2, L_3, and L_4	1.5 mH
Outer dc-link capacitances C_1, and C_4	2350 μF
Inner dc-link capacitances C_2, and C_3	1410 μF
Three-phase RL load	R = 20 Ω/10 Ω, L = 4 mH
Voltage difference band of NPV v_{dif_band}	3 V

is denoted as NPV-Method-2. An additional resistor (1.5 kΩ) is connected in parallel with the lower capacitor (C_3) to create a dc unbalance of NPV. In this research, the voltage difference band is selected as 3 V, which is smaller than 4% of the voltage across dc-link capacitor. The dynamic results in non-boost and boost operations are shown in Fig. 3.50. The function of NPV balance control is enabled at t = 1 s, and it is apparently seen that both NPV balance methods can suppress the dc unbalance of NPV resulted from the additional resistor. The voltage oscillations of NPV-Method-1 are lower than those of NPV-Method-2, because NPV-Method-1 obtains the appropriate duty cycle of the small vector by regulating the voltage difference between two dc-link capacitors in a closed-loop form, which can avoid the over-regulation of capacitor voltages to the opposite direction. These comparative results coincide with the above analyses. Thus, NPV-Method-1 is adopted in the proposed HSVM scheme.

2. **Experimental results**

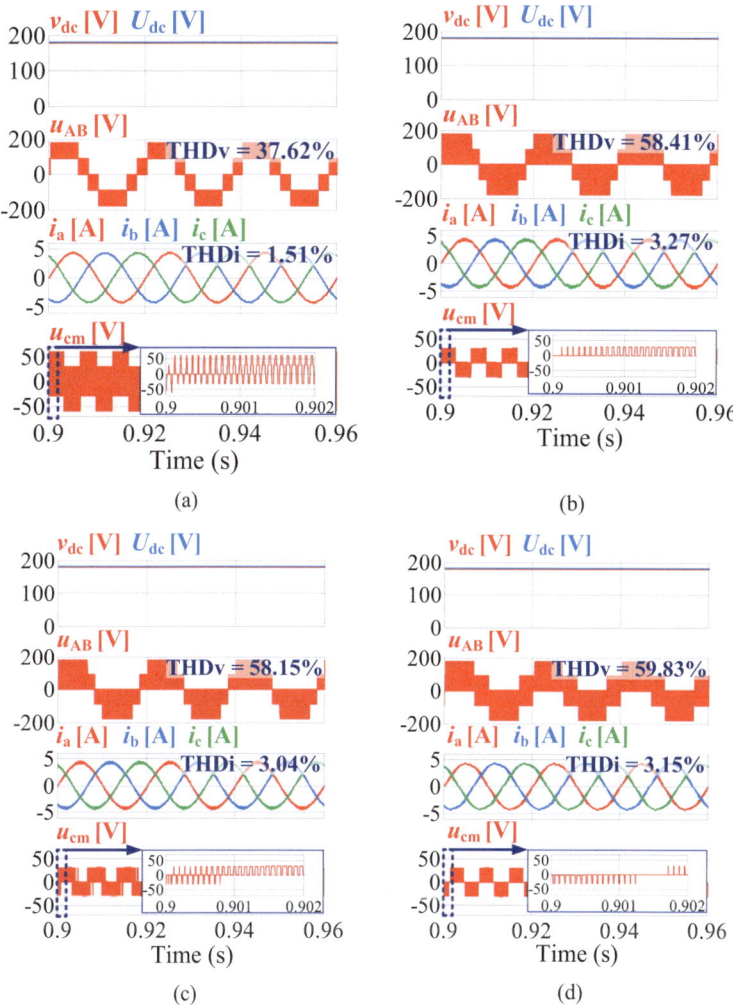

Fig. 3.48 Simulated results of the non-boost operation mode. **a** Conventional SVM method. **b** LMZ-ST method. **c** LMZS-ST method. **d** HSVM method

Five cases are included in the experimental results for convenience of readers to understanding the effect of CMV suppression using different modulation methods. The five cases are case 1 non-boost operation mode, case 2 boost operation mode, and case 3 measurement of CMV spectra and leakage current, case 4 verification of neutral-point balance ability, and case 5 transient performance of HSVM modulation method, which will be elaborated as follows.

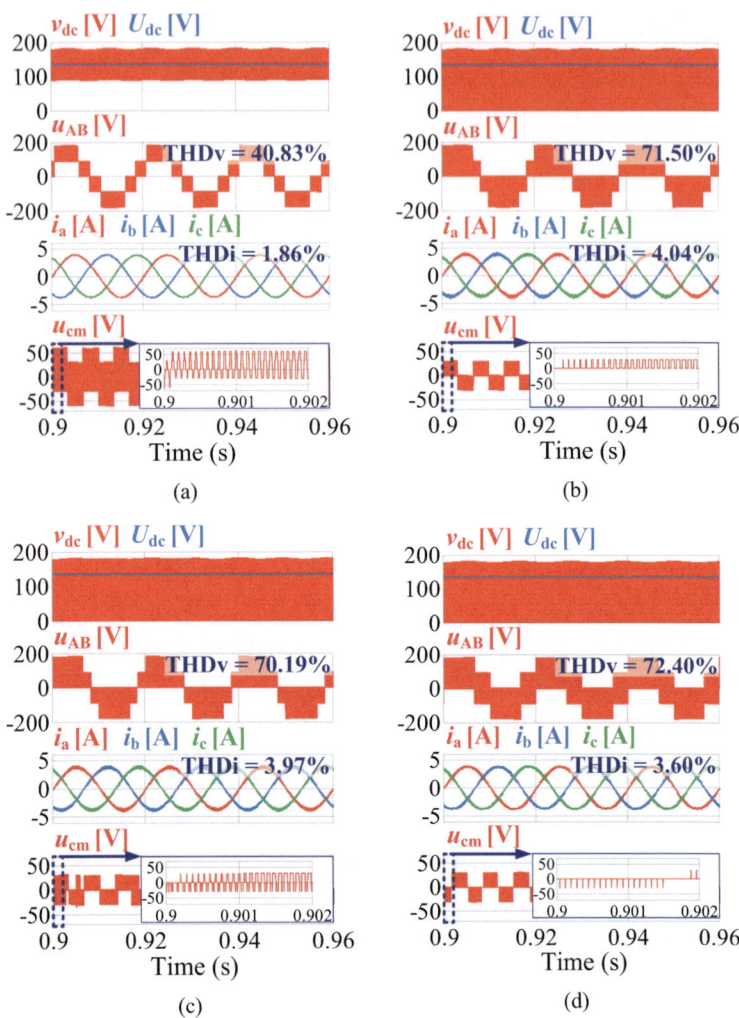

Fig. 3.49 Simulated results of the boost operation mode. **a** Conventional SVM method. **b** LMZ-ST method. **c** LMZS-ST method. **d** HSVM method

Case 1: Non-boost Operation Mode

In this case, the modulation index and shoot-through duty cycle are the same as those in simulation results. The obtained experimental waveforms for different methods are provided in Fig. 3.51. It is clearly observed that the dc-link voltage is equal to the dc input voltage. As displayed in Fig. 3.51a, the CMV magnitude generated by the conventional SVM method is as high as one-third of the dc-link voltage magnitude. Compared with Method-1, the other methods can suppress the CMV by half, as shown in Fig. 3.51b–d. Since the LMZ-ST method does not utilize small vectors, the NPV balance cannot be actively controlled. Compared with the LMZ-ST method, the

3.3 Experimental Verifications

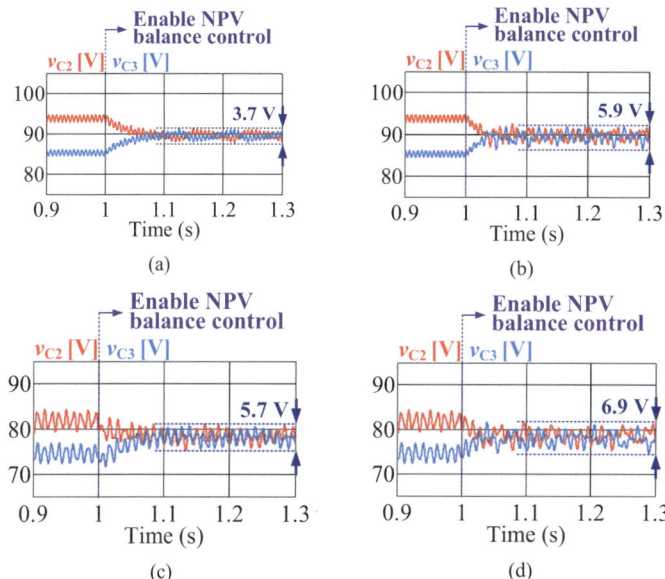

Fig. 3.50 Dynamic waveforms of NPV balance control. **a** Non-boost operation and NPV-Method-1. **b** Non-boost operation and NPV-Method-2. **c** Boost operation and NPV-Method-1. **d** Boost operation and NPV-Method-2

LMZS-ST method additionally utilizes six small vectors, which however increases the frequency of CMV. When the HSVM method is applied, the number of CMV changes in one switching period is equal to one when the P-type or N-type small vector is selected. Otherwise, the CMV is kept at zero. Thus, the CMV magnitude and frequency are simultaneously suppressed, as depicted in Fig. 3.51d.

Case 2: Boost Operation Mode

To test the boost operation mode of the QZS-3LT^{2}I, different modulation indices and shoot-through duty cycle are set. When the input voltage of the dc power source is reduced to 135 V, and the modulation index and shoot-through duty cycle are given as 0.84 and 0.125, respectively, Fig. 3.52 provides the corresponding waveforms. It is clearly observed that the dc-link voltages of the four methods have become pulse waveforms, whose magnitudes are boosted to 180 V. It is proved that the function of voltage boosting can be realized.

The conventional SVM method adopts the nearest three vectors to generate the ac output voltage, which results in adjacent level switching. Therefore, the line-to-line voltage of the conventional SVM method changes from $U_{dc}/2$ to U_{dc} at the top part, as shown in Fig. 3.52a. However, the CMV magnitude is as high as $V_{dc}/3$. For the other three methods, the small vectors with high CMV magnitudes are not utilized, and the CMV magnitudes are reduced to $V_{dc}/6$. Moreover, the line-to-line voltages of these three methods change from 0 to U_{dc} at the top part, which increases the harmonics in the output waveforms to some extent. When the HSVM method is

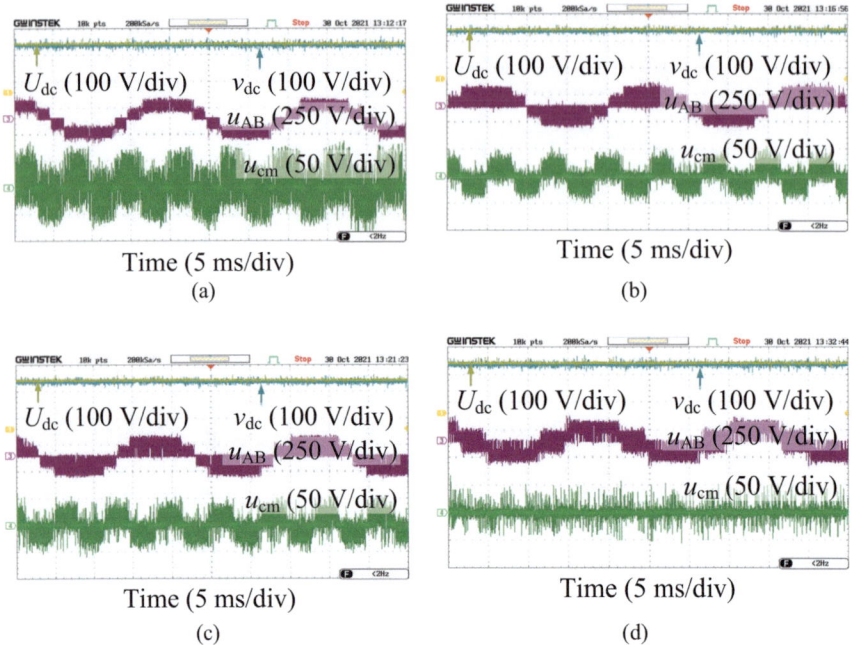

Fig. 3.51 Experimental results of the non-boost operation mode. **a** Conventional SVM method. **b** LMZ-ST method. **c** LMZS-ST method. **d** HSVM method

applied, the voltages across capacitors C_2 and C_3 in the boost case are shown in Fig. 3.52b, which proves that the NPV balance is not deteriorated by the insertion of shoot-through states.

Next, the input voltage of the dc power source is further reduced to 90 V to validate the high-boost operation for the HSVM scheme. The modulation index and shoot-through duty cycle are given as 0.70 and 0.25, respectively. The obtained experimental waveforms are depicted in Fig. 3.53. It is observed that the magnitude of the dc-link voltage is twice the dc input voltage, which proves the feasibility of high-boost operation for the proposed scheme.

Case 3: Measurement of CMV Spectra and Leakage Current

Since the leakage current is influenced by the CMV magnitude and frequency, the spectra and RMS values of CMV with different modulation methods are measured and compared in different operation conditions. Then the leakage currents with different modulation methods are tests.

1. **Spectra of CMV**

    As indicated by [20], the frequency characteristics of CMV can be reflected by the corresponding spectrum. Therefore, the CMV spectra for different schemes have been tested experimentally. The data for CMV waveforms are stored by adopting the digital storage oscilloscope, and the CMV spectra for different

3.3 Experimental Verifications

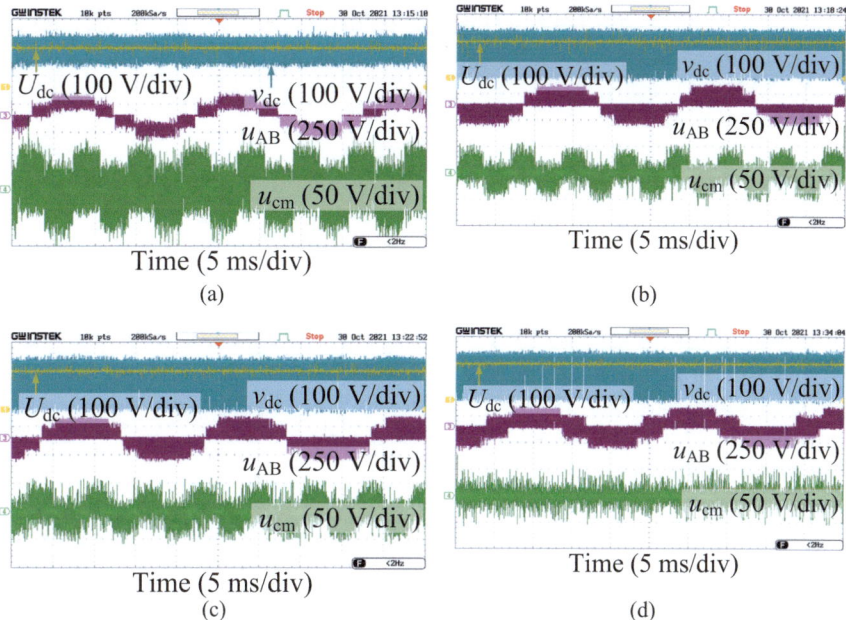

Fig. 3.52 Experimental results of the boost operation mode. **a** Conventional SVM method. **b** LMZ-ST method. **c** LMZS-ST method. **d** HSVM method

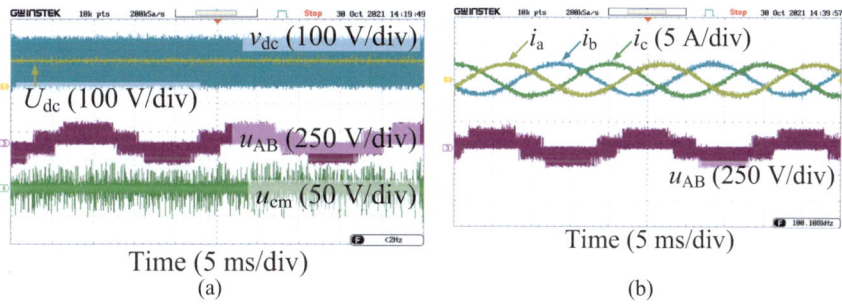

Fig. 3.53 Experimental results of the high-boost operation for HSVM scheme ($U_{dc} = 90$ V). **a** DC input voltage, dc-link voltage, line-to-line voltage, and CMV. **b** Three-phase output currents

schemes are subsequently obtained by using MATLAB software, as shown in Fig. 3.54. When the conventional SVM method is used, the CMV magnitude around the switching frequency (12.5 kHz) and double switching frequency (25 kHz) are high, which can induce large leakage currents and CM EMI. The LMZ-ST and LMZS-ST methods can reduce the CMV magnitude around the switching frequency (12.5 kHz), but the CMV magnitudes around the double switching frequency (25 kHz) are still high. Fortunately, the HSVM method can effectively reduce the high-frequency components of CMV. Besides, the fundamental magnitude of CMV for the HSVM method is the lowest among the four methods.

2. **RMS value of CMV**

   To illustrates the capability of CMV reduction of the LMZ-ST method, the CMV RMS values of different modulation indices are tested using the conventional SVM and LMZ-ST methods. In this case, the dc input voltage and shoot-through duty cycle are set as 360 V and 0.05, respectively. The CMV RMS values of different modulation indices are shown in Fig. 3.55. For the conventional SVM method, when the modulation index is around 0.5, the dwell times of small voltage vectors with high CMV are increased, and the CMV RMS value is the highest. For the LMZ-ST method, as the modulation index increases, the CMV RMS values increase. The CMV RMS value of the LMZ-ST method is lower than that of the conventional SVM method.

3. **Leakage current**

   In [21], a novel machine model for high frequencies was developed, which revealed that the leakage current is directly proportional to the CMV frequency. A parasitic capacitance is connected between the common point of the three-phase load and the negative point of the dc power source, and the leakage current is measured at the line where the parasitic capacitance is connected, as shown in Fig. 3.56. The parasitic capacitance C_{PV} is chosen as 4.7 µF.

   The measured waveforms of leakage currents are shown in Fig. 3.57. It is observed that the leakage current of the conventional SVM method is the highest among four methods, which results from the effects of high CMV. The LMZ-ST and LMZS-ST methods can suppress the leakage currents to some extent. By reducing the magnitude and high-frequency components of CMV, the HSVM method can further reduce the leakage current. These experimental results are well consistent with the theoretical analyses.

Table 3.14 summarizes the numerical comparisons for different schemes. It is mentioned that the maximum frequency of the harmonics is selected as 30 kHz when analyzing the values of total harmonic distortion (THD). It is found that the conventional SVM method has the best quality of output waveforms, since this method adopts nearest three vectors to generate the ac output voltage. When the HSVM method is adopted, the THD values of the line-to-line voltage (THDv) and output current (THDi) are higher than those of conventional SVM method, but the magnitude and RMS value of CMV can be effectively reduced. Moreover, the HSVM method can effectively suppress the leakage current by reducing the magnitude and high-frequency components of CMV. More importantly, a relatively low THDi is obtained due to the adoption of the three-level topology. Compared with the LMZ-ST and LMZS-ST methods, the HSVM method can effectively reduce the RMS values of CMV and leakage current.

3.3 Experimental Verifications

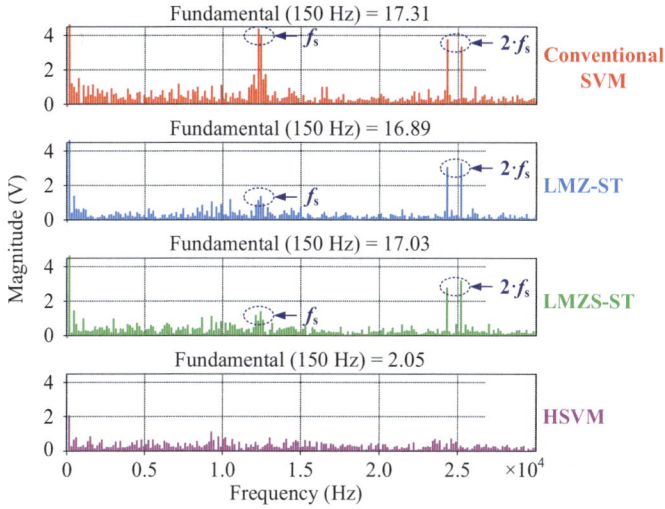

Fig. 3.54 Experimental results of CMV spectra for different modulation methods

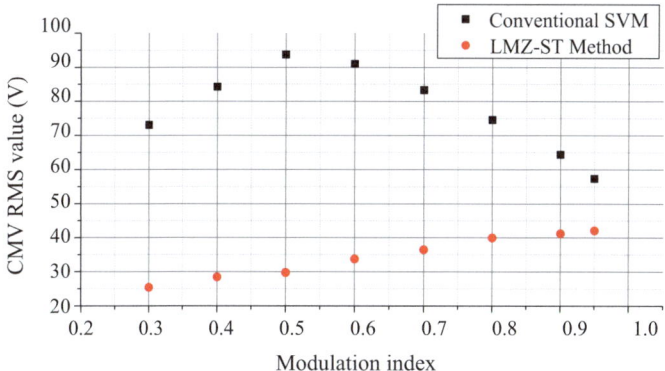

Fig. 3.55 CMV RMS values of different modulation indices

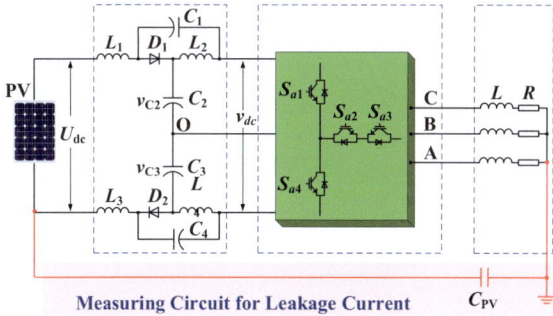

Fig. 3.56 The measuring circuit for leakage current

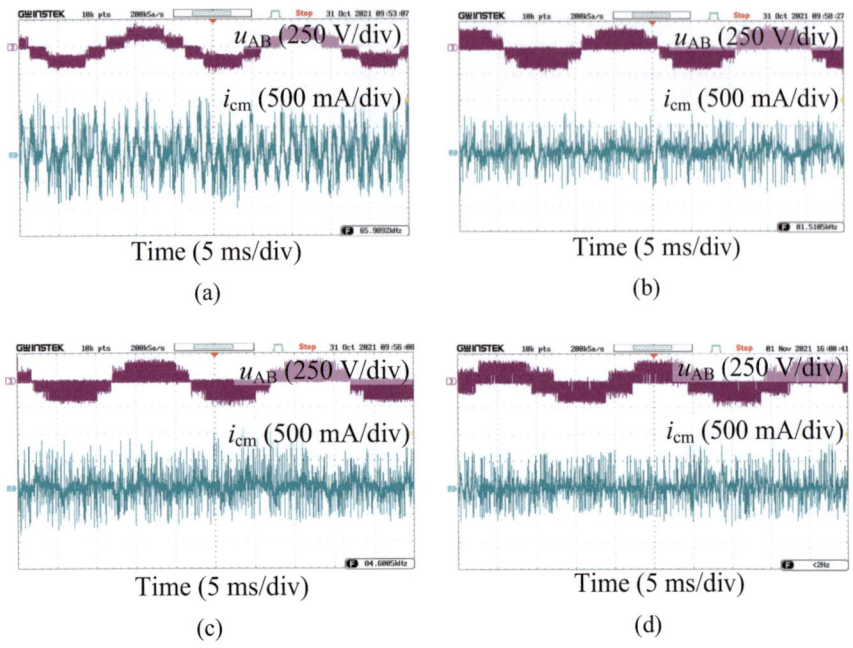

Fig. 3.57 Experimental waveforms of leakage currents for different methods. **a** Conventional SVM method. **b** LMZ-ST method. **c** LMZS-ST method. **d** HSVM method

Table 3.14 Numerical comparisons of performance indices for different schemes

Methods	Conventional method	LMZ-ST method	LMZS-ST method	HSVM method
Line-to-line voltage THD	40.89%	71.59%	70.36%	71.82%
Load current THD	2.37%	4.11%	4.05%	3.94%
Magnitude of CMV	60 V	30 V	30 V	30 V
RMS value of CMV	35.7 V	17.1 V	19.5 V	7.8 V
RMS value of i_{cm}	283.3 mA	95.4 mA	114.1 mA	36.6 mA
Neutral-point control ability	Yes	No	Yes	Yes

3.3 Experimental Verifications

Case 4: Verification of NPV Balance Ability

In this part, the HSVM method is used to verify the ability of NPV balance. An additional 470 Ω resistor is added in parallel with capacitor C_3 to simulate the non-ideal conditions of the DC side. At the beginning of the tests, the active NPV control based on small vector selection is activated. During the system operation, the active NPV control is disabled. Figure 3.58 gives the experimental waveforms under different conditions. When the NPV balance control strategy is disabled, the voltage across capacitor C_2 is increased, while the voltage across capacitor C_3 is decreased, because the current that flows through the additional resistor results in the NPV unbalance. These results demonstrate that the active NPV balance control of the proposed HSVM scheme is effective in both non-boost and boost operations.

Case 5: Transient Performance of HSVM Modulation Method

The transient performance of the HSVM scheme is shown in Fig. 3.59, where a step change in the modulation index is conducted. In Fig. 3.59a, the QZS-3LT²L is initially operated at a modulation index of 0.64. When the modulation index is changed from

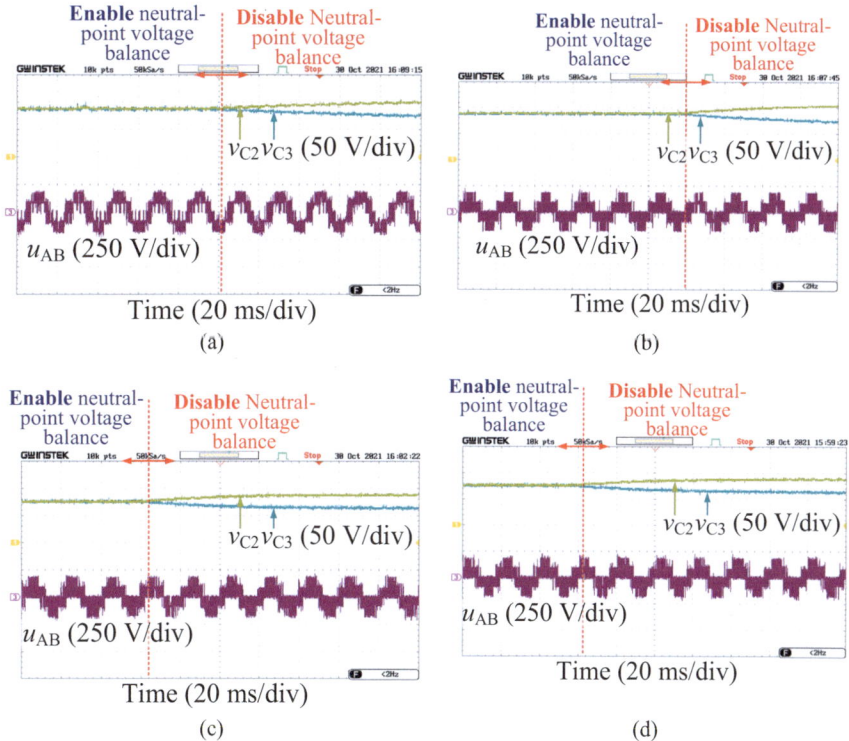

Fig. 3.58 Experimental waveforms of NPV balance control ability using HSVM method. **a** Non-boost operation mode and $m = 0.95$. **b** Non-boost operation mode and $m = 0.64$. **c** Boost operation mode and $m = 0.84$. **d** Boost operation mode and $m = 0.64$

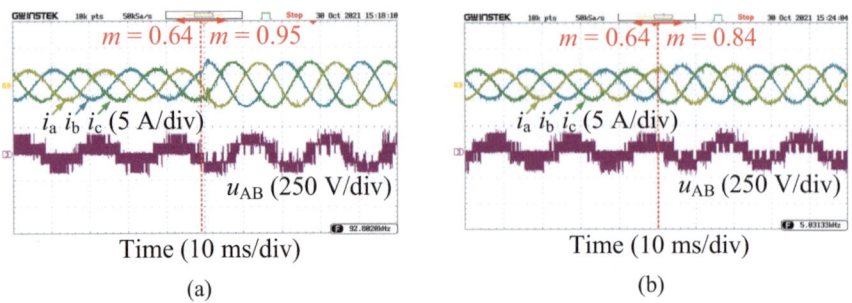

Fig. 3.59 Transient performance of HSVM modulation method under different modulation index. **a** Non-boost operation. **b** Boost operation

0.64 to 0.95, the current flowing through the load is increased. It is clear that the load current is proportional to the modulation index. When the modulation index is 0.64, the THDi is 5.91%. When the modulation index is increased to 0.95, the THDi is decreased to 3.19%. In boost operation, when the modulation index is changed from 0.64 to 0.84, the experimental waveforms of the line-to-line voltage and three-phase load currents are shown in Fig. 3.59b. When the modulation index is 0.64, the THDi is 5.80%. When the modulation index is increased to 0.84, the THDi is decreased to 3.94%. These results demonstrate that the THDi decreases as the modulation index increases.

3.4 Summary

In PV power generation systems, the parasitic capacitor inevitably leads to the leakage current. The leakage current causes electromagnetic inferences, distorts the output currents quality and threatens people safety. Thus, the leakage current suppression methods are elaborated in this chapter. The generation mechanism of leakage current is analyzed and the corresponding models are established firstly. Then according to the theoretical knowledge, two types of leakage current suppression methods are elaborated. The first type method is proposed from the perspective of CM circuit, and the second type method is proposed from the perspective of exciting source suppression. The detailed implementation processes of these methods are given step by step, which provides much guidance for readers working with the subject. Finally, the experimental results are given to facilitate readers to understand the characterizes of different leakage current suppression methods.

References

1. Automatic disconnection device between a generator and the public low-voltage grid, VDE V 0126-1-1 (2006)
2. X. Guo, X. Jia, Hardware-based cascaded topology and modulation strategy with leakage current reduction for transformerless PV systems. IEEE Trans. Ind. Electron. **63**(12), 7823–7832 (2016)
3. L. Zhou, F. Gao, T. Xu, A family of neutral-point-clamped circuits of single-phase PV inverters: Generalized principle and implementation. IEEE Trans. Power Electron. **32**(6), 4307–4319 (2017)
4. C.T. Morris, D. Han, B. Sarlioglu, Reduction of common mode voltage and conducted EMI through three-phase inverter topology. IEEE Trans. Power Electron. **32**(3), 1720–1724 (2017)
5. M. Schweizer, J.W. Kolar, Design and implementation of a highly efficient three-level T-type converter for low-voltage applications. IEEE Trans. Power Electron. **28**(2), 899–907 (2013)
6. X. Li, X. Xing, C. Zhang, A. Chen, C. Qin, G. Zhang, Simultaneous common-mode resonance circulating current and leakage current suppression for transformerless three-level t-type PV inverter system. IEEE Trans. Ind. Electron. **66**(6), 4457–4467 (2019)
7. Z. Hu, X. Xing, C. Liu, R. Zhang, F. Blaabjerg, A modified discontinuous PWM method for three-level inverters with the improved LCL filter. IEEE Trans. Power Electron. **39**(5), 5498–5509 (2024)
8. Z. Hu, X. Xing, H. Zhang, F. Blaabjerg, Modeling and suppression method of low order harmonics for three-level inverter with small capacitance value. IEEE Trans. Ind. Electron. **72**(1), 470–480 (2025)
9. C. Bao, X. Ruan, X. Wang et al., Step-by-step controller design for LCL-type grid-connected inverter with capacitor-current-feedback active-damping. IEEE Trans. Power Electron. **29**(3), 1239–1253 (2014)
10. K. Li, M. Wei, C. Xie et al., Triangle carrier-based DPWM for three-level NPC inverters. IEEE J. Emerg. Sel. Topics Power Electron. **6**(4), 1966–1978 (2018)
11. F. Liu, K. Xin, Y. Liu, An adaptive discontinuous pulse width modulation (DPWM) method for three phase inverter, in *2017 IEEE Applied Power Electronics Conference and Exposition (APEC)* (2017), pp. 1467–1472
12. S. Lei, H. Hu, B. Ling, Design of harmonic current suppression strategy based on resonance control. Power Electron. **56**(7), 37–40 (2022)
13. O. Husev et al., Comparison of impedance-source networks for two and multilevel buck-boost inverter applications. IEEE Trans. Power Electron. **31**(11), 7564–7579 (2016)
14. M.C. Cavalcanti, K.C. de Oliveira, A.M. de Farias et al., Modulation techniques to eliminate leakage currents in transformerless three-phase photovoltaic systems. IEEE Trans. Ind. Electron. **57**(4), 1360–1368 (2010)
15. Y. Liu, B. Ge, H. Abu-Rub et al., Hybrid pulsewidth modulated single-phase quasi-Z-source grid-tie photovoltaic power system. IEEE Trans. Ind. Inform. **12**(2), 621–632 (2016)
16. C. Qin, C. Zhang, A. Chen et al., A space vector modulation scheme of the quasi-Z-source three-level T-type inverter for common-mode voltage reduction. IEEE Trans. Ind. Electron. **65**(10), 8340–8350 (2018)
17. C. Qin, C. Zhang, X. Xing et al., Simultaneous common-mode voltage reduction and neutral-point voltage balance scheme for the quasi-Z-source three-level T-type inverter. IEEE Trans. Ind. Electron. **67**(3), 1956–1967 (2020)
18. C. Qin, X. Xing, Y. Jiang, Hybrid space vector modulation scheme to reduce common-mode voltage magnitude and frequency in three-level quasi-Z-source inverter. IEEE J. Emerg. Sel. Topics Power Electron. **10**(6), 6810–6821 (2022)
19. E. Makovenko, O. Husev, E. Romero-Cadaval et al., Single phase three-level neutral-point-clamped quasi-Z-source inverter. IET Power Electron. **8**(1), 1–10 (2015)
20. A. Janabi, B. Wang, Hybrid SVPWM scheme to minimize the common-mode voltage frequency and amplitude in voltage source inverter drives. IEEE Trans. Power Electron. **34**(2), 1595–1610 (2019)

21. O. Magdun, A. Binder, High-frequency induction machine modeling for common mode current and bearing voltage calculation. IEEE Trans. Ind. Appl. **50**(3), 1780–1790 (2014)

Open Access This chapter is licensed under the terms of the Creative Commons Attribution 4.0 International License (http://creativecommons.org/licenses/by/4.0/), which permits use, sharing, adaptation, distribution and reproduction in any medium or format, as long as you give appropriate credit to the original author(s) and the source, provide a link to the Creative Commons license and indicate if changes were made.

The images or other third party material in this chapter are included in the chapter's Creative Commons license, unless indicated otherwise in a credit line to the material. If material is not included in the chapter's Creative Commons license and your intended use is not permitted by statutory regulation or exceeds the permitted use, you will need to obtain permission directly from the copyright holder.

Chapter 4
Control Technology of Photovoltaic Generation Systems for Maximizing Power Generation

For the photovoltaic (PV) generation systems, the output power is one of the important performance indices for users, which is directly affected by the utilization of the PV array. It is well known that the utilization of the PV array is not only determined by its internal characteristics, but also the external environment. The internal characteristics mainly refer to the "Photovoltaic Effect", which is determined by the materials. The external environment includes light irradiance, load condition, and environmental temperature and so on. PV array has different I-V curve with different environment conditions. Usually, there is one maximum power point (MPP) in each I-V curve. If the PV array is always operating in the MPP, the maximum utilization of PV array can be maintained, thus obtaining the maximum output power.

Therefore, for PV power generation systems, the key of improving output power is to find the best MPP and adopt appropriate control technology to ensure that the PV array runs at that point [1]. The control technology that ensures the operation of PV array in MPP is called maximum power point tracking (MPPT) technology.

Additionally, for the large-scale PV applications, the laying range of PV array is very wide, and partial shading and panel mismatch is common, which leads to the I-V curve and MPP of each PV panel are different. The commonly used single MPPT technology, which is performed by a central inverter, results in reduced power conversion efficiency. The main reason is that all PV panels are operated at a common MPPT algorithm that might not be effective for all of them.

Thus, this chapter focuses on the control technology that employed in PV generation systems for output power improvement.

First, for generalized PV applications, the conventional MPPT technology is illustrated. Theoretically, when the output impedance of the PV array is equal to the equivalent impedance of PV inverter, the maximum output power of the PV array can be attained. The essence of MPPT technology is that by adjusting the output impedance of the PV array to match the equivalent impedance of PV inverter, which is integrated into outer loop for power control. To be specifically, the output voltage

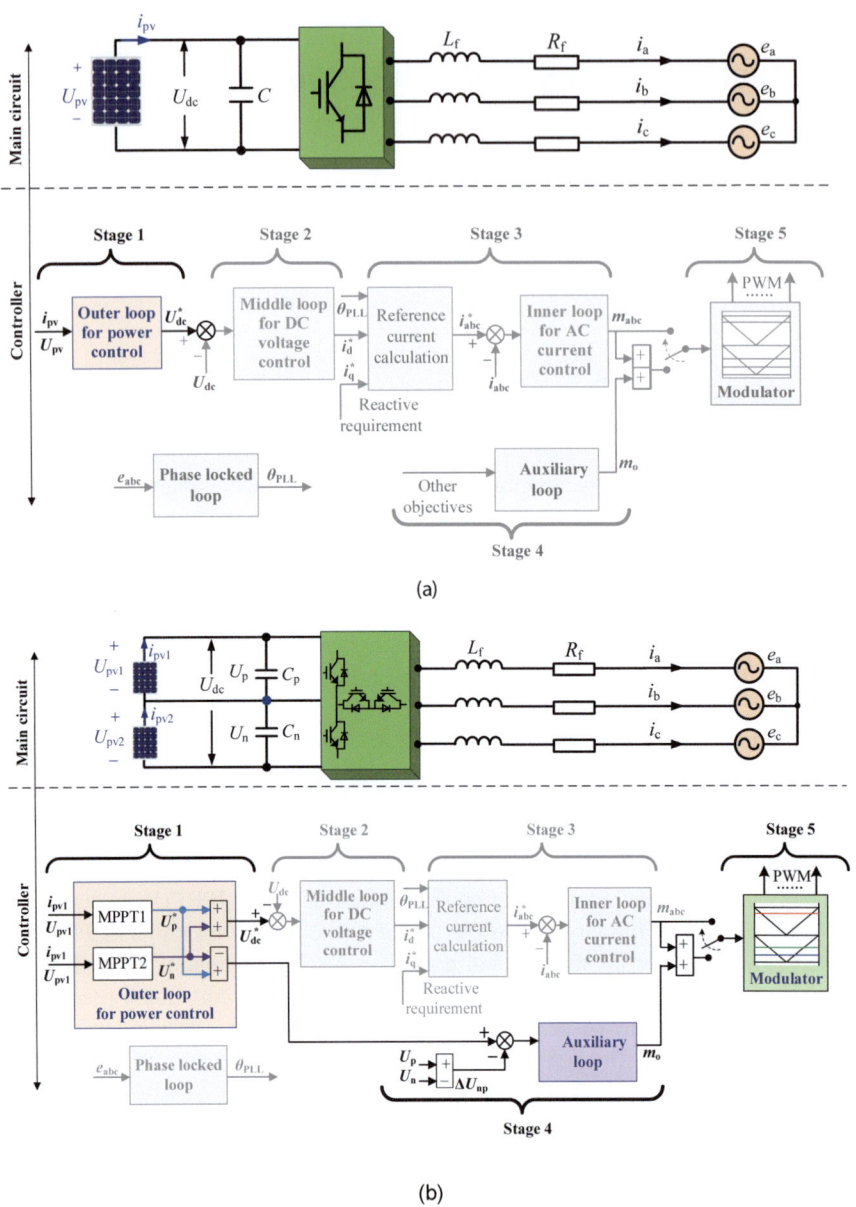

Fig. 4.1 The modules associated with maximum power point operation in overall control diagram **a** conventional MPPT technology, **b** separate MPPT technology

and current of the PV array are sampled and controlled in real-time in outer loop. The control scheme is shown in Fig. 4.1a.

Then, for large-scale PV applications, to further improve the output power, the separate MPPT technology is presented for high-power three-level T-type inverter ($3LT^2I$). With the separate MPPT technology, the total MPPTs efficiency can be improved under the partial shading and panel mismatch conditions. The control scheme of separate MPPT technology is shown in Fig. 4.1b. Four key points need to be emphasized. (1) In outer loop, two MPPT control modules are adopted to calculate the maximum point of power and adjust PV panels, individually. (2) The output of each MPPT control module is summed up and considered as the command of middle loop for DC voltage control, which ensures the $3LT^2I$ tracking the total power generation of the inverter based on the MPPTs. (3) The difference of the output of MPPT control modules is considered as the command of neutral-point voltage (NPV) controller, which ensures the voltages of upper and lower capacitors in $3LT^2I$ to track the output reference voltage of each MPPT control modules. The NPV controller is integrated into auxiliary loop. (4) The appropriate modulation methods of $3LT^2I$ are proposed, which provide balance line voltage whatever the NPV balanced or unbalanced. Thus, the high quality of AC output current is maintained. The appropriated modulation methods are integrated into PWM modulator.

Theoretical analysis, implementation process and control results of these two control technologies are elaborated as follows.

4.1 Conventional Maximum Power Point Tracking (MPPT) Technology

According to the single diode model of PV panel, the I-V curves and P-U curves are obtained under different irradiation and temperature.

When the temperature is kept constant, the I-V curves and P-U curves with the change of irradiance are shown in Fig. 4.2. It can be seen from the I-V curve that as irradiance decreases, the open circuit voltage (U_{oc}) hardly changes, while the short circuit current (I_{sc}) rapidly decreases. The P-U curves show that the MPP varies with different irradiance, and power at the MPP decreases with the decrease of irradiance.

When the irradiance is constant and the temperature changes, the I-V curves and P-U curves are shown in Fig. 4.3. It can be seen from the I-V curve that as irradiance decreases, the short circuit current changes a little, while the open circuit voltage increases linearly with the decrease of temperature. Meanwhile, the MPP varies with different temperature.

It can be seen that the operation condition of PV array is seriously affected by external environment, and presents typical nonlinear characteristics. It should be pointed out that the output voltage and current of PV array change with the change of external environment. Therefore, the equivalent output impedance of PV array varies with the external environment.

According to the theory of power circuit, when the equivalent output impedance of PV array at MPP is equal to the load impedance, the maximum output power of the

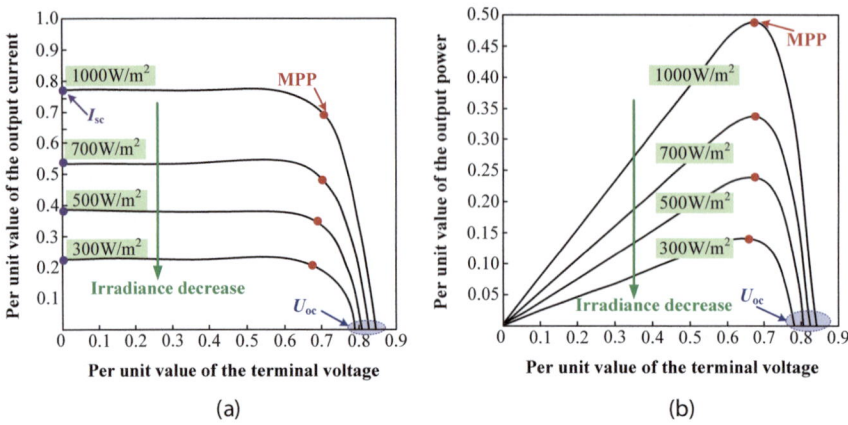

Fig. 4.2 The output characteristics of PV panel with the change of irradiance **a** the *I-V* curves, **b** *P-U* curves

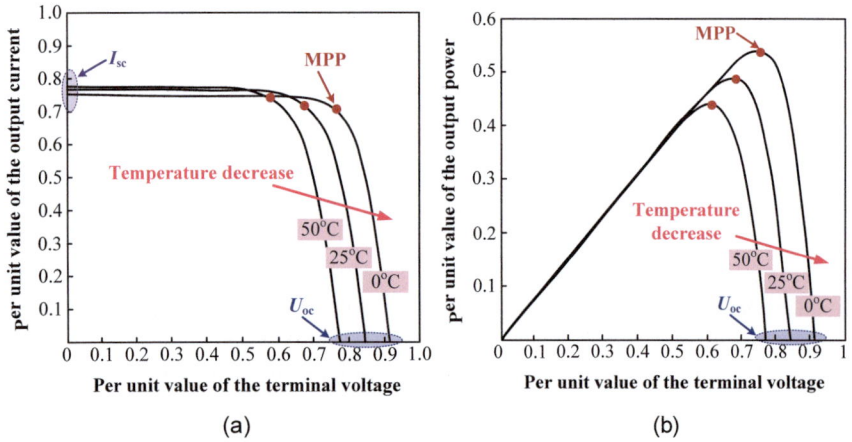

Fig. 4.3 The output characteristics of PV panel with the change of temperature **a** the *I-V* curves, **b** *P-U* curves

PV array can be attained. Obviously, the process of MPPT is the process for matching the equivalent output impedance of PV array at MPP and the equivalent impedance of PV inverter. As the above analyses domonstrate, the equivalent output impedance of PV array at MPP is heavily dependent on environmental conditions. Therefore, if the appropriate control technology is used to adjust the equivalent impedance of PV inverter in real time to match the equivalent output impedance of PV array at MPP, MPPT can be achieved.

The equivalent impedance matching process of PV array and PV inverter is shown in Fig. 4.4. The equivalent impedance curve of PV inverter is displayed by the blue line. If the PV array operates in irradiance 1 condition, the *I-V* curve and blue line

4.1 Conventional Maximum Power Point Tracking (MPPT) Technology

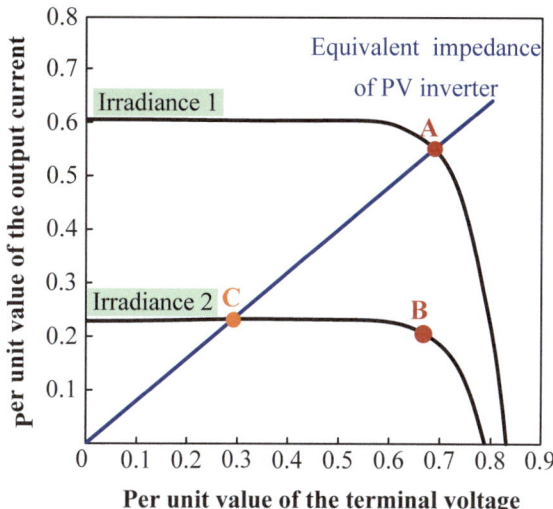

Fig. 4.4 The impedance matching process of PV array and PV inverter

intersect at point A. Moreover, point A is the MPP under the condition of irradiance 1. Thus, the equivalent impedance of PV inverter matches with that of PV array at MPP, and the maximum output power can be maintained. When irradiance 2 occurs, the MPP is the B. If the equivalent impedance curve of PV inverter is constant, the *I-V* curve and blue line intersect at point C. Fortunately, the MPP is B instead of C, which reduces the output power. Therefore, the equivalent impedance curve of PV inverter should be adjusted by appropriate control methods in real time to match the equivalent impedance matching of the PV array at MPP.

As introduced before, the MPPT methods are responsible for capturing the maximum possible amount of energy from the PV panels, which are integrated into outer loop for power control of the PV inverter. In both single-stage and two-stage cases, the grid-tied inverter control scheme is in most cases based on the well-known voltage-oriented control (VOC) technique normally using the synchronous dq frame (note that this is not the case in single-phase inverters). The MPPT method is executed by the inverter in the single-stage case. In the two-stage case, the MPPT is achieved by the dc-dc converter while the inverter accurately controls the dc-link voltage which is a fixed value.

There are several MPPT methods found in practice and in the literature that can reach up to 99% of the MPP efficiency. A classification of the most common MPPT methods is given in Fig. 4.5 [2]. Among these methods, the hill-climbing and fractional methods are widely adopted in commercial PV systems.

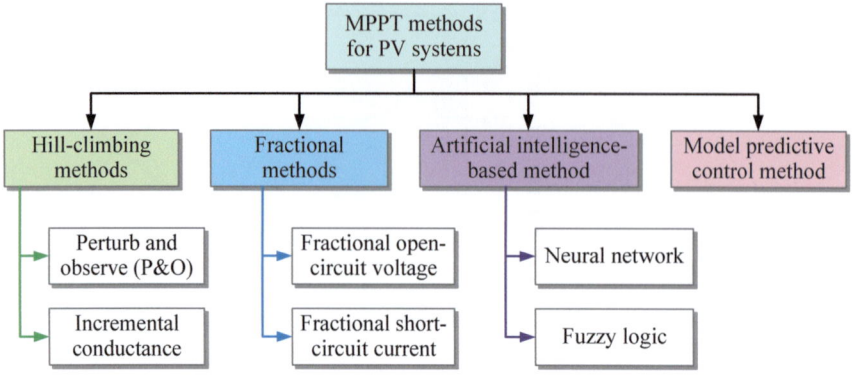

Fig. 4.5 Classification of MPPT methods

4.1.1 Perturb and Observe Algorithm

The Perturb and Observe (P&O) algorithm is a widely used MPPT method, and Fig. 4.6 displays the flowchart of this algorithm. The fundamental steps involved in the P&O algorithm are summarized as follows. At first, the algorithm is started with an initial operating voltage and current from the solar panel. The voltage and current of the solar panel are measured to calculate the generated power. Next, the operating voltage is slightly changed. Typically, this involves increasing or decreasing the voltage by a small amount. After perturbing the voltage, wait for a brief moment to allow the system to stabilize, then measure the new power output. The decision making is conducted as follows. If the power increases after the perturbation, the same direction of perturbation is continued in subsequent cycles (i.e., if the voltage was increased and power increased, continue to increase the voltage). If the power decreases, reverse the direction of perturbation (i.e., if the voltage was increased and power decreased, then decrease the voltage). Continuously repeat the above steps to find the MPP dynamically.

The advantages of the P&O algorithm are summarized as follows:

1. Simplicity: The algorithm is straightforward to implement and requires no complex models of the PV system.
2. Effectiveness: It works well under varying environmental conditions (e.g., changing sunlight levels) and can track the MPP in real-time.
3. Low-cost: It requires minimal additional hardware and processing power.

The P&O algorithm also has several shortcomings as follows:

1. Oscillation around MPP: Near the MPP, the algorithm can cause oscillations, leading to suboptimal power output.
2. Slow response: During sudden changes in environmental conditions, P&O may not respond quickly enough, potentially missing the MPP.

4.1 Conventional Maximum Power Point Tracking (MPPT) Technology

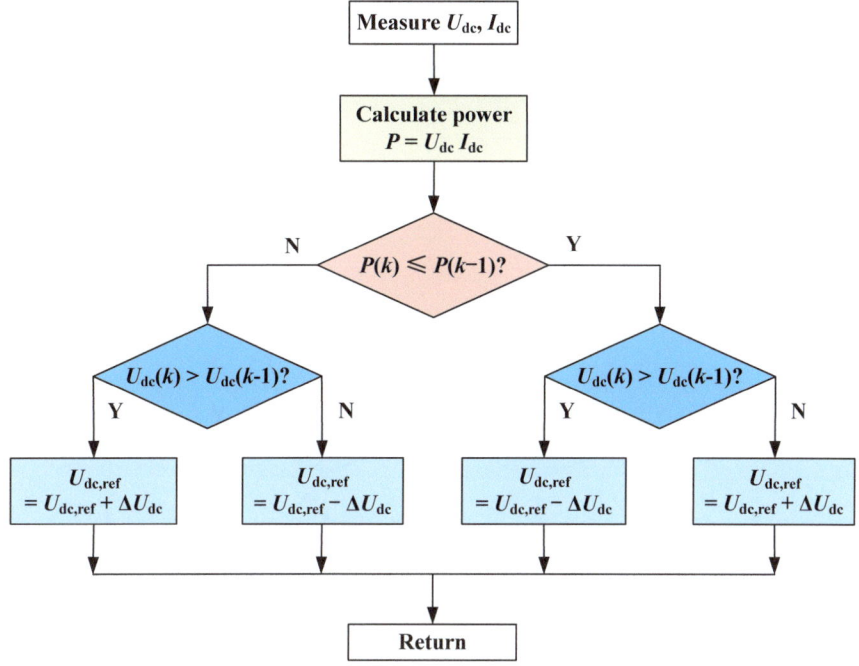

Fig. 4.6 Flowchart of the perturb and observe algorithm

3. Dependence on perturbation size: The choice of perturbation size affects the speed and accuracy of tracking. A small perturbation may track slowly, while a large perturbation might overshoot and cause oscillation.

In summary, the P&O algorithm is a popular MPPT technique due to its simplicity and effectiveness in various operating conditions, allowing renewable energy systems to optimize their performance efficiently.

4.1.2 Incremental Conductance Algorithm

The incremental conductance (INC) algorithm is another MPPT method used in PV generation systems, which improves upon traditional methods such as the P&O algorithm by providing faster and more accurate tracking of the MPP. Figure 4.7 depicts the corresponding flowchart of this algorithm. It is particularly effective under varying environmental conditions, such as changes in solar irradiance and temperature.

The principle behind the INC algorithm is to determine the MPP by calculating the instantaneous conductance and the incremental conductance of the PV system.

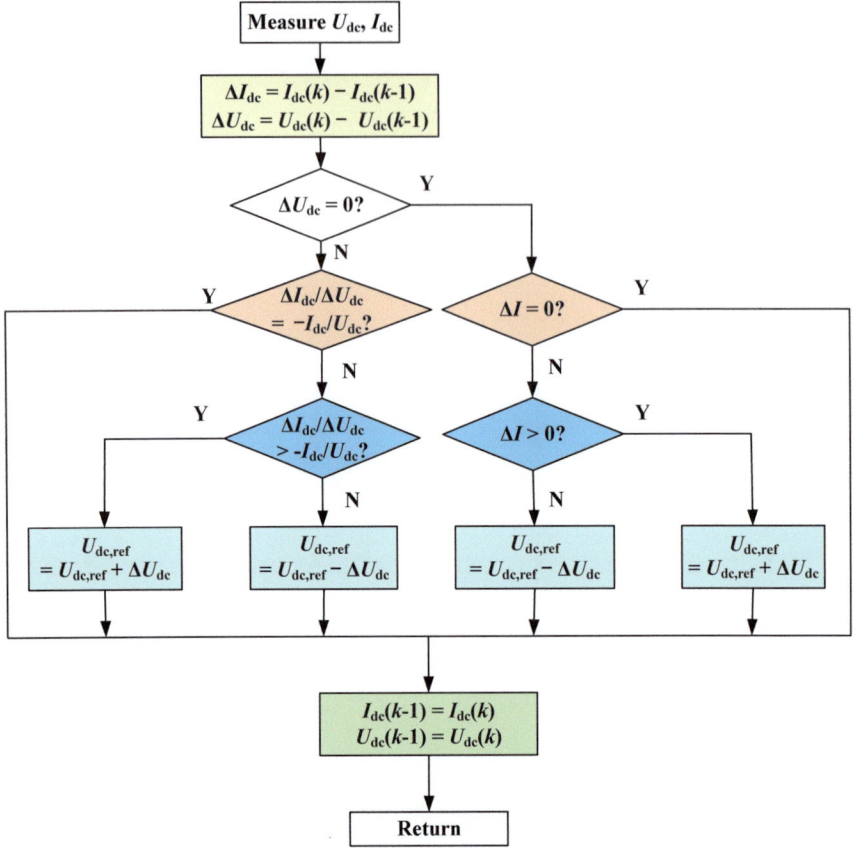

Fig. 4.7 Flowchart of the incremental conductance algorithm

At the MPP, the derivative of the power with respect to voltage is zero. The conditions established by this algorithm are explained as follows.

When the MPP is reached, $dP/dU = 0$. This indicates that the incremental conductance is equal to the instantaneous conductance.

When the operating point is to the left of the MPP, $dP/dU > 0$. This indicates that the incremental conductance is greater than the instantaneous conductance.

When the operating point is to the right of the MPP, $dP/dU < 0$. This indicates that the incremental conductance is less than the instantaneous conductance.

The advantages of the INC algorithm are summarized as follows:

1. Improved convergence speed: It reacts more rapidly to changes in environmental conditions compared to traditional methods.
2. Robust in non-ideal conditions: It works well in partially shaded conditions or rapidly changing light conditions.
3. Higher accuracy: Achieves better performance by precisely identifying the MPP.

4.1 Conventional Maximum Power Point Tracking (MPPT) Technology

The INC algorithm also has several disadvantages as follows:

1. Complexity: The algorithm implementation is more complex than simpler techniques like P&O.
2. Computational load: Requires more calculations and faster processors to handle dynamic changes in real-time.

To summarize, the INC algorithm is a powerful MPPT technique that helps optimize the performance of PV systems under changing environmental conditions. Its efficiency makes it a preferred choice in many solar energy applications.

4.1.3 Fractional Open-Circuit Voltage Method

The fractional open-circuit voltage method is a well-established technique used for MPPT in PV systems, which is based on the relationship between the open-circuit voltage and the MPP voltage of a solar panel. This method provides a straightforward and effective way to estimate the MPP without requiring complex measurements of current and voltage in real time.

The basic principle behind the fractional open-circuit voltage method is that the MPP voltage (U_{MPP}) of a solar panel is approximately a fixed fraction of its open-circuit voltage (U_{oc}), that is, $U_{\text{MPP}} = k_1 U_{\text{oc}}$. Typically, this fraction (k_1) can be a value between 0.7 and 0.85, depending on the characteristics of the particular solar panel. This value can be derived from manufacturer data or empirical observation. The operating point of the inverter or load is adjusted to match this estimated MPP voltage. This process is repeated during operation to ensure that the PV system operates close to the MPP as conditions change.

The advantages of the fractional open-circuit voltage method are summarized as follows. First, this method is relatively easy to implement, since it does not require continuous measurement and real-time calculations of both current and voltage. Compared to more complex MPPT algorithms, this method is less computationally intensive, leading to lower power consumption overall. Moreover, it can effectively track the MPP under stable and moderate conditions without significant oscillations or complexity.

It is worth noting that the fractional open-circuit voltage method also has some limitations. This method may not respond quickly to rapid changes in irradiance which could lead to reduced energy capture in dynamic conditions. The choice of the fraction (k_1) can vary with temperature and sunlight conditions; if it is incorrectly set, the system may operate sub-optimally. Fine-tuning of the fraction k_1 for different environmental conditions or different types of solar panels can be challenging.

4.1.4 Fractional Short-Circuit Current Method

The fractional short-circuit current method is another MPPT technique, which is similar to the fractional open-circuit voltage method. However, instead of utilizing the open-circuit voltage, the fractional short-circuit current method derives the MPP from the short-circuit current (I_{sc}) of the solar panel.

The fractional short-circuit current method is based on the principle that the MPP current (I_{MPP}) of a solar panel is approximately a fixed fraction of its short-circuit current (I_{sc}), that is, $I_{MPP} = k_2 I_{sc}$. This relationship is used to estimate the MPP current without the need to continuously measure the voltage and current of the PV system. The coefficient k_2 is a constant that typically ranges from 0.7 to 0.85, based on the characteristics of the specific solar panel. The operating point of the inverter is regulated to track or maintain this estimated MPP current (I_{MPP}). It is worth noting that the short-circuit current (I_{sc}) is periodically measured to update the MPP current as environmental conditions change.

Like the fractional open-circuit voltage method, the fractional short-circuit current method is also straightforward to implement, relying on simple measurements and calculations. In addition, this method is less computationally intensive compared to dynamic MPPT methods, which helps in conserving energy.

Similar to the fractional open-circuit voltage method, the response to rapid changes in sunlight may not be immediate, which leads to the potential energy loss. Besides, the effectiveness of the fractional short-circuit current method highly depends on the accurate selection of the fraction (k_2). This value can vary due to temperature fluctuations, irradiance levels, and specific PV panel characteristics. Another problem associated with this method is the risk during measurement. To be specific, while measuring the short-circuit current, care must be taken not to exceed the panel's current rating or short the system for extended periods, which may lead to damage.

4.1.5 Advanced MPPT Methods

With the development of modern control theory and artificial intelligence technology, several advanced MPPT methods have been proposed, and some typical solutions are briefly introduced here.

The model predictive control (MPC) strategy is an advanced MPPT method that uses a mathematical model of the PV system to predict the future power output based on the current operating conditions. It then determines the optimal operating point that maximizes the power output.

Methods based on artificial intelligence (AI), such as neural networks and fuzzy logic, can be used for MPPT by learning from past data and making predictions based on patterns. These methods can adapt to changing environmental conditions and optimize the power output accordingly, but the calculation complexity is increased compared to conventional MPPT methods.

It is important to note that the choice of MPPT method depends on the specific requirements, cost considerations, and environmental conditions of the PV power generation system. Different methods may have their advantages and disadvantages in terms of accuracy, efficiency, and complexity.

4.2 Separate Maximum Power Point Tracking Technology

The cost of PV panels dropped in 2008 by around 40% to levels under one euro per watt, which promoted the rapid development of the PV power generation industry [3, 4]. In particular, more and more large-scale PV power stations have been built.

The centralized inverter is preferred in large-scale PV applications for connecting all PV panels to the grid. In this structure, all PV strings are arranged in parallel and connected to a common central inverter, to reach a high-power level as shown in Fig. 4.8a. Moreover, all strings are operated based on a common MPPT algorithm. This configuration has the advantages of causes lesser complexity, lower investment costs, and higher efficiency. However, for large-scale PV applications, the laying range of the PV array is very wide, and the partial shading and temperature difference is common. Consequently, the output characteristics of all PV panels are difficult to be consistent. Figure 4.8b shows the *I-V* curves of normal and shaded panels, respectively. It can be seen that even if the external irradiance is the same, the *I-V* curves and MPPs will be different due to the presence of shadows. If all strings are operated based on a common MPPT algorithm, the impedance matching between the PV panels and PV inverter cannot be maintained. Thus, the MPPT performance is degraded.

To improve the MPPT performance, the DC-DC converters are connected to each string, which is called a multistring inverter [5]. Each string is separately connected to the interface converter, and is operated based on its own MPPT algorithm (shown in Fig. 4.9), which facilitates overall MPPT performance of the whole system. However, investment costs are relatively high and also the system efficiency is decreased due to applying multiple converters. Further, the control and implementation complexity are increased due to the additional dc–dc converter stages. Therefore, the improvement of the MPPT control without dc–dc converter is a hot topic.

Hereby, centralized topology with separated MPPTs can be considered as the best topology since investment costs and complexity are reduced, while efficiency is increased and MPPTs are strictly followed. Inherently, the three-level inverter has a split dc-link, whose voltages can be controlled separately according to practical applications by appropriate control technology. If the split dc-link voltage structure and control technology are applied to the split PV connection, separate MPPTs on the PV string become possible without an increment in the conversion stage. Thus, the control technology for three-level PV inverter with separate MPPTs is presented in this section.

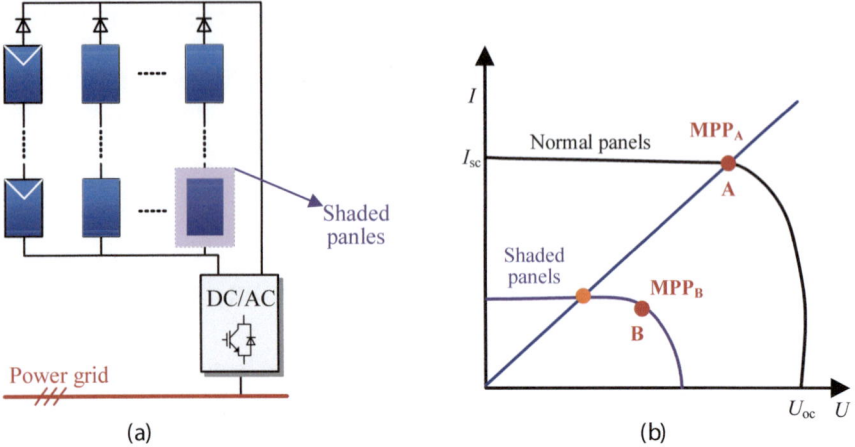

Fig. 4.8 The topology for large-scale PV power stations **a** centralized PV inverter, **b** the I-V curves of normal and shaded panels

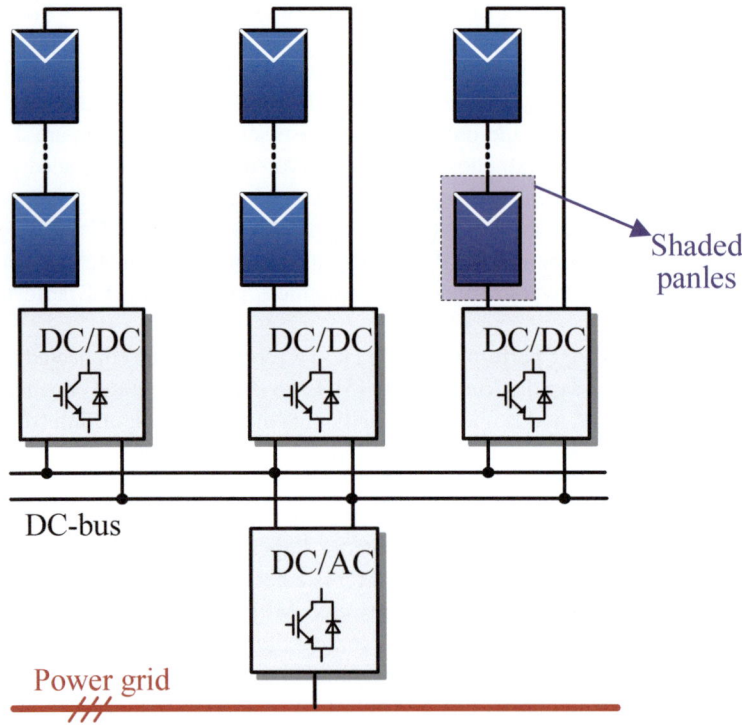

Fig. 4.9 Multistring inverter topology for large-scale PV power stations

4.2 Separate Maximum Power Point Tracking Technology

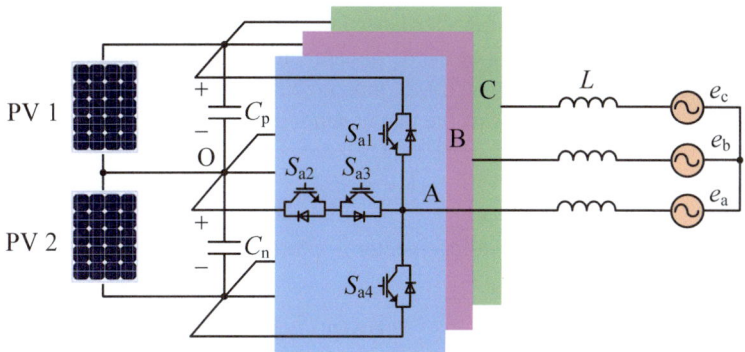

Fig. 4.10 The topology of centralized three-level inverter with separate PV modules

In a centralized three-level PV inverter system, separate dc power sources can be configured for two dc-link capacitors, which can improve the system efficiency through controlling the capacitor voltages separately [6]. In some hybrid energy storage systems such as the solar PV and battery storage integration system, asymmetric control of the dc-link voltages is also necessary to utilize maximal power from different sources [7]. However, unbalanced NPV conditions negatively affect the operation of the three-level inverter, and three-phase output currents are distorted when adopting the traditional modulation method. To cope with this severe problem, various methods have been put forward for the three-level inverter with unbalanced NPV conditions [8–11], which will be analyzed in this section.

Figure 4.10 shows the simplified circuit diagram of the centralized three-level PV inverter system. It is seen that separate PV modules are connected in parallel with two dc-link capacitors (C_p and C_n). The split dc-link capacitors can be controlled asymmetrically to achieve the separate MPPTs in PV module, which is connected to each capacitor. The operating status of one phase leg can be represented by three switching states: [P], [O], and [N]. Select the dc-link neutral-point as the reference, the output pole voltages of state [P], [O], and [N] are U_p, 0, and $-U_n$, respectively.

In order to indicate the voltage difference between two dc-link capacitors (C_p and C_n), we define the dc unbalancing coefficient (k) as

$$k = \frac{U_p - U_n}{U_p + U_n} = \frac{U_p - U_n}{U_{dc}} \tag{4.1}$$

where U_p and U_n are voltages across the upper and lower dc-link capacitors C_p and C_n, respectively. U_{dc} is the total dc-link voltage, that is, $U_{dc} = U_p + U_n$. The voltages across the upper and the lower dc-link capacitors can be expressed by means of the dc unbalancing coefficient k and the total dc-link voltage U_{dc}, as expressed by

$$\begin{cases} U_p = \frac{1+k}{2} \cdot U_{dc} \\ U_n = \frac{1-k}{2} \cdot U_{dc} \end{cases} \tag{4.2}$$

When k is equal to zero, the NPV balance is achieved. Since the sum of U_p and U_n is equal to $U_{dc}/2$, the range of the dc unbalancing coefficient can be obtained as follows. In practical applications, the dc unbalancing degree is usually restricted by considering the block voltages of the power semiconductor switches.

$$-1 < k < 1 \tag{4.3}$$

The basic definition of space voltage vector is expressed as follows:

$$V = \frac{2}{3}\left[u_{AO}(t)e^{j0} + u_{BO}(t)e^{j\frac{2\pi}{3}} + u_{CO}(t)e^{j\frac{4\pi}{3}}\right] \tag{4.4}$$

where $e^{jx} = \cos x + j\sin x$ and $x = 0, 2\pi/3$, or $4\pi/3$. $u_{AO}(t)$, $u_{BO}(t)$, and $u_{CO}(t)$ are the instantaneous output phase voltages.

For active switching state [POO], the generated output phase voltages are

$$\begin{cases} u_{AO}(t) = U_p = \frac{1+k}{2} \cdot U_{dc} \\ u_{BO}(t) = 0 \\ u_{CO}(t) = 0 \end{cases} \tag{4.5}$$

The corresponding space vector V_{POO} can be obtained as follows:

$$V_{POO} = \frac{2}{3}U_p = \frac{1+k}{3}U_{dc} \tag{4.6}$$

By utilizing the same methodology, the expressions of other basic vectors in Sector 1 are provided as follows:

$$V_{PNN} = \frac{2}{3}U_{dc} \tag{4.7}$$

$$V_{PPN} = \frac{1}{3}U_{dc} + j\frac{\sqrt{3}}{3}U_{dc} \tag{4.8}$$

$$V_{PON} = \frac{2U_p + U_n}{3} + j\frac{\sqrt{3}U_n}{3} = \frac{3+k}{6}U_{dc} + j\frac{\sqrt{3}(1-k)}{6}U_{dc} \tag{4.9}$$

$$V_{ONN} = \frac{1-k}{3}U_{dc} \tag{4.10}$$

4.2 Separate Maximum Power Point Tracking Technology

$$V_{\text{PPO}} = \frac{1+k}{6}U_{\text{dc}} + j\frac{\sqrt{3}(1+k)}{6}U_{\text{dc}} \tag{4.11}$$

$$V_{\text{OON}} = \frac{1-k}{6}U_{\text{dc}} + j\frac{\sqrt{3}(1-k)}{6}U_{\text{dc}} \tag{4.12}$$

$$V_{\text{OOO}} = V_{\text{PPP}} = V_{\text{NNN}} = 0 \tag{4.13}$$

From the expressions of basic vectors in Sector 1, one can find that the positions of large and zero vectors the same as the balanced case, whereas the positions of medium and small vectors are changed by the unbalanced NPV conditions. What's more, the following equations are satisfied:

$$V_{\text{PPN}} - V_{\text{PON}} = -\frac{1+k}{6}U_{\text{dc}} + j\frac{\sqrt{3}(1+k)}{3}U_{\text{dc}} \tag{4.14}$$

$$V_{\text{PNN}} - V_{\text{PON}} = \frac{1-k}{6}U_{\text{dc}} + j\frac{\sqrt{3}(k-1)}{6}U_{\text{dc}} \tag{4.15}$$

We can further obtain the algebraic relation as follows:

$$V_{\text{PPN}} - V_{\text{PON}} = (V_{\text{PNN}} - V_{\text{PON}}) \cdot \frac{k+1}{k-1} \tag{4.16}$$

It is revealed that the medium vectors not only change in length but also in direction. However, it should be noted that the medium vector [PON] stays on the connecting line of the two adjacent large vectors ([PNN] and [PPN]) if the total dc-link voltage keeps constant. A similar conclusion is also applicable for the medium vectors in other sectors.

For the three-level inverter with unbalanced NPV conditions, the revised space vector diagrams are obtained, as shown in Fig. 4.11. It is clear that these space vector diagrams become complicated and asymmetric due to the effects of unbalanced NPV conditions.

The revised space vector diagram is divided into six sectors by six medium vectors. The positions of the zero vectors and the large vectors keep the same as the balanced case whether the dc unbalancing coefficient is negative or positive. The overlapping small vectors are split and cannot be considered as redundant vectors anymore. One of them becomes longer and the other shorter, depending on the sign of the dc unbalancing coefficient. The medium vectors not only change in length but also in direction; however, it should be noted that the medium vector stays on the connecting line of the two adjacent large vectors.

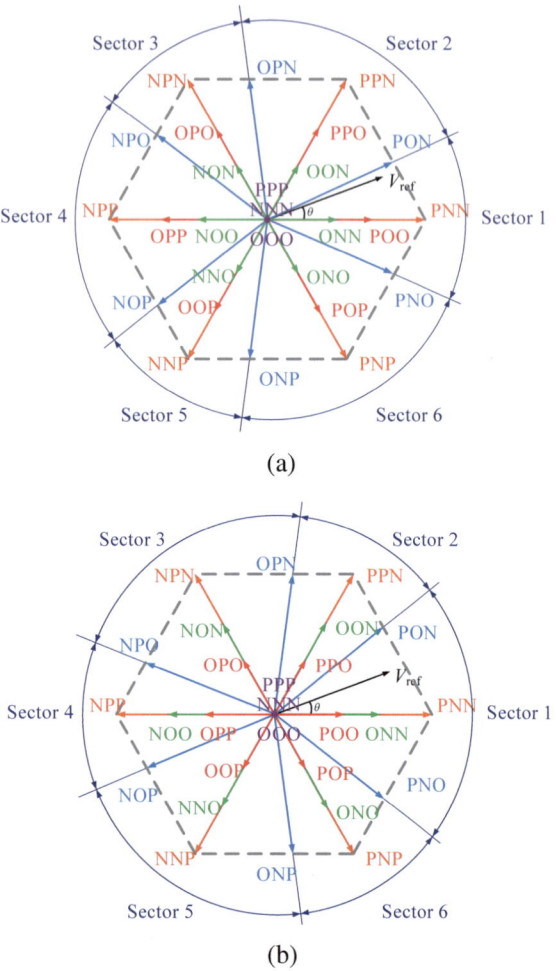

Fig. 4.11 Revised space vector diagrams of the three-level inverter with unbalanced NPV conditions **a** positive unbalancing coefficient, **b** negative unbalancing coefficient

Another issue associated with the three-level inverter is the common-mode voltage (CMV). The CMV is generated by the switching of power devices in the inverter. The high magnitude, high frequency, and slew rate of CMV cause several unwanted drawbacks. In the motor drive system, the CMV causes excessive bearing current and premature failure of bearing. In PV generation systems, the CMV gives rise to leakage current and increases the common-mode electromagnetic interference. The CMV of the inverter is calculated as follows [12]:

$$u_{cm} = \frac{u_{AO} + u_{BO} + u_{CO}}{3} \quad (4.17)$$

where u_{AO}, u_{BO}, and u_{CO} are three-phase output voltages. With unbalanced NPV conditions, the CMVs calculated by the above expression for all the basic voltage

4.2 Separate Maximum Power Point Tracking Technology

Table 4.1 Basic voltage vectors and CMVs with unbalanced NPV conditions

Vectors	State	CMV	State	CMV	State	CMV
Large	[PNN]	$\frac{U_p-2U_n}{3}$	[PPN]	$\frac{2U_p-U_n}{3}$	[NPN]	$\frac{U_p-2U_n}{3}$
	[NPP]	$\frac{2U_p-U_n}{3}$	[NNP]	$\frac{U_p-2U_n}{3}$	[PNP]	$\frac{2U_p-U_n}{3}$
Medium	[PON]	$\frac{U_p-U_n}{3}$	[OPN]	$\frac{U_p-U_n}{3}$	[NPO]	$\frac{U_p-U_n}{3}$
	[NOP]	$\frac{U_p-U_n}{3}$	[ONP]	$\frac{U_p-U_n}{3}$	[PNO]	$\frac{U_p-U_n}{3}$
P-type	[POO]	$\frac{U_p}{3}$	[PPO]	$\frac{2U_p}{3}$	[OPO]	$\frac{U_p}{3}$
	[OPP]	$\frac{2U_p}{3}$	[OOP]	$\frac{U_p}{3}$	[POP]	$\frac{2U_p}{3}$
N-type	[ONN]	$-\frac{2U_n}{3}$	[OON]	$-\frac{U_n}{3}$	[NON]	$-\frac{2U_n}{3}$
	[NOO]	$-\frac{U_n}{3}$	[NNO]	$-\frac{2U_n}{3}$	[ONO]	$-\frac{U_n}{3}$
Zero	[PPP]	U_p	[OOO]	0	[NNN]	U_n

vectors are summarized in Table 4.1. It can be seen that the CMV magnitudes of voltage vectors with shade are relatively high. If the modulation method avoids using these basic vectors and takes the unbalanced NPV conditions into consideration, the reference voltage vector can be correctly synthesized with reduced CMV, and three-phase sinusoidal output currents can be obtained. Based on these analyses, the CMV reduction method for the three-level inverter with unbalanced NPV conditions will be elaborated in Sect. 4.2.1.3.

4.2.1 Based on Continuous Modulation Methods

The continuous modulation methods for the three-level inverter with unbalanced NPVs mainly include hybrid modulation method [13], virtual space vector modulation method [14, 15], and common-mode voltage reduction method [16], etc.

Hybrid Modulation Method

The hybrid modulation method for the three-level inverter with unbalanced NPV conditions was investigated in [13], which is a combination of the space vector modulation (SVM) and carrier-based modulation (CBM). Specifically, a novel sector division method for revised space vector diagrams with unbalanced NPV conditions is first given to identify the type of switching sequence for each phase. Instead of calculating the durations of the adopted basic vectors, the hybrid modulation method adopts the CBM with calculated zero-sequence voltage to produce the pulses directly, which greatly simplifies the implementation process.

For Sector 1, the basic vectors [OOO] ([NNN], [PPP]), [POO], [ONN], [PNN], [PON], and [PNO] are included. According to the basic vectors included in Sector 1, it is found that the switching states of phase A can only be [O] and [P], and the switching states of both phase B and phase C can only be [N] and [O].

As the conventional SVM method, the center-pulse double-edged switching sequences can be classified into two types, as shown in Fig. 4.12. According to the working principle of the three-level inverter, the switching states of the four switches of phase j (j = a, b, c) are also given. The switching states of the first type switching sequence include [O] and [P], while those of the second type switching sequence include [N] and [O]. The voltage-second balance equation for the first and second type switching sequence can be written, where $v_{j,\text{ref}}$ represents the sinusoidal command voltage. v_z is defined as the zero-sequence voltage, and t_j is the duration time as defined in Fig. 4.6.

$$\left(v_{j,\text{ref}} + v_z\right) \cdot T_s = U_p \cdot t_j \tag{4.18}$$

$$\left(v_{j,\text{ref}} + v_z\right) \cdot T_s = -U_n \cdot \left(T_s - t_j\right) \tag{4.19}$$

For the hybrid modulation method, a sector division method for the revised space vector diagrams with unbalanced NPV condition is first given to identify the switching sequence type for each phase. In the sector division method, the revised space vector diagrams are divided into 6 sectors by the medium voltage vectors, and the principles of identifying the sector in which the reference voltage vector is located in is provided as follows:

1. if $-\alpha \leqslant \theta < \alpha$, Sector = 1;
2. if $\alpha \leqslant \theta < (2\pi/3 - \alpha)$, Sector = 2;
3. if $(2\pi/3 - \alpha) \leqslant \theta < (2\pi/3 + \alpha)$, Sector = 3;
4. if $(2\pi/3 + \alpha) \leqslant \theta < \pi$ or $-\pi \leqslant \theta < (-2\pi/3 - \alpha)$, Sector = 4;
5. if $(-2\pi/3 - \alpha) \leqslant \theta < (-2\pi/3 + \alpha)$, Sector = 5;
6. if $(-2\pi/3 + \alpha) \leqslant \theta < (-\alpha)$, Sector = 6.

The variable θ represents the angle of the reference voltage vector. According to the geometrical characteristics of the revised space vector diagram, the angle α is calculated as follows:

$$\alpha = \begin{cases} \frac{2\pi}{3} - \arcsin\left(\frac{1}{\sqrt{1+\frac{k^2}{12}}}\right) & \text{For } k < 0 \\ \arcsin\left(\frac{1}{\sqrt{1+\frac{k^2}{12}}}\right) - \frac{\pi}{3} & \text{For } k > 0 \end{cases} \tag{4.20}$$

The position of the medium vector changes with the unbalancing coefficient, however, it has been proven that the medium vector stays on the connecting line of the two adjacent large vectors for any dc unbalancing coefficient. Thus, the switching sequence

4.2 Separate Maximum Power Point Tracking Technology

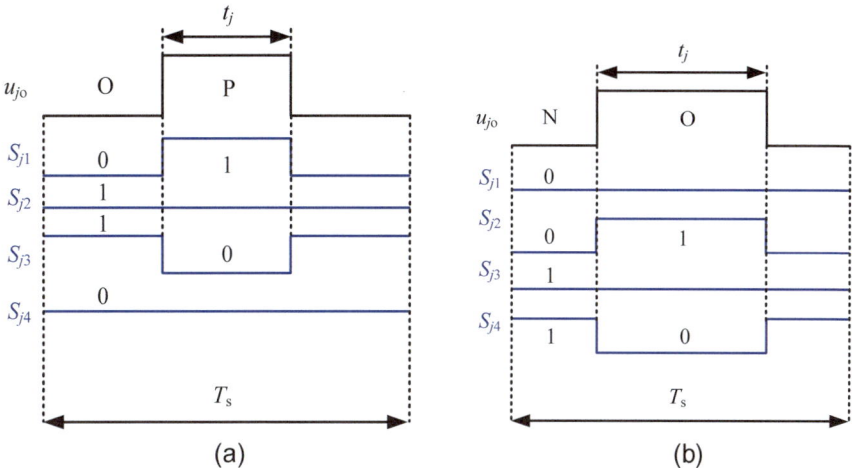

Fig. 4.12 Switching sequence type **a** type-I, **b** type-II

determining method is applicable for all the dc unbalancing coefficients ranging from -1 to 1 owing to the same geometrical feature. This sector division method is easy to calculate and there is no need to calculate the subsectors. Once the sector is calculated, the types of switching sequences can be obtained, which are summarized in Table 4.2.

When Sector $= 1$, the corresponding voltage-second balance equations are given as follows, and the intervals defined as t_a, t_b, and t_c are calculated subsequently.

$$\begin{cases} (v_{a,\text{ref}} + v_z) \cdot T_s = U_p \cdot t_a \\ (v_{b,\text{ref}} + v_z) \cdot T_s = -U_n \cdot (T_s - t_b) \\ (v_{c,\text{ref}} + v_z) \cdot T_s = -U_n \cdot (T_s - t_c) \end{cases} \quad (4.21)$$

$$\begin{cases} t_a = \frac{(v_{a,\text{ref}} + v_z) \cdot T_s}{U_p} \\ t_b = \frac{(v_{b,\text{ref}} + v_z) \cdot T_s}{U_n} + T_s \\ t_c = \frac{(v_{c,\text{ref}} + v_z) \cdot T_s}{U_n} + T_s \end{cases} \quad (4.22)$$

Table 4.2 The types of switching sequences in different sectors

Sector	1	2	3	4	5	6
Phase a	I	I	II	II	II	I
Phase b	II	I	I	I	II	II
Phase c	II	II	II	I	I	I

It can be seen that the dwell time can be obtained by calculating the zero-sequence voltage. The calculated dwell time must satisfy $0 \leq t_j \leq T_s$, because the voltage-second balance should not be broken. Doing so, the satisfactory output waveforms can be guaranteed.

Generally, the intervals t_j ($j =$ a, b, c) for various sector can be calculated, where κ_j is equal to 1 for the Type I sequence and -1 for another case.

$$t_j = \left(\frac{1 - \kappa_j}{2}\right) \cdot T_s + \frac{2(v_{a,\text{ref}} + v_z) \cdot T_s}{(1 + \kappa_j \cdot k) \cdot U_{dc}} \tag{4.23}$$

The allowable range of the zero-sequence voltage for each phase in Sector 1 are obtained as follows:

$$\begin{cases} -v_{a,\text{ref}} \leq v_z \leq -v_{a,\text{ref}} + \frac{1+k}{2} U_{dc} \\ -v_{b,\text{ref}} - \frac{1-k}{2} U_{dc} \leq v_z \leq -v_{b,\text{ref}} \\ -v_{c,\text{ref}} - \frac{1-k}{2} U_{dc} \leq v_z \leq -v_{c,\text{ref}} \end{cases} \tag{4.24}$$

The minimal and maximal values of the zero-sequence voltage for three-phase system can be obtained as

$$v_{z,\text{min}} = \max\left\{-v_{a,\text{ref}}, -v_{b,\text{ref}} - \frac{1-k}{2} U_{dc}, -v_{c,\text{ref}} - \frac{1-k}{2} U_{dc}\right\} \tag{4.25}$$

$$v_{z,\text{max}} = \min\left\{-v_{a,\text{ref}} + \frac{1+k}{2} U_{dc}, -v_{b,\text{ref}}, -v_{c,\text{ref}}\right\} \tag{4.26}$$

The minimal and maximal values of the zero-sequence voltage in other sectors can be obtained by using the similar approach. If other targets are not considered, the zero-sequence voltage for the hybrid modulation method can be selected as the average value of $v_{z,\text{min}}$ and $v_{z,\text{max}}$ for simplicity.

After calculating t_j, the drive pulses for the four switches to produce the corresponding switching sequences can be designed. The drive pulses for S_{j1} and S_{j3} are complementary, so as S_{j2} and S_{j4}. To produce the drive pulses, a modulation wave generator is needed to calculate the modulation signals (m_{j1} and m_{j2}). The modulation signal m_{j1} is used to produce the drive pulses for S_{j1} and S_{j3}, while, m_{j2} is used to produce the drive pulses for S_{j2} and S_{j4}. According to the relationship between the switching states and the switch statuses, the modulation signals m_{j1} and m_{j2} can be calculated as follows:

$$m_{j1} = \begin{cases} 1 - \frac{v_{a,\text{ref}} + v_z}{\frac{1+k}{2} \cdot U_{dc}}, & \text{if } \kappa_j = 1 \\ 1, & \text{if } \kappa_j = -1 \end{cases} \tag{4.27}$$

$$m_{j2} = \begin{cases} 0, & \text{if } \kappa_j = 1 \\ -\frac{v_{a,\text{ref}} + v_z}{\frac{1-k}{2} \cdot U_{dc}}, & \text{if } \kappa_j = -1 \end{cases} \tag{4.28}$$

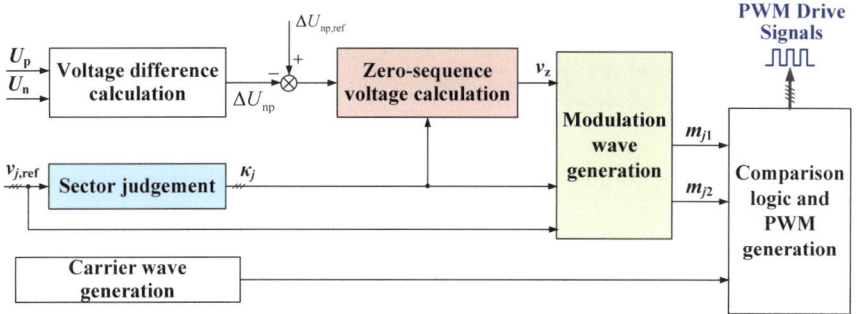

Fig. 4.13 Control diagram of the hybrid modulation method

The drive pulses are directly obtained according to the size-relationship between the modulation waves and a single carrier, as shown in Fig. 4.6. If the modulation wave is larger than the carrier, the corresponding switch turns ON; otherwise, it turns OFF.

The control diagram of the hybrid modulation strategy is shown in Fig. 4.13. It can be viewed as a combination of the SVPWM and CBM methods, i.e., it identifies the switching sequence through a simplified sector division method as SVPWM, while the pulses for power semiconductor switches are directly obtained according to the CBM method instead of calculating the duty ratios of the utilized basic voltage vectors.

It has been well recognized that the performance of modulation strategy is affected by the zero-sequence voltage, and the zero-sequence voltage calculation principle to realize the asymmetric control of the dc links is studied as follows. The average value of the current flowing through the neutral-point ($i_{\mathrm{np,av}}$) is determined by the intervals of state [O] and the phase currents of the three legs. If the switching sequence type is known, $i_{\mathrm{np,av}}$ can be calculated as

$$i_{\mathrm{np,av}} = \frac{\sigma}{U_{\mathrm{dc}}} \cdot v_z \tag{4.29}$$

where σ and ζ are respectively given as

$$\sigma = -\sum_{j=a,b,c} \frac{i_j}{k + \kappa_j} \tag{4.30}$$

$$\zeta = -\frac{1}{U_{\mathrm{dc}}} \sum_{j=a,b,c} \frac{v_{j,\mathrm{ref}} \cdot i_j}{k + \kappa_j} \tag{4.31}$$

Hence, the asymmetric control of the capacitor voltages can be realized by adjusting $i_{\mathrm{np,av}}$ to charge or discharge the neutral-point according to the difference between the voltage difference of dc-side capacitors and their reference. It is a simple way that the reference of $i_{\mathrm{np,av}}$ is obtained by a closed-loop control, and a PI regulator

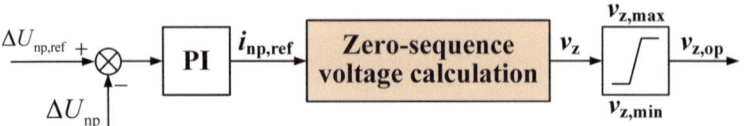

Fig. 4.14 Schematic of the asymmetric control of the dc links with the hybrid modulation method

can be employed, as shown in Fig. 4.14. The control parameters of the PI regulator can be designed by referring to the methodology of typical control system. Then, the optimized zero-sequence voltage can be calculated.

$$v_z = \frac{U_{dc}}{\sigma}(i_{np,av} - \zeta) \tag{4.32}$$

Virtual Space Vector Modulation Method

For the three-level inverter, the virtual space vector modulation (VSVM) method has been proposed for eliminating the low-frequency oscillations of NPVs at full modulation indices and power factors [17]. This part will investigate the VSVM method for the three-level inverter with unbalanced NPV conditions, which can adjust the unbalancing degree with a high degree of precision, high modulation index and low power factor.

First, the VSVM method for the three-level inverter with balanced NPV conditions is briefly reviewed. It is known that the average current must be zero to avoid a variation of the NPV. The appropriate combination of switching states must be selected for the small vectors in order to achieve this goal, but this is not possible when the modulation index is high and the power factor is low. This is due to the fact that in these conditions, the current introduced by the medium vectors cannot be fully compensated by the current introduced by the small vectors.

To achieve full control of the NPV, a set of new virtual vectors are defined as a linear combination of the vectors corresponding to certain switching states. The new virtual vectors, shown in Fig. 4.15 for Sector 1 of the space vector diagram, have an associated switching average equal to zero.

When the unbalanced NPV conditions occur, the definition of virtual vectors is explained as follows. Based on the revised space vector diagrams, the new virtual vectors that neither charge nor discharge the NPV are defined. It should be pointed out that the influence of the given space vector on the NPV with unbalanced NPV conditions is the same as with balanced NPV conditions, owing to the fact that the load connection state of the given vector is the same for any unbalancing degree. For example, although the medium vector [PON] shifts in the space vector diagrams with unbalanced NPV conditions, it connects the currents of phase B to the neutral-point for both the balanced and unbalanced NPV conditions. That is to say, we can define

4.2 Separate Maximum Power Point Tracking Technology

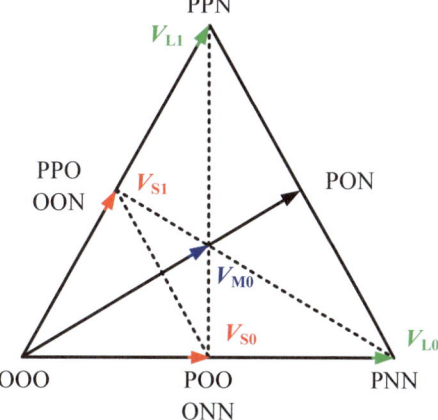

Fig. 4.15 Virtual space vectors for the Sector 1 of the space vector diagram

the new virtual vectors according to the above-mentioned principles and the results in Sector 1 are given as follows:

$$\begin{cases} V'_Z = V_{OOO} \\ V'_{S0} = \frac{1}{2}(V_{ONN} + V_{POO}) \\ V'_{S1} = \frac{1}{2}(V_{PPO} + V_{OON}) \\ V'_{M0} = \frac{1}{3}(V_{PON} + V_{PPO} + V_{ONN}) \\ V'_{L0} = V_{PNN} \\ V'_{L1} = V_{PPN} \end{cases} \quad (4.33)$$

By substituting the expressions of basic vectors into the definitions of virtual vectors, the positions of the defined virtual vectors with unbalanced NPV conditions can be acquired as follows:

$$\begin{cases} V'_Z = 0 \\ V'_{S0} = \frac{1}{3}U_{dc} \\ V'_{S1} = \frac{1}{6}U_{dc} + j\frac{\sqrt{3}}{6}U_{dc} \\ V'_{M0} = \frac{1}{3}U_{dc} + j\frac{\sqrt{3}}{9}U_{dc} \\ V'_{L0} = \frac{2}{3}U_{dc} \\ V'_{L1} = \frac{1}{3}U_{dc} + j\frac{\sqrt{3}}{3}U_{dc} \end{cases} \quad (4.34)$$

After investigating the expressions of virtual vectors with balanced NPV conditions and those with unbalanced NPV conditions, it is interesting to find that the positions of the defined virtual vectors are identical for both cases and independent of the unbalancing degree. Adopting the same method, the interesting finding is also accurate for other sectors, and thus, it can be concluded that the virtual space vector diagrams for the three-level inverter with unbalanced NPV conditions are exactly the same as

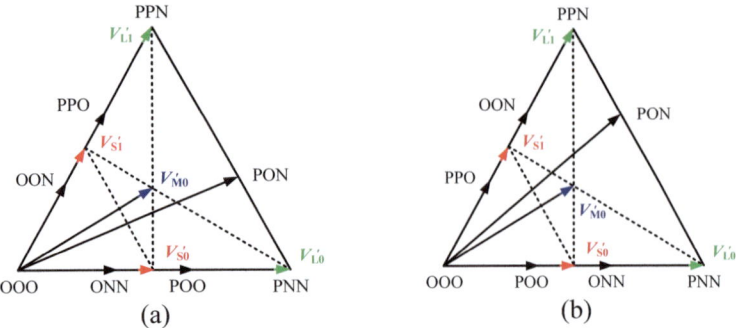

Fig. 4.16 Virtual space vector diagram for the three-level inverter with unbalanced NPV conditions **a** $k > 0$, **b** $k < 0$

the virtual space vector diagram of the conventional virtual space vector modulation strategy, while the real vector diagrams of these two cases are very different. Figure 4.16 depicts the virtual space vector diagrams for the three-level inverter with unbalanced NPV conditions.

Based on the above-mentioned principle, the processes of virtual vectors selection, switching states sequence arrangement, and duty ratios calculation for the VSVPWM with unbalanced NPV conditions have no difference between the methods for balanced case, and the fast-processing implementation strategy can be also applied without any modifications [18–21]. To be specific, the original modulation signals are modified to obtain space vector modulation patterns in order to achieve the maximum range for linear operation mode as follows:

$$\begin{cases} v'_{a,\text{ref}} = v_{a,\text{ref}} + v_z \\ v'_{b,\text{ref}} = v_{b,\text{ref}} + v_z \\ v'_{c,\text{ref}} = v_{c,\text{ref}} + v_z \end{cases} \quad (4.35)$$

where $v_z = -\dfrac{\max\{v_{a,\text{ref}}, v_{b,\text{ref}}, v_{c,\text{ref}}\} + \min\{v_{a,\text{ref}}, v_{b,\text{ref}}, v_{c,\text{ref}}\}}{2}$.

The two modified modulation signals for each phase will be obtained from these signals, which must accomplish

$$\begin{cases} v'_{a,\text{ref}} = v_{ap} + v_{an} \\ v'_{b,\text{ref}} = v_{bp} + v_{bn} \\ v'_{c,\text{ref}} = v_{cp} + v_{cn} \end{cases} \quad (4.36)$$

where $v_{ip} \geq 0$ and $v_{in} \leq 0$, with $i = $ a, b, c. The signals with subscript p will only cross the upper carrier, and the signals with subscript n will only cross the lower one carrier.

To maintain the locally averaged voltages on the dc-link capacitors constant, the new modulation signals for three phases can be constructed as follows:

4.2 Separate Maximum Power Point Tracking Technology

$$\begin{cases} v_{ap} = \frac{v_{a,ref} - \min\{v_{a,ref}, v_{b,ref}, v_{c,ref}\}}{2} \\ v_{an} = \frac{v_{a,ref} - \max\{v_{a,ref}, v_{b,ref}, v_{c,ref}\}}{2} \end{cases} \quad (4.37)$$

$$\begin{cases} v_{bp} = \frac{v_{b,ref} - \min\{v_{a,ref}, v_{b,ref}, v_{c,ref}\}}{2} \\ v_{bn} = \frac{v_{b,ref} - \max\{v_{a,ref}, v_{b,ref}, v_{c,ref}\}}{2} \end{cases} \quad (4.38)$$

$$\begin{cases} v_{cp} = \frac{v_{c,ref} - \min\{v_{a,ref}, v_{b,ref}, v_{c,ref}\}}{2} \\ v_{cn} = \frac{v_{c,ref} - \max\{v_{a,ref}, v_{b,ref}, v_{c,ref}\}}{2} \end{cases} \quad (4.39)$$

The algorithm can be easily implemented using the scheme that is shown in Fig. 4.17. Figure 4.18 shows the waveforms for this example.

In addition, the switching sequence with unbalanced NPV conditions can be arranged by referencing the one designed for the balanced case, so as to attain the minimum switching events. The duty ratios d_{aP} can be calculated for any dc unbalancing coefficients with the reference voltages. The expression for d_{aN} is the same but phase shifted by 180°. The duty ratio expressions for phases B and C are the same as for phase A, but phase shifted by 120° and 240°, respectively.

$$d_{aP} = \begin{cases} \frac{\sqrt{3}}{2} m \cos(\theta - \frac{\pi}{6}), & 0 \le \theta < \frac{2\pi}{3} \\ 0, & \frac{2\pi}{3} \le \theta < \frac{4\pi}{3} \\ \frac{\sqrt{3}}{2} m \cos(\theta + \frac{\pi}{6}), & \frac{4\pi}{3} \le \theta < 2\pi \end{cases} \quad (4.40)$$

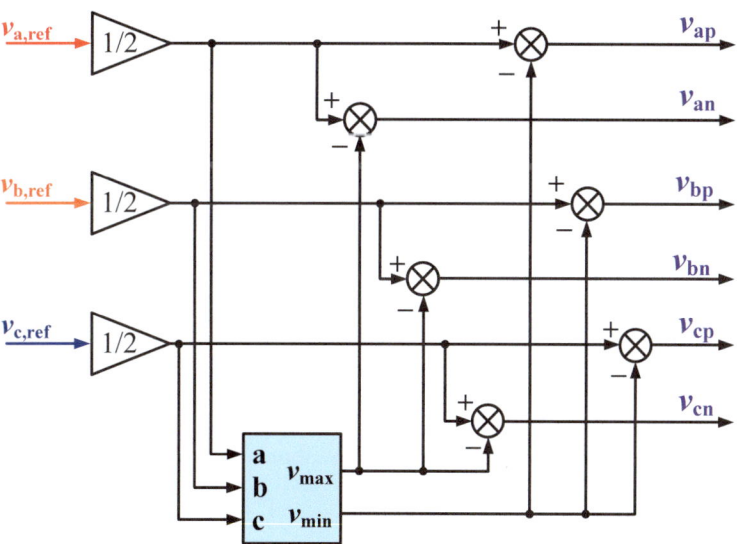

Fig. 4.17 Scheme for the generation of v_{ip} and v_{in} ($i =$ a, b, c)

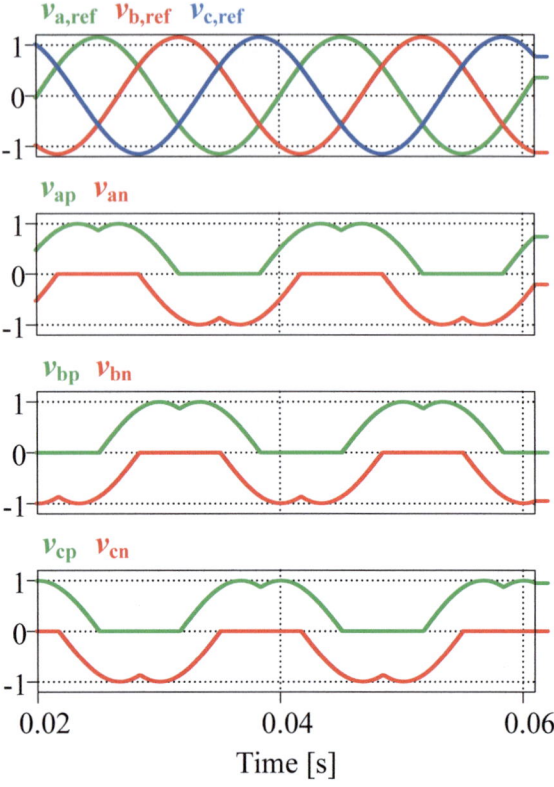

Fig. 4.18 Example of three-phase sinusoidal modulation signals and modified signals for VSVM strategy

As introduced before, the asymmetric control of the upper and lower dc-link voltages is helpful to enhance the overall efficiency in centralized PV generation system. The asymmetric control strategy with the VSVM method will be introduced as follows. Similar to the hybrid modulation method, the dc unbalancing coefficient can be controlled by adjusting the neutral-point current. However, the neutral-point current keeps at zero for the VSVM method, and modifications should be conducted to the VSVM method in order to realize this control objective.

As a typical example, the switching sequence in Sector 1 shown in Fig. 4.19 is analyzed to explain how to adjust the NP current. The average value of NP current in a single carrier period for the switching sequence in Fig. 4.19 can be expressed as follows, and it is theoretically 0 without considering the dead-time and nonlinearity of the power semiconductor switches, because three duty ratios (d_{aO}, d_{bO}, and d_{cO}) are the same values.

$$i_{np,av} = \sum_{j=a,b,c} i_j d_{jO} \qquad (4.41)$$

4.2 Separate Maximum Power Point Tracking Technology

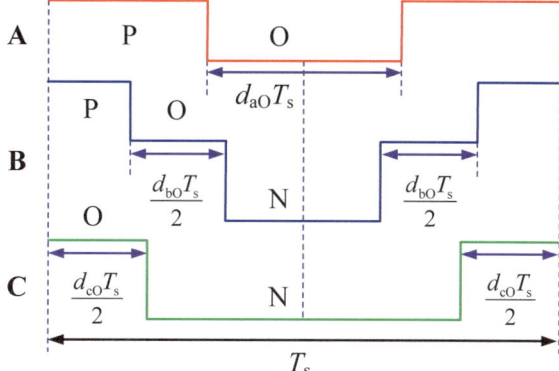

Fig. 4.19 The switching sequence in Sector 1 with the negative dc unbalancing coefficient

To ensure the simple and intuitive property of the algorithm, the asymmetric control of the dc-link voltages is conducted without varying the injected zero-sequence voltage. Instead, only the duty ratios of the phase verifying $d_{jP} > 0$ and $d_{jN} > 0$ are perturbed. As shown in Fig. 4.19, the voltage-second balancing principle will be broken by adjusting d_{aO} or d_{cO}, if the injected zero-sequence voltage keeps unchanged, and only d_{bO} can be perturbed without breaking the voltage-second balance owing to the volt-second variation caused by perturbing d_{bO} can be compensated by appropriately adjusting d_{bP} and d_{bN} as

$$\begin{cases} d'_{bO} = d_{bO} + d_{comp} \\ d'_{bP} = d_{bP} - \frac{1-k}{2} d_{comp} \\ d'_{bN} = d_{bN} - \frac{1+k}{2} d_{comp} \end{cases} \quad (4.42)$$

Then, the neutral-point current is varied as

$$i_{np,av} = i_b \cdot d_{comp} \quad (4.43)$$

The variation of d_{bO} can be calculated, and its maximal and minimal values should be limited, so as to ensure the values of duty cycles (d_{bO}, d_{bP}, and d_{bN}) are all between zero and one.

$$d_{comp} = \frac{i_{np,ref}}{i_b} \quad (4.44)$$

$$\begin{cases} d_{comp,max} = \min\left(1 - d_{bO}, \frac{2d_{bP}}{1-k}, \frac{2d_{bN}}{1+k}\right) \\ d_{comp,min} = \max\left(-d_{bO}, \frac{2(d_{bP}-1)}{1-k}, \frac{2(d_{bN}-1)}{1+k}\right) \end{cases} \quad (4.45)$$

For other sectors, the phase being perturbed is the one verifying $d_{jP} > 0$ and $d_{jN} > 0$. In practical applications, the reference of the NP current can be obtained by a proportional-integral controller of which the input is the error between the given and real unbalancing degree.

Common-Mode Voltage Reduction Method

From the perspective of space vector modulation, the hybrid modulation and the virtual SVM strategies utilize all the basic vectors to synthesize the reference vector and produce the AC output voltage, which induce a high magnitude of the CMV. The high magnitude, high frequency, and slew rate (dv/dt) of the CMV cause several unwanted drawbacks. In a PV generation system, the CMV gives rise to leakage current and increases the common-mode electromagnetic interference. In the motor drive system, the CMV causes excessive bearing current and premature failure of bearing. Therefore, it is desired to reduce the CMV.

To this end, the CMV reduction method for the three-level inverter with separate MPPTs will be elaborated in this part. At first, the space vector diagrams that include the basic vectors with low CMV magnitudes are briefly introduced, which helps to understand the principle of the CMV reduction method.

Figure 4.20 provides the space vector diagrams of different dc unbalancing coefficients for reducing the CMV, in which the basic vectors with high CMV magnitudes are removed. The whole space vector diagram is divided into 12 sectors by the large and medium vectors, the small and zero vectors are not redundant any more. In concrete terms, for the zero vector, only the vector state [OOO] is included. Moreover, six small vectors with low CMV magnitudes are utilized. The outer hexagon formed by large vectors remains unchanged, and the maximum magnitude of the reference vector is equal to the radius of the inscribed circle of the outer hexagon. Therefore, the dc-link voltage utilization is not affected. It is clear that the space vector diagrams become complicated and asymmetric due to the effects of unbalanced NPV conditions. Figure 4.20c displays the space vector diagram with a balanced NPV condition. It is obviously observed that the balanced NPV condition can be regarded as a typical case of the unbalanced NPV conditions.

Based on the above-mentioned analyses, the detailed working principle of the CMV reduction method is explained as follows. Basically, the carrier-based approach with the configuration of the carrier is developed, so as to cope with the difficulties resulted from the unbalanced NPV conditions. In three-level inverters using carrier-based modulation method, three sinusoidal reference voltages are used to determine three-phase output voltages of the inverter. These reference voltages for each phase are expressed as

$$\begin{cases} v_{a,\text{ref}} = m \cdot \frac{U_{dc}}{2} \cdot \sin(\omega_1 t) \\ v_{b,\text{ref}} = m \cdot \frac{U_{dc}}{2} \cdot \sin\left(\omega_1 t - \frac{2\pi}{3}\right) \\ v_{c,\text{ref}} = m \cdot \frac{U_{dc}}{2} \cdot \sin\left(\omega_1 t + \frac{2\pi}{3}\right) \end{cases} \quad (4.46)$$

4.2 Separate Maximum Power Point Tracking Technology

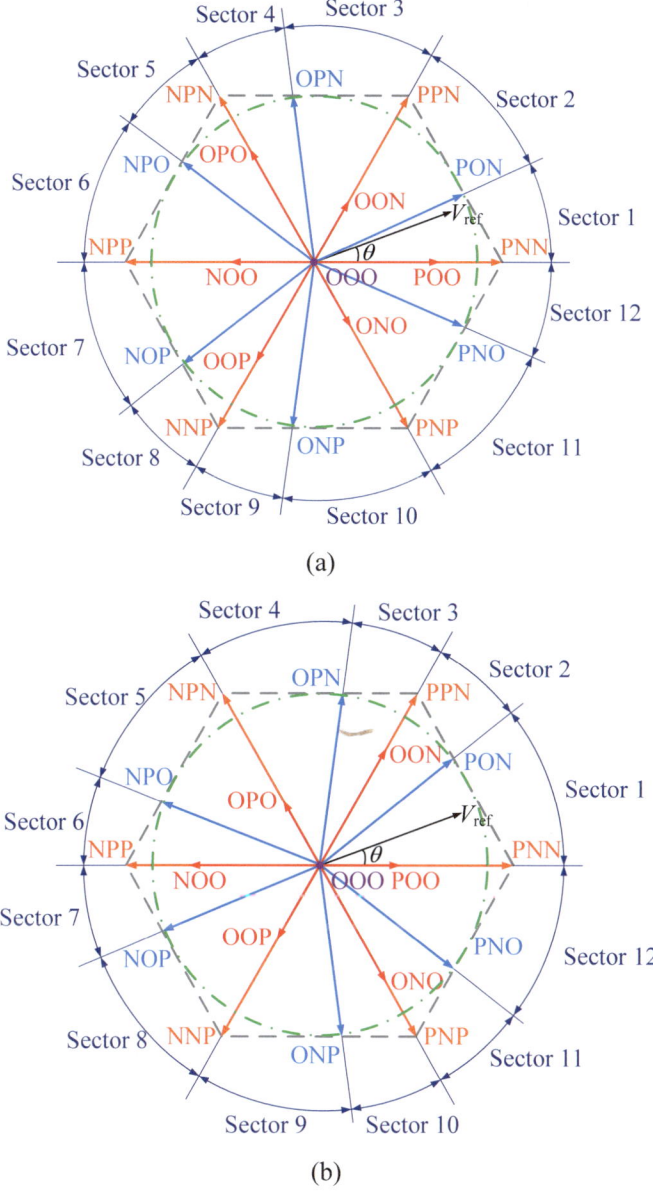

Fig. 4.20 Space vector diagrams of the three-level inverter with separate MPPTs for reducing CMV **a** $k > 0$, **b** $k < 0$, **c** $k = 0$

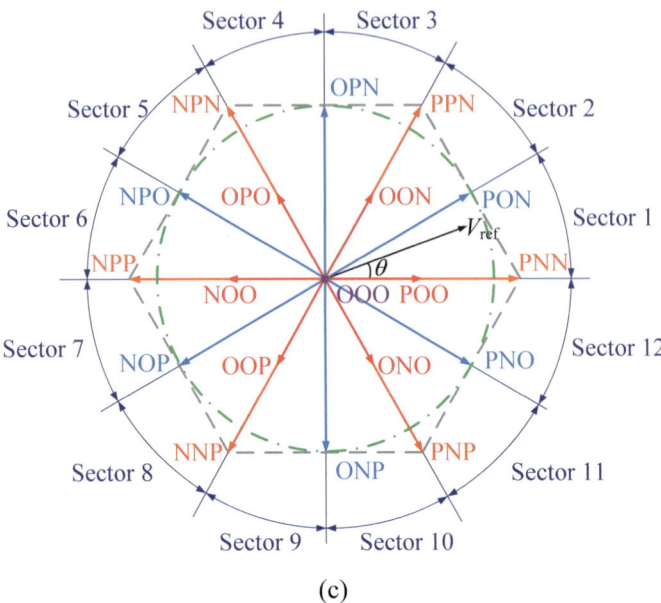

Fig. 4.20 (continued)

where ω_1 is the fundamental angular frequency, and m is the modulation index, which can be defined as follows:

$$m = \frac{2 \cdot V_m}{U_{dc}} \quad (4.47)$$

where V_m is the peak value of the reference voltage, and U_{dc} is the total dc-link voltage, that is $U_{dc} = U_p + U_n$.

The generalized form of the voltage-second balance equations in one carrier period can be written by considering the injected zero-sequence voltage component.

$$\begin{cases} (v_{a,\text{ref}} + v_z) \cdot T_s = \int_0^{T_s} v_{ao}(t)dt \\ (v_{b,\text{ref}} + v_z) \cdot T_s = \int_0^{T_s} v_{bo}(t)dt \\ (v_{c,\text{ref}} + v_z) \cdot T_s = \int_0^{T_s} v_{co}(t)dt \end{cases} \quad (4.48)$$

where $v_{a,\text{ref}}$, $v_{b,\text{ref}}$, and $v_{c,\text{ref}}$ represent three-phase reference voltages. v_z is the zero-sequence voltage.

To further simplify the calculation process, the normalized reference voltages are introduced as follows:

4.2 Separate Maximum Power Point Tracking Technology

$$\begin{cases} v_a = m \cdot \sin(\omega_1 t) \\ v_b = m \cdot \sin(\omega_1 t - \frac{2\pi}{3}) \\ v_c = m \cdot \sin(\omega_1 t + \frac{2\pi}{3}) \end{cases} \quad (4.49)$$

The maximum, minimum, and medium values of three-phase normalized modulation waveforms are calculated by using the simple sorting algorithm, as expressed in the following equation:

$$\begin{cases} v_{\max} = \max\{v_a, v_b, v_c\} \\ v_{\min} = \min\{v_a, v_b, v_c\} \\ v_{\mathrm{mid}} = \sum\limits_{j=a,b,c} v_j - v_{\max} - v_{\min} \end{cases} \quad (4.50)$$

The basic considerations of the CMV reduction method are described as follows. First, the carrier waveforms should be determined. As introduced before, in order to reduce the CMV, only the basic vectors with low CMV magnitudes can be employed to synthesize the reference vector. It is not difficult to conclude that the alternative phase opposition disposition (APOD) scheme can satisfy this requirement. Furthermore, to avoid the switching events between switching cycles, each switching sequence is started and ended by the zero vector [OOO]. Therefore, the carriers that should be used are shown in Fig. 4.21, in which the upper and lower carriers are the concave and convex carrier, respectively.

For this configuration of two carriers, the state transitions in one switching cycle include two types. One type is [O] → [P] → [O] (denoted as Type-1 state transition), and the other type is [O] → [N] → [O] (denoted by Type-2 state transition), as shown in Fig. 4.21a, b, respectively. To facilitate the following analyses, the state transition function of Phase j (j = a, b, c) is defined as follows:

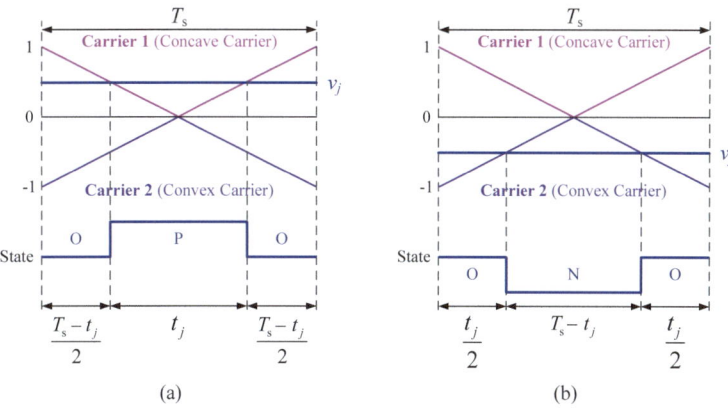

Fig. 4.21 Switching sequence for Phase j (j = a, b, c) **a** [O] → [P] → [O], **b** [O] → [N] → [O]

$$n_j = \begin{cases} 1 & \text{For Type-1 state transition} \\ -1 & \text{For Type-2 state transition} \end{cases} \quad (4.51)$$

In addition, for the Type-1 state transition, the dwell time (t_j) of Phase j (j = a, b, c) is defined as the duration of state [P]. For the Type-2 state transition, the dwell time (t_j) of Phase j is defined as the duration of state [O].

Sector 1 in the space vector diagram is considered as a representative example to illustrate the CMV reduction method. In this case, the reference voltage of Phase A is positive, whereas the reference voltages of Phases B and C are negative. Therefore, the state transition of Phase A is [O] → [P] → [O], and the state transitions of Phases B and C are [O] → [N] → [O]. The values of the state transition functions are $n_a = 1$, $n_b = -1$, and $n_c = -1$. The voltage-second balance equations in Sector 1 can be written as

$$\begin{cases} (v_{a,\text{ref}} + v_z) \cdot T_s = U_p \cdot t_a \\ (v_{b,\text{ref}} + v_z) \cdot T_s = -U_n \cdot (T_s - t_b) \\ (v_{c,\text{ref}} + v_z) \cdot T_s = -U_n \cdot (T_s - t_c) \end{cases} \quad (4.52)$$

The voltage-second balance equations in Sector 1 can be further written as follows:

$$\begin{cases} (v_a + v_z) \cdot \frac{U_{dc}}{2} \cdot T_s = \frac{1+k}{2} \cdot U_{dc} \cdot t_a \\ (v_b + v_z) \cdot \frac{U_{dc}}{2} \cdot T_s = -\frac{1-k}{2} \cdot U_{dc} \cdot (T_s - t_b) \\ (v_c + v_z) \cdot \frac{U_{dc}}{2} \cdot T_s = -\frac{1-k}{2} \cdot U_{dc} \cdot (T_s - t_c) \end{cases} \quad (4.53)$$

where v_z is the normalized zero-sequence component. It is noted that this variable can be used to control the NPV with reduced CMV.

Three-phase dwell times in Sector 1 is obtained as follows:

$$\begin{cases} t_a = (v_a + v_z) \cdot T_s \cdot \frac{1}{1+k} \\ t_b = T_s + (v_b + v_z) \cdot T_s \cdot \frac{1}{1-k} \\ t_c = T_s + (v_c + v_z) \cdot T_s \cdot \frac{1}{1-k} \end{cases} \quad (4.54)$$

When the reference voltage vector is located in other sectors, three-phase dwell time can be calculated by using similar method. Tables 4.3 and 4.4 summarize state transition functions and three-phase dwell time in all sectors. It can be seen that three-phase dwell time is calculated by three-phase modulation waveforms and the unbalancing coefficient.

Next, the injected zero-sequence component needs to be determined. It is obvious that the maximum value of three-phase modulation waveforms (v_{max}) is always positive, and the minimum value of three-phase modulation waveforms (v_{min}) is always negative. The medium value of three-phase modulation waveforms (v_{mid}) can be positive or negative, and the sign of the medium modulation waveform cannot be changed after injecting the zero-sequence component. To guarantee the linear modulation range and avoid over-modulation, the modulation waveforms cannot exceed

4.2 Separate Maximum Power Point Tracking Technology

Table 4.3 State transition functions in all sectors

Sector	v_a	v_b	v_c	n_a	n_b	n_c
12, 1	+	−	−	1	−1	−1
2, 3	+	+	−	1	1	−1
4, 5	−	+	−	−1	1	−1
6, 7	−	+	+	−1	1	1
8, 9	−	−	+	−1	−1	1
10, 11	+	−	+	1	−1	1

Table 4.4 Three-phase dwell time in all sectors

Sector	t_a	t_b	t_c
12, 1	$(v_a+v_z)\cdot T_s \cdot \frac{1}{1+k}$	$T_s + (v_b+v_z)\cdot T_s \cdot \frac{1}{1-k}$	$T_s + (v_c+v_z)\cdot T_s \cdot \frac{1}{1-k}$
2, 3	$(v_a+v_z)\cdot T_s \cdot \frac{1}{1+k}$	$(v_b+v_z)\cdot T_s \cdot \frac{1}{1+k}$	$T_s + (v_c+v_z)\cdot T_s \cdot \frac{1}{1-k}$
4, 5	$T_s + (v_a+v_z)\cdot T_s \cdot \frac{1}{1-k}$	$(v_b+v_z)\cdot T_s \cdot \frac{1}{1+k}$	$T_s + (v_c+v_z)\cdot T_s \cdot \frac{1}{1-k}$
6, 7	$T_s + (v_a+v_z)\cdot T_s \cdot \frac{1}{1-k}$	$(v_b+v_z)\cdot T_s \cdot \frac{1}{1+k}$	$(v_c+v_z)\cdot T_s \cdot \frac{1}{1+k}$
8, 9	$T_s + (v_a+v_z)\cdot T_s \cdot \frac{1}{1-k}$	$T_s + (v_b+v_z)\cdot T_s \cdot \frac{1}{1-k}$	$(v_c+v_z)\cdot T_s \cdot \frac{1}{1+k}$
10, 11	$(v_a+v_z)\cdot T_s \cdot \frac{1}{1+k}$	$T_s + (v_b+v_z)\cdot T_s \cdot \frac{1}{1-k}$	$(v_c+v_z)\cdot T_s \cdot \frac{1}{1+k}$

the bounds of the carriers. In addition, the zero-sequence component (v_z) obtained by the NPV controller can be positive or negative. Depending on the combination of the signs of v_{mid} and v_z, four cases can be determined to calculate the boundaries of the zero-sequence component as follows.

1. Case 1 ($v_z > 0$ and $v_{\text{mid}} > 0$): The medium modulation waveform is positive, and the three-phase modulation waves move up. The inequalities should be satisfied as

$$\begin{cases} \frac{v_{\max}+v_z}{1+k} < 1 \\ \frac{v_{\min}+v_z}{1-k} < 0 \\ \frac{v_{\text{mid}}+v_z}{1+k} + \frac{v_{\min}+v_z}{1-k} < 0 \end{cases} \quad (4.55)$$

The bounds of the zero-sequence component in Case 1 are obtained as follows:

$$0 < v_z < \min\left\{1+k-v_{\max},\ -v_{\min},\ -\frac{(1-k)\cdot v_{\text{mid}} + (1+k)\cdot v_{\min}}{2}\right\} \quad (4.56)$$

2. Case 2 ($v_z > 0$ and $v_{\text{mid}} < 0$): The medium modulation waveform is negative, and three-phase modulation waveforms move up. The inequalities should be satisfied as

$$\begin{cases} \frac{v_{\max}+v_z}{1+k} < 1 \\ \frac{v_{\mathrm{mid}}+v_z}{1-k} < 0 \end{cases} \tag{4.57}$$

The bounds of the zero-sequence component in Case 2 are shown as follows:

$$0 < v_z < \min\{1+k-v_{\max}, -v_{\mathrm{mid}}\} \tag{4.58}$$

3. Case 3 ($v_z < 0$ and $v_{\mathrm{mid}} > 0$): In this case, the inequalities should be satisfied as

$$\begin{cases} \frac{v_{\mathrm{mid}}+v_z}{1+k} > 0 \\ \frac{v_{\min}+v_z}{1-k} > -1 \end{cases} \tag{4.59}$$

Then, we can obtain the bounds of the zero-sequence component in Case 3, as expressed in the following inequality:

$$\max\{-v_{\mathrm{mid}}, k-1-v_{\min}\} < v_z < 0 \tag{4.60}$$

4. Case 4 ($v_z < 0$ and $v_{\mathrm{mid}} < 0$): In this case, the inequalities expression should be satisfied

$$\begin{cases} \frac{v_{\max}+v_z}{1+k} > 0 \\ \frac{v_{\min}+v_z}{1-k} > -1 \\ \frac{v_{\max}+v_z}{1+k} + \frac{v_{\mathrm{mid}}+v_z}{1-k} > 0 \end{cases} \tag{4.61}$$

Then, we can obtain the bounds of the zero-sequence component in Case 4, as expressed in the following inequality:

$$\max\left\{-v_{\max}, k-1-v_{\min}, -\frac{(1-k)v_{\max}+(1+k)v_{\mathrm{mid}}}{2}\right\} < v_z < 0 \tag{4.62}$$

In summary, the zero-sequence component should be restricted by considering the above inequalities in order not to generate basic vectors with high CMV magnitudes. After conducting the saturation process, the final zero-sequence component (denoted as v'_z) can be obtained, and the final modulation waveform of Phase j ($j=$ a, b, c) is calculated as

$$m_j = \begin{cases} (v_j+v'_z) \cdot \frac{1}{1+k} & \text{For } n_j = 1 \\ (v_j+v'_z) \cdot \frac{1}{1-k} & \text{For } n_j = -1 \end{cases} \tag{4.63}$$

where v_j, v_z, k, and n_j are the original modulation waveform of Phase j ($j=$ a, b, c), the final zero-sequence component, the unbalancing coefficient, and the state transition function, respectively.

4.2 Separate Maximum Power Point Tracking Technology

The asymmetric control of dc-link voltages with the common-mode voltage reduction method will be discussed as follows, which is designed based on the deadbeat controller with fast dynamic performance. As introduced before, the voltages across two dc-link capacitors can be separately regulated by controlling the NP current, which is expressed as follows:

$$i_{np} = C_{dc} \cdot \left(\frac{dU_p}{dt} - \frac{dU_n}{dt} \right) = C_{dc} \cdot \frac{d\Delta U_{np}}{dt} \quad (4.64)$$

where $\Delta U_{np} = U_p - U_n$ is the voltage difference between two dc-link capacitors, and C_{dc} is the dc-link capacitance.

The discrete-time model of the neutral-point current can be obtained as

$$i_{np}(k) = C_{dc} \cdot \frac{\Delta U_{np}(k+1) - \Delta U_{np}(k)}{T_s} \quad (4.65)$$

where T_s is the sampling period.

Next, in order to control the neutral-point current by injecting the appropriate zero-sequence component, the relationship between the zero-sequence component and the neutral-point current should be obtained. When the reference voltage vector is located in Sector 1, according to the definition of the three-phase dwell time, the durations of state [O] in Phases A, B, and C are $T_s - t_a$, t_b, and t_c, respectively. Since state [O] affects the neutral-point current, the neutral-point current in Sector 1 is calculated as

$$i_{np} = \frac{i_a \cdot (T_s - t_a) + i_b \cdot t_b + i_c \cdot t_c}{T_s} \quad (4.66)$$

Substituting the three-phase dwell time into the above equation gives the following relationship:

$$i_{np} = -\left(\frac{v_a \cdot i_a}{k+1} + \frac{v_b \cdot i_b}{k-1} + \frac{v_c \cdot i_c}{k-1} \right) - v_z \cdot \left(\frac{i_a}{k+1} + \frac{i_b}{k-1} + \frac{i_c}{k-1} \right) \quad (4.67)$$

The neutral-point currents in other sectors can be calculated by using similar analytical method, and the obtained results are listed in Table 4.5.

According to the state transition functions and the expressions of the neutral-point current, we can derive the generalized form of the neutral-point current in all sectors, as expressed by

$$i_{np} = -\sum_{j=a,b,c} \frac{v_j \cdot i_j}{k+n_j} - v_z \cdot \sum_{j=a,b,c} \frac{i_j}{k+n_j} \quad (4.68)$$

Table 4.5 Calculation of NP currents in all sectors

Sector	i_{np}
12, 1	$-\left(\frac{v_a \cdot i_a}{k+1} + \frac{v_b \cdot i_b}{k-1} + \frac{v_c \cdot i_c}{k-1}\right) - v_z \cdot \left(\frac{i_a}{k+1} + \frac{i_b}{k-1} + \frac{i_c}{k-1}\right)$
2, 3	$-\left(\frac{v_a \cdot i_a}{k+1} + \frac{v_b \cdot i_b}{k+1} + \frac{v_c \cdot i_c}{k-1}\right) - v_z \cdot \left(\frac{i_a}{k+1} + \frac{i_b}{k+1} + \frac{i_c}{k-1}\right)$
4, 5	$-\left(\frac{v_a \cdot i_a}{k-1} + \frac{v_b \cdot i_b}{k+1} + \frac{v_c \cdot i_c}{k-1}\right) - v_z \cdot \left(\frac{i_a}{k-1} + \frac{i_b}{k+1} + \frac{i_c}{k-1}\right)$
6, 7	$-\left(\frac{v_a \cdot i_a}{k-1} + \frac{v_b \cdot i_b}{k+1} + \frac{v_c \cdot i_c}{k+1}\right) - v_z \cdot \left(\frac{i_a}{k-1} + \frac{i_b}{k+1} + \frac{i_c}{k+1}\right)$
8, 9	$-\left(\frac{v_a \cdot i_a}{k-1} + \frac{v_b \cdot i_b}{k-1} + \frac{v_c \cdot i_c}{k+1}\right) - v_z \cdot \left(\frac{i_a}{k-1} + \frac{i_b}{k-1} + \frac{i_c}{k+1}\right)$
10, 11	$-\left(\frac{v_a \cdot i_a}{k+1} + \frac{v_b \cdot i_b}{k-1} + \frac{v_c \cdot i_c}{k+1}\right) - v_z \cdot \left(\frac{i_a}{k+1} + \frac{i_b}{k-1} + \frac{i_c}{k+1}\right)$

Supposing the expecting voltage difference to its reference value at $(k+1)T$ and ignoring the delay caused by sampling and calculation, the reference value of neutral-point current can be derived as

$$i_{np,ref} = \frac{C_{dc} \cdot [\Delta U_{np,ref} - \Delta U_{np}(k)]}{T_s} \quad (4.69)$$

The generalized form of the zero-sequence component is obtained as

$$v_z = \frac{i_{np,ref} + \sum_{j=a,b,c} \frac{v_j \cdot i_j}{k+n_j}}{-\sum_{j=a,b,c} \frac{i_j}{k+n_j}} \quad (4.70)$$

The overall control block diagram is shown in Fig. 4.22. The current control scheme is designed in the synchronous reference frame. According to the reference value of the difference between the two capacitor voltages, the reference NP current is calculated using the deadbeat controller. Then, the optimized zero-sequence component for asymmetric control of dc-link voltages is calculated by considering three-phase modulation waveforms, three-phase output currents, state transition function, and unbalancing coefficient. The zero-sequence component is restricted by considering the ranges in different cases, and the final zero-sequence component can be obtained. By comparing the final modulation waveforms and two carriers with APOD, the PWM signals can be generated and used to control the power semiconductor switches in the three-level inverter.

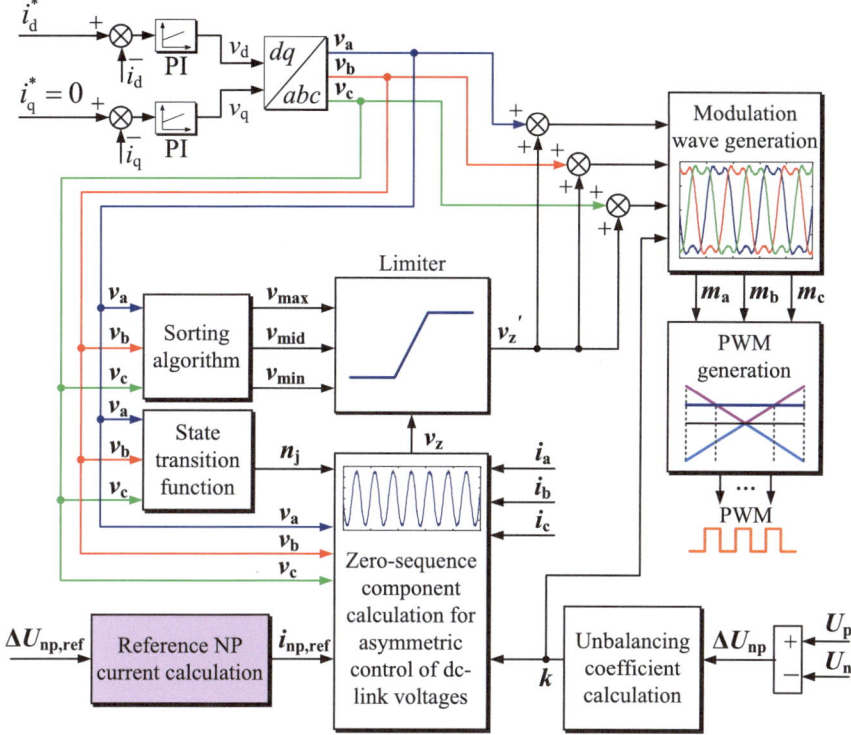

Fig. 4.22 Control block diagram of the CMV reduction method for the three-level inverter for separate MPPTs

4.2.2 Based on Discontinuous Modulation Method

To realize the objective of switching losses reduction and system efficiency enhancement, the discontinuous modulation method has been proposed for the three-level inverter with balanced NPV conditions. When unbalanced NPVs occur, the implementation process becomes complicated, especially the design of switching sequences, because the subsector calculation is time consuming in the asymmetric space vector diagrams [22].

Fortunately, the hybrid modulation method in Sect. 4.2.1.1 can be extended to discontinuous modulation method with a few modifications, so as to realize the objective of switching losses reduction with unbalanced NPV conditions. In concrete terms, the target of bus clamping is realized by selecting the zero-sequence voltage as $v_{z,min}$ or $v_{z,max}$.

When the reference vector lies in Sector 1, the switching sequences of the discontinuous modulation strategy are given in Tables 4.6 and 4.7. It can be seen that the clamping styles of the discontinuous modulation strategy in Sector 1 can be divided into six cases according to the selected zero-sequence voltage and the size relationship of $v_{z,min}$ and $v_{z,max}$, and the switching sequences in other sectors can also be analyzed by the same method.

To sum up, the discontinuous modulation strategy can reduce one third of the average switching frequency owing to the fact that one switching sequence of the three phases is clamped to the positive, negative dc bus, or the neutral-point within each carrier period. The switching losses are reduced by the discontinuous modulation strategy owing to the reduction of the average switching frequency. It should be noted that the calculating processes are the same for different dc unbalancing coefficients ranging from -1 to 1. Thus, the discontinuous modulation method can be applied for all the unbalanced cases. However, the dc unbalancing coefficient is usually restricted in practical applications by considering the block voltages of the power semiconductor switches.

Next, the asymmetric control of the dc-links for discontinuous modulation method is described as follows. The zero-sequence voltage for the discontinuous modulation can only be selected from two values, i.e., $v_{z,max}$ and $v_{z,min}$. Considering the fact that only two different values are available, a simple hysteresis control strategy is developed to select the optimal v_z to ensure the real value of the dc unbalancing coefficient follows its reference value.

Table 4.6 Switching sequences for $v_z = v_{z,min}$

v_z	$v_{z,min} = -v_a$	$v_{z,min} = -U_n - v_b$	$v_{z,min} = -U_n - v_c$
Switching sequence	(waveform)	(waveform)	(waveform)

Table 4.7 Switching sequences for $v_z = v_{z,max}$

v_z	$v_{z,max} = U_p - v_a$	$v_{z,min} = -v_b$	$v_{z,min} = -v_c$
Switching sequence	(waveform)	(waveform)	(waveform)

4.2 Separate Maximum Power Point Tracking Technology

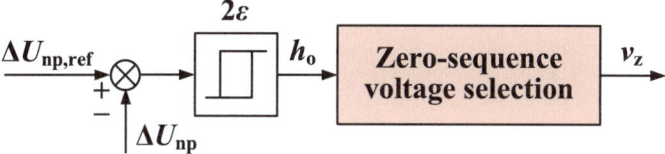

Fig. 4.23 Control diagram of the asymmetric control strategy for the discontinuous modulation method

The block diagram of the discontinuous modulation method is shown in Fig. 4.23, where the hysteresis bandwidth is 2ε, and the output of the hysteresis controller (h_o) changes to "1" if "$k_{ref} - k > \varepsilon$" and "0" if "$k_{ref} - k < -\varepsilon$".

The sector is firstly calculated, and then the polarity of the variable σ is obtained. The optimal zero-sequence voltage v_z can be selected according to the principles as follows:

1. if $h_o = 1$ and $\sigma > 0$, then v_z is selected as $v_{z,max}$;
2. if $h_o = 1$ and $\sigma < 0$, then v_z is selected as $v_{z,min}$;
3. if $h_o = 0$ and $\sigma > 0$, then v_z is selected as $v_{z,min}$;
4. if $h_o = 0$ and $\sigma < 0$, then v_z is selected as $v_{z,max}$.

To understand the above principles, an example for principle (1) is given here. In this case, h_o is equal to 1 and the real value of the dc unbalancing coefficient is smaller than its reference, thus, the zero-sequence voltage of which the corresponding value of $i_{np,av}$ is bigger should be selected to ensure the discharge of the lower capacitor is maximal to stop or decrease the speed of dc unbalancing coefficient becoming smaller than its reference. It can be seen that $i_{np,av}$ is a linear function of v_z. If the value of σ is positive, $i_{np,av}$ is an increasing function of v_z and thus the optimal v_z is $v_{z,max}$; otherwise, it is $v_{z,min}$ as principle (2). The hysteresis bandwidth can be selected as proper values according to the parameters of the main circuit and working conditions.

4.2.3 Based on Fault-Tolerant Control Method

The unbalanced NPV easily leads to the fault of power semiconductor switches. In order to simultaneously realize the fault-tolerant control (FTC) and unbalanced NPV control (UNPVC), a half cycle FTC (HC-FTC) strategy and a whole cycle FTC (WC-FTC) strategy are proposed in [23]. The carrier is modified to avoid using the impossible switching states and realize FTC. Meanwhile, the mathematical models of NP current under both normal and fault conditions are established based on volt-second balance principle, and thereby the zero-sequence component (ZSC) derived by the model can be used to realize the UNPVC whether the power semiconductor switch is faulty or not. In addition, the ZSC is selected in detail to improve the NPV control ability in HC-FTC. Furthermore, the application range of HC-FTC is obtained

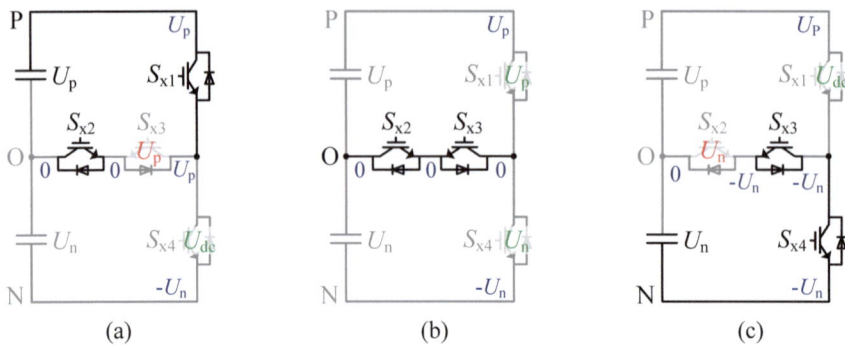

Fig. 4.24 The voltage stress of power semiconductor switches **a** switching state [P], **b** switching state [O], **c** switching state [N]

and the WC-FTC with wider application range is analyzed. Two strategies can be selected according to the requirements. The detailed principles and implementation will be elaborated in this section, which includes four parts.

Part 1: Analysis of OCF with Unbalanced NPV

Generally, there are two cases of OCF in the 3LT^{2}I: inner power semiconductor switch fault and outer power semiconductor switch fault. When an outer power semiconductor switch fault, the modulation index should be reduced to realize FTC, which will cause the inverter to be off-grid or the power rating to be reduced.

In addition, the unbalanced NPV severely aggravates the voltage stress of inner power semiconductor switches. The voltage stress of power semiconductor switches in different switching states is shown in Fig. 4.24, where the blue word, green word and red word represent electrical potential, normal voltage stress and abnormal voltage stress, respectively.

Table 4.8 summarizes the voltage stress of different power semiconductor switches under balanced and unbalanced conditions. As shown in Table 4.8, the voltage stress of all power semiconductor switches is U_{dc} under unbalanced conditions. Outer power semiconductor switches can withstand the voltage stress as they are designed with the higher rated power. However, the inner power semiconductor switches are designed with the lower rated power and are likely to overvoltage with unbalanced NPV. Therefore, this paper mainly focuses on the OCF occurring in inner power semiconductor switches regardless of outer switches.

Phase A is taken as the fault phase and the current paths in phase A are shown in Fig. 4.25, where the solid green line and dashed red line represent current paths under normal and fault conditions, respectively. When S_{a2} is faulty and the switching state is [O], the switching state [O] is replaced by [N] when $i_a > 0$ as shown in Fig. 4.25a. However, the current path does not change when $i_a < 0$ as shown in Fig. 4.25b. Consequently, the switching state [O] is impossible only when S_{a2} fault and $i_a > 0$. When S_{a3} is faulty, the switching state [O] is impossible only when $i_a < 0$ as shown in Fig. 4.25c, d. The switching state [O] is impossible in the whole fundamental cycle

4.2 Separate Maximum Power Point Tracking Technology

Table 4.8 Voltage stress of different power semiconductor switches

Switch	Voltage stress under balanced conditions	Voltage stress under unbalanced conditions
S_{x1}	U_{dc}	U_{dc}
S_{x2}	$U_{dc}/2$	U_{dc}
S_{x3}	$U_{dc}/2$	U_{dc}
S_{x4}	U_{dc}	U_{dc}

when OCF occurs in S_{a2} and S_{a3} as shown in Fig. 4.25e, f. S_{a2} fault is taken as an example, and other inner switches fault can be analyzed by the same method.

Part 2: Carrier-Based FTC

Carrier-based PWM (CBPWM) is widely used due to its easy implementation. In a T-type three-level inverter, the PWM signals can be generated by comparing the modulation waveforms m_x (x = a, b, c) with two carriers with amplitude of 1, which are defined as C_{r1} and C_{r2} as shown in Fig. 4.26. When m_x is greater than 0, the output pole voltage is [P] and [O] (denoted by type A). When m_x is smaller than 0, the output pole voltage is [O] and [N] (denoted by type B). For type A, the dwell time

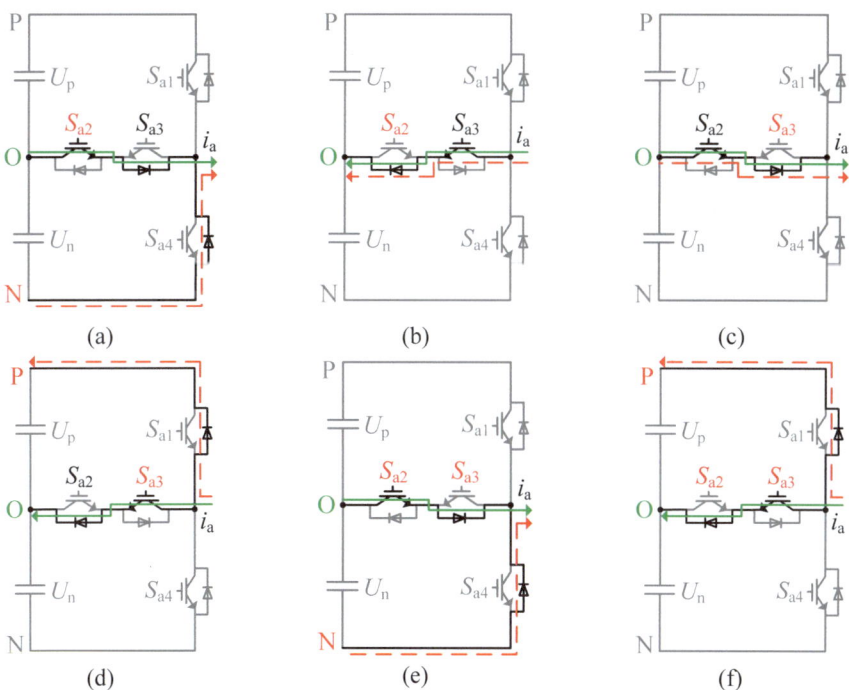

Fig. 4.25 Current paths in phase A **a** S_{a2} fault, $i_a > 0$, **b** S_{a2} fault, $i_a < 0$, **c** S_{a3} fault, $i_a > 0$, **d** S_{a3} fault, $i_a < 0$, **e** S_{a2} and S_{a3} fault, $i_a > 0$, **f** S_{a2} and S_{a3} fault, $i_a < 0$

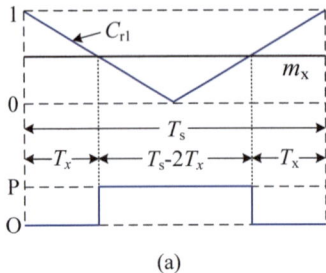 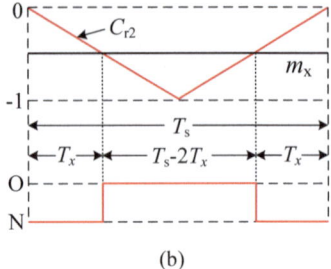

Fig. 4.26 Two carrier types **a** type A, **b** type B

T_x is defined as the half of switching state [O] duration. For type B, T_x is defined as the half of switching state [N] duration. T_s is the sampling time.

The symbolic function of three-phase is defined as

$$n_x = \begin{cases} 1, & \text{type A} \\ -1, & \text{type B} \end{cases} \quad (4.71)$$

The duty cycle of switching state [O] is defined as d_x, which can be expressed as

$$d_x = \begin{cases} \frac{2T_x}{T_s}, & \text{type A} \\ \frac{(T_s - 2T_x)}{T_s}, & \text{type B} \end{cases} \quad (4.72)$$

The output current will be distorted if the reference voltage is synthesized by impossible switching states caused by the OCF occurs in S_{a2}. As analyzed above, the switching state [O] is impossible when $i_a > 0$ but is possible when $i_a < 0$. Therefore, a fundamental cycle can be divided into two half cycles: the healthy half cycle and the fault half cycle. If the switching state [O] of phase A is not used to synthesize the reference voltage in the fault half cycle, the distortion will be eliminated.

Hence, the C_{r1} and C_{r2} are also adopted in the healthy half cycle to generate three-level pole voltages. While a single carrier with an amplitude of 2 (denoted by C_{r3}) should be adopted in the fault half cycle as shown in Fig. 4.27. At this time, the reference voltage is synthesized only by using [P] and [N] instead of [P], [O] and [N] in the fault half cycle. That is to say, in this case, d_a is equal to 0. While in the healthy half cycle, [P], [O] and [N] are all used to synthesize the reference voltage. Since this strategy will not generate switching state [O] in phase A in the half cycle, it is named as HC-FTC strategy.

Part 3: UNPVC in HC-FTC Strategy

The three-phase reference modulation waveforms are

4.2 Separate Maximum Power Point Tracking Technology

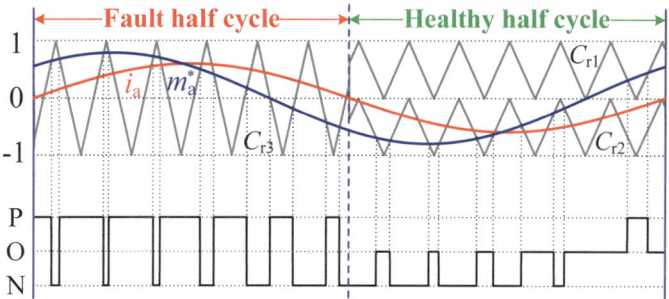

Fig. 4.27 The modified carrier of HC-FTC strategy in phase A

$$\begin{cases} m_a^* = m \cdot \cos(\omega_1 t) \\ m_b^* = m \cdot \cos(\omega_1 t - \frac{2\pi}{3}) \\ m_c^* = m \cdot \cos(\omega_1 t + \frac{2\pi}{3}) \end{cases} \quad (4.73)$$

where ωt is the phase angle of phase A. The three-phase currents can be expressed as

$$\begin{cases} i_a = I_m \cdot \cos(\omega_1 t - \varphi) \\ i_b = I_m \cdot \cos(\omega_1 t - \frac{2\pi}{3} - \varphi) \\ i_c = I_m \cdot \cos(\omega_1 t + \frac{2\pi}{3} - \varphi) \end{cases} \quad (4.74)$$

where I_m is the peak value of the phase current and φ is the *pf* angle.

The UNPVC can be implemented based on CBPWM with proper ZSC injection and the modulation waveforms after injecting ZSC can be expressed as

$$m'_x = m_x^* + m_z \quad (4.75)$$

where m_z represents the ZSC. And the key to realize UNPVC is to calculate the proper m_z.

According to the volt-second balance principle, there is

$$m'_x \cdot \frac{U_{dc}}{2} \cdot T_s = n_x \cdot \frac{U_{dc} \cdot (1 + k \cdot n_x)}{2} \cdot (1 - d_x) \cdot T_s \quad (4.76)$$

The three-phase dwell times in all sectors are summarized in Table 4.9.

The phase current flowing from the inverter to the load is defined as the positive direction, and the NP current flowing out of the NP is defined as positive direction. The required unbalanced NPV can be controlled by an appropriate NP current. Since the NP current is only influenced by the switching state [O] in one phase, the mathematical model of NP current is divided into two cases.

Case I: In the healthy half cycle, the three phases all have switching state [O] and the NP current can be expressed as

Table 4.9 Three-phase dwell times in all sectors

Sector	T_a	T_b	T_c
I	$\frac{T_s}{2} - \frac{m'_a \cdot T_s}{2(k+1)}$	$\frac{m'_b \cdot T_s}{2(k-1)}$	$\frac{m'_c \cdot T_s}{2(k-1)}$
II	$\frac{T_s}{2} - \frac{m'_a \cdot T_s}{2(k+1)}$	$\frac{T_s}{2} - \frac{m'_b \cdot T_s}{2(k+1)}$	$\frac{m'_c \cdot T_s}{2(k-1)}$
III	$\frac{m'_a \cdot T_s}{2(k-1)}$	$\frac{T_s}{2} - \frac{m'_b \cdot T_s}{2(k+1)}$	$\frac{m'_c \cdot T_s}{2(k-1)}$
IV	$\frac{m'_a \cdot T_s}{2(k-1)}$	$\frac{T_s}{2} - \frac{m'_b \cdot T_s}{2(k+1)}$	$\frac{T_s}{2} - \frac{m'_c \cdot T_s}{2(k+1)}$
V	$\frac{m'_a \cdot T_s}{2(k-1)}$	$\frac{m'_b \cdot T_s}{2(k-1)}$	$\frac{T_s}{2} - \frac{m'_c \cdot T_s}{2(k+1)}$
VI	$\frac{T_s}{2} - \frac{m'_a \cdot T_s}{2(k+1)}$	$\frac{m'_b \cdot T_s}{2(k-1)}$	$\frac{T_s}{2} - \frac{m'_c \cdot T_s}{2(k+1)}$

$$i_{\text{nph}} = d_a \cdot i_a + d_b \cdot i_b + d_c \cdot i_c \tag{4.77}$$

Meanwhile, the three-phase currents satisfy

$$i_a + i_b + i_c = 0 \tag{4.78}$$

The NP current in the healthy half cycle can be generalized as

$$i_{\text{nph}} = -\sum_{x=a,b,c} \frac{m_x^* \cdot i_x}{k + n_x} - m_{\text{zh}} \cdot \sum_{x=a,b,c} \frac{i_x}{k + n_x} \tag{4.79}$$

where m_{zh} represents the ZSC in the healthy half cycle and can be obtained by the reference NP current i_{NP}^*. Therefore, m_{zh} can be calculated as

$$m_{\text{zh}} = -\frac{i_{\text{NP}}^* + \sum_{x=a,b,c} \frac{m_x^* \cdot i_x}{k + n_x}}{\sum_{x=a,b,c} \frac{i_x}{k + n_x}} \tag{4.80}$$

Case II: In the fault half cycle, the switching state [O] in phase A is not used to synthesize the reference voltage and the NP current will not be influenced by phase A. In this case, the NP current is derived as

$$i_{\text{npf}} = d_b \cdot i_b + d_c \cdot i_c \tag{4.81}$$

The NP current in the fault half cycle can be generalized as

$$i_{\text{npf}} = -i_a - \sum_{x=b,c} \frac{m_x^* \cdot i_x}{k + n_x} - m_{\text{zf}} \cdot \sum_{x=b,c} \frac{i_x}{k + n_x} \tag{4.82}$$

where m_{zf} is the ZSC in the fault half cycle, which can be calculated as

4.2 Separate Maximum Power Point Tracking Technology

$$m_{zf} = -\frac{i_{np}^* + i_a + \sum_{x=b,c} \frac{m_x^* \cdot i_x}{k+n_x}}{\sum_{x=b,c} \frac{i_x}{k+n_x}} \tag{4.83}$$

Meanwhile, to avoid the overmodulation, m_z should be limited in the following range:

$$m_{zmin} \leq m_z \leq m_{zmax} \tag{4.84}$$

where $m_{z\,min}$ and $m_{z\,max}$ can be expressed as

$$\begin{cases} m_{zmin} = \max\left\{\frac{(n_x-1)(k \cdot n_x+1)}{2} - m_x^*\right\} \\ m_{zmax} = \min\left\{\frac{(n_x+1)(k \cdot n_x+1)}{2} - m_x^*\right\} \end{cases} \tag{4.85}$$

It can be concluded that m_{zmin} and m_{zmax} are related to m, $\omega_1 t$ and k. Fig. 4.28 illustrates m_{zmin} and m_{zmax} under different m and $\omega_1 t$ conditions when $k = 0$ and $k = 0.25$. The right side of Fig. 4.28 shows the ZSC and NPV in a fundamental cycle when $m = 0.9$. As shown in the shaded region, the ZSC is less than the lower limit and m_{zf} is set to m_{zmin} in this case, which makes the NPV shift to the unexpected direction. If the same ZSC m_{zf} is injected in the healthy half cycle, the NPV deviation will be eliminated because m_{zf} will be set to m_{zmax} due to the symmetry of ZSC. However, as m_{zh} is injected in the healthy half cycle and m_{zf} is injected in the fault half cycle, the symmetry of ZSC is broken and the NPV deviation cannot be eliminated. Hence, it is necessary to change the ZSC in the healthy half cycle to improve the NPV control ability.

Sector III is taken as an example, and the NP current model in the healthy half cycle can be expressed as

$$i'_{nph} = -\sum_{x=a,b,c} \frac{m_x^* \cdot i_x}{k+n_x} - m'_{zh} \cdot \sum_{x=a,b,c} \frac{i_x}{k+n_x} = -\sum_{x=a,b,c} \frac{m_x^* \cdot i_x}{k+n_x} - \frac{2i_b}{1-k^2} \cdot m'_{zh} \tag{4.86}$$

As $1 - k^2 > 0$, m'_{zh} is depended on the polarity of i_b. It is evident that the ZSC has the best NPV control ability when it is clamped to maximum or minimum. Therefore, m_{zh} in Sector III can be expressed as

$$m'_{zh} = \begin{cases} m_{zh}, & \Delta U_{np,ref} - V_{TH} \leq \Delta U_{np} \leq \Delta U_{np,ref} + V_{TH} \\ m_{zmax}, & \Delta U_{np} > \Delta U_{np,ref} + V_{TH} \ \& \ i_b > 0 \text{ or} \\ & \Delta U_{np} < \Delta U_{np,ref} - V_{TH} \ \& \ i_b \leq 0 \\ m_{zmin}, & \Delta U_{np} > \Delta U_{np,ref} + V_{TH} \ \& \ i_b \leq 0 \text{ or} \\ & \Delta U_{np} < \Delta U_{np,ref} - V_{TH} \ \& \ i_b > 0 \end{cases} \tag{4.87}$$

where ΔU_{np} is the NPV, $\Delta U_{np,ref}$ is the reference NPV and ΔV_{TH} is the threshold of NPV deviation. m'_{zh} in other sectors can be analyzed by the same method. Among

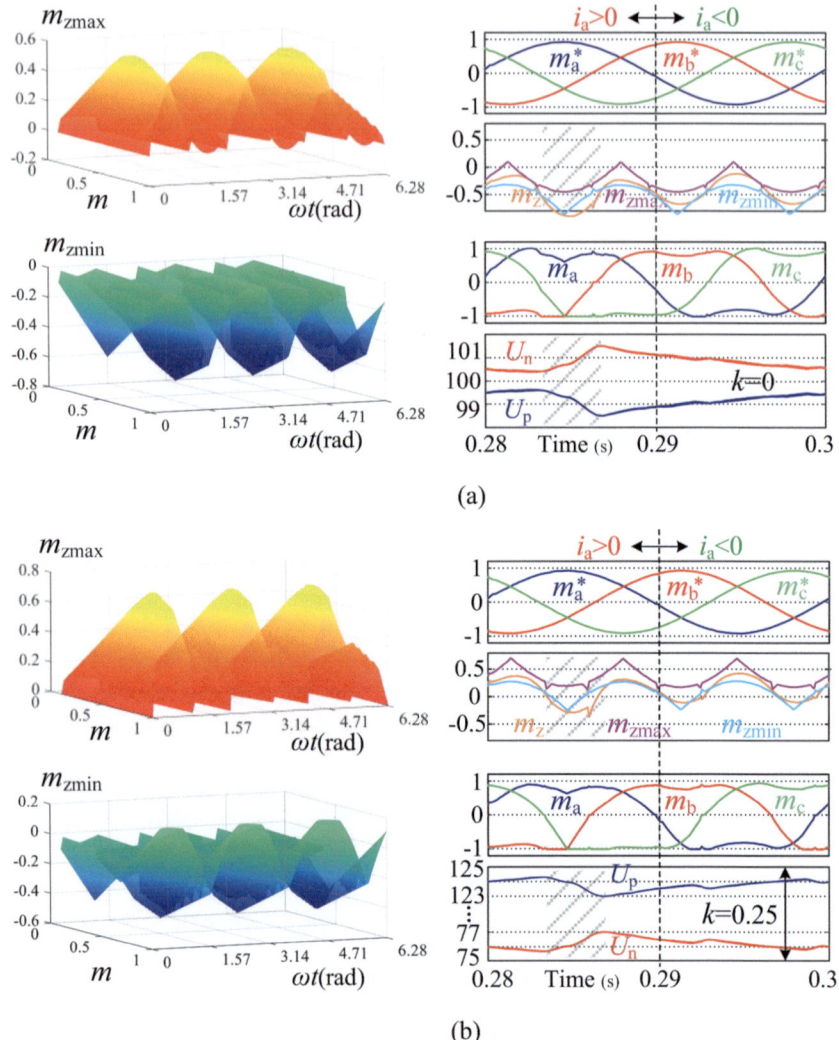

Fig. 4.28 The ZSC limitations and NPV **a** $k = 0$, **b** $k = 0.25$

them, Sector I and IV are related to i_a, Sector III and VI are related to i_b and Sector II and V are related to i_c. After adopting this method, the NPV control ability increases significantly as shown in Fig. 4.29. It can be seen that the NPV is controlled to the expected value at a very fast speed.

The final modulation waveforms m_x are obtained as

$$m_x = \begin{cases} \frac{m_x^* + m_{zf}}{1 + k \cdot n_x}, & i_a > 0 \\ \frac{m_x^* + m_{zh}'}{1 + k \cdot n_x}, & i_a < 0 \end{cases} \tag{4.88}$$

4.2 Separate Maximum Power Point Tracking Technology

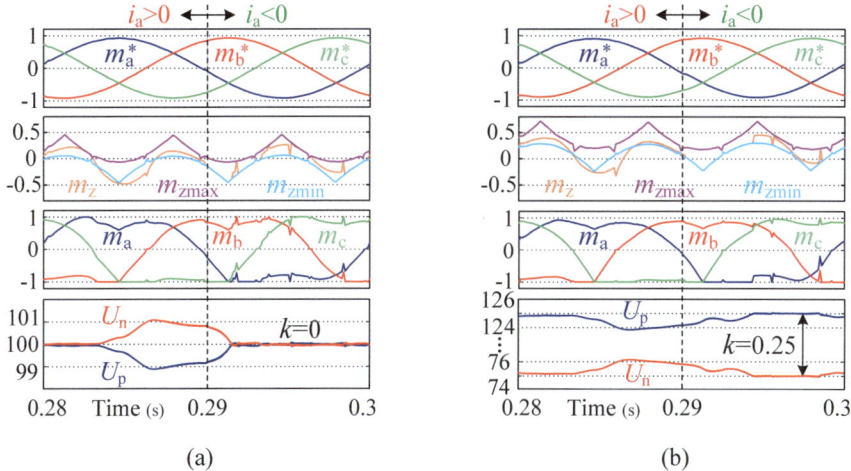

Fig. 4.29 Simulation results after changing the ZSC **a** $k = 0$, **b** $k = 0.25$

Figure 4.30 shows the overall control block diagram of the HC-FTC strategy. First, a PI controller is used to track the output currents, which generates the reference modulation waveforms m^*_x. Second, the NPV is also controlled by a PI controller that generates the i^*_{np}. Then, m'_{zh}, m_{zf}, $m_{z\,min}$ and $m_{z\,max}$ are calculated. The selected ZSC is limited and injected into m^*_x to obtain the final modulation waveforms m_x. Finally, m_x are compared with the modified carriers to generate the PWM signals.

Part 4: Application Range and the WC-FTC Strategy

In order to simplify the analysis, the NP current analysis is performed in the manner of single DC source. To get the application ranges of the HC-FTC strategy under different m and pf conditions, the variation of NPV in a fundamental cycle should be considered. Let $C_1 = C_2 = C$ and ignore the initial voltage of the capacitor. According to the Kirchhoff current law (KCL), the NP current can be expressed as

$$i_{np} = C\frac{dU_p}{dt} - C\frac{dU_n}{dt} = C\frac{\Delta U_{np}}{T_s} \quad (4.89)$$

In a fundamental cycle, the maximum and minimum of NPV deviations can be expressed as

$$\begin{cases} \Delta U_{npmax} = \frac{1}{C}\int_0^{2\pi} i_{npmax}\, d\omega t \\ \Delta U_{npmin} = \frac{1}{C}\int_0^{2\pi} i_{npmin}\, d\omega t \end{cases} \quad (4.90)$$

where i_{npmax} represents the NP current takes the maximum value in each switching cycle and i_{npmin} represents the NP current takes the minimum value in each switching

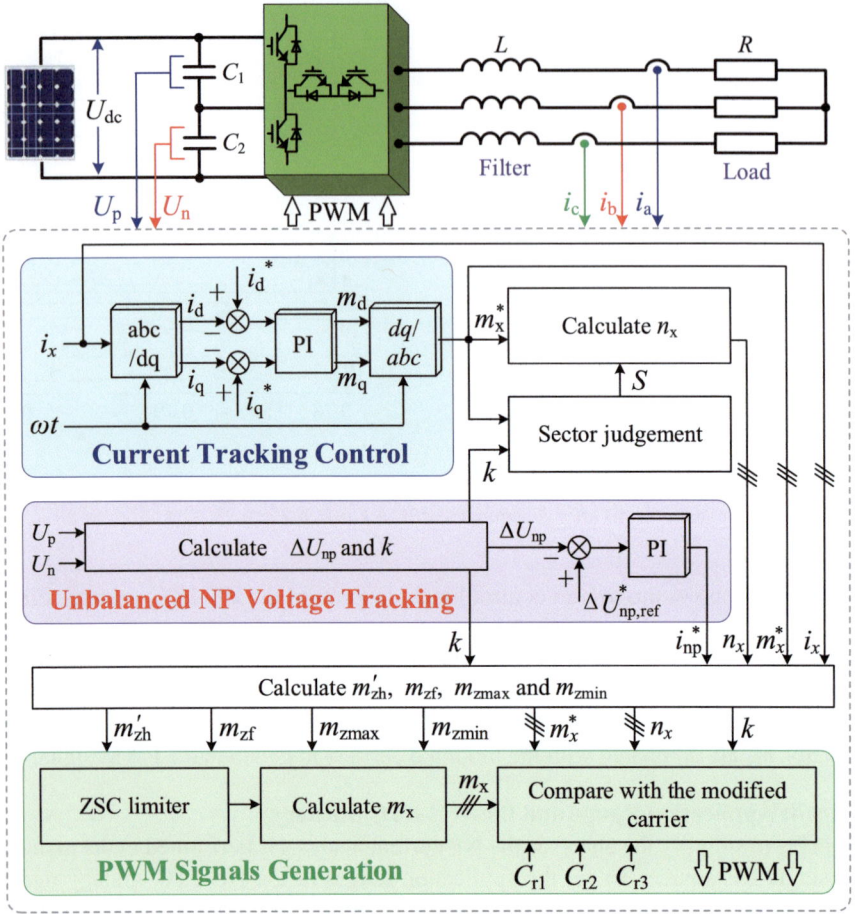

Fig. 4.30 The overall control diagram of the HC-FTC strategy

cycle. As the NP current is piecewise linear, the i_{npmax} and i_{npmin} will be obtained at the calculated optimal ZSC or boundary value. Therefore, in the healthy half cycle, the NP currents with m_{zh}, $m_{z\,max}$ and $m_{z\,min}$ injection are compared, respectively. And in the fault half cycle, the NP currents with m_{zf}, $m_{z\,max}$ and $m_{z\,min}$ injection are compared, respectively.

Combining the above analysis, the ΔU_{npmax} and ΔU_{npmin} are related to m, pf, k, I_m and C. Since the I_m and C are constants, the ΔU_{npmax} and ΔU_{npmin} under different m and pf are shown in Fig. 4.31.

If $\Delta U_{npmin} < 0 < \Delta U_{npmax}$, there is at least one value to control the NPV, which is a prerequisite to guarantee the reliable and safe operation. If $\Delta U_{npmin} < \Delta U_{npmax} < 0$ or $0 < \Delta U_{npmin} < \Delta U_{npmax}$, the NPV will be uncontrollable. The application ranges of the HC-FTC strategy are shown in Fig. 4.32. It is shown that there are some uncontrollable regions no matter under balanced or unbalanced conditions. Therefore, in

4.2 Separate Maximum Power Point Tracking Technology

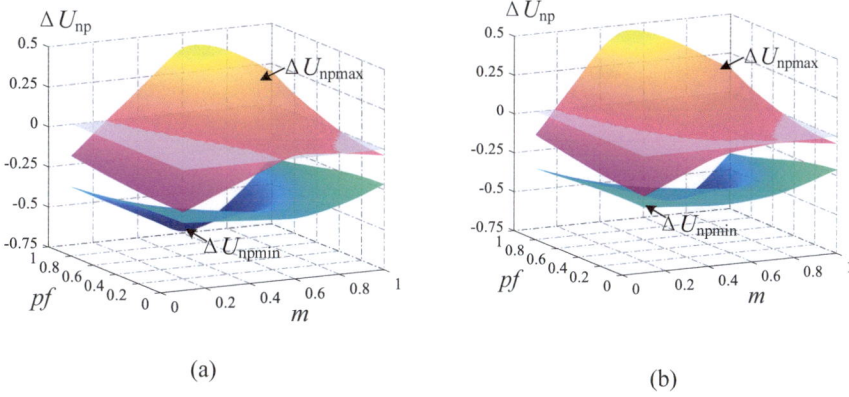

Fig. 4.31 The NPV variation in a fundamental cycle of the HC-FTC strategy **a** $k = 0$, **b** $k = 0.25$

these uncontrollable regions, another strategy with wider application range needs to be adopted.

Moreover, the HC-FTC strategy is only applicable to the case of single inner switch fault. When OCF occurs both in S_{a2} and S_{a3}, the switching state [O] is impossible in phase A in a fundamental cycle as analyzed above. Therefore, the WC-FTC strategy is provided, where the C_{r3} is used in the whole fundamental cycle to avoid using the impossible switching states in the fault phase and eliminate the output current distortion. In this case, m_{zf} is injected in the whole fundamental cycle to realize the FTC and UNPVC, simultaneously. In order to obtain the application ranges of the WC-FTC strategy, Fig. 4.33 shows the ΔU_{npmax} and ΔU_{npmin} under different m and pf. It can be seen that $\Delta U_{npmin} < 0 < \Delta U_{npmax}$ is always satisfied under balanced or unbalanced conditions. Therefore, the WC-FTC strategy is applicable in any operating conditions.

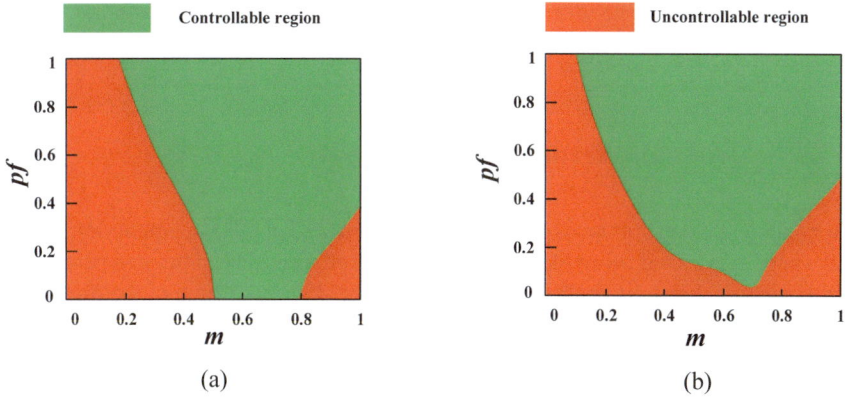

Fig. 4.32 The application ranges of the HC-FTC strategy **a** $k = 0$, **b** $k = 0.25$

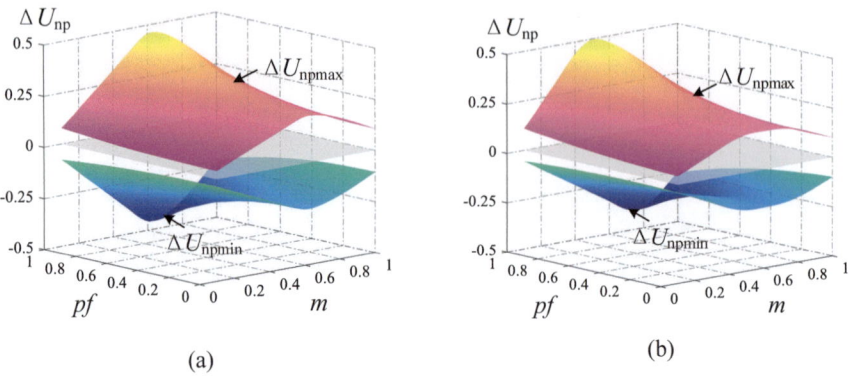

Fig. 4.33 The NPV variation in a fundamental cycle of the WC-FTC strategy **a** $k = 0$, **b** $k = 0.25$

To summarize, the centralized three-level inverter with separate MPPTs is the best choice topology for large-scale PV power generation system, since investment costs and complexity are reduced, while efficiency is increased and MPPTs are strictly followed. However, the NPV imbalance is inevitably, which leads to three-phase output currents are distorted when adopting the traditional modulation method. To cope with this severe problem, varieties of methods are presented in this section.

4.3 Simulation and Experimental Validations

To validate the effectiveness and correctness of the presented technology, comprehensive simulation and experimental studies have been conducted. It should be noted that conventional MPPT technology for generalized PV applications has been extensively studied by scholars at home and abroad, and relatively mature. Thus, the simulation and experimental results are mainly obtained from the separate MPPT technology.

4.3.1 Simulation Results and Discussions

To evaluate the performance of different separate MPPT methods, simulation studies have been conducted using MATLAB/Simulink software. The simulated parameters are given in Table 4.10. It should be noted that the dc-link voltage of the three-level inverter is supplied by a single dc power source, and the voltages across the two capacitors are controllable, which is used to validate the effectiveness of the modulation method and asymmetric control strategy of dc-link voltages simultaneously. The peak value of output currents is controlled at 10 A. Three different methods for three-level inverters under unbalanced NPV conditions are compared. Method-1

4.3 Simulation and Experimental Validations

Table 4.10 Parameters for simulated tests

Parameters	Values
DC input voltage (U_{dc})	350 V
DC-link capacitors (C_{dc})	940 µF
Fundamental ac output frequency (f_1)	50 Hz
Switching frequency (f_{sw})	10 kHz
Sampling period (T_s)	100 µs
Dead-time (t_d)	2 µs
Filter inductance (L)	3 mH
Load resistance (R_{load})	16 Ω

is the virtual SVM method. Method-2 is the SVM method. Method-3 is the CMV reduction method with unbalanced NPV conditions.

With a Positive Unbalancing Coefficient

In the first simulation, the voltage difference between the two dc-link capacitors is controlled to be 60 V, and the unbalancing coefficient is equal to 0.1714. The simulated waveforms are shown in Fig. 4.34, and the detailed comparisons of different methods are summarized in Table 4.11. The asymmetric control of dc-link voltages is achieved. Since Method-1 and Method-2 use all the basic vectors to synthesize the reference voltage vector, the maximum CMV magnitudes are 136.7 and 136.5 V, respectively. Among three methods, the current THD value of Method-2 is the minimum. Method-4 can restrict the CMV magnitude and obtain sinusoidal output currents simultaneously. Since Method-4 avoids using the basic voltage vectors with high CMV magnitudes, the current THD value is slightly higher than that of Method-2. Fortunately, the current THD value of Method-3 remains at an acceptable level.

To measure the leakage current, a leakage resistor and a leakage capacitor are connected between the common point of three-phase load and the negative point of dc source. The leakage current is measured at the line where the leakage resistor and the leakage capacitor are connected, as shown in Fig. 4.35, where $R_{leakage}$ and $C_{leakage}$ are the leakage resistor and the leakage capacitor, respectively.

When the adopted leakage capacitor and leakage resistor are 0.015 µF and 15 Ω, the zoom-in waveforms of CMV and leakage current are shown in Fig. 4.36. The RMS values of leakage currents for different methods are summarized in Table 4.11. Since the CMV magnitudes of Method-1 and Method-2 are high, the leakage currents become large. Method-4 takes the unbalanced NPV conditions of three-level inverter into consideration, and good-quality output currents can be obtained. Compared with Method-1 and Method-2, the leakage current of Method-3 can be reduced to some extent.

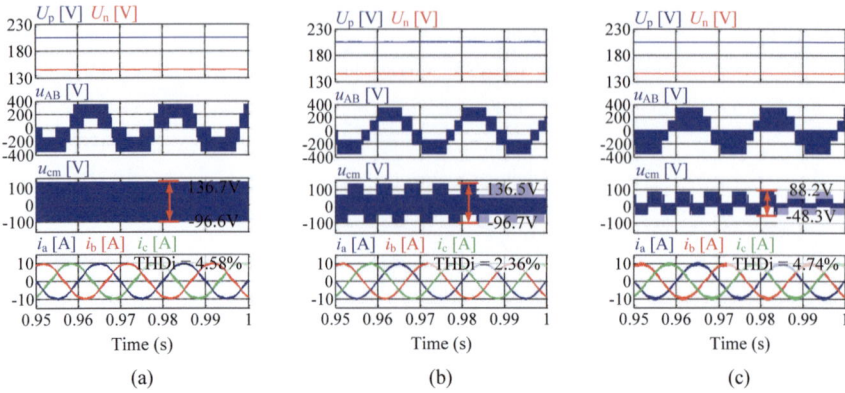

Fig. 4.34 Simulation results of positive unbalancing coefficient **a** Method-1, **b** Method-2, **c** Method-3

Table 4.11 Comparisons of different methods for positive unbalancing coefficient

Method	Range of CMV (V)	THDi (%)	i_{cm_RMS} (mA)
Method-1	− 96.6–136.7	4.58	241
Method-2	− 96.7–136.5	2.36	236
Method-3	− 48.3–88.2	4.74	209

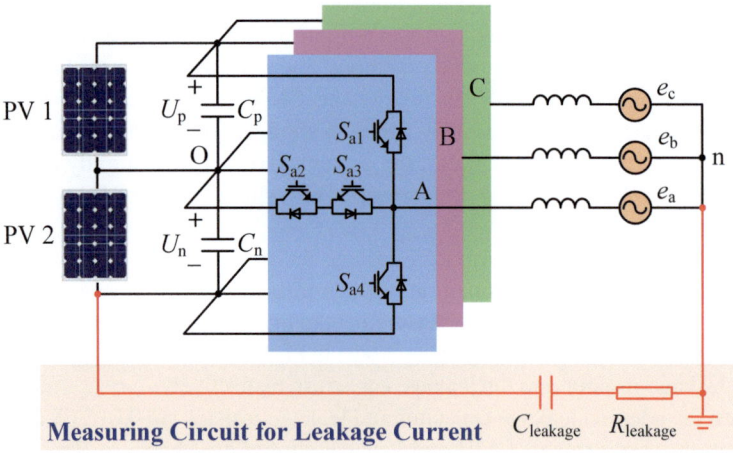

Fig. 4.35 Measuring circuit for leakage current

With a Negative Unbalancing Coefficient

Next, consider the case that the voltage difference between two dc-link capacitors is negative. In this case, the voltage difference is controlled to be − 60 V, and

4.3 Simulation and Experimental Validations

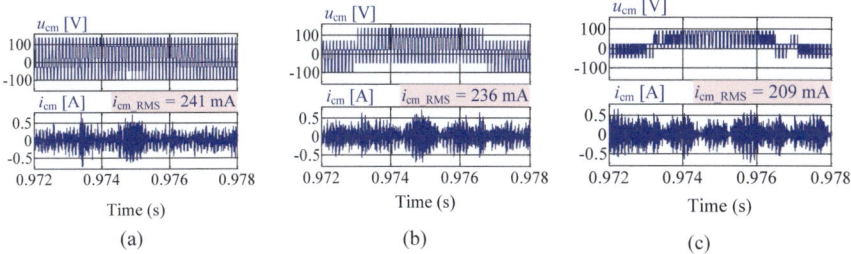

Fig. 4.36 Zoom-in simulated waveforms of CMV and leakage current for positive unbalancing coefficient **a** Method-1, **b** Method-2, **c** Method-3

the unbalancing coefficient is equal to -0.1714. The simulated results are shown in Fig. 4.37, and the detailed comparisons are listed in Table 4.12. The zoom-in simulated waveforms of CMV and leakage current for the negative unbalancing coefficient ($k = -0.1714$) are shown in Fig. 4.38. Since all the basic vectors are employed in Method-1 and Method-2, the generated CMVs are relatively high. When Method-3 is adopted, output currents of good quality are obtained, and the CMV is suppressed at the same time.

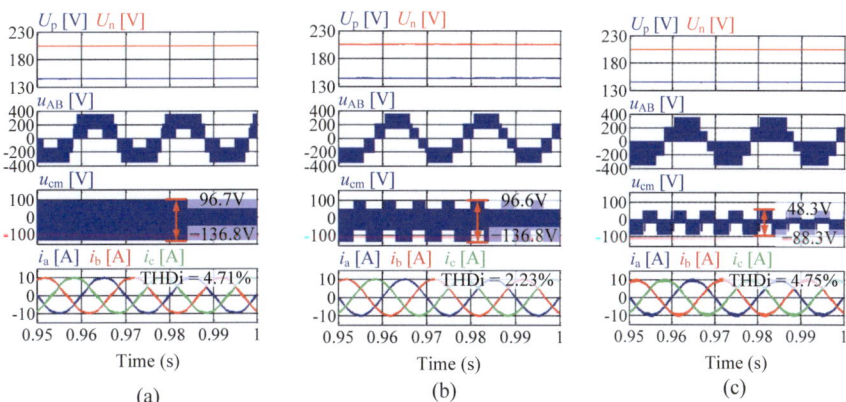

Fig. 4.37 Simulation results of negative unbalancing coefficient **a** Method-1, **b** Method-2, **c** Method-3

Table 4.12 Comparisons of different methods for negative unbalancing coefficient

Method	Range of CMV (V)	THDi (%)	i_{cm_RMS} (mA)
Method-1	-136.8–96.7	4.71	225
Method-2	-136.8–96.6	2.23	223
Method-3	-88.3–48.3	4.75	206

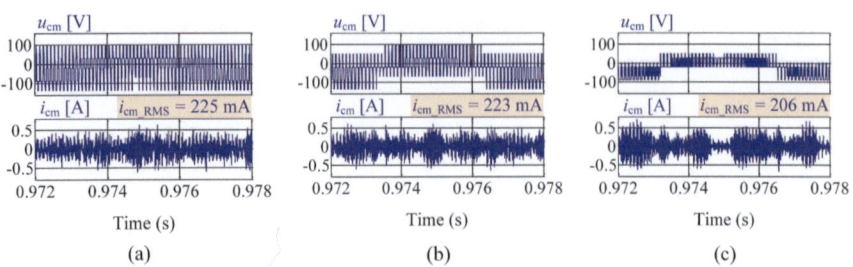

Fig. 4.38 Zoom-in simulated waveforms of CMV and leakage current for negative unbalancing coefficient **a** Method-1, **b** Method-2, **c** Method-3

Asymmetric Control of Capacitor Voltages

Figures 4.39 and 4.40 provide the simulation results of asymmetric control of capacitor voltages for the hybrid modulation method and common-mode voltage reduction method, respectively. It is clear that the asymmetric control of dc-link voltages can be achieved flexibly. A sinusoidal output current can be obtained whether the NPV is balanced or not. In comparison with the hybrid modulation strategy, the common-mode voltage reduction method can suppress the common-mode voltage magnitude by half.

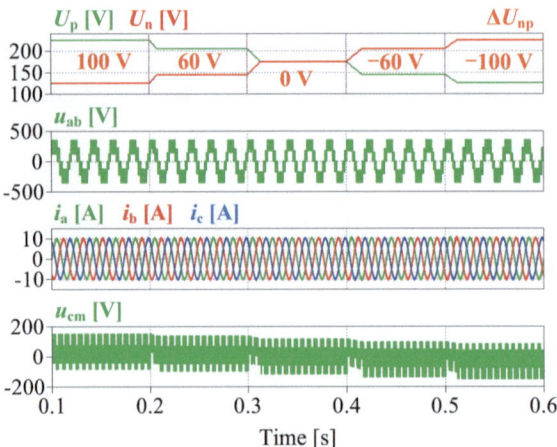

Fig. 4.39 Simulation results of asymmetric control of capacitor voltages for the hybrid modulation method

4.3 Simulation and Experimental Validations

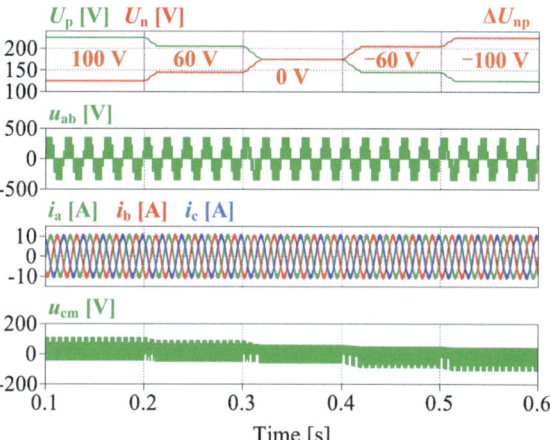

Fig. 4.40 Simulation results of asymmetric control of capacitor voltages for the common-mode voltage reduction method

4.3.2 Experimental Results and Discussions

An experimental platform was designed and built in the laboratory, as shown in Fig. 4.41. extensive experimental results are provided to demonstrate the performances of different control methods. Parameters for experimental tests are identical to those for simulations. It should be noted that the dc-link voltage of the three-level inverter is supplied by a single dc power source, and the voltages across the two capacitors are controllable, which is used to validate the effectiveness of the modulation method and asymmetric control strategy of dc-link voltages simultaneously. The peak value of output currents is controlled at 10 A. Four different methods are compared. Method-1 is the virtual SVM method. Method-2 is the SVM method. Method-3 is the CMV reduction method with unbalanced NPV conditions.

With a Positive Unbalancing Coefficient

The experimental results of positive unbalancing coefficient are shown in Figs. 4.42, 4.43, and 4.44, where the voltage difference between the two dc-link capacitors is controlled to be 60 V. The CMVs of Method-1 and Method-2 are relatively high, because the basic vectors with high CMV magnitudes are utilized by these two methods. When Method-3 is used, three-phase sinusoidal output currents can be guaranteed. Compared with Method-1 and Method-2, the CMV magnitude of Method-3 is reduced by half.

The experimental waveforms of the modulation waveforms, the zero-sequence component, and its corresponding limitations are shown in Fig. 4.45. It is clearly seen that the zero-sequence voltage can be restricted within its lower bound ($v_{z,min}$) and upper bound ($v_{z,max}$) in order not to generate the basic voltage vectors with high CMV magnitudes.

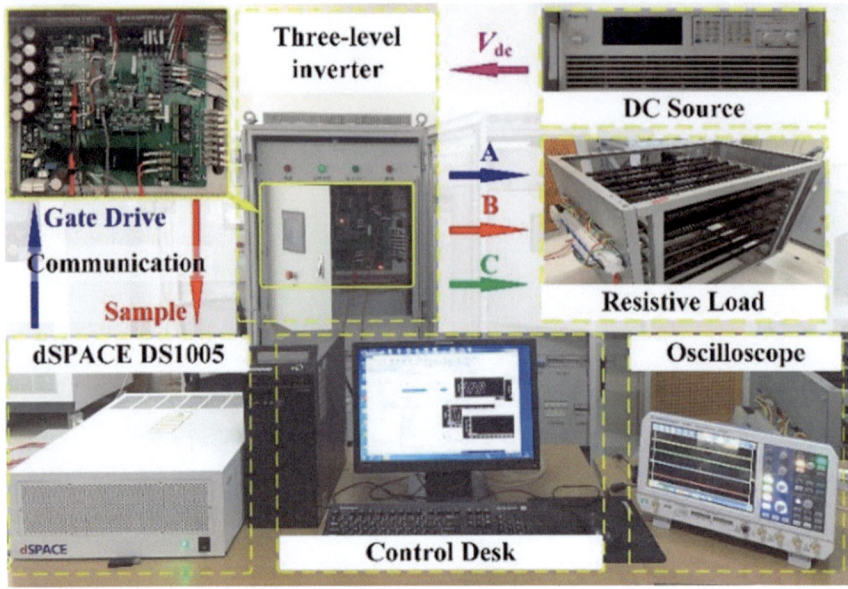

Fig. 4.41 Photograph of experimental platform

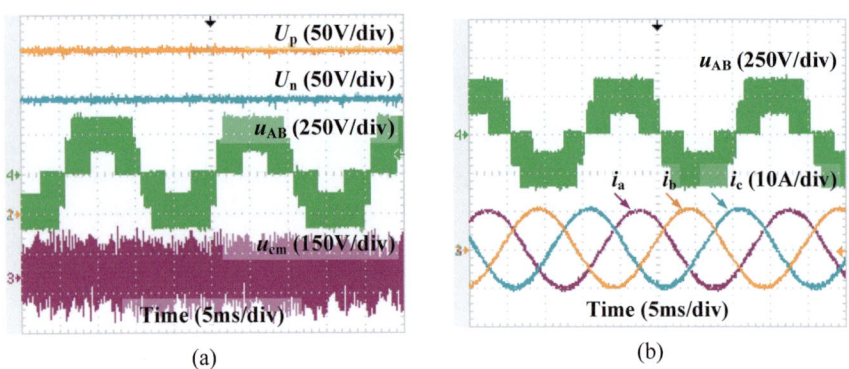

Fig. 4.42 Experimental results of positive unbalancing coefficient for Method-1 **a** capacitor voltage, line voltage, and CMV, **b** three-phase output currents

With a Negative Unbalancing Coefficient

When the voltage difference between the two dc-link capacitors is equal to -60 V, the experimental waveforms are shown in Figs. 4.46, 4.47, and 4.48. In comparison with Method-1 and Method-2, Method-3 can effectively suppress the CMV by half. In addition, three-phase sinusoidal output currents with good quality can be guaranteed

4.3 Simulation and Experimental Validations 201

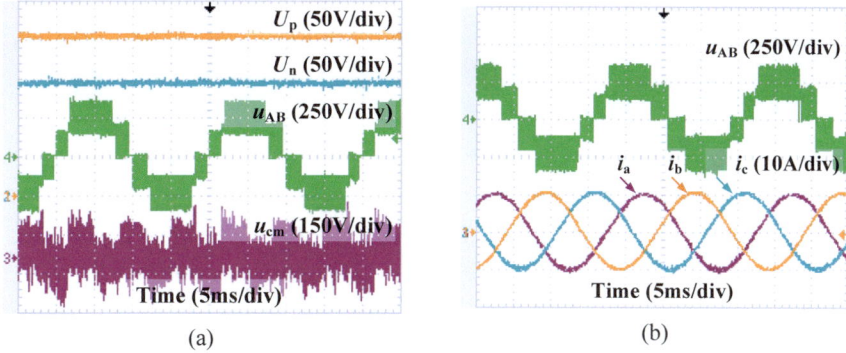

Fig. 4.43 Experimental results of positive unbalancing coefficient for Method-2 **a** capacitor voltages, line voltage, and CMV, **b** three-phase output currents

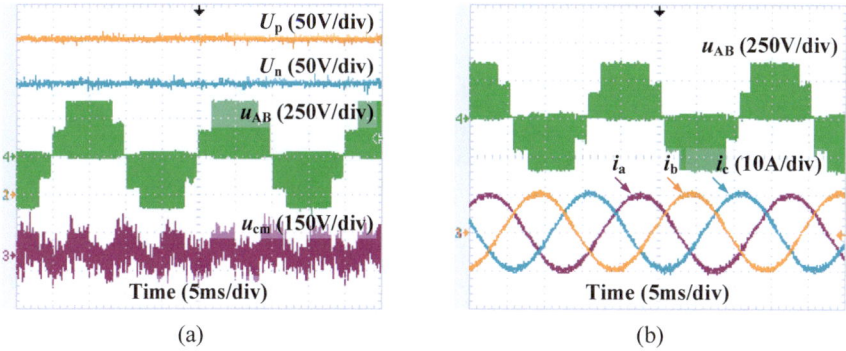

Fig. 4.44 Experimental results of positive unbalancing coefficient for Method-3 **a** capacitor voltages, line voltage, and CMV, **b** three-phase output currents

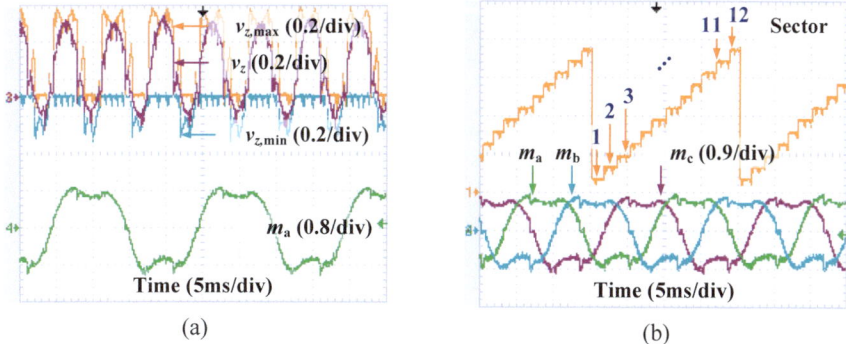

Fig. 4.45 Experimental results of the modulation waveforms, zero-sequence component, and the corresponding limitations for positive unbalancing coefficient ($k = 0.1714$) **a** zero-sequence component and the corresponding limitations, **b** three-phase modulation waveforms

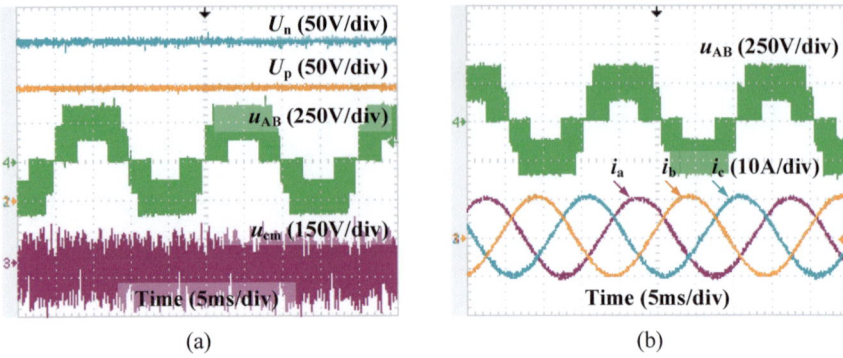

Fig. 4.46 Experimental results of negative unbalancing coefficient for Method-1 **a** capacitor voltage, line voltage, and CMV, **b** three-phase output currents

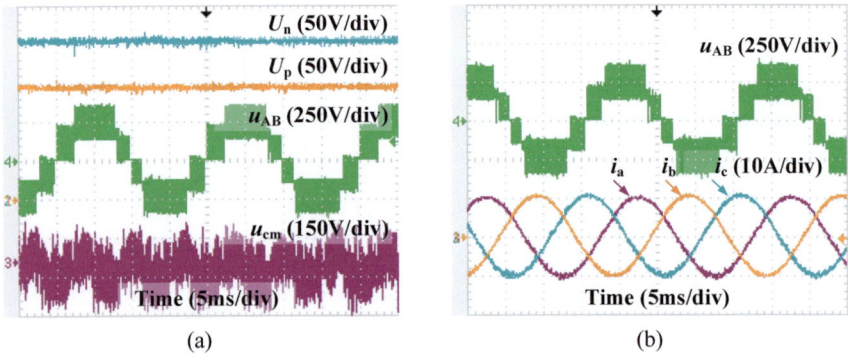

Fig. 4.47 Experimental results of negative unbalancing coefficient for Method-2 **a** capacitor voltage, line voltage, and CMV, **b** three-phase output currents

at the same time, since the unbalanced dc links are considered to synthesize the reference vector and produce the AC output voltage.

The experimental results of modulation waveforms, the zero-sequence component, and its limitations are shown in Fig. 4.49. Since the limitations are taken into consideration, the injected zero-sequence component v_z has been restricted within $v_{z,min}$ and $v_{z,max}$, which can restrict the CMV magnitude.

Under the Balanced NPV Condition

In order to demonstrate the applicable NPV conditions, the experimental tests under a balanced NPV condition have been conducted, and the obtained results are shown in Fig. 4.50. When the reference value of capacitor voltage difference is equal to zero, balanced NPV can be achieved, and the CMV can be reduced in comparison to

4.3 Simulation and Experimental Validations

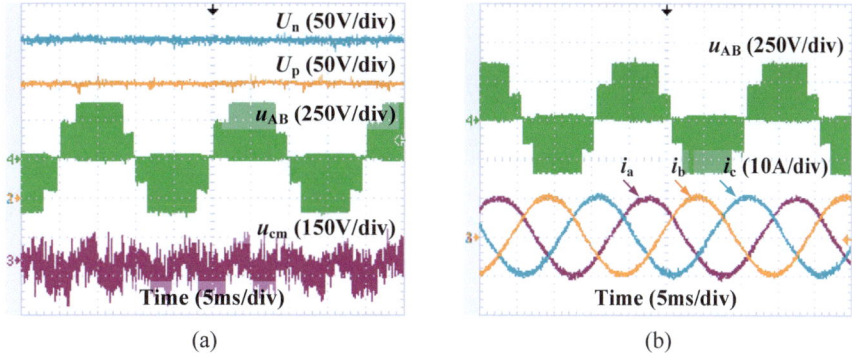

Fig. 4.48 Experimental results of negative unbalancing coefficient for Method-3 **a** capacitor voltage, line voltage, and CMV, **b** three-phase output currents

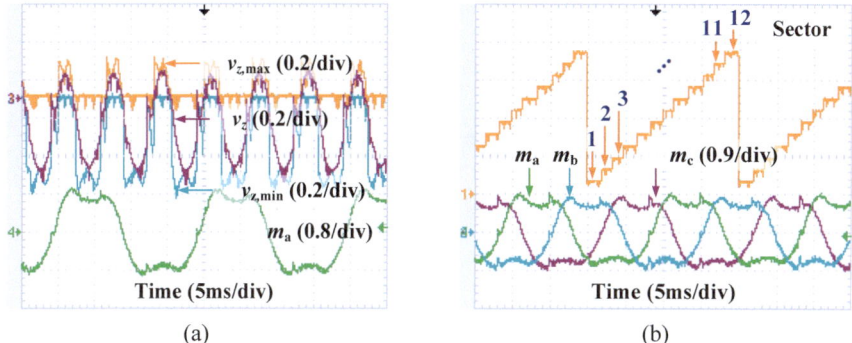

Fig. 4.49 Experimental results of the modulation waveforms, zero-sequence component, and the corresponding limitations for negative unbalancing coefficient ($k = -0.1714$) **a** zero-sequence component and the corresponding limitations, **b** three-phase modulation waveforms

the conventional SVM method. These experimental waveforms reveal that the CMV reduction method can be used under unbalanced and balanced NPV conditions. The experimental results of modulation waveforms, the zero-sequence component, and its limitations with balanced NPV condition are shown in Fig. 4.51. It is clear that the zero-sequence component does not exceed its upper bound and lower bound, which can realize the objective of CMV reduction.

Asymmetric Control of Capacitor Voltages

Next, the asymmetric control of dc links is tested. Figure 4.52a, b show the results of the conventional PI control strategy, whereas Fig. 4.52c, d shows the results of the deadbeat control strategy. It is clear that the asymmetric control of dc-link voltages

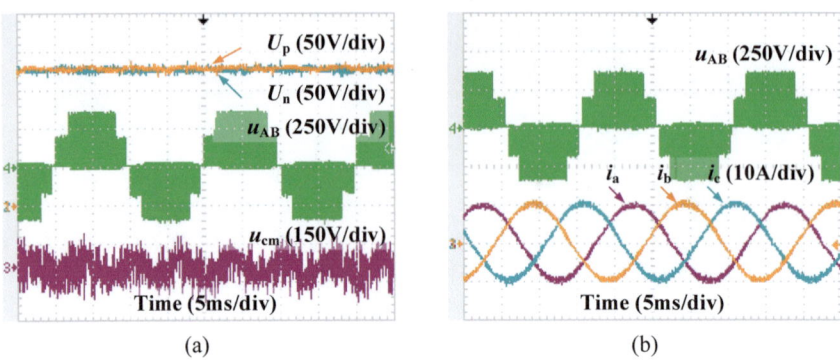

Fig. 4.50 Experimental results with balanced NPV condition for Method-3 **a** capacitor voltage, line voltage, and CMV, **b** three-phase output currents

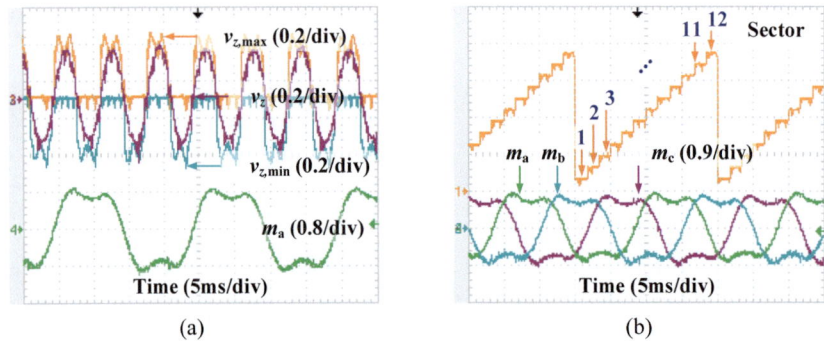

Fig. 4.51 Experimental results of the modulation waveforms, zero-sequence component, and the corresponding limitations with balanced NPV condition **a** zero-sequence component and the corresponding limitations, **b** three-phase modulation waveforms

can be achieved flexibly, and the deadbeat control strategy has a faster dynamic response. Compared with the conventional PI control strategy, the settling time is reduced from 20 to 7 ms. The sinusoidal output current can be obtained whether the NPV is balanced or not.

4.3 Simulation and Experimental Validations

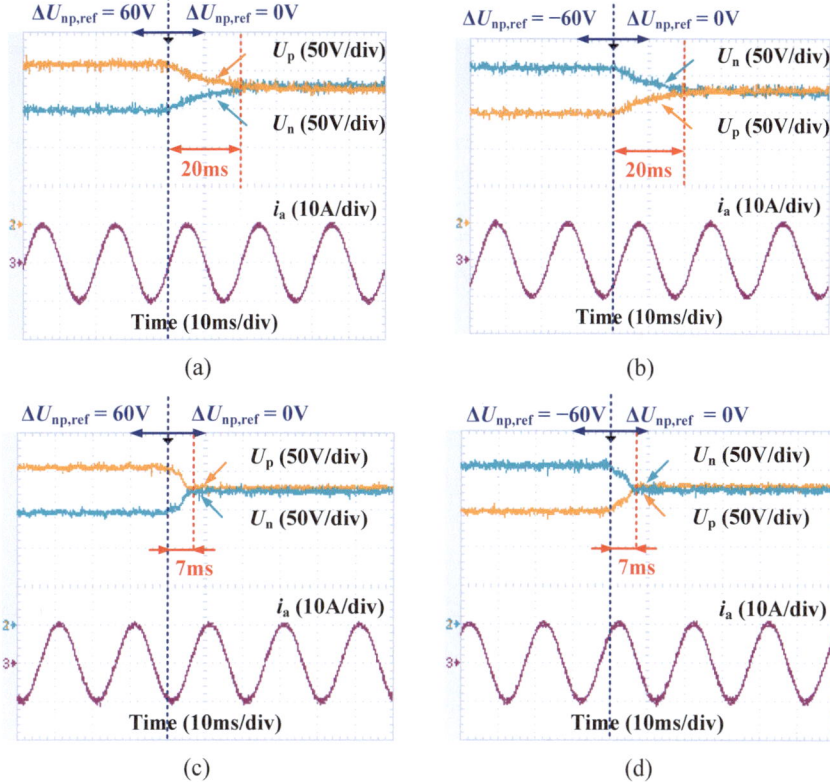

Fig. 4.52 Experimental waveforms of asymmetric control of dc-link voltages **a**, **b** conventional PI control strategy, **c**, **d** deadbeat control strategy

Fault-Tolerant Control Method

This part provides the experimental results of the fault-tolerant control method. The gate driving signal is forced to 0 to simulate the power semiconductor switch fault. Four strategies for the three-level inverter are compared. Strategy-1 represents the FTC strategy with NPV oscillations suppression proposed in [24]. Strategy-2 represents the UNPVC strategy under power semiconductor switch normal operating conditions proposed in [16]. Strategy-3 represents the WC-FTC strategy and strategy-4 represents the HC-FTC strategy. Three cases are included in this part: (1) Steady results, dynamic results and comparison among the four strategies are given in this part.

1. **Steady results**

    Figure 4.53 shows the steady experimental results of the output current of phase A, pole voltage of phase A and the voltages of dc-link capacitors in the four strategies when $k = 0.25$ and $m = 0.9$. As shown in Fig. 4.53a, the NPV cannot be maintained at the desired unbalanced value as strategy-1 was proposed under the balanced NPV conditions. After a period of time, the NPV becomes balanced, which proves that strategy-1 can realize FTC and NPV balance control but cannot realize UNPVC. As shown in Fig. 4.53b, the UNPVC can be perfectly realized in strategy-2 under the normal power semiconductor switch conditions. However, the output current is severely distorted when OCF occurs as shown in Fig. 4.53c, the reason is the mathematical model of NP current is no longer suitable under the power semiconductor switch fault conditions. Figure 4.53d, e show the experimental results of strategy-3 and strategy-4 when S_{a2} fault. Both strategies can

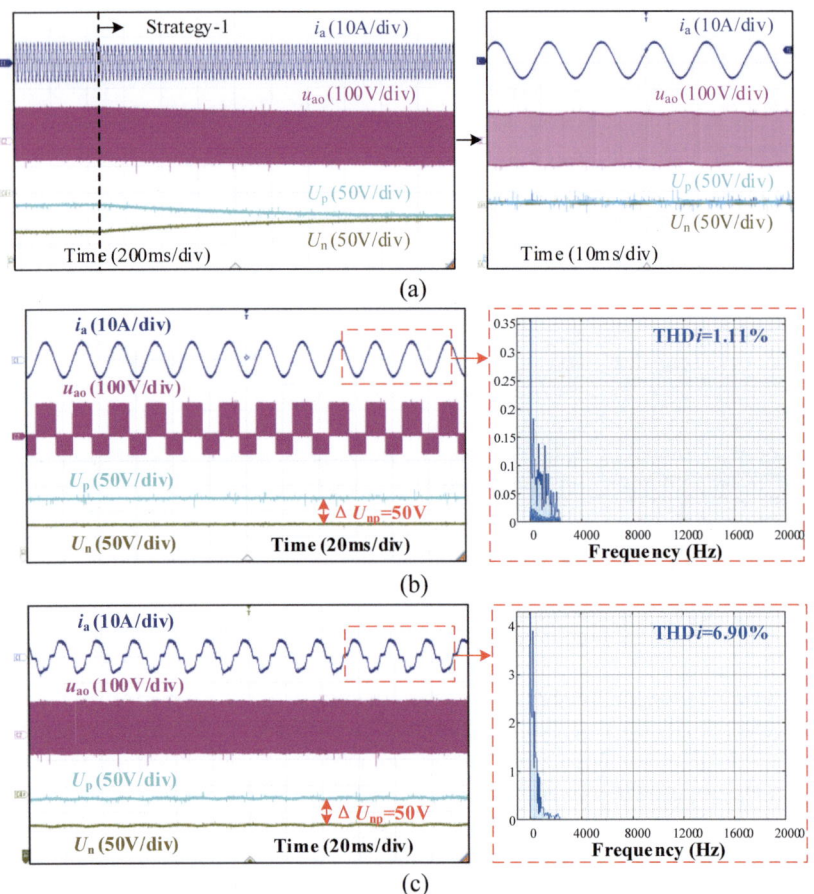

Fig. 4.53 Experimental results when $k = 0.25$ and $m = 0.9$ **a** strategy-1, fault condition, **b** strategy-2, normal condition; **c** strategy-2, fault condition, **d** strategy-3, fault condition, **e** strategy-4, fault condition

4.3 Simulation and Experimental Validations

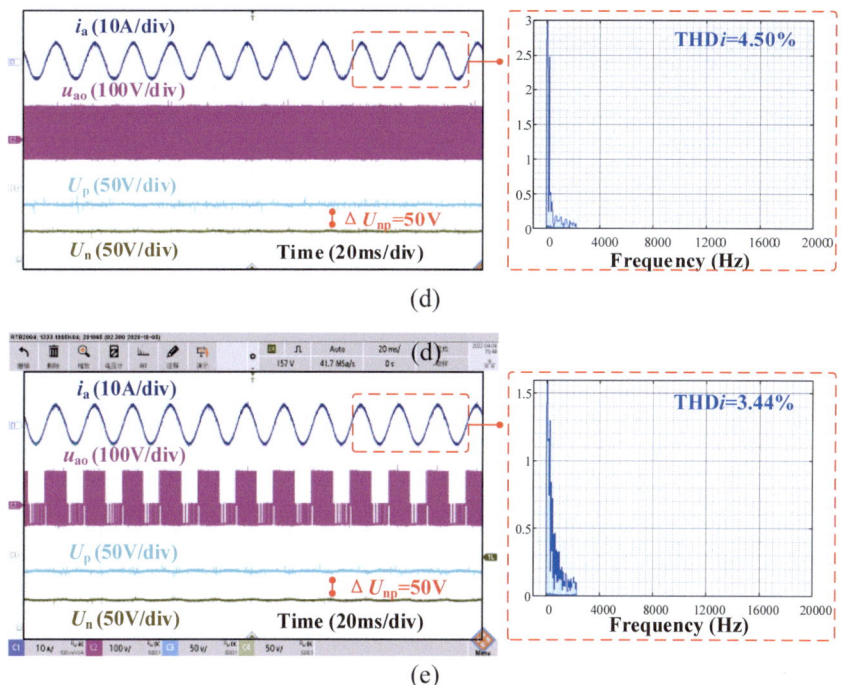

Fig. 4.53 (continued)

realize the FTC and UNPVC, simultaneously. It should be noted that in a fundamental cycle, the pole voltage changes between switching state [P] and [N] when strategy-3 is adopted. While it is the half cycle that the pole voltage changes between switching state [P], [O] and [N] when strategy-4 is adopted, which is benefit to improve the performance of output current.

In order to quantitatively compare the current quality of different strategies, the output phase current total harmonic distortion (THDi) is calculated by

$$\text{THD}i = \sqrt{\sum_{n=2}^{\infty} \left(\frac{I_n}{I_1}\right)^2} \tag{4.91}$$

where I_1 and I_n are the fundamental and nth-order harmonic component of output phase current. As shown in Fig. 4.51, the THDi is 1.11% of strategy-2 under the normal conditions, while it increases to 6.91% when OCF occurs. In strategy-3 and strategy-4, the THDi is reduced to 4.50% and 3.44%, respectively. It is obvious that the THDi of strategy-3 and strategy-4 are both significantly lower than strategy-2 when OCF occurs and NPV is unbalanced. It is noted that the

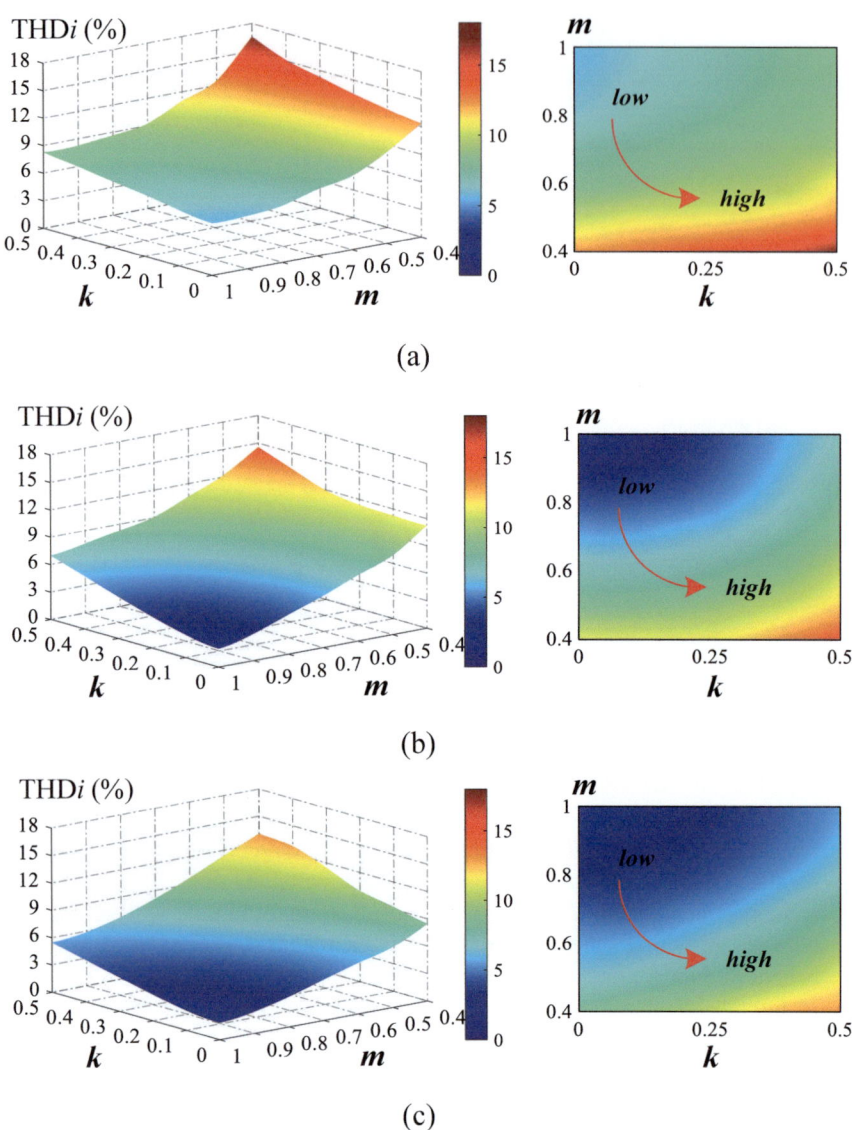

Fig. 4.54 The measured THDi under different k and m **a** strategy-2, **b** strategy-3, **c** strategy-4

THDi of strategy-1 is meaningless, as this strategy cannot maintain unbalanced NPV.

To further comprehensively compare the three strategies in output current performance, Fig. 4.54 illustrates the THDi of the three strategies under different m and k. It is clearly seen that the THDi are all increased with the increase of k and the decrease of m. It is shown that strategy-2 always has the highest THDi

4.3 Simulation and Experimental Validations

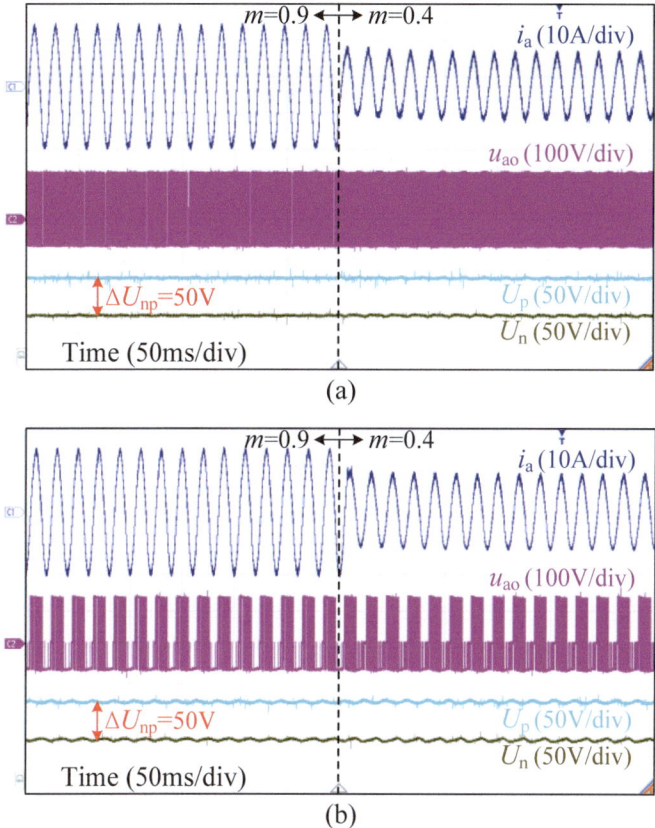

Fig. 4.55 Experimental results when m changes from 0.9 to 0.4 **a** strategy-3, **b** strategy-4

as the current is severely distorted. Furthermore, the THDi of strategy-4 is lower than that of strategy-3 at any operating conditions since the available switching state [O] in the healthy half cycle is adopted, which is consistent with theoretical analysis.

2. **Dynamic results**

    In order to verify the dynamic response ability of the strategies under different operating conditions, dynamic experimental results are shown in Figs. 4.55, 4.56 and 4.57. The dynamic experimental results when $k = 0.25$ and m changes from 0.9 to 0.4 are shown in Fig. 4.55. It can be seen that the phase current always remains sinusoidal and the NPV is maintained at the desired unbalanced value, which proves that the strategies are also effective under low modulation index.

    Besides, in order to compare the dynamic response performance of the two strategies, the dynamic experimental results when k changes from 0.25 to -0.15 and $m = 0.9$ are shown in Fig. 4.56. It can be seen that the strategies are also

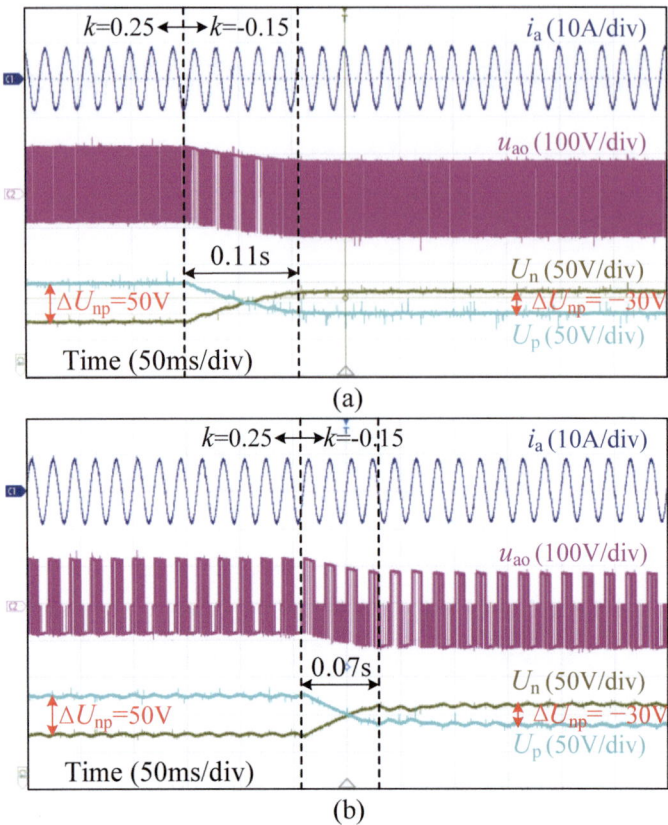

Fig. 4.56 Experimental results when k changes from 0.25 to -0.15 **a** strategy-3, **b** strategy-4

valid when $k < 0$. And the dynamic response time of strategy-3 is 0.11 s, while the dynamic response time of strategy-4 is reduced to 0.07 s.

Figure 4.57 shows the dynamic experimental results when k changes from 0.25 to 0 and $m = 0.9$. The dynamic response time is about 0.08 s in strategy-3. While the dynamic response time is reduced to 0.05 s when strategy-4 is adopted. More importantly, it proves that the strategies are valid whether the NPV is balanced or unbalanced as the mathematical model of NP current is established by the voltage-second balance principle.

For a more detailed comparison of the dynamic response performance in the two strategies, the dynamic response time under different variations of k is shown in Fig. 4.58. It can be seen that strategy-4 has a better dynamic response performance than strategy-3 due to the ZSC selection method in the healthy half cycle, which is consistent with the theoretical analysis.

3. **Comparison among the four strategies**

4.3 Simulation and Experimental Validations

In order to make a comprehensive comparison among the four strategies, Table 4.13 is given. It can be found that the two strategies can simultaneously realize UNPVC and FTC, which have not been realized by existing strategies. Strategy-2 has the highest THDi as the current is severely distorted. Furthermore, the strategy-4 has a better performance in output current and dynamic response compared with strategy-3 due to the additional switching states adoption and the ZSC selection method in the healthy half cycle. Strategy-3 is easier to implement as the current direction judgment and ZSC selection are not needed. More importantly, strategy-4 becomes invalid under some operating conditions as analyzed before, and only strategy-3 can be adopted when OCF occurs both in S_{a2} and S_{a3}. Therefore, when strategy-4 is not applicable, strategy-3 with wider application range is adopted to ensure the safe and reliable operation of the three-level inverter system. While strategy-4 is adopted in order to obtain better current quality and dynamic response ability when it is applicable.

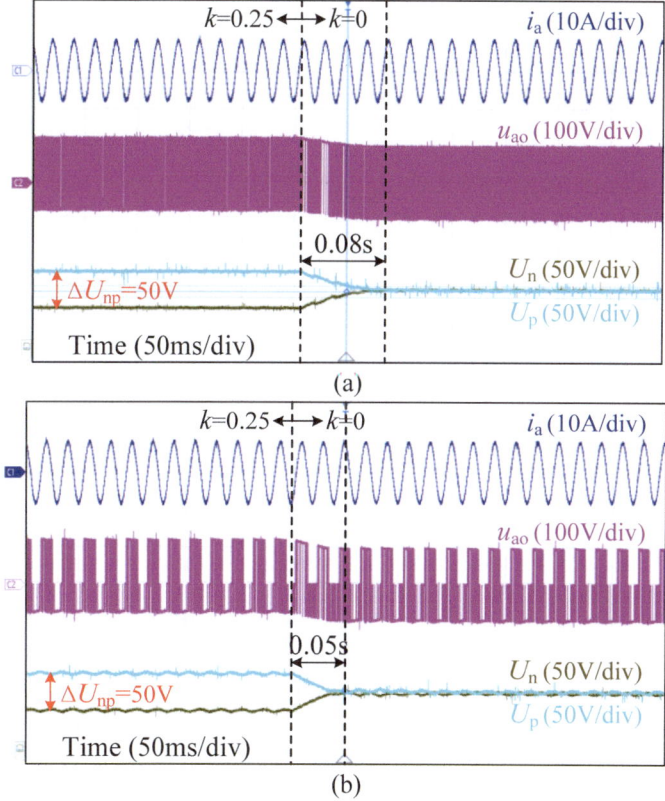

Fig. 4.57 Experimental results when k changes from 0.25 to 0 **a** strategy-3, **b** strategy-4

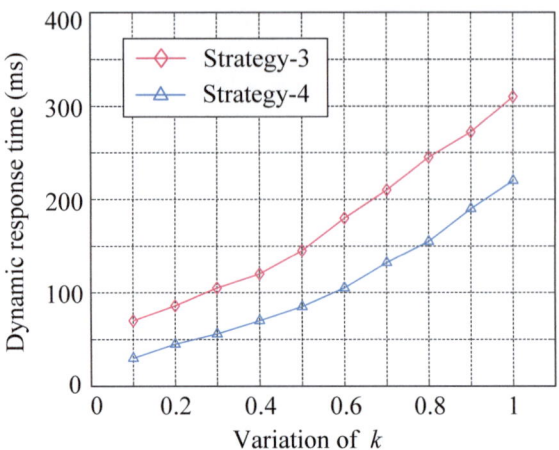

Fig. 4.58 Dynamic response time under different variation of k

Table 4.13 Comparison among the four strategies

	Strategy-1	Strategy-2	Strategy-3	Strategy-4
UNPVC	No	Yes	Yes	Yes
FTC	Yes	No	Yes	Yes
THDi	–	High	Moderate	Low
Dynamic response	–	–	Slow	Fast
Application range	–	–	Wide	Narrow
Complexity	Complex	Simple	Simple	Moderate

4.4 Summary

In this chapter, the control technologies of PV generation systems for maximizing power generation are elaborated, which consist of the conventional MPPT technology and separate MPPT technology. First, for generalized PV applications, the basic principles, classification, and typical strategies of the conventional MPPT technology are introduced. Then, for large-scale PV applications, the separate MPPT technology methods are presented for high-power 3LT^{2}I to further improve the output power. Especially, for the problems caused by imbalanced NPV, various control and modulation methods are presented, which include continuous modulation, discontinuous modulation, and fault-tolerant modulation. The implementation process of each method is elaborated step by step, which provides much guidance for readers. Besides, extensive simulated results based on MATLAB/Simulink software and experimental results based on the prototype system are provided to demonstrate the operating performances with the corresponding control technology. The simulation and experimental results facilitate readers to understand the characteristics of control technology.

References

1. H. Xiao, X. Wang, *Transformerless Photovoltaic Grid-Connected Inverters* (Springer, Singapore, 2020)
2. H. Abu-Rub, M. Malinowski, K. Al-Haddad, *Power Electronics for Renewable Energy Systems, Transportation and Industrial Applications* (Wiley-IEEE Press, 2014)
3. R. Teodorescu, M. Liserre, P. Rodriquez, *Grid Converters for Photovoltaic and Wind Power System* (Wiley, 2011)
4. EPIA, *Global Market Outlook for Photovoltaics 2014–2018* (2014)
5. J.S. Lee, K.B. Lee, Variable DC-link voltage algorithm with a wide range of MPPT for a two-string PV system. Energies **6**(1), 58–78 (2013)
6. Q. Changwei, *Research on High-Performance Modulation and Control Technology of Three-Level Photovoltaic Inverter System* (Shandong University, 2019)
7. H.R. Teymour, D. Sutanto, K.M. Muttaqi, P. Ciufo, Solar PV and battery storage integration using a new configuration of a three-level NPC inverter with advanced control strategy. IEEE Trans. Energy Convers. **29**(2), 354–365 (2014)
8. U. Choi, F. Blaabjerg, K. Lee, Control strategy of two capacitor voltages for separate MPPTs in photovoltaic systems using neutral-point-clamped inverters. IEEE Trans. Ind. Appl. **51**(4), 3295–3303 (2015)
9. Z. Zhang, A. Chen, X. Xing, C. Zhang, Space vector modulation based control strategy of three-level inverter for separate MPPTs in photovoltaic system, in *2016 IEEE 8th International Power Electronics and Motion Control Conference (IPEMC-ECCE Asia)* (2016), pp. 2394–2400
10. Z. Ye, Y. Xu, X. Wu, G. Tan, X. Deng, Z. Wang, A simplified PWM strategy for a neutral-point-clamped (NPC) three-level converter with unbalanced dc links. IEEE Trans. Power Electron. **31**(4), 3227–3238 (2016)
11. Z. Chu, C. Qin, X. Li, L. Yang, C. Hou, H. Bai, DC-link current ripple reduction method for the reduced switch count three-level inverter with unbalanced neutral-point voltages. IEEE J. Emerg. Sel. Topics Power Electron. https://doi.org/10.1109/JESTPE.2025.3525464
12. C. Qin, X. Li, Improved control scheme for simultaneous reduction of common-mode voltage and current harmonic distortion of the Vienna-type rectifier with balanced and unbalanced neutral-point voltages. ISA Trans. **131**, 415–426 (2022)
13. X. Wu, G. Tan, G. Yao, C. Sun, G. Liu, A hybrid PWM strategy for three-level inverter with unbalanced DC links. IEEE J. Emerg. Sel. Topics Power Electron. **6**(1), 1–15 (2018)
14. X. Wu, G. Tan, Z. Ye, G. Yao, Z. Liu, G. Liu, Virtual-space-vector PWM for a three-level neutral-point-clamped inverter with unbalanced DC-links. IEEE Trans. Power Electron. **33**(3), 2630–2642 (2018)
15. C. Qin, X. Li, Improved virtual space vector modulation scheme for the reduced switch count three-level inverter with balanced and unbalanced neutral-point voltage conditions. IEEE Trans. Power Electron. **38**(2), 2092–2104 (2023)
16. C. Qin, X. Li, X. Xing, C. Zhang, G. Zhang, Common-mode voltage reduction method for three-level inverter with unbalanced neutral-point voltage conditions. IEEE Trans. Ind. Informat. **17**(10), 6603–6613 (2021)
17. S. Busquets-Monge, J. Bordonau, D. Boroyevich, S. Somavilla, The nearest three virtual space vector PWM—a modulation for the comprehensive neutral-point balancing in the three-level NPC inverter. IEEE Power Electron. Lett. **2**(1), 11–15 (2004)

18. J. Pou, J. Zaragoza, P. Rodriguez, S. Ceballos, V. Sala, R. Burgos, D. Boroyevich, Fast-processing modulation strategy for the neutral-point-clamped converter with total elimination of the low-frequency voltage oscillations in the neutral point. IEEE Trans. Ind. Electron. **54**(4), 2288–2299 (2007)
19. J. Wang, Y. Gao, W. Jiang, A carrier-based implementation of virtual space vector modulation for neutral-point-clamped three-level inverter. IEEE Trans. Ind. Electron. **64**(12), 9580–9586 (2017)
20. W. Jiang, L. Wang, J. Wang, X. Zhang, P. Wang, A carrier-based virtual space vector modulation with active neutral-point voltage control for a neutral-point-clamped three-level inverter. IEEE Trans. Ind. Electron. **65**(11), 8687–8696 (2018)
21. C. Qin, X. Li, C. Hou, C. Su, A fast-processing implementation method of virtual space vector modulation for the impedance-source three-level inverter. IEEE Trans. Circuits Syst. II, Exp. Briefs **71**(7), 3568–3572 (2024)
22. M.M. Hashempour, M. Yang, T. Lee, An adaptive control of DPWM for clamped-three-level photovoltaic inverters with unbalanced neutral-point voltage. IEEE Trans. Ind. Appl. **54**(6), 6133–6148 (2018)
23. X. Pang, A. Chen, X. Li, Z. Zhong, X. Liu, Carrier-based fault tolerant control strategies for three-level T-type inverter with unbalanced neutral-point voltage. IEEE J. Emerg. Sel. Topics Power Electron. **12**(4), 4154–4164 (2024)
24. J. Chen, C. Zhang, A. Chen, X. Xing, Fault-tolerant control strategies for T-type three-level inverters considering neutral-point voltage oscillations. IEEE Trans. Ind. Electron. **66**(4), 2837–2846 (2019)

Open Access This chapter is licensed under the terms of the Creative Commons Attribution 4.0 International License (http://creativecommons.org/licenses/by/4.0/), which permits use, sharing, adaptation, distribution and reproduction in any medium or format, as long as you give appropriate credit to the original author(s) and the source, provide a link to the Creative Commons license and indicate if changes were made.

The images or other third party material in this chapter are included in the chapter's Creative Commons license, unless indicated otherwise in a credit line to the material. If material is not included in the chapter's Creative Commons license and your intended use is not permitted by statutory regulation or exceeds the permitted use, you will need to obtain permission directly from the copyright holder.

Chapter 5
Advanced Technology of Photovoltaic Inverters for High Robustness and Performance

With the increase of Photovoltaic (PV) industry, the power generation capacity and installed capacity of PV stations are constantly increasing. Under this background, PV equipment suppliers and users from all over the world not merely focus on indicators such as leakage current and MPPT. The high robustness and high performance of PV inverters are gradually becoming the important indicators. Thus, in this chapter, the advanced control technology is proposed to satisfy these newly emerging requirements.

First, the nonlinear control technology for robustness improvement is given, which improves the anti-interference capability of the PV inverter. The nonlinear control technology is integrated into middle loop and inner loop to regulate the dc-link voltage and grid currents. With the nonlinear control technology, the enhanced dynamic and steady-state response of PV inverters is achieved under load disturbance and parameter uncertainties. Then the control technologies for AC-side current harmonic and ripple suppression and DC-side neutral-point voltage (NPV) oscillation suppression are proposed, which improves the transmitted power quality and system safety. The control technologies to achieve the above high performances are integrated into auxiliary loop and PWM modulator. The auxiliary loop adjusts the output variables in real time according to the content of current harmonic/ripple and deviation value of NPV. The output variables are fed into PWM modulator. The PWM modulator adjusts the carrier's phase or modulation waveforms according to the input variables. The modules that achieve the above high robustness and high performances are depicted in Fig. 5.1 with highlighted color.

Moreover, in PV generation system, the storage battery is gradually collocated in dc-link of PV inverter to increase the absorption rate and to suppress fluctuation of power. The configuration of PV generation system with a storage battery is shown in Fig. 5.2. In this system, the power can flow in both directions. The direction of power flowing is determined by the output power of PV panels, the state of battery, the electricity consumption of AC load and the electricity price of the power grid. Typically, when the irradiance of PV panels is sufficient, the PV inverter transfers

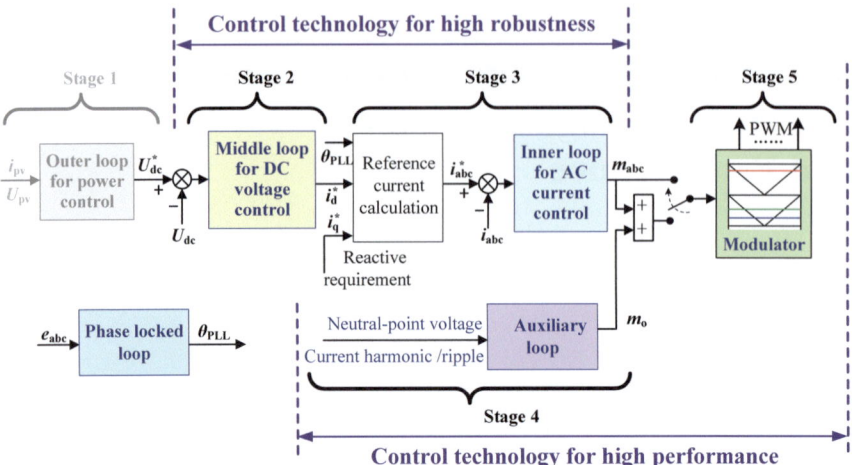

Fig. 5.1 The modules associated with high robustness and high performance in overall control diagram

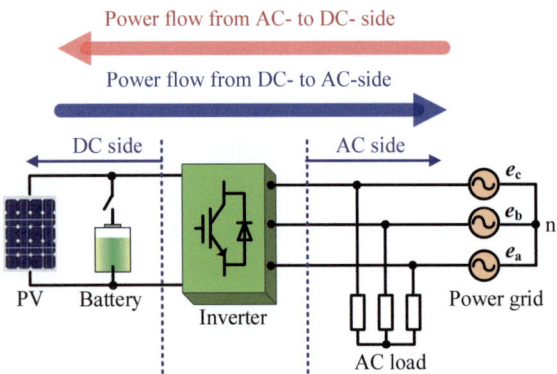

Fig. 5.2 The configuration of PV generation system with storage battery

the energy from PV panel to the load or the power grid. In this case, the power flow from DC side to AC side. At night, the electricity price of the power grid is greatly reduced since the electricity consumption of AC load is dramatically decreased. To reduce the electricity cost and the power fluctuation, the storage battery is usually charged by the power grid. In this case, the power flow from AC side to DC side.

It should be noted that both power flowing directions and topology of PV inverters directly affect the control effect. To comprehensively explain the control technology for robustness and performance improvement of the PV inverter system, two power flowing directions and two typical topologies are adopted to elaborate corresponding control technologies. Two power flowing directions include from AC- to DC-side and from DC- to AC-side. Two typical topologies include two-level inverter and three-level T-type inverter ($3LT^2I$).

5.1 Nonlinear Control Technology for High Robustness

The control strategy directly determines the achievement of control objectives and system performance of the PV inverter when the modulation strategies are fixed. However, PV inverter is a strongly nonlinear system, and its control performance is affected by multiple sources of disturbances. On one hand, parameter uncertainties lead to deviations between the system's mathematical model and the actual system. On the other hand, the PV inverter faces disturbances from both the DC- and AC-sides, such as dc-link current variations and grid voltage fluctuations on the AC side. Under these disturbances, traditional linear PI control methods result in slow dynamic response, poor disturbance rejection, weak robustness, and low control accuracy, potentially even causing system instability. Therefore, researching advanced nonlinear control strategies for PV inverters to improve robustness has significant scientific research value and practical relevance.

In this section, two advanced nonlinear robust control strategies are introduced for PV inverters, which are an improved finite-time control (IFTC) method and a disturbance observer-based finite-time control (DOFTC) method, respectively. The IFTC method belongs to the dual loop voltage and direct current control structure. In this control structure, current loop is used to realize fast and accurate current tracking, and the voltage loop is used to counteract load disturbances and maintain the dc voltage stable. This control structure is clear and intuitive, which is widely used in engineering. The DOFTC method belongs to the dual loop voltage and indirect current control structure. In this control structure, voltage loop is still used to maintain the dc voltage stable, while the indirect current loop is used to realize current tracking control by controlling active and reactive power. Different from the dual-loop voltage and direct current control structure, the dual-loop voltage and indirect current control structure directly regulates power without the need for a phase-locked loop (PLL). The theoretical derivation and implementation process of these two control strategies are elaborated in the following sections.

In this section, the specific structure of the PV generation system shown in Fig. 5.3 is taken as an example to illustrate the control strategies for robustness improvement. In this specific structure, the topology is $3LT^2I$ and the power flowing direction is from AC side to DC side. In this case, the current flowing from AC side to DC side is defined as positive.

In this structure, the capacitors C_1 and C_2 are adopted to filter voltage ripples and reduce voltage variations. Usually, these two capacitors are generally chosen to have the same capacitance value C. On the AC side, the $3LT^2I$ is connected to the power grid through the L filter. R denotes the equivalent series resistance of the filter and the transmission line. U_{dc} and e_{abc} denote the dc-link voltage and grid voltages,

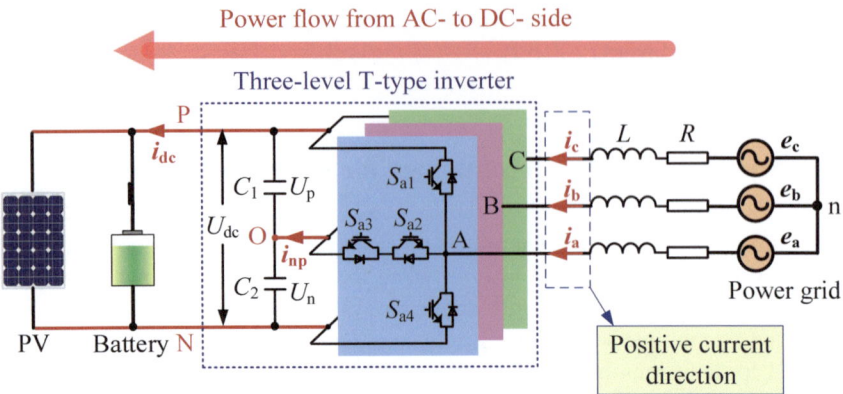

Fig. 5.3 The topology of 3LT²I with the power flowing from AC side to DC side

respectively. For the convenience of modeling and analysis, the gird currents are i_a, i_b, and i_c, and dc-link current is i_{dc}.

5.1.1 Improved Finite-Time Control Method

Based on the dual-loop voltage and direct current control structure, the author presents an IFTC strategy for PV inverter to enhance the robustness and guarantee the control performance under system uncertainties [1]. Specifically, the voltage loop employs a finite-time adaptive controller that can counteract dc-link disturbances without relying on current sensors. In the direct-current tracking loop, the control objective is the current, and the finite-time controllers combined with a command filter are constructed to obtain fast and accurate current tracking performance. The command filter is utilized to avoid calculating the derivative of current references. Theoretical analysis demonstrates the effectiveness of the IFTC strategy. The models of the 3LT²I with system uncertainties are first established. Then, based on the models, the design process of IFTC is given step by step. Modeling and controller design details are described in the following sections.

Modeling of 3LT²I With System Uncertainties

Before starting the modeling, some standard assumptions are imposed on the 3LT²I, including ideal grid voltages, balanced NPV, and neglected power losses.

Using the Kirchhoff's voltage law, the AC-side mathematical model in dq rotating coordinate frame is built as

5.1 Nonlinear Control Technology for High Robustness

$$\begin{cases} L\dot{i}_d = -Ri_d + \omega L i_q + e_d - \frac{U_{dc}}{2}m_d \\ L\dot{i}_q = -Ri_q - \omega L i_d + e_q - \frac{U_{dc}}{2}m_q \end{cases} \quad (5.1)$$

where e_{dq}, $i_{dq,n}$ and m_{dq} are grid voltages, grid currents and averaged duty cycles in the dq rotating coordinate frame, respectively. ω represents the grid frequency.

In practice, the filter inductance varies with operating conditions. On one hand, due to the nonlinearity of the magnetic core, the inductance of the inductor varies with the flowing current. On the other hand, the temperature of the magnetic core also influences the value of inductance. Therefore, the value of the filter inductance is formalized as

$$L = L_0 + \Delta L \quad (5.2)$$

where L_0 is the nominal value of the filter inductance and ΔL is the uncertain part. Besides, the line resistance R is difficult to measure and is also considered as the uncertain parameter. Considering the aforementioned parameter uncertainties, the AC-side dynamic model is rewritten as

$$\begin{cases} \dot{i}_d = \omega i_q + \frac{1}{L_0}e_d - \frac{U_{dc}}{2L_0}m_d + d_1(t) \\ \dot{i}_q = -\omega i_d + \frac{1}{L_0}e_q - \frac{U_{dc}}{2L_0}m_q + d_2(t) \end{cases} \quad (5.3)$$

where d_1, d_2 are the lumped disturbances of AC side caused by parameter uncertainties, which are bounded disturbances and expressed as

$$\begin{cases} d_1(t) = \frac{\Delta L}{L_0}\omega i_q - \frac{\Delta L}{L_0}\dot{i}_d - \frac{1}{L_0}Ri_d \\ d_2(t) = -\frac{\Delta L}{L_0}\omega i_d - \frac{\Delta L}{L_0}\dot{i}_q - \frac{1}{L_0}Ri_q \end{cases} \quad (5.4)$$

At the dc-link, using the law of power conservation, it is obtained that

$$\frac{1}{2}CU_{dc}\dot{U}_{dc} = \frac{3}{2}e_d i_d - U_{dc}i_{dc} \quad (5.5)$$

The above expression formalized as

$$\begin{cases} \dot{U}_{dc} = 3\frac{e_d}{CU_{dc}}i_d - 2U_{dc}\rho \\ \rho = \frac{i_{dc}}{CU_{dc}} \end{cases} \quad (5.6)$$

where ρ is taken as the unknown dc-link disturbances caused by dc-link current variations.

Combining the AC-side and dc-link models, it is revealed that the 3LT²I is essentially a multivariable nonlinear system affected by dc-link disturbances and parameter uncertainties. Therefore, it's meaningful to investigate an advanced robust control strategy that ensures high-quality output performance.

Design Improved Finite-Time Controller

To enhance the robustness, an IFTC strategy based on dual-loop voltage and direct current control structure is proposed for the $3LT^2I$. This section includes three steps. Step 1: In the voltage loop, a finite-time adaptive controller is designed to ensure the control performance under dc-link disturbances. Step 2: In the direct-current loop, the finite-time controller, combined with a command filter, is designed for fast and accurate current tracking performance. Step 3: Theoretical analysis demonstrates the effectiveness and stability of this method.

Step 1: Design of Voltage Loop with Finite-Time Adaptive Controller

In the voltage regulation loop, the task is to regulate the dc-link voltage to reach maintain in its reference value. Its main obstacle exists in dc-link disturbances that bring damaging active power impulses and cause large dc-link voltage fluctuations. In terms of control design, the challenge lies in designing a disturbance observer technique with a simple structure to estimate and suppress the effects of dc-link disturbances. Furthermore, combining such a disturbance observer technique with the finite-time control poses challenges in the stability analysis. To address these issues, a finite-time adaptive controller is constructed to estimate and compensate for dc-link disturbances, which guarantees the control performance under dc-link disturbances. Details are given as follows.

For the $3LT^2I$, the current dynamics are much faster than the dc-link voltage dynamics. Using singular perturbation theory, the dc-link dynamic is reduced to

$$\dot{U}_{dc} = 3\frac{e_d}{CU_{dc}}i_d^* - 2U_{dc}\rho \tag{5.7}$$

where i_d^* denotes the reference of i_d and is also the introduced virtual controller for regulating dc-link voltage.

Define the voltage regulation error and the disturbance estimation error $\tilde{\rho}$ as

$$\begin{cases} z_1 = U_{dc} - U_{dc}^* \\ \tilde{\rho} = \rho - \hat{\rho} \end{cases} \tag{5.8}$$

and choose a Lyapunov function candidate as

$$V_1 = \frac{1}{2}z_1^2 + \frac{1}{2r}\tilde{\rho}^2 \tag{5.9}$$

where $\hat{\rho}$ is the estimation of ρ, and r is a positive constant value. Then, the derivative of this Lyapunov function is expressed as

$$\dot{V}_1 = z_1\dot{z}_1 - \frac{1}{r}\tilde{\rho}\dot{\hat{\rho}} = z_1\left(3\frac{e_d}{CU_{dc}}i_d^* - 2U_{dc}\rho\right) - \frac{1}{r}\tilde{\rho}\dot{\hat{\rho}} \tag{5.10}$$

5.1 Nonlinear Control Technology for High Robustness

If one constructs $\hat{\rho}$ and i_d^* such that $\dot{V}_1$ meets Lemma 2 in [2], then the disturbance estimation error $\tilde{\rho}$ and the voltage regulation error z_1 tend to an arbitrarily small neighborhood of zero in a finite time. Under this guidance, the adaptive law and the finite-time voltage controller are constructed as

$$\begin{cases} \dot{\hat{\rho}} = -2rz_1 U_{dc} - \sigma\hat{\rho} - \sigma\hat{\rho}^\mu \\ i_d^* = \frac{1}{3e_d} U_{dc}[-k_1 z_1 - s_1 \eta(z_1) + 2CU_{dc}\hat{\rho}] \end{cases} \quad (5.11)$$

where σ, k_1 and s_1 are positive control parameters. $\mu = \mu_1/\mu_2$, where μ_1 and μ_2 are positive odd integers with $\mu_1 < \mu_2$. $\eta(z_1)$ is a smooth switch function described by

$$\eta(z_1) = \begin{cases} z_1^\mu, & |z_1| \geq \kappa \\ \gamma_1 z_1 + \gamma_2 z_1^3, & |z_1| < \kappa \end{cases} \quad (5.12)$$

where $\gamma_1 = (1/2)(3-\mu)\kappa^{\mu-1}$, $\gamma_2 = (1/2)(\mu-1)\kappa^{\mu-3}$, and κ is a small positive constant. Using this smooth switch function, the constructed voltage controller satisfies C_1 continuous, and its derivative exists and is bounded [3].

Substituting the constructed adaptive law and finite-time voltage controller into $\dot{V}_1$ yields

$$\begin{aligned} \dot{V}_1 &= -\frac{1}{C}k_1 z_1^2 - \frac{1}{C}s_1 z_1 \eta(z_1) - 2z_1 V_{dc}\tilde{\rho} - \frac{1}{r}\tilde{\rho}\dot{\hat{\rho}} \\ &\leq -\frac{1}{C}k_1 z_1^2 - \frac{1}{C}s_1 z_1^{\mu+1} + s_1 \kappa^{\mu+1} + \frac{\sigma}{r}\tilde{\rho}\hat{\rho} + \frac{\sigma}{r}\tilde{\rho}\hat{\rho}^\mu \end{aligned} \quad (5.13)$$

Based on the Young's inequality and Lemma 3 in [2], the following inequalities hold that

$$\begin{cases} \tilde{\rho}\hat{\rho} \leq -\frac{1}{2}\tilde{\rho}^2 + \frac{1}{2}\rho^2 \\ \tilde{\rho}\hat{\rho}^\mu \leq -c\tilde{\rho}^{\mu+1} + d\rho^{\mu+1} \end{cases} \quad (5.14)$$

where $c > 0$ and $d > 0$. Then, we have

$$\begin{aligned} \dot{V}_1 &\leq -\frac{1}{C}k_1 z_1^2 - \frac{1}{C}s_1 z_1^{\mu+1} - \frac{\sigma}{2r}\tilde{\rho}^2 - \frac{\sigma c}{r}\tilde{\rho}^{\mu+1} \\ &\quad + \frac{1}{C}s_1 \kappa^{\mu+1} + \frac{\sigma}{2r}\rho^2 + \frac{\sigma d}{r}\rho^{\mu+1} \\ &\leq -a_1 V_1 - b_1 V_1^\gamma + \delta_1 \end{aligned} \quad (5.15)$$

where $a_1 = \min\{2k_1/C, \sigma\}$, $b_1 = \min\{2^\gamma s_1/C, 2^\gamma c\sigma r^{\gamma-1}\}$, $\delta_1 = s_1\kappa^{\mu+1}/C + \sigma/2r\rho^2 + \sigma d/r\rho^{\mu+1}$, and $\gamma = (\mu+1)/2$. Based on this, the fast finite-time stability criterion in [2] is satisfied. The main results are presented as follows.

For the dc-link subsystem with dc-link disturbances, using the finite-time adaptive voltage controller, $\tilde{\rho}$ and z_1 tend to an arbitrarily small neighborhood of zero in a finite time, achieving fast and accurate voltage control under dc-link disturbances. Its proof will be given later.

Step 2: Design of Inner Loop with Finite-Time Controller Combined with a Command Filter

In the current tracking loop, the task is to drive i_d and i_q to track their references. Firstly, define current tracking errors as

$$\begin{cases} z_2 = i_d - i_d^* \\ z_3 = i_q - i_q^* \end{cases} \tag{5.16}$$

where i^*q is the reference of i_q. Then, choose a Lyapunov function candidate as

$$V_2 = \frac{1}{2}z_2^2 + \frac{1}{2}z_3^2 \tag{5.17}$$

Taking the derivative of V_2 yields

$$\begin{aligned} \dot{V}_2 &= z_2 \dot{z}_2 + z_3 \dot{z}_3 \\ &= z_2 \left[\omega i_q + \frac{1}{L_0} e_d - \frac{U_{dc}}{2L_0} m_d + d_1(t) - \dot{i}_d^* \right] \\ &\quad + z_3 \left[-\omega i_d + \frac{1}{L_0} e_q - \frac{U_{dc}}{2L_0} m_q + d_2(t) - \dot{i}_q^* \right] \end{aligned} \tag{5.18}$$

As per the above equation, the derivatives of i^*d and i^*q are needed for constructing the current tracking controller, which brings the difficulty. Usually, i^*q is set as a constant, while i^*d is a time-varying signal generated by the voltage regulation loop. To this end, a command filter is introduced to approximate the derivative of i^*d, and its structure is given as

$$\begin{cases} \dot{\varphi}_1 = \varphi_2 \\ \dot{\varphi}_2 = \frac{1}{\zeta^2}\left[-\rho_1 \arctan(\varphi_1 - i_d^*) - \rho_2 \arctan(\zeta \varphi_2)\right] \end{cases} \tag{5.19}$$

where φ_1 and φ_2 are filter output signals with i^*d being the input. $\zeta > 0$ is a filter parameter. $\rho_1 > 0$ and $\rho_2 > 0$ are utilized for increasing the adjustment freedom.

Based on Lemma 2 in [4], it is obtained that the filter error satisfies

$$\left|\varphi_2 - \dot{i}_d^*\right| \leq \mathcal{O} \tag{5.20}$$

after a finite-time transient process t_f, where $\mathcal{O}$ is bounded and can be made small enough by selecting proper ζ. In particular, the smaller ζ is, the smaller $\mathcal{O}$ is.

Similarly, a finite-time current controller is constructed as

$$\begin{cases} m_d = -\frac{2}{U_{dc}}\left(-k_2 z_2 - s_2 z_2^\mu - \omega L_0 i_q - e_d + L_0 \varphi_2\right) \\ m_q = -\frac{2}{U_{dc}}\left(-k_2 z_3 - s_2 z_3^\mu + \omega L_0 i_d - e_q\right) \end{cases} \tag{5.21}$$

5.1 Nonlinear Control Technology for High Robustness

where k_2 and s_2 are positive control parameters.

Substituting the finite-time current controller, it is obtained that

$$\begin{aligned} \dot{V}_2 &= -\frac{k_2}{L_0} \sum_{i=2}^{3} z_i^2 - \frac{s_2}{L_0} \sum_{i=2}^{3} z_i^{\mu+1} + z_2 d_1 + z_3 d_2 - z_2 (\varphi_2 - \dot{i}_d^*) \\ &\leq -\left(\frac{k_2}{L_0} - 1\right) \sum_{i=2}^{3} z_i^2 - \frac{s_2}{L_0} \sum_{i=2}^{3} z_i^{\mu+1} + \tfrac{1}{2} d_1^2 + \tfrac{1}{2} d_2^2 + \tfrac{1}{2} \mathcal{O}^2 \\ &\leq -a_2 V_2 - b_2 V_2^{\gamma} + \delta_2 \end{aligned} \quad (5.22)$$

with $a_2 = 2 k_2/L_0 - 2$, $b_2 = 2^{\gamma} s_2/L_0$, and $\delta_2 = 1/2 d2\ 1 + 1/2 d2\ 2 + 1/2\ \mathcal{O}^2$. The main results are presented as follows.

For the AC-side subsystem with parameter uncertainties, using the finite-time current controller with the command filter, z_2 and z_3 tend to an arbitrarily small neighborhood of zero in a finite time. Its proof will be given later.

Figure 5.4 displays the control block diagram of the proposed IFTC strategy. It integrates a dual-loop voltage and direct current control structure to regulate the dc-link voltage and grid currents. Additionally, the difference of the dc-link capacitor voltages should be maintained around zero to balance the NPV. For the purpose of achieving this goal simply and efficiently, a classical PI method is utilized.

Step 3: Theoretical Analyses for Effectiveness and Stability of IFTC

Due to the coupling between the voltage regulation loop and the current tracking loop, the system stability needs to consider both loops together. Thus, choose a Lyapunov function as

$$V = V_1 + V_2 \quad (5.23)$$

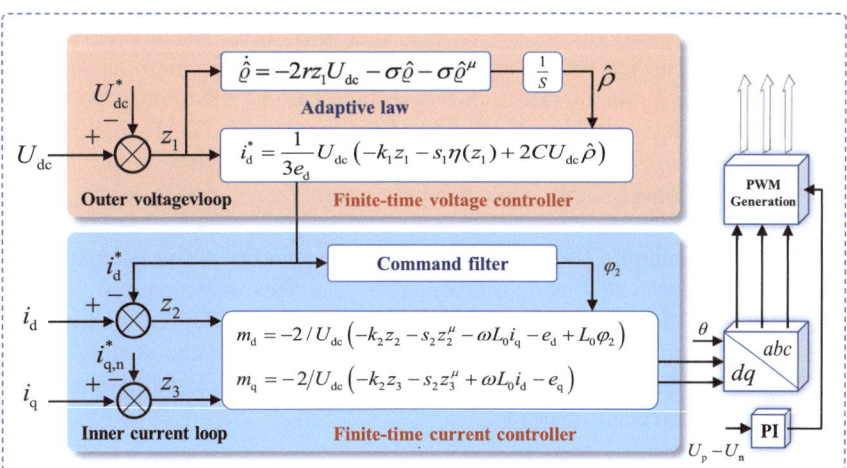

Fig. 5.4 Block diagram of the IFTC strategy based on dual-loop voltage and direct current control structure

Then, it is obtained that

$$\dot{V} \leq -aV - bV^{\gamma} + \delta \tag{5.24}$$

with $a = \min\{a_1, a_2\}$, $b = \min\{b_1, b_2\}$, and $\delta = \delta_1 + \delta_2$.

Choosing the control parameters such that $a > 0$ and $b > 0$, we have

$$\begin{cases} |z_i| \leq \sqrt{\frac{2\delta}{a-\tau}} \\ |\tilde{\rho}| \leq \sqrt{\frac{2\delta}{a-\tau}} \end{cases} \tag{5.25}$$

within a finite time, $0 < \tau < a$ [2]. The radiuses of the region of tracking errors are adjustable and are small enough when selecting large a. This completes the proof.

In summary, this section investigates a robust IFTC strategy of the $3LT^2I$ based on the dual-loop voltage and direct current control structure. The voltage loop designs a finite-time adaptive controller to enhance control performance in the presence of dc-link disturbances. By integrating the current controller with a command filter, the current tracking loop achieves fast and accurate current tracking performance. Theoretical analysis demonstrates the IFTC strategy's effectiveness.

5.1.2 Disturbance Observer-Based Finite-Time Control Method

In the dual-loop voltage and direct current control structure, PLL controller is required to obtain the phase of grid voltage, so that the mathematical model in dq rotating coordinate frame is established. Usually the performance of dc-link voltage and grid current is also affected by the PLL controller. Thus, lots of factors should be considered to design the structure and parameters of PLL controller, which is a very complicated process. Thus, in this section, based on the dual-loop voltage and indirect current control structure, the DOFTC method for $3LT^2I$ is presented by the author's researcher team [4]. In this control structure, voltage loop is used to maintain the dc voltage stable, while the inner control loop is used to realize current tracking control by controlling active and reactive power, which is defined as the indirect current control. This method enhances the robustness and guarantees the control performance under load disturbance and parameter uncertainties.

In DOFTC method, a finite-time disturbance observer (FTDO) is designed to fast and accurately estimate disturbances. Then, the disturbance estimations are compensated in the feedback control loop. Finally, the stability of the closed-loop system is analyzed and theoretically proved.

5.1 Nonlinear Control Technology for High Robustness

Mathematical Modeling for DOBFTC

Reviewing the content of above section, the dynamic model of the dc-link is expressed as

$$\begin{cases} \dot{U}_{dc} = 2\frac{P}{CU_{dc}} + d_1 \\ d_1 = -2\frac{i_{dc}}{C} \end{cases} \quad (5.26)$$

where d_1 is taken as the unknown dc-link disturbances.

In the $\alpha\beta$ stationary coordinate frame, the instantaneous active power P and reactive power Q of the AC side are defined as

$$\begin{cases} P = \frac{3}{2}e_\alpha i_\alpha + \frac{3}{2}e_\beta i_\beta \\ Q = \frac{3}{2}e_\beta i_\alpha - \frac{3}{2}e_\alpha i_\beta \end{cases} \quad (5.27)$$

where i_α and i_β are the $\alpha\beta$-axis components of i_{abc}. To directly regulate the instantaneous power of the 3LT^{2}I without any PLL involved, Hu et al. [5] proposed an AC side model in the $\alpha\beta$ stationary coordinate frame as

$$\begin{cases} \dot{P} = -\frac{3}{2L}e_\alpha u_\alpha - \frac{3}{2L}e_\beta u_\beta - \omega Q - \frac{R}{L}P + \frac{3}{2L}\left(e_\alpha^2 + e_\beta^2\right) \\ \dot{Q} = \frac{3}{2L}e_\alpha u_\beta - \frac{3}{2L}e_\beta u_\alpha + \omega P - \frac{R}{L}Q \end{cases} \quad (5.28)$$

where u_α and u_β are the $\alpha\beta$-axis components of u. ω is the grid angle frequency and R is the equivalent resistance of the power transmission line.

Then, an improved AC-side model is built as

$$\begin{cases} \dot{P} = -\frac{3}{2L}u_1 - \omega Q - \frac{R}{L}P + \frac{3}{2L}\left(e_\alpha^2 + e_\beta^2\right) + d_2 \\ \dot{Q} = \frac{3}{2L}u_2 + \omega P - \frac{R}{L}Q + d_3 \end{cases} \quad (5.29)$$

in which d_1 and d_2 are lumped disturbances with respect to the AC-side model. The introduced new system control inputs are defined as

$$\begin{bmatrix} u_1 \\ u_2 \end{bmatrix} = \begin{bmatrix} e_\alpha & e_\beta \\ -e_\beta & e_\alpha \end{bmatrix} \begin{bmatrix} u_\alpha \\ u_\beta \end{bmatrix} \quad (5.30)$$

and we define $A = [e_\alpha, e_\beta, -e_\beta, e_\alpha]$. The new control inputs u_1 and u_2 are introduced to decouple original control inputs u_α and u_β [6]. Based on this, we can directly and independently regulate active and reactive power.

Design of Disturbance Observer-Based Finite-Time Controller

For the 3LT^{2}I with disturbance d_i, the DOFTC strategy is proposed, which consists of three parts: the finite-time controller, the FTDO, and the stability analysis of the closed-loop system.

Part 1: Finite-Time Controller

The continuous finite-time controller is designed on account of the practical finite-time Lyapunov stability criterion and the backstepping design procedure. According to the mathematical model of 3LT^{2}I and the basic idea of backstepping technique, the cascade structure is obtained by introducing the virtual controller P^*, where the dc-link voltage U_{dc} is adjusted when P tends to P^*, and the power P and Q are controlled by the designed u_1 and u_2. Subsequently, the proposed finite-time controller inherently has dual-loop structure consisting of a voltage loop for voltage control and an inner loop for power control, as shown in Fig. 5.5.

Each stage of backstepping design only needs to deal with a relatively simple system, so that the constructed controller is flexible and simple. Define the closed-loop tracking errors of the dynamic system as

$$\begin{cases} z_1 = U_{dc} - U_{dc}^* \\ z_2 = P - P^* \\ z_3 = Q - Q^* \end{cases} \quad (5.31)$$

in which U^*dc, P^* and Q^* are the references of U_{dc}, P and Q, respectively. Naturally, the control objective is to ensure that z_i tends to a small neighborhood of zero in finite time, $i = 1, 2, 3$.

To achieve the abovementioned control objectives and avoid the chattering problem, continuous finite-time controllers are constructed as

$$\begin{cases} P^* = \frac{1}{2}CU_{dc}\left(-k_1 z_1 - s_1 z_1^\gamma - \hat{d}_1\right) \\ u_1 = -\frac{2}{3}L\left(-k_2 z_2 - s_2 z_2^\gamma - f_1 - \hat{d}_2 - \frac{2z_1}{CU_{dc}} + \hat{\dot{P}}^*\right) \\ u_2 = \frac{2}{3}L\left(-k_3 z_3 - s_3 z_3^\gamma - f_2 - \hat{d}_3\right) \end{cases} \quad (5.32)$$

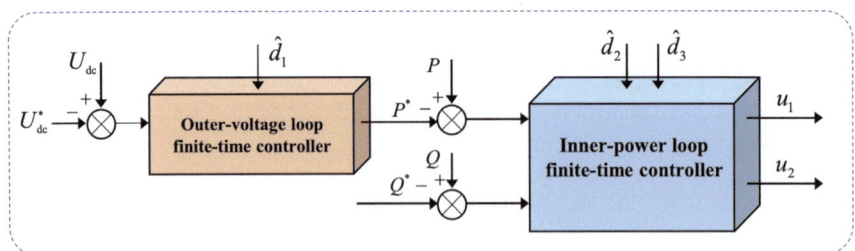

Fig. 5.5 Block diagram of DOFTC with a dual-loop voltage and indirect current control structure

5.1 Nonlinear Control Technology for High Robustness

where $0 < \gamma < 1$. The control parameters k_i and s_i are positive constants. $\hat{d}_i$ and $\hat{\dot{P}}^*$ are estimated values of d_i and $\dot{P}^*$, respectively, and they will be obtained by the subsequent finite-time disturbance observer. The detailed design processes of finite-time controller are given as follows.

Step 1: For the dc-voltage subsystem, the Lyapunov function V_1 is designed as

$$V_1 = \frac{1}{2}z_1^2 \tag{5.33}$$

Based on the definitions of error variables, taking the derivative of V_1 yields

$$\dot{V}_1 = z_1 \dot{z}_1 = z_1 \left(\frac{2P^*}{CU_{dc}} + \frac{2z_2}{CU_{dc}} + d_1 \right) \tag{5.34}$$

By substituting the constructed virtual controller P^*, it follows that

$$\dot{V}_1 = -k_1 z_1^2 - s_1 z_1^{\gamma+1} + z_1 \tilde{d}_1 + z_2 \frac{2z_1}{CU_{dc}} \tag{5.35}$$

with disturbance estimation errors $\tilde{d}_i = d_i - \hat{d}_i$, where "~" represents the estimation error.

Step 2: Similarly, the second Lyapunov function V_2 is designed as

$$V_2 = V_1 + \frac{1}{2}z_2^2 \tag{5.36}$$

Taking the derivative of V_2 yields

$$\begin{aligned}\dot{V}_2 &= \dot{V}_1 + z_2 \dot{z}_2 = -k_1 z_1^2 - s_1 z_1^{\gamma+1} + z_1 \tilde{d}_1 + z_2 \frac{2z_1}{CU_{dc}} + z_2 \dot{z}_2 \\ &= -k_1 z_1^2 - s_1 z_1^{\gamma+1} + z_1 \tilde{d}_1 + z_2 \left(-\frac{3}{2L} u_1 + f_1 + d_2 - \dot{P}^* + \frac{2z_1}{CU_{dc}} \right) \end{aligned} \tag{5.37}$$

By substituting the constructed u_1, the derivative of V_2 is rewritten as

$$\dot{V}_2 = \sum_{i=1}^{2} \left(-k_i z_i^2 - s_i z_i^{\gamma+1} + z_i \tilde{d}_i \right) - z_2 \tilde{\dot{P}}^* \tag{5.38}$$

with $\tilde{\dot{P}}^* = \dot{P}^* - \hat{\dot{P}}^*$.

Step 3: The final Lyapunov function is formed by summing up the Lyapunov functions associated with each subsystem, which guarantees the stability of the whole closed-loop inverter system. The final Lyapunov function is designed as

$$V = V_2 + \frac{1}{2}z_3^2 \tag{5.39}$$

Then, taking the derivative of V yields

$$\dot{V} = \sum_{i=1}^{2}\left(-k_i z_i^2 - s_i z_i^{\gamma+1} + z_i \tilde{d}_i\right) - z_2 \tilde{P}^* + z_3 \dot{z}_3$$

$$= \sum_{i=1}^{2}\left(-k_i z_i^2 - s_i z_i^{\gamma+1} + z_i \tilde{d}_i\right) - z_2 \tilde{P}^* + z_3 \left(\frac{3}{2L}u_2 + f_2 + d_3\right) \tag{5.40}$$

By substituting u_2, the derivative of V is rewritten as

$$\dot{V} = \sum_{i=1}^{3}\left(-k_i z_i^2 - s_i z_i^{\gamma+1} + z_i \tilde{d}_i\right) - z_2 \tilde{P}^* \tag{5.41}$$

If the estimation errors $\tilde{d}_i$ and $\tilde{P}^*$ tend to a small neighborhood of zero, the practical finite-time stability condition in [2] is guaranteed by the designed finite-time controller and the control objectives are achieved.

Part 2: Finite-Time Disturbance Observer

At first, the estimated model of the 3LT²I is constructed as

$$\begin{cases} \dot{\hat{U}}_{dc} = 2\frac{P}{CU_{dc}} \\ \dot{\hat{P}} = -\frac{3}{2L}u_1 + f_1(P,Q) \\ \dot{\hat{Q}} = \frac{3}{2L}u_2 + f_2(P,Q) \end{cases} \tag{5.42}$$

Obviously, the lumped disturbances are derived as

$$\begin{cases} d_1 = \dot{\tilde{U}}_{dc} \\ d_2 = \dot{\tilde{P}} \\ d_3 = \dot{\tilde{Q}} \end{cases} \tag{5.43}$$

where

$$\begin{cases} \tilde{U}_{dc} = U_{dc} - \hat{U}_{dc} \\ \tilde{P} = P - \hat{P} \\ \tilde{Q} = Q - \hat{Q} \end{cases} \tag{5.44}$$

In general, differential signals are directly calculated by numerical difference. If signals $\tilde{U}_{dc}$, $\tilde{P}$, and $\tilde{Q}$ have random noise components, the direct differentiation

5.1 Nonlinear Control Technology for High Robustness

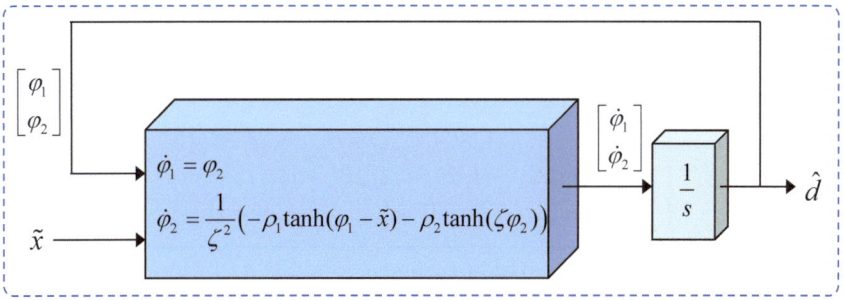

Fig. 5.6 Structure block diagram of the finite-time differentiator

cannot be applied in controller design, because of the explicit differentiation sensitivities to noises, which may lead to system instability. So, it is inevitable to construct a differentiator.

For the uniform notation, defining $d_4 = \dot{P}^*$ and $\hat{d}_4 = \hat{\dot{P}}^*$. The detailed structure of the finite-time differentiator is

$$\begin{cases} \dot{\varphi}_1 = \varphi_2 \\ \dot{\varphi}_2 = \frac{1}{\zeta^2}\left[-\rho_1 \tanh(\varphi_1 - \tilde{x}) - \rho_2 \tanh(\zeta \varphi_2)\right] \end{cases} \quad (5.45)$$

where $\varphi_2 = \hat{d} = \left[\hat{d}_1, \hat{d}_2, \hat{d}_3, \hat{d}_4\right]^T$ are the differentiator output signals with $\tilde{x} = \left[\tilde{U}_{dc}, \tilde{P}, \tilde{Q}, P^*\right]^T$ acting as the input signals. Figure 5.6 shows the structure of the finite-time differentiator.

By using Lemma 2 in [4], the estimation error satisfies

$$|\tilde{d}_i| = |d_i - \hat{d}_i| \leq \varepsilon, \quad i = 1, 2, 3, 4 \quad (5.46)$$

for $t \geq T_f$, i.e., d_i is accurately identified.

Part 3: Stability Analysis
Consider the 3LT^{2}I with uncertain disturbances. According to the continuous finite-time controllers and the FTDO, the tracking error z_i converges to a small neighborhood of the origin within finite time. Meanwhile, the closed-loop system of the 3LT^{2}I is practical finite-time stable. Its proof is given as follows.

Proof According the Young's inequality, the inequality holds that

$$z_j \tilde{d}_j \leq \frac{1}{2}z_j^2 + \frac{1}{2}\varepsilon^2, \quad -z_2\tilde{\dot{P}}^* \leq \frac{1}{2}z_2^2 + \frac{1}{2}\varepsilon^2 \quad (5.47)$$

for $t \geq T_f$. Then, we have

$$\dot{V} \le -k_1 z_1^2 + \tfrac{1}{2} z_1^2 - s_1 z_1^{2\tau} - k_2 z_2^2 + z_2^2 - s_2 z_2^{2\tau} \\ - k_3 z_3^2 + \tfrac{1}{2} z_3^2 - s_3 z_3^{2\tau} + 2\varepsilon^2 \le -\lambda_1 V - \lambda_2 V^\tau + \eta \tag{5.48}$$

where

$$\begin{cases} \lambda_1 = \min\{2k_1 - 1, 2k_2 - 2, 2k_3 - 1\} \\ \lambda_2 = \min\{2^\tau s_1, 2^\tau s_2, 2^\tau s_3\}, \eta = 2\varepsilon^2, \tau = \frac{(\gamma+1)}{2} \end{cases} \tag{5.49}$$

If the control parameters are chosen such that $\lambda_1 > 0$ and $\lambda_2 > 0$, the closed-loop system of the 3LT^{2}I satisfies the practical finite-time stability condition [2], and the following equation holds that

$$|z_i| \le \sqrt{\frac{2\eta}{\lambda_1 - \mu}} \tag{5.50}$$

within a finite time, $0 < \mu < \lambda_1$. Meanwhile, the closed-loop system of the 3LT^{2}I is practical finite-time stable. It is obtained that U_{dc} and Q accurately track their reference signals even with uncertain disturbances. This completes the proof.

In summary, two nonlinear control methods are proposed in this section. The IFTC method belongs to dual-loop voltage and direct current control structure. The voltage loop considers dc-link loads as unknown external disturbances and designs a finite-time adaptive controller to enhance control performance in the presence of load disturbances without using current sensors. In the inner loop, the current controller integrates a command filter, which achieves fast and accurate current tracking performance. The DOFTC method belongs to dual-loop voltage and indirect current control structure. In this method, the FTDO is proposed to fast and accurately estimate uncertain disturbances for the 3LT^{2}I, and the practical finite-time stability analysis of the closed-loop system is discussed and proved. It is proved that the DOFTC can achieve active disturbance-rejection ability, high tracking precision, and quick dynamic response for the 3LT^{2}I.

5.2 Control Technology for AC-Side Current Quality Improvement

As one of the main control objectives of inverters, the AC-side current performance has always attracted extensive attention. For example, the IEC 61727 [7] requires that current THD limits shall not exceed 5% in the rated PV system output. Usually, the current THD can be reduced by increasing the switching frequency. However, in high-power PV generation system, in order to improve the inverter efficiency and reduce the system loss, the switching frequency is usually set very low. The lower switching frequency will increase the harmonic contents of the inverter output current. To overcome this drawback, the AC-side current quality improvement technology has been proposed, including current harmonic elimination method and current ripple suppression method.

5.2.1 AC-Side Current Harmonic Elimination Method

For a three-phase inverter, the main components of harmonic contents are low-order harmonics such as the 5th, 7th, 11th, 13th, 17th, 19th, etc. Selective harmonic elimination pulse width modulation (SHEPWM) method has the outstanding performance in accurately eliminating low-order harmonics of the output voltage at low switching frequency. In return, the output current's quality will also be remarkably improved. That is to say, the SHEPWM method can eliminate low-order harmonics as well as increase system efficiency by reducing switching losses. For SHEPWM, one key issue is to obtain switching angles by solving nonlinear transcendental equations, in which complicated calculation seems to be inevitable in the practical implementation of SHEPWM.

Many optimization algorithms are used to solve the equations, for instance, genetic algorithm, gray wolf optimization algorithm, and other swarm intelligence algorithms. The common disadvantages of the above algorithms are the complexity of implementation and large amount of computation. Particle swarm optimization (PSO) algorithm [8] has merits such as ease to understand and simplicity to implement, which has become a superior solution to solve this problem and is widely used in SHEPWM to solve switching angles. Although the convergence speed of PSO algorithm is relatively fast, it cannot change its speed step size to adjust according to the actual situation in the search space, resulting in premature convergence and stagnation of local optimal solution. An improved linear inertia weight PSO (LIW-PSO) algorithm was utilized in [9, 10] to calculate the switching angles for SHEPWM in multilevel inverters. This improvement makes the search step size of particles decrease linearly with the increase of iterations, which can prevent particles from falling into local extremum to a certain extent, but once trapped, it is still difficult to jump out. Therefore, how to improve the algorithm and enhance performance when using PSO to solve switching angles in SHEPWM for multilevel inverters still needs further investigation.

Thus, to accurately eliminate low-order harmonics with low switching frequency, an improved particle swarm optimization (IPSO) based on SHEPWM method is proposed by the author's team [11]. Using this method, the output current quality will be remarkably improved. The proposed method consists of three parts: constructing harmonic elimination equations, solving switching angles using the IPSO algorithm, and validation results for current harmonic elimination method. The details of each part will be given in the following section.

Constructing Harmonic Elimination Equations

It should be noted that the harmonic elimination equations are directly related to the output phase voltage waveform. That is to say, when the output phase voltage waveform has different levels, the harmonic elimination equations are different. In this section, the $3LT^2I$ topology and the power flowing from AC side to DC side is taken as the example to illustrate the implementation of the proposed IPSO based on SHEPWM. For ease of analysis and readers' understanding, the $3LT^2I$ topology is given in Fig. 5.7.

The output waveform of three-level inverter by using SHEPWM method usually has 1/4 period symmetry and 1/2 period symmetry about 180°. Taking odd switching angles N in a quarter period as an example, the $3LT^2I$ output voltages are depicted in Fig. 5.8.

According to symmetry of waveforms and Dirichlet theorem, the Fourier series of the output voltages u_{JO} (J = A, B, C) of $3LT^2I$ are expressed as

$$u_{JO}(\omega t) = \sum_{n=1}^{\infty} b_n \sin n\omega t \quad (5.51)$$

The sine term coefficient is expressed as

$$b_n = \begin{cases} 0, & n \text{ is even} \\ \frac{2U_{dc}}{n\pi} \sum_{j=1}^{N} (-1)^{j+1} \cos(n\alpha_j), & n \text{ is odd} \end{cases} \quad (5.52)$$

where n = 1, 2, 3, ..., ω is the fundamental angular frequency, and α_j (j = 1, 2, ..., N) indicates switching angles satisfying

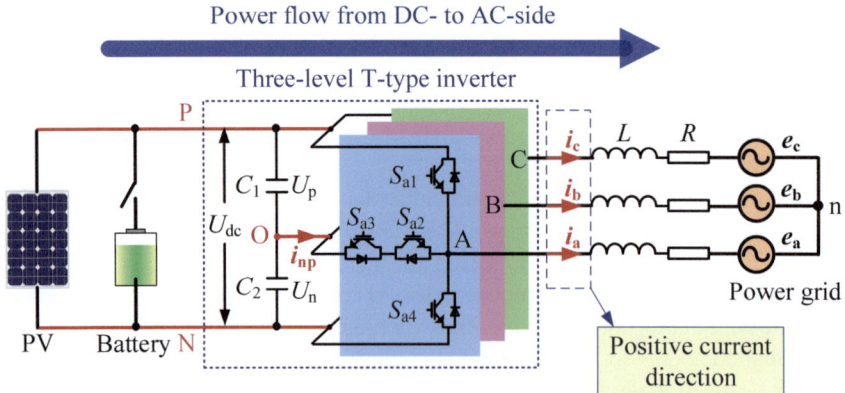

Fig. 5.7 The structure of $3LT^2I$ with the power flowing from DC- to AC-side

5.2 Control Technology for AC-Side Current Quality Improvement

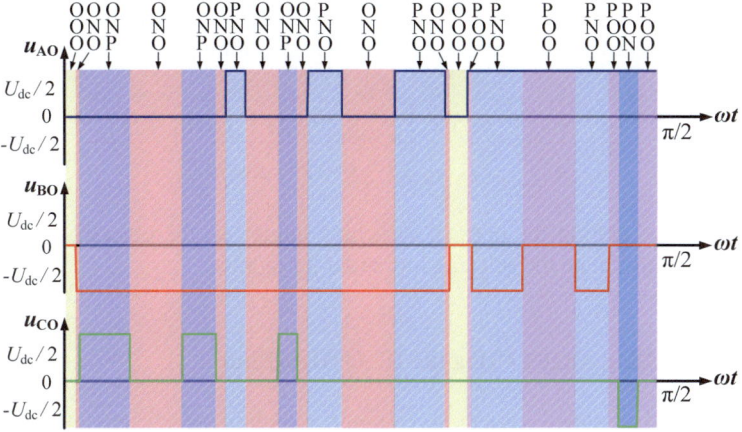

Fig. 5.8 Vectorization diagram of SHEPWM three-phase output voltages

$$0 < \alpha_1 < \alpha_2 < \cdots < \alpha_N < \frac{\pi}{2} \tag{5.53}$$

Define the modulation index as

$$m = \frac{2V_R}{U_{dc}} \tag{5.54}$$

where V_R is the magnitude of inverter output voltage.

Since tripled harmonics are automatically removed in output line voltage for three-phase three-wire systems, and even harmonics are also naturally eliminated, the harmonic order eliminated by the three-phase three-level SHEPWM method is $6k \pm 1$ ($k = 1, 2, \ldots$).

Consequently, the harmonic elimination equations are

$$\begin{cases} \frac{4}{\pi} \sum_{j=1}^{N} (-1)^{j+1} \cos \alpha_j = m \\ \frac{4}{5\pi} \sum_{j=1}^{N} (-1)^{j+1} \cos(5\alpha_j) = 0 \\ \frac{4}{7\pi} \sum_{j=1}^{N} (-1)^{j+1} \cos(7\alpha_j) = 0 \\ \cdots \\ \frac{4}{(3N-2)\pi} \sum_{j=1}^{N} (-1)^{j+1} \cos\left[(3N-2)\alpha_j\right] = 0 \end{cases} \tag{5.55}$$

Here, the odd N is taken as an example. Besides, the maximum order of harmonic that can be eliminated is $3N - 2$ when N is odd and is $3N - 1$ when N is even.

In order to solve the switching angles conveniently, the harmonic elimination equations are transformed into a constrained optimization problem.

$$F(\alpha_1, \alpha_2, \ldots, \alpha_N) = \left[\frac{4}{\pi}\sum_{j=1}^{N}(-1)^{j+1}\cos\alpha_j - m\right]^2 \\ + \sum_{i=5,7,\ldots,3N-2}\left[\frac{4}{i\pi}(-1)^{j+1}\cos(i\alpha_j)\right]^2 \quad (5.56)$$

For the sake of getting the optimal switching angles α_j, the objective function is minimized to eliminate the selected low-order harmonics.

Solving Switching Angles Using the IPSO Algorithm

The harmonics elimination equation system is a multivariable transcendental equation system with complex nonlinear characteristics and a huge amount of calculation, which is extremely difficult to solve by conventional methods. As a representative swarm intelligence optimization algorithm, the PSO algorithm is easy to understand and simple to implement, becoming an excellent choice for solving the harmonic elimination equations.

However, in the standard PSO algorithm, the inertia weight ω is a constant, which cannot dynamically adjust the search speed step size according to the specific situation. Thus, it falls into a local optimum easily, which leads to premature convergence and stagnation of the algorithm, resulting in the optimization performance degradation of the algorithm.

In order to overcome these drawbacks, the IPSO algorithm is proposed, which has a fast convergence rate and high solution accuracy.

Suppose D is the dimension of the search space, and the particle amount is n. Denote $X_i = (x_{id})_{1 \times D}$ and $P_i = (p_{id})_{1 \times D}$ ($d = 1, 2, \ldots, D$) as the current and best position of the ith particle, respectively, the latter of which is known as the individual optimal position. $P_g = (p_{gd})_{1 \times D}$ represents the best position of n particles in the population, which is called the global optimal position. $V_i = (v_{id})_{1 \times D}$ indicates the velocity of the ith particle. The update formulas for position and velocity of the ith particle's dth component at the $(t + 1)$th iteration are as follows:

$$v_{id}^{t+1} = \omega_{id}^t \cdot v_{id}^t + c_{1id}^t \cdot \text{rand}_1 \cdot (p_{id} - x_{id}^t) \\ + c_{2id}^t \cdot \text{rand}_2 \cdot (p_{gd} - x_{id}^t) \quad (5.57)$$

$$x_{id}^{t+1} = x_{id}^t + v_{id}^{t+1} \quad (5.58)$$

where $d = 1, 2, \ldots, D$, $i = 1, 2, \ldots, n$, c_{1id}^t and c_{2id}^t are cognitive and social learning factors, rand_1 and rand_2 are random numbers between 0 and 1, and ω_{id}^t is the inertia weight factor of the ith particle's dth component at tth iteration.

It can be seen from the above equations that the appropriate choice of inertia weight can make the particles have balanced exploitation and exploration ability [12],

5.2 Control Technology for AC-Side Current Quality Improvement

which largely determines the implementation effect of the algorithm. Therefore, the research on inertia weight is particularly important.

In view of this, the LIW-PSO algorithm was proposed [13] and applied in [9, 10], in which the improvement of ω was

$$\omega(t) = \omega_{\max} - \frac{(\omega_{\max} - \omega_{\min})t}{t_{\max}} \qquad (5.59)$$

where $\omega_{\min}$ and $\omega_{\max}$ denote the minimum and maximum values of ω, t and $t_{\max}$ signify the current and maximum iteration numbers, respectively. This improvement can avoid particles falling into local extremum to a certain extent, but once trapped, it is still difficult to jump out. Moreover, the change of ω in this improvement is only linearly related to the current iteration number, which cannot better adapt to the optimization problems with complex nonlinear characteristics.

To better dynamically adjust the local and global searching capabilities for the proposed IPSO algorithm, the inertia weight of the ith particle's dth component at tth iteration in this article is defined as

$$\omega_{id}^{t} = \exp\left(\frac{-t}{|F^{t}(x_{\text{best},d}) - F^{t}(x_{id})|t_{\max}}\right) \qquad (5.60)$$

where $F^{t}(x_{id})$ and $F^{t}(x_{\text{best},d})$ are the fitness values of the ith and optimal particles' dth component at the tth iteration, respectively.

At the same time, the cognitive and social learning factors are modified as

$$c_{1id}^{t} = c_{1,\max} - \frac{(c_{1,\max} - c_{1,\min})t}{t_{\max}} \qquad (5.61)$$

$$c_{2id}^{t} = c_{2,\max} - \frac{(c_{2,\max} - c_{2,\min})t}{t_{\max}} \qquad (5.62)$$

where $c_{1,\min}$, $c_{1,\max}$, $c_{2,\min}$, and $c_{2,\max}$ are the minimum and maximum values of the cognitive and social learning factors.

When solving the harmonic elimination equations, according to the nonlinear characteristics of the equations, a negative exponential inertia weight is introduced in the inertia weight of the ith particle's dth component at tth iteration in the whole iteration process of the proposed IPSO algorithm, and ω_{id}^{t} decreases nonlinearly with the increase of iteration number. $|F^{t}(x_{\text{best},d}) - F^{t}(x_{id})|$ can control the search step in the optimization process. According to the property of exponential function, ω_{id}^{t} can be limited to [0,1]. At the beginning of the iteration, the value of ω_{id}^{t} is large. It can be seen from the expression of update formulas for position and velocity and inertia weight that a large ω_{id}^{t} will make the particles spread across the entire search space with a large velocity to determine the approximate range of the optimal value, which is conducive to guiding the particles out of the trap of local optimum and improving the global searching ability. With the increase of iterations, $|F^{t}(x_{\text{best},d}) -$

$F^t(x_{id})$ decreases and ω_{id}^t also decreases nonlinearly. The search range of most particles gradually decreases and concentrates in the neighborhood of the optimal value. A small ω_{id}^t can prevent missing the optimal solution. This is beneficial to raise the particles' convergence rate and searching accuracy. The improvement in inertia weight makes excellent use of the ability of ω_{id}^t to balance local and global search. The strong points of the proposed IPSO algorithm are simple form, less computation, and easy implementation. It can significantly increase the convergence rate, and particles can quickly jump out of the local optimal position.

The steps of the proposed IPSO algorithm to solve the harmonic optimization problem are listed below. The flowchart is shown in Fig. 5.9.

Step 1: Set control parameters and initialize switching angles' position and velocity.

Step 2: According to the optimization problem, calculate the fitness value of switching angles to search for the individual and global optimal position.

Step 3: Update ω_{id}^t, c_{1id}^t, and c_{2id}^t.

Step 4: Update the switching angles' velocity and position and calculate the corresponding fitness value to update the individual and global optimal extremum.

Step 5: Determine whether the termination criteria is met. If it is satisfied, the iterative operation ends; if it is not satisfied, the next iterative operation is entered.

Validation Results for the Current Harmonic Elimination Method

The performance of IPSO is compared with the standard PSO and LIW-PSO, [11] in terms of the capability of jumping out of local optimum and total number of iterations to obtain accurate switching angles. Different modulation indices of 0.80, 0.85, 0.90, and 0.95 are used to cover the whole variation range. The comparisons of variation for the objective function value of the three methods are shown in Fig. 5.10. Taking $m = 0.90$ as an example, when the termination condition is satisfied, the number of iterations of the proposed IPSO is 281, while that of the standard PSO and LIW-PSO algorithm is 464 and 568, respectively. Furthermore, it can be seen from the circle parts that the proposed IPSO can jump out of local optimum much faster than the standard PSO and LIW-PSO, thus reducing the number of iterations and greatly accelerating the convergence speed.

Using the proposed IPSO to solve the constrained optimization problem, the switching angle curves with different modulation indices in four cases are obtained in Fig. 5.11, which indicates the applicability of the proposed IPSO to the constrained optimization problem with different numbers of switching angles. The switching angle curves are stored in the memory and can be searched online through the controller.

All in all, the IPSO algorithm is proposed for solving harmonic elimination equations of SHEPWM to obtain switching angles, which defines a nonlinear negative exponential inertia weight that considers the current and optimal fitness values and, thus, can dynamically adjust the local and global search speed and step size in real

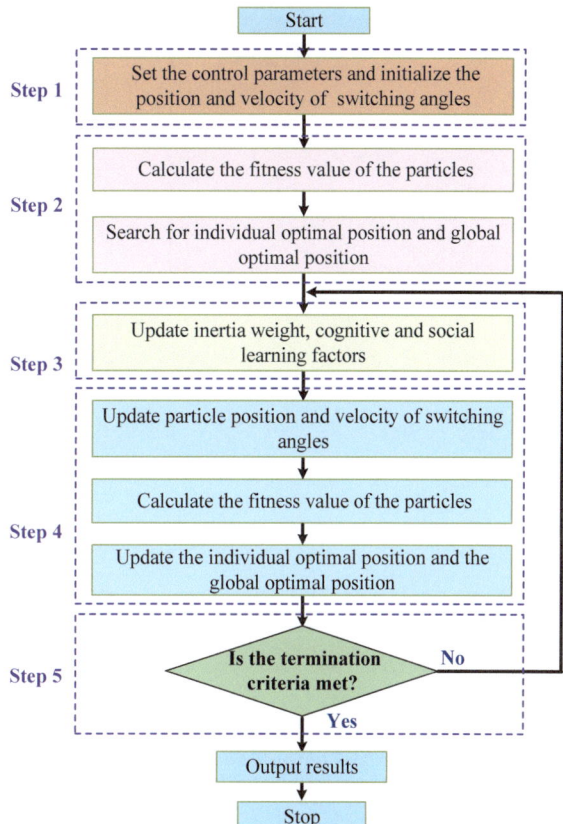

Fig. 5.9 Flowchart of the proposed IPSO algorithm

time according to actual working conditions. It overcomes the difficulty that the particle's velocity step size cannot be adjusted according to the actual situation in the search space or cannot adapt to the optimization problem with complex nonlinear characteristics when the standard PSO algorithm or LIW-PSO algorithm is used to solve the harmonic elimination equations. This difficulty makes the algorithm easy to fall into local optimum, causing premature convergence and stagnation, which results in performance degradation of the algorithm. The proposed IPSO algorithm not only improves the convergence speed and search accuracy of particles but also raises the global search ability. In addition, it has a simple form, low computational cost, and easy implementation, which can be used to calculate the switching angles quickly and accurately. Simulation results verify that the proposed IPSO algorithm can significantly improve the convergence speed and jump out of the local optimal position highly fast.

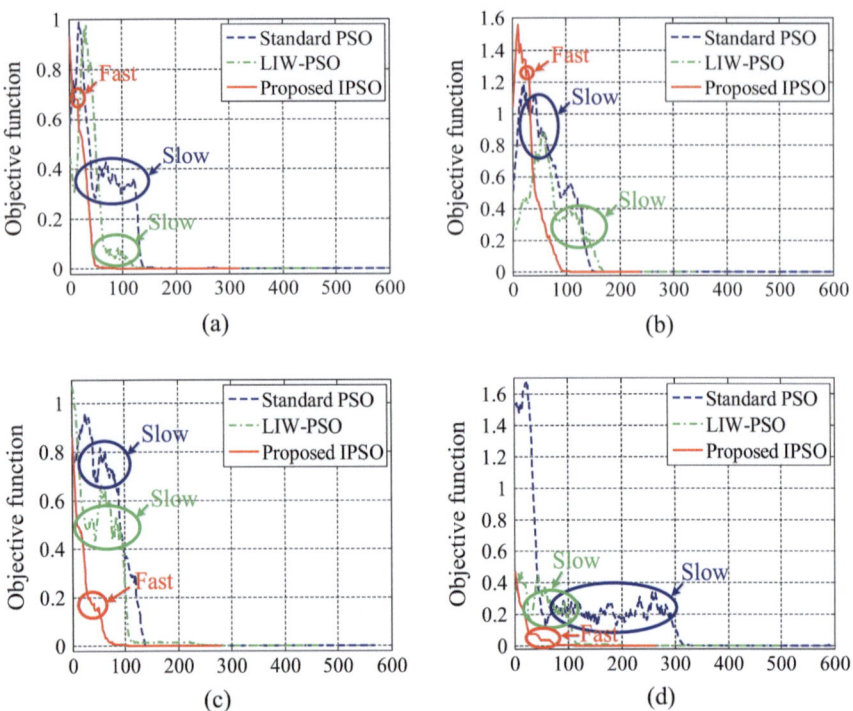

Fig. 5.10 Comparison variation of objective function for three PSO algorithms with different modulation indices **a** $m = 0.80$ **b** $m = 0.85$ **c** $m = 0.90$ **d** $m = 0.95$

5.2.2 AC-Side Current Ripple Suppression Methods

Recently, the International Energy Agency envisioned that the share of renewable energy in power generation will increase from 29% in 2020 to 60% in 2030 and close to 90% in 2050 [14], and large-scale renewable energy power plants (such as photovoltaic (PV) plants [15]) will become dominant. In the high-power and current rating PV applications, using a single inverter typically cannot meet the requirements due to the capacity limit of semiconductor devices. Furthermore, it is difficult for a single inverter with a single maximum power point tracking to achieve optimal efficiency in high-power rating applications. Thus, the paralleled architectures of inverters have been a common approach and drawn increased attention in high-power rating applications due to the advantages of high-power ratings, high efficiency, high reliability, and system redundancy [16]. In PV applications, the current harmonics of the total output current of parallel inverters in the point of common coupling are strictly constrained. The IEC 61727 and IEEE Std. 1547 give the limitations of current THD in the PV system [17]. In high-power rating applications, due to the limited switching frequency in each parallel inverter, large filters are usually

5.2 Control Technology for AC-Side Current Quality Improvement

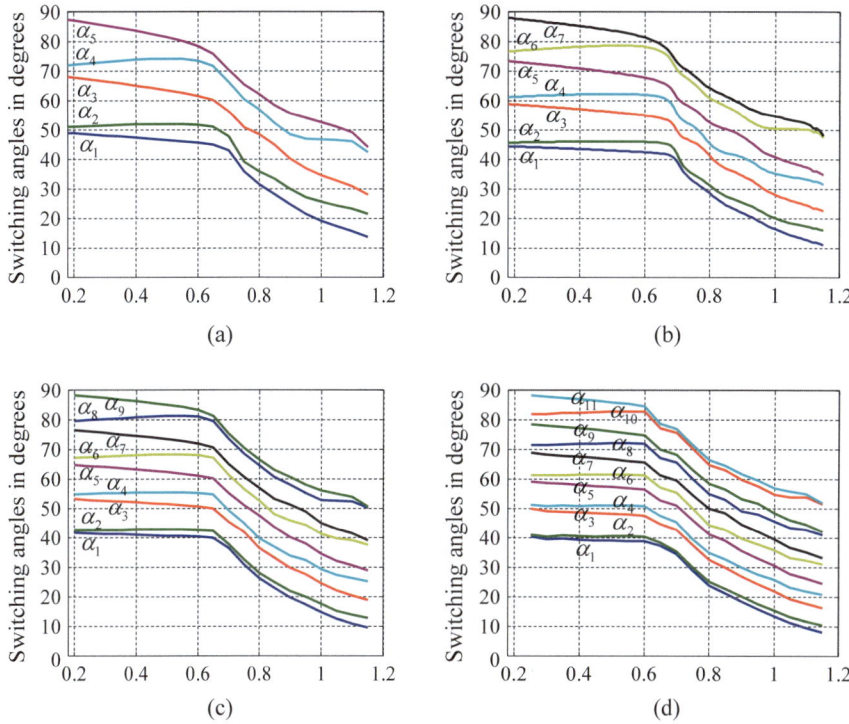

Fig. 5.11 Switching angle curves with respect to the modulation index for different numbers of switching angles: **a** $N = 5$; **b** $N = 7$; **c** $N = 9$; **d** $N = 11$

required to reduce the THD in the total output current, which increases the size, cost, and weight of the overall system inevitably.

To overcome these drawbacks and to meet the high current quality requirement in high-power rating photovoltaic applications, the total current ripple suppression methods for parallel inverters are researched extensively by the author's team and illustrated in this section. The current ripple suppression methods are categorized into two types. The fist type is the improved carrier-based modulation (ICBM) strategy, which is based on zero-sequence injection method and derived from conventional carrier-based modulation strategies. The second type is the double phase-shifted pulse width modulation (DPS-PWM) strategy, in which additional modified angles are added in the initial phase-shifted angles of carriers in the conventional phase-shifted pulse width modulation (PS-PWM) strategy. To verify the effectiveness of the proposed two methods, the experimental results are given.

ICBM Method Based on Zero-Sequence Injection

To improve the grid-connected current quality, the ICBM current ripple suppression strategy based on zero-sequence injection is proposed by the author in [18]. Based on root mean square value analysis of the output current ripple, a proper zero-sequence component is injected into three-phase modulation waves in the PS-PWM to achieve a low total current ripple. Three parts are included to elaborate the proposed method in this section. (1) The analyses of output characteristic with different zero-sequence component modulation waves. (2) The principle of total current ripple reduction for the two paralleled two-level inverters. (3) The principle of total current ripple reduction for the two paralleled multilevel inverters.

(1) **The analyses of output characteristic with different zero-sequence component modulation waves**

The two parallel three-phase two-level inverters and the power flowing from AC side to DC side shown in Fig. 5.12 are selected as the objects to clarify the proposed method. In this structure, the DC side of each inverter is connected to a separate PV module, and the AC side is connected to the AC grid or the load. Note that only an L filter is used here for simplicity, and the neutral-point O of the dc-link capacitance is chosen as the reference point.

Assuming the paralleled two-level inverters work under similar working conditions. The three-phase modulation waves (m_a, m_b, m_c) for each inverter are the same and normalized as

$$\begin{cases} m_x = m_x^* + m_z \quad x = a, b, c \\ m_a^* = m \times \cos(2\pi f_1 t) \\ m_b^* = m \times \cos\left(2\pi f_1 t - \frac{2\pi}{3}\right) \\ m_c^* = m \times \cos\left(2\pi f_1 t + \frac{2\pi}{3}\right) \end{cases} \quad (5.63)$$

where m_z represents the zero-sequence component modulation waves injection, m represents the modulation index, and f_1 represents the fundamental frequency.

Figure 5.13 shows the PS-PWM applied in the two paralleled two-level inverters, where u_{x1} and u_{x2} represent the output phase voltage of inverter 1 and inverter 2, U_{dc} represents the dc-link voltage.

The carrier of inverter 2 (C_{r2}) lags 180° behind the carrier of inverter 1 (C_{r1}). When the modulation wave is lower than the carrier, the output phase voltage is $-U_{dc}/2$, otherwise, it is $U_{dc}/2$. To simplify the description, the voltages $U_{dc}/2$, 0, $-U_{dc}/2$ are called states [P], [O], [N]. The output phase voltage states of each inverter vary between [P] and [N] in one carrier period.

For two paralleled inverters, the equivalent total output phase voltage u_x can be calculated as

$$u_x = \frac{u_{x1} + u_{x2}}{2} \quad x = a, b, c \quad (5.64)$$

5.2 Control Technology for AC-Side Current Quality Improvement

The equivalent total output phase voltage states of two parallel two-level inverters will vary between [O] and [N], or [O] and [P] in one carrier period. More specially, if the equivalent total output phase voltage states of two parallel inverters vary in one carrier period, the voltage states will start with [O] when the PS-PWM is applied.

Since the two parallel two-level inverters can provide three types of voltage states, the two parallel two-level inverters can be regarded as one three-level inverter. The space voltage vectors diagram shown in Fig. 5.14 for the three-level inverter is used here to further analyze the total output waveform and harmonic characteristics of the two parallel two-level inverters.

It is worth noticing that the voltage vector sequence for the paralleled inverters with PS-PWM is different from that for the typical three-level inverter with SVM, since the varied voltage states should start with [O] in the PS-PWM. When the reference voltage vector (V_{ref}) is located at Sector 1, there are seven kinds of possible voltage vector sequences that start with [OOO] and are listed in Table 5.1 to synthesize the V_{ref} in half a carrier period.

It can be seen that by changing the value range of the zero-sequence component m_z, the reference voltage vector (V_{ref}) can be synthesized by one of the seven kinds of vector sequences. The zero-sequence components $m_{z1} \sim m_{z8}$ are expressed as

$$\begin{cases} m_{z1} = -1 - m_{min} & m_{z2} = -m_{max} \\ m_{z3} = -\frac{m_{max}+m_{mid}}{2} & m_{z4} = -\frac{m_{max}+m_{min}}{2} \\ m_{z5} = -m_{mid} & m_{z6} = -\frac{m_{mid}+m_{min}}{2} \\ m_{z7} = -m_{min} & m_{z8} = 1 - m_{max} \end{cases} \quad (5.65)$$

where $m_{max}, m_{mid}, m_{min}$ represent the maximum, medium, and minimum values of instantaneous three-phase modulation waves (m_a^*, m_b^*, m_c^*) without m_z injection. It should also be noted that the range of m_z is strictly limited to [m_{z1}, m_{z8}] in the PS-PWM. Thus, some of the voltage vector sequences with the m_z out of [m_{z1}, m_{z8}] are not considered in this section.

Furthermore, when the voltage vector sequence is determined, by determining the value of m_z, the dwell time of the applied voltage vectors can also be obtained. For example, when the vector sequence 4 is applied, the dwell times of voltage vectors in half of one carrier period T_s can be shown in Fig. 5.15, where t_1, t_2, t_3 and t_4 represent half the dwell times of the first, second, third, and fourth voltage vector in the corresponding vector sequence. They are calculated as

$$\begin{cases} t_1 = \left(\frac{1}{2} - \frac{m_{max}}{2} - \frac{m_z}{2}\right)\frac{T_s}{2} \\ t_2 = \left(\frac{m_{max}}{2} + \frac{m_{min}}{2} + m_z\right)\frac{T_s}{2} \\ t_3 = \left(\frac{m_{mid}}{2} - \frac{m_{min}}{2}\right)\frac{T_s}{2} \\ t_4 = \left(-\frac{m_{mid}}{2} - \frac{m_z}{2}\right)\frac{T_s}{2} \end{cases} \quad (5.66)$$

Therefore, the zero-sequence component m_z is selected as a key parameter to further optimize the voltage vector sequence and the corresponding vectors' dwell times, which directly impact the total current ripple. Based on the above analyses, the principle for total current ripple reduction methods in two paralleled inverter system is given as follows.

(2) **Principle of total current ripple reduction for the two paralleled two-level inverters**

Firstly, the principle of total current ripple reduction methods for two paralleled two-level inverter system is illustrated in this section. Figure 5.16 shows the single-phase equivalent circuit on the AC side, which is derived to analyze the total output current harmonics.

where u_{sx} is the output fundamental frequency voltage, u_x is the total output phase voltage, and u_{nO} represents the voltage between the middle point n of the three-phase grid/load and reference point O

The filter inductor voltage which corresponds to the output current ripple slope (k_x) can be calculated as

$$L \frac{di_x}{dt} = u_x - u_{sx} - u_{nO} = L \times k_x \tag{5.67}$$

For phase x, the output fundamental frequency voltage u_{sx} can be approximated as

$$u_{sx} = m_x^* \times \frac{U_{dc}}{2} \tag{5.68}$$

The u_{nO} can be calculated as

$$u_{nO} = \frac{1}{3}(u_a + u_b + u_c) \tag{5.69}$$

When the reference voltage vector (V_{ref}) is located at Sector 1 and the vector sequence 4 is applied, the inductor voltages under different voltage vectors and the current ripple of phase x in vector sequence 4 are illustrated in Fig. 5.17, where i_x is the instantaneous value of total output current, and $i_{x,avg}$ represents the averaged value of i_x in half of the carrier period T_s. It should be noticed that the current ripple of i_x is symmetrical in half of a carrier period.

Therefore, the instantaneous value of the total output phase current in the quarter carrier period is used to describe the current ripple quantitatively. For phase x, the total output current ripple i_{px} is calculated as

$$i_{px} = i_x - i_{x,avg} \tag{5.70}$$

The mean square values ($RMS(i_{px})^2$) of i_{px} in one carrier period can be obtained as

5.2 Control Technology for AC-Side Current Quality Improvement

$$RMS(i_{px})^2 = \frac{4}{T_s} \begin{Bmatrix} \int_0^{t_1}(k_{1x}t)^2 dt + \int_{t_1}^{t_1+t_2}[k_{1x}t_1 + k_{2x}(t-t_1)]^2 dt \\ + \int_{t_1+t_2}^{t_1+t_2+t_3}[k_{1x}t_1 + k_{2x}t_2 + k_{3x}(t-t_1-t_2)]^2 dt+ \\ \int_{t_1+t_2+t_3}^{t_1+t_2+t_3+t_4}[k_{1x}t_1 + k_{2x}t_2 + k_{3x}t_3 + k_{4x}(t-t_1-t_2-t_3)]^2 dt \end{Bmatrix}$$
(5.71)

where k_{1x}, k_{2x}, k_{3x}, k_{4x} represent the current ripple slope of the first, second, third, and fourth voltage vector in vector sequence. More specifically, the k_{1x}, k_{2x}, k_{3x}, k_{4x} in Fig. 5.17 correspond to vectors [OOO], [POO], [PON], [PNN], respectively.

Furthermore, the average value ($I_{\text{aver\_RMS}}^2$) of $RMS(i_{\text{pa}})^2$, $RMS(i_{\text{pb}})^2$, $RMS(i_{\text{pc}})^2$ can be expressed as

$$I_{\text{aver\_RMS}}^2 = \frac{1}{3}\left[RMS(i_{\text{pa}})^2 + RMS(i_{\text{pb}})^2 + RMS(i_{\text{pc}})^2\right]$$
(5.72)

The dwell times of each voltage vector are deterministic and related to the m_z when the vector sequence is deterministic. Thus, the average mean square value of three-phase total output current ripples in different vector sequences can be expressed as a function of m_z. The $f_i(m_z)$ represents the average mean square value of three-phase total output current ripples when the voltage vector sequence i ($i = 1, ..., 7$) is applied, and can be calculated as

$$I_{\text{aver\_RMS}}^2 = f_i(m_z) \; i = 1, ..., 7$$
(5.73)

According to the above equation, by selecting a proper m_z, the average mean square value of three-phase total output current ripples can be minimized. Since the $f_i(m_z)$ is a continuous function, the optimal m_z ($m_{z,\text{opt}}$) which minimizes $f_i(m_z)$ can be simply found in either the boundaries of m_z range or the extreme points of $f_i(m_z)$ in the m_z range. To simplify the calculation, the extreme points of $f_i(m_z)$ in the m_z range are not considered here.

When the boundary values of m_z range are applied, the average mean square values of three-phase total output current ripples in Sector 1 are plotted in Fig. 5.18a, c, e, g, while the modulation index (m) and the $2\pi f_1 t$ are varied. By comparing the magnitude of the average mean square values $I_{\text{aver\_RMS}}^2$, the $m_{z,\text{opt}}$ can be obtained as Fig. 5.18b, d, f, h. The $I_{\text{aver\_RMS}}^2$ corresponding to $m_{z,\text{opt}}$ is the lowest among those corresponding to $m_{z1} \sim m_{z8}$.

Similarly, $m_{z,\text{opt}}$ in all ranges of Sector 1 can be obtained as Fig. 5.19a, and be represented in the space vector diagram as Fig. 5.19b.

Furthermore, according to the symmetry of the three-phase modulation waves, the $m_{z,\text{opt}}$ in all sectors can be shown in Fig. 5.20. Note that m_{z1} and m_{z8} are used here for simplicity. The $m_{z,\text{opt}}$ can be obtained when the modulation waves (m_{a*}, m_{b*}, m_{c*}) are determined. By injecting the $m_{z,\text{opt}}$ into the modulation waves of each inverter, a lower THD in the total output current can be obtained.

(3) **Principle of total current ripple reduction for the two paralleled multilevel inverters**

The current ripple evaluation in the multilevel inverter can be carried out as an extension of the analysis of the two-level inverter. Thus, the proposed ICBM can be further extended to apply to two paralleled multilevel inverters by using a similar principle. The two paralleled three-level inverters are selected as an example to clarify the proposed method in this section, and the two paralleled multilevel inverters can be deduced in the same way.

Regard the two paralleled three-level inverters as a five-level inverter. The five-level space voltage vectors diagram and partial voltage vectors are shown in Fig. 5.21. The equivalent total output phase voltage levels can be divided into five styles: $U_{dc}/2$, $U_{dc}/4$, 0, $-U_{dc}/4$, $-U_{dc}/2$. And these five styles are remarked as states $[2, 1, 0]$, $[-1]$, $[-2]$.

By subtracting the particular vectors from the equivalent voltage vector (V_{ref}), the five-level space voltage vectors diagram can be converted to the three-level space voltage vectors diagram. For example, when the equivalent voltage vector (V_{ref}) is located in S1, the particular vector is $[1\ -1\ -1]$. For all sectors of the five level space voltage vectors diagram, the particular vectors are listed in Table 5.2. When the particular vector is subtracted from the equivalent voltage vector (V_{ref}), the three-phase modulation waves also need to be changed, which is expressed as

$$m_{nx} = 2(m_x^* + m_{sx}) \quad x = a, b, c \tag{5.74}$$

where the m_{nx} represents the modulation waves applied in the three-level space voltage vectors diagram, and the m_{sx} is also listed in Table 5.2.

When the five-level space voltage vectors diagram is converted to the three-level space voltage vectors diagram, the m_z calculated by ICBM for two paralleled two-level inverters can be applied to reduce the total current ripple.

Furthermore, to obtain the modulation waves applied to each three-level inverter in parallel, the three-level space voltage vectors diagram should be restored to the five-level space voltage vectors diagram by adding an offset value (m_{offx}) in the modulation wave m_x^*, which is expressed as

$$m_{tx} = m_x^* + \frac{1}{3} \sum_{x=a,b,c} \left[\frac{1}{2}(m_{nx} + m_z) + m_{offx} \right] \quad x = a, b, c \tag{5.75}$$

where m_{tx} represents the modulation waves applied in each three-level inverter in parallel.

The diagram of the proposed ICBM in two parallel three-level inverters is shown in Fig. 5.22. Firstly, by using the vector shift principle, the three-phase reference modulation waves (m_x^*) are converted to modulation waves for the three-level space voltage vectors diagram (m_{nx}). And then, based on m_{nx}, the m_z calculated by ICBM for two parallel two-level inverters is added to the m_{nx}.

5.2 Control Technology for AC-Side Current Quality Improvement

Finally, the offset value (m_{offx}) is added to the m_x^* to obtain the modulation waves for the three-level inverters in parallel.

In addition, for the two parallel multilevel inverters or an even number of two-level inverters in parallel, the parallel inverters can also be regarded as a multilevel inverter (such as seven-level, nine-level…), and convert the multilevel space voltage vectors diagram to three-level space voltage vectors diagram. Based on m_z calculated by ICBM in two parallel two-level inverters, the lower current ripple can also be achieved in the above paralleled inverters.

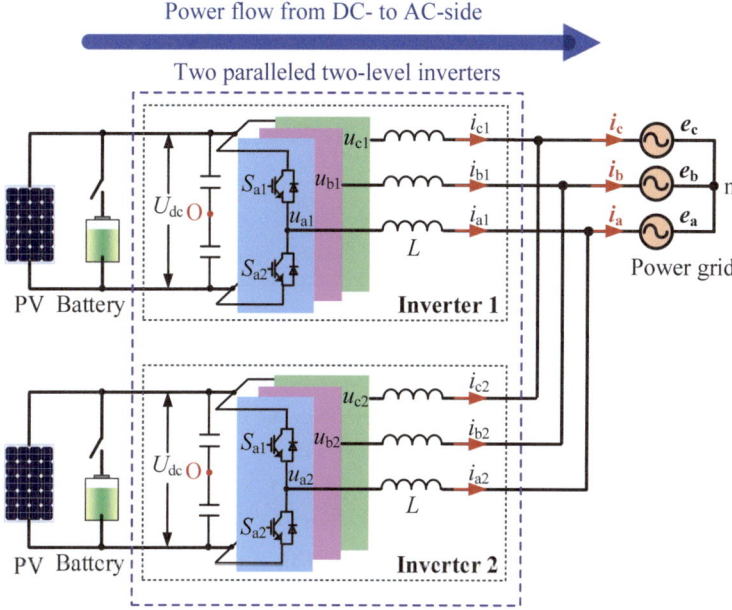

Fig. 5.12 Typical structure of two paralleled two-level inverters

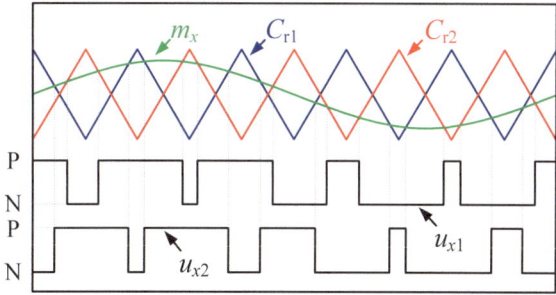

Fig. 5.13 The principle of PS-PWM

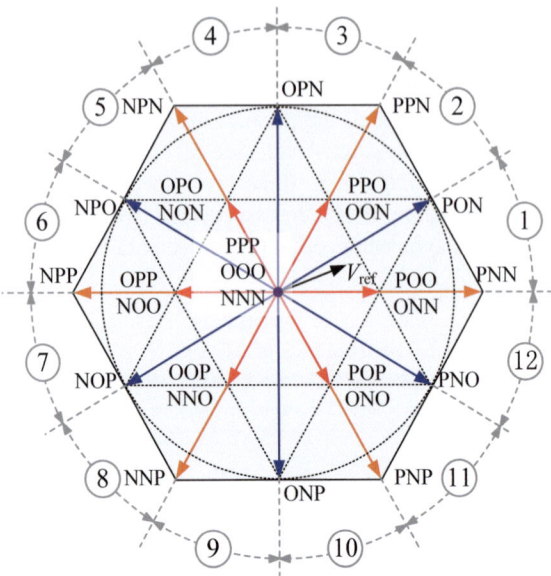

Fig. 5.14 Space voltage vectors diagram for three-level inverter

Table 5.1 The possible voltage vector sequences in PS-PWM

Order	Voltage vector sequence	Range of m_z
1	OOO-OON-ONN-NNN-ONN-OON-OOO	$[m_{z1}, m_{z2}]$
2	OOO-OON-ONN-PNN-ONN-OON-OOO	$[m_{z2}, m_{z3}]$
3	OOO-OON-PON-PNN-PON-OON-OOO	$[m_{z3}, m_{z4}]$
4	OOO-POO-PON-PNN-PON-POO-OOO	$[m_{z4}, m_{z5}]$
5	OOO-POO-PON-PPN-PON-POO-OOO	$[m_{z5}, m_{z6}]$
6	OOO-POO-PPO-PPN-PPO-POO-OOO	$[m_{z6}, m_{z7}]$
7	OOO-POO-PPO-PPP-PPO-POO-OOO	$[m_{z7}, m_{z8}]$

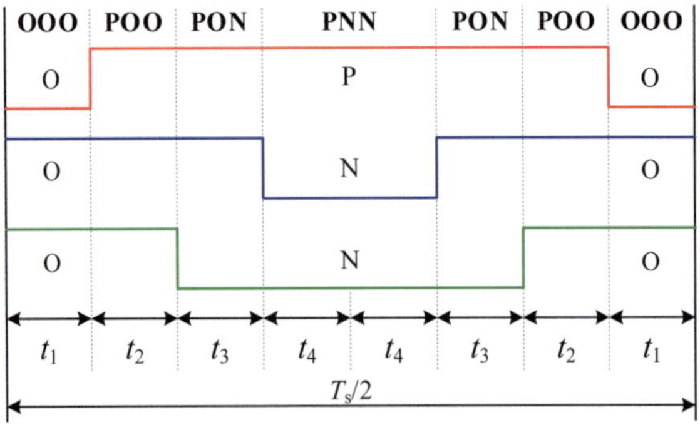

Fig. 5.15 The dwell times of voltage vectors in sequence 4

5.2 Control Technology for AC-Side Current Quality Improvement

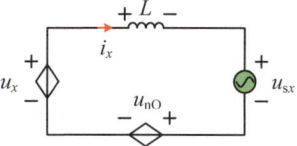

Fig. 5.16 Single phase equivalent circuit in AC side

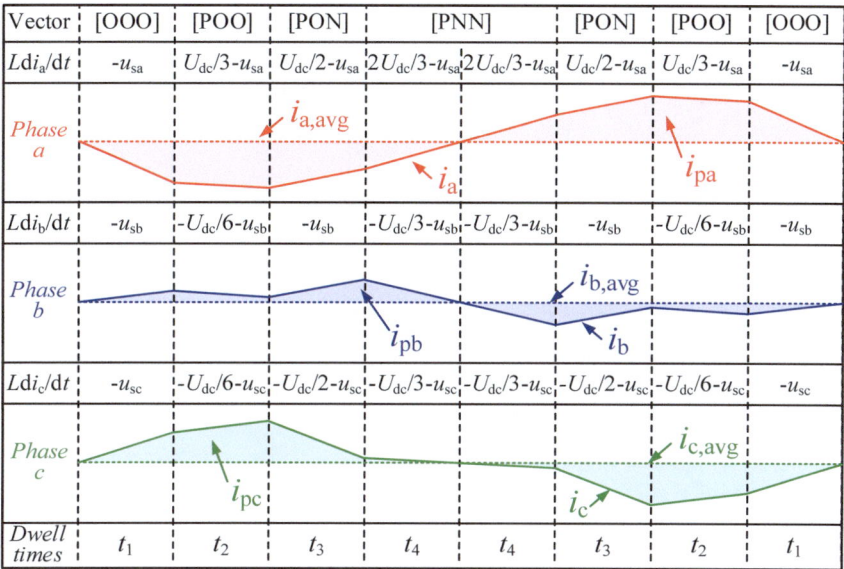

Fig. 5.17 The three-phase currents, when sequence 4 is applied

DPS-PWM Method Based on Phase-Shifted Angles

Aiming at improving the quality of the total output current in PV applications and is easy to extend/implement in multiple parallel inverters, a double phase-shifted PWM (DPS-PWM) strategy based on phase-shifted angles is proposed by the author in [19]. The phase-shifted angles of carriers are modified in each inverter twice to avoid the case that the total output line voltage changes between three voltage levels in one carrier period. Compared with PS-PWM, the proposed DPS-PWM can provide a better total output current and line voltage quality and can be easily realized and expand in practical applications as PS-PWM. To elaborate the proposed method, three parts are included in this section. (1) The analyses of output characteristic with phase-shifted angles. (2) Principle of DPS-PWM for two parallel two-level inverters. (3) Principle of DPS-PWM for multiple parallel two-level inverters.

(1) **The analyses of output characteristic with phase-shifted angles**

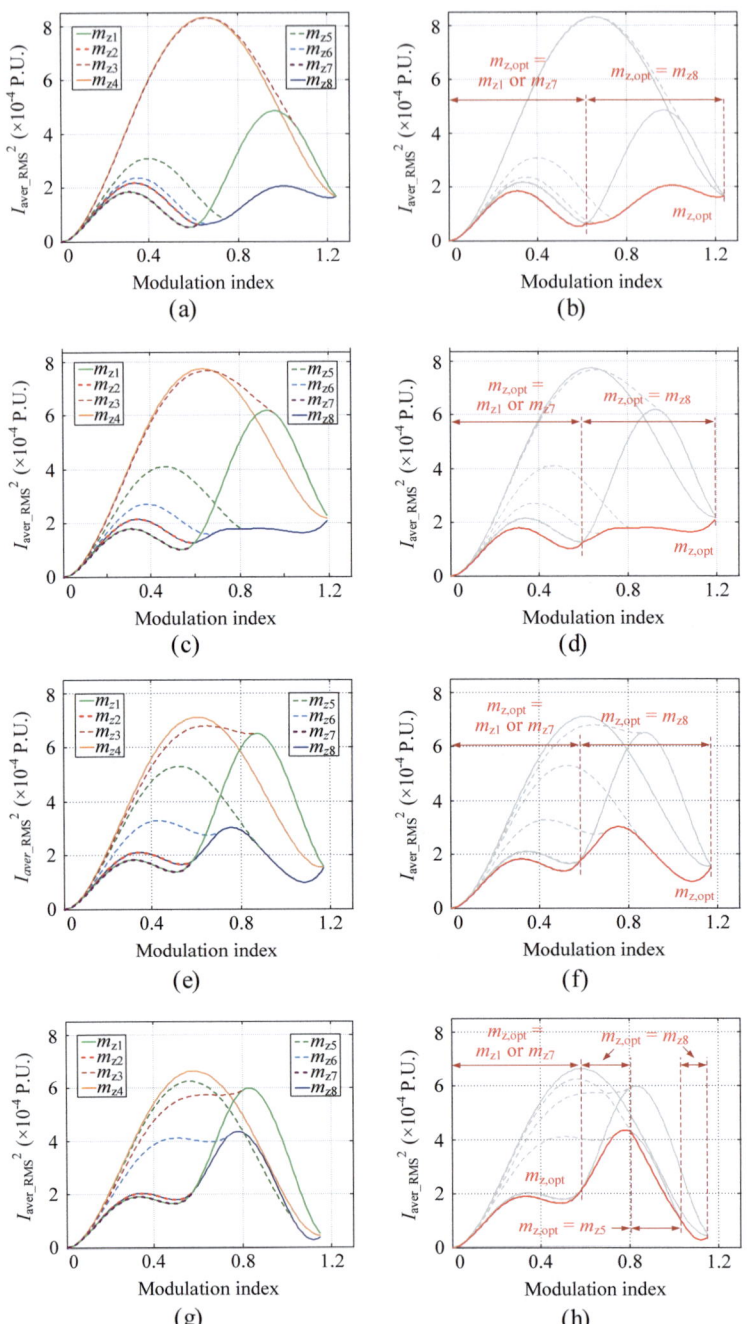

Fig. 5.18 The average means square values of current ripples in Sector 1: **a** $2\pi f_1 t = 0.15$; **b** $2\pi f_1 t = 0.15$; **c** $2\pi f_1 t = 0.25$; **d** $2\pi f_1 t = 0.25$; **e** $2\pi f_1 t = 0.35$; **f** $2\pi f_1 t = 0.35$; **g** $2\pi f_1 t = 0.45$; **h** $2\pi f_1 t = 0.45$

5.2 Control Technology for AC-Side Current Quality Improvement 249

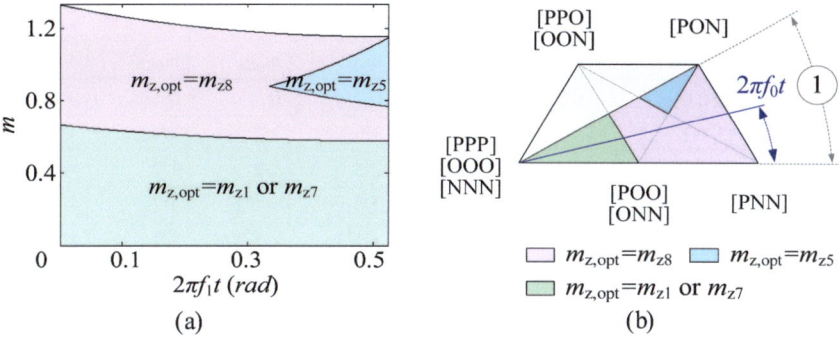

Fig. 5.19 $m_{z,opt}$ in Sector 1: **a** the value of $m_{z,opt}$; **b** $m_{z,opt}$ in Sector 1

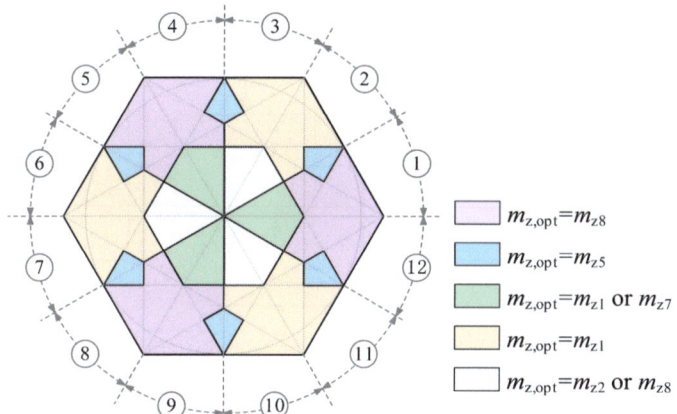

Fig. 5.20 The $m_{z,opt}$ in all sectors

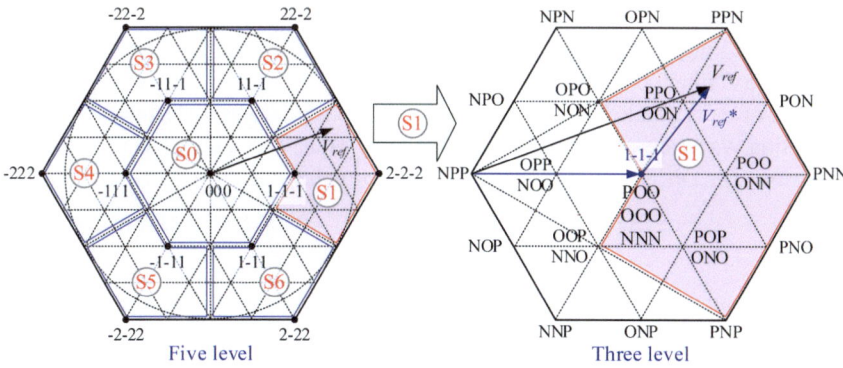

Fig. 5.21 The five-level and three-level space voltage vectors diagram

Table 5.2 The particular vectors for five-level SVM diagram

Sector	Vector	m_{sa}	m_{sb}	m_{sc}	m_{offa}	m_{offb}	m_{offc}
S0	[0 0 0]	0	0	0	0	0	0
S1	[1 −1 −1]	−2/3	1/3	1/3	1/2	−1/2	−1/2
S2	[1 1 −1]	−1/3	−1/3	2/3	1/2	1/2	−1/2
S3	[−1 1 −1]	1/3	−2/3	1/3	−1/2	1/2	−1/2
S4	[−1 1 1]	2/3	−1/3	−1/3	−1/2	1/2	1/2
S5	[−1 −1 1]	1/3	1/3	−2/3	−1/2	−1/2	1/2
S6	[1 −1 1]	−1/3	2/3	−1/3	1/2	−1/2	1/2

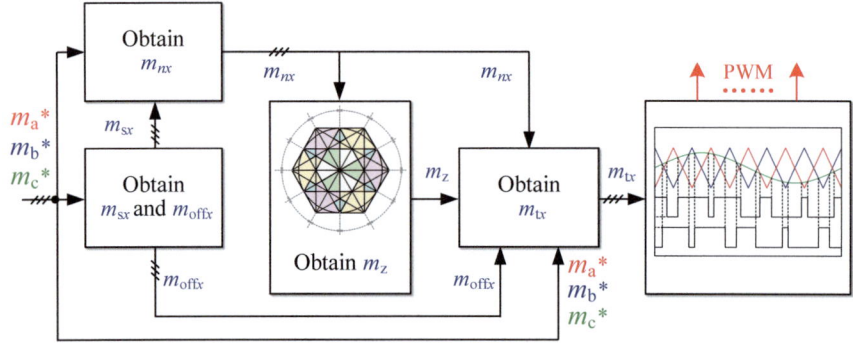

Fig. 5.22 The diagram of ICBM strategy based on zero-sequence injection method

The three-phase two-level inverters in PV applications are selected as the application objects to clarify the proposed method. A general structure of the parallel system is shown in Fig. 5.23. The DC side of each inverter is directly connected, and the neutral point (O) of the dc-link voltage is chosen as the reference point. The AC side is connected to the AC grid or a load through a filter. Note that only a L filter (L_l) is used here for simplicity.

For each parallel inverter in PV applications, the output current and modulation wave are expected to be sinusoidal. Thus, the three-phase modulation waves (m_a, m_b, m_c) are approximatively normalized as

$$\begin{cases} m_a = m \times \cos(2\pi f_1 t) \\ m_b = m \times \cos(2\pi f_1 t - \frac{2\pi}{3}) \\ m_c = m \times \cos(2\pi f_1 t + \frac{2\pi}{3}) \end{cases} \quad (5.76)$$

The single-phase equivalent circuit in AC side is derived as Fig. 5.24, where u_{xi} and i_{xi} (x = a, b, c; i = 1,…, N) represent the output phase voltage and phase current of phase x in the ith inverter. Since the load current and grid voltage are sinusoidal in the PV applications, and the power factor in PV applications is

5.2 Control Technology for AC-Side Current Quality Improvement

high, an equivalent impedance is applied to describe the load/grid. Z_x represents the equivalent impedance of load/grid in phase x, u_{lx} represents the voltage on Z_x, u_{Og} represents the voltage between point O and g.

The total output current of phase x can be expressed as

$$\frac{L_l}{N}\frac{di_x}{dt} + Z_x i_x = \frac{L_l}{N}\frac{d}{dt}\sum_{i=1}^{N} i_{xi} + Z_x \sum_{i=1}^{N} i_{xi} = \frac{1}{N}\sum_{i=1}^{N}(u_{xi} - u_{Og}) \quad (5.77)$$

where

$$u_{Og} = \frac{1}{N}\sum_{i=1}^{N}\frac{1}{3}(u_{ai} + u_{bi} + u_{ci}) \quad (5.78)$$

The sum of phase voltages for phase x in each parallel inverter is defined as the total phase voltage V_{Px}. The difference between two of the three-phase total phase voltage V_{Px} is defined as the total line voltage V_{LLT}. Note that the difference between V_{Pa} and V_{Pb} is regarded as the V_{LLT} in this paper for simplicity. Define the total output voltage V_{gTx} for phase x. These voltages are expressed as

$$\begin{cases} V_{Px} = \sum_{i=1}^{N} u_{xi} \\ V_{LLT} = \sum_{i=1}^{N}(u_{ai} - u_{bi}) = V_{Pa} - V_{Pb} \\ V_{gTx} = \sum_{i=1}^{N}[(u_{ki} - u_{Og})] = V_{Px} - \frac{1}{3}\sum_{x=a,b,c} V_{Px} \end{cases} \quad (5.79)$$

From the above expressions, it is noted that the harmonic characteristics of the total output current in parallel inverters are determined by the total output voltage V_{gTx}. Meanwhile, similar conclusions can be drawn by using the same above principle when the different types of filters are applied.

Furthermore, referring to the article [20], the spectrum of pulse width modulated phase voltage V_{Px} can be approximately expressed as

$$\begin{aligned} V_{Pk}(t) = &\frac{A_{00}}{2} + \sum_{m=1}^{\infty}[A_{m0}\cos(mx) + B_{m0}\sin(mx)] \\ &+ \sum_{n=1}^{\infty}[A_{0n}\cos(ny) + B_{0n}\sin(ny)] \\ &+ \sum_{m=1}^{\infty}\sum_{\substack{n=-\infty \\ (n\neq 0)}}^{\infty}\left[\begin{array}{c}A_{mn}\cos(mx+ny) \\ +B_{mn}\sin(mx+ny)\end{array}\right] \end{aligned} \quad (5.80)$$

where

$$\begin{cases} w = 2\pi f_c t + \theta_c \\ y = 2\pi f_1 t + \theta_1 \\ C_{mn} = A_{mn} + jB_{mn} = \frac{1}{2\pi^2} \int_{-\pi}^{\pi} \int_{-\pi}^{\pi} f(w,y) e^{j(mw+ny)} dw dy \end{cases} \quad (5.81)$$

and $f(w,y)$ represents the contour plot of V_{Px}, A_{mn} and B_{mn} represent the harmonic amplitude in frequency $(nf_1 + mf_c)$.

Combining voltages expressions, the spectrum of V_{gTx} and V_{LLT} are approximately expressed as

$$V_{LLT}(t) = \sum_{n=1}^{\infty} \begin{bmatrix} -2A_{0n} \sin\left(\frac{n\pi}{3}\right) \sin\left(ny - \frac{n\pi}{3}\right) \\ + 2B_{0n} \sin\left(\frac{n\pi}{3}\right) \cos\left(ny - \frac{n\pi}{3}\right) \end{bmatrix}$$
$$+ \sum_{m=1}^{\infty} \sum_{\substack{n=-\infty \\ (n \neq 0)}}^{\infty} \begin{bmatrix} -2A_{mn} \sin\left(\frac{n\pi}{3}\right) \sin\left(mw + ny - \frac{n\pi}{3}\right) \\ + 2B_{mn} \sin\left(\frac{n\pi}{3}\right) \cos\left(mw + ny - \frac{n\pi}{3}\right) \end{bmatrix} \quad (5.82)$$

$$V_{gTx}(t) = \sum_{n=1}^{\infty} \begin{bmatrix} \frac{4}{3} A_{0n} \sin\left(\frac{n\pi}{3}\right)^2 \cos(ny) \\ + \frac{4}{3} B_{0n} \sin\left(\frac{n\pi}{3}\right)^2 \sin(ny) \end{bmatrix}$$
$$+ \sum_{m=1}^{\infty} \sum_{\substack{n=-\infty \\ (n \neq 0)}}^{\infty} \begin{bmatrix} \frac{4}{3} A_{mn} \sin\left(\frac{n\pi}{3}\right)^2 \cos(mw + ny) \\ + \frac{4}{3} B_{mn} \sin\left(\frac{n\pi}{3}\right)^2 \sin(mw + ny) \end{bmatrix} \quad (5.83)$$

It is worth noticing that the harmonic amplitude of V_{gTx} is proportional to that of V_{LLT} at the same frequency. Thus, in the proposed method, the V_{LLT} is selected to reflect the total output characteristics in parallel inverters, since the V_{LLT} is easier to measure and calculate in practical applications than V_{gTx}.

(2) **Principle of DPS-PWM for two parallel two-level inverters.**

Take two parallel two-level inverters as an example. When the PS-PWM is applied in the parallel two-level inverters, the carrier waveforms and total phase voltages in one carrier period are shown in Fig. 5.25, where C_{r1} and C_{r2} represent the carrier waveforms for inverter 1 and 2, and C_{r2} lags C_{r1} π rad since PS-PWM is applied.

The V_{Pa} and V_{Pb} represent the total phase voltage corresponding to m_a and m_b, ([N], [O], [P]) represent the voltage level $(-U_{dc}, 0, U_{dc})$. When the modulation waves (m_a and m_b) are higher than zero, the V_{Pa} and V_{Pb} vary between [P] and [O]. Otherwise, the V_{Pa} and V_{Pb} vary between [O] and [N]. Thus, the V_{LLT} which represents the difference between V_{Pa} and V_{Pb} also varies between adjacent voltage levels in one carrier period. Furthermore, as shown in Fig. 5.25a, d, when m_a and m_b are lower or higher than zero, the V_{LLT} will also jump between [P] and [O]. And in Fig. 5.25b, c, when $m_a > 0$, $m_b < 0$, the V_{LLT} will switch between three voltage levels (2[P], [P] and [O]). Compared with Fig. 5.25a, d,

V_{LLT} in Fig. 5.25b, c contain higher voltage harmonic amplitude. That means a higher voltage harmonic amplitude in the total output voltage V_{gTx} will result in a higher THD in the total output current i_x. To further improve the output current quality of parallel inverters with PS-PWM, a DPS-PWM strategy is proposed in the proposed method. By modifying the phase-shifted angles of carriers in each inverter twice, the higher voltage harmonic amplitude can be avoided.

To be specific, the phase-shifted angles of carriers in each inverter consist of two parts, namely initial angle and modified angle. As shown in Fig. 5.26, the y_a axis of carriers (the carrier amplitude direction) is divided into two sectors and named s_1 and s_2. In s_1 and s_2, the initial angles of carrier 1 and carrier 2 are 0 rad and π rad, which is in accordance with PS-PWM. The modified angles of each carrier are the same in each sector. In s_1, the modified angles of each carrier are 0 rad, while those in s_2 are $\pi/2$ rad or $-\pi/2$ rad. It is worth noticing that the same effect of V_{Pb} can be achieved when the modified angles in s_2 are $\pi/2$ rad or $-\pi/2$ rad.

And only the modified angles in s_2 are $\pi/2$ rad as shown in Fig. 5.27 for simplicity. When the modified angles in s_2 are $\pi/2$ rad, by modifying the phase-shifted angles of carriers as the sum of the initial angle and modified angle, it can be noted that the total phase voltage lags by $\pi/2$ rad in s_2, while the total phase voltage is consistent with that of PS-PWM in s_1. That makes the higher voltage level state of V_{Pa} and V_{Pb} centered, and the total line voltage V_{LLT} can be changed accordingly.

In Fig. 5.27a, d, when the proposed DPS-PWM is applied, the V_{LLT} varies between [P] and [O], the same as in PS-PWM. And when $m_a > 0$, $m_b < 0$, the V_{LLT} will only switch between two adjacent voltage levels ([P] and [O]) due to the total phase voltage V_{Pb} lagging by $\pi/2$ rad, as shown in Fig. 5.27b, c. Since the total line voltage V_{LLT} can jump between two adjacent voltage levels and avoid the jumping between three voltage levels, the voltage harmonic amplitude in V_{LLT} and V_{gTx} are effectively reduced.

(3) **Principle of DPS-PWM for multiple parallel two-level inverters.**

To be more specific, as shown in Fig. 5.28, when the PS-PWM is applied, the initial voltage levels of each total output voltage (V_{pa}, V_{pb}, V_{pc}) are [O]. That results in the total line voltage V_{LLT} possibly jumping in three adjacent voltage levels. And the equivalent output voltage vectors for synthesizing the reference voltage vector are [OOO], [POO], [PON], [PNN].

Figure 5.29 shows the three-level voltage vector diagram, and the voltage vector V_{ref} is the reference voltage vector corresponding to (m_a, m_b, m_c). It can be noted that, when the PS-PWM is applied, the voltage vector [OOO] with significant vector errors is used for reference vector synthesis. That inevitably leads to significant harmonics in V_{LLT}, V_{gTx} and the total output current i_x.

When the DPS-PWM is applied, the high voltage levels of each total output voltage (V_{pa}, V_{pb}, V_{pc}) are centered. That results in the total line voltage V_{LLT} jumping in two adjacent voltage levels. And the equivalent output voltage vectors for synthesizing reference voltage vector are [POO], [PON], [PNN], [ONN]. Consistent with the nearest-three-vectors three-level SVPWM, since only the nearest three voltage vectors are applied to synthesize the reference vector in the DPS-PWM, the harmonics in V_{LLT} and V_{gTx} can be effectively reduced, which means an effective reduction of output current harmonics.

Based on the above analysis, the proposed DPS-PWM can also be extended to meet the requirements of multiple-parallel two-level inverters. And the main principle is modifying the phase-shifted angles of carriers in each inverter twice to make the higher voltage level state of three-phase total phase voltage V_{Px} centered, so that the case that V_{LLT} varies between three voltage states is avoided.

Assuming there are N inverters in parallel, the y_a axis of carriers is divided to N sectors ($s_1 \sim s_N$), and the sector number j ($j = 1,..., N$) is calculated based on the modulation wave m_x), as shown as

$$j = N - \frac{N}{2}\left[(m_x + 1) - (m_x + 1) \bmod\left(\frac{2}{N}\right)\right] \quad (5.84)$$

For the inverter i ($i = 1, ..., N$), the phase-shift angle θ_{dp_ij} of DPS-PWM in s_j ($j = 1, ..., N$) consists of two parts, where θ_{p1_i} represents the initial angle, θ_{p2_j} represents the modified angle.

$$\theta_{dp\_ij} = \theta_{p1\_i} + \theta_{p2\_j} \quad (5.85)$$

Similar to PS-PWM, the initial carrier angle of the ith inverter can be obtained as

$$\theta_{p1\_i} = \frac{(i-1) \times 2\pi}{N} \quad (5.86)$$

On the basis of initial angle θ_{p1_i}, when the modulation wave m_x is located in s_j of y_a axis, the phase-shift angle θ_{dp_ij} of each carrier should be further modified by θ_{p2_j}. The modified angle θ_{p2_j} is obtained as

$$\theta_{p2\_j} = \pm\frac{(j-1) \times \pi}{N} \quad (5.87)$$

When the $\theta_{p2\_j} < 0$, the carrier waveforms in DPS-PWM for all sectors are shown in Fig. 5.30a. When the $\theta_{p2\_j} > 0$, the corresponding carrier waveform is shown in Fig. 5.30b. Where C_{r1} to C_{rN} represent the carriers applied in inverter 1 to inverter N.

More specifically, based on the above description, the proposed method for three parallel two-level inverters is selected as an example to further elaborate the proposed DPS-PWM.

As shown in Fig. 5.31, when there are three inverters in parallel, the y_a axis of carriers is divided into s_1, s_2 and s_3. In s_1, s_2 and s_3, the initial angles θ_{p1i} of carrier 1, carrier 2 and carrier 3 are 0 rad, $2\pi/3$ rad and $4\pi/3$ rad, the same as PS-PWM. Since the y_a axis of carriers is divided into three parts, the modified

5.2 Control Technology for AC-Side Current Quality Improvement

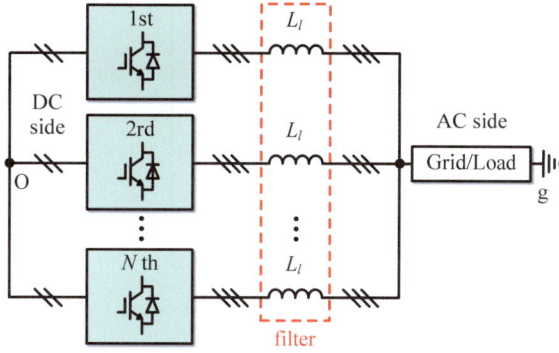

Fig. 5.23 The general structure of parallel inverters with DC side directly connected

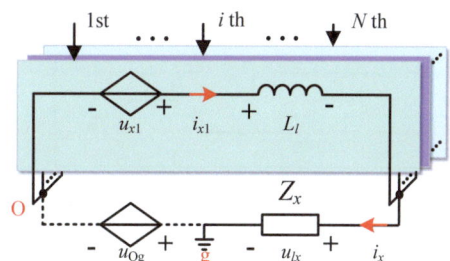

Fig. 5.24 Single-phase equivalent circuit in the ith inverter

angles θ_{p2j} change correspondingly. And the modified angles θ_{p2j} of s_1, s_2 and s_3 can be 0 rad, $\pi/3$ rad and $2\pi/3$ rad, respectively. Also, the same effect can be achieved when the modified angles θ_{p2j} of s_1, s_2 and s_3 are 0 rad, $-\pi/3$ rad and $-2\pi/3$ rad.

It can be noted that the higher voltage level states of V_{Pa} and V_{Pb} are centered when DPS-PWM is applied, which means a better quality of V_{LLT}. Accordingly, a lower THD in total output current can be achieved.

Validation Results for the Current Ripple Suppression

To verify the superiority and effectiveness of the proposed ICBM and DPS-PWM for the current ripple suppression of parallel inverters systems, the experimental results are given in this section. The experimental parameters for each inverter are listed as follows. The dc-link voltage U_{dc} is 200 V, the current filter inductor is 5 mH, the equivalent load resistance is 12 Ω, the carrier frequency f_c is 2500 Hz, the fundamental frequency f_1 is 50 Hz.

(1) **Experimental results for ICBM method based on zero-sequence injection**

Three methods are compared in this section. Method-1 is SVM, Method-2 is the ISVM, which is based on conventional PS-PWM, and Method-3 represents the proposed ICBM based on zero-sequence injection.

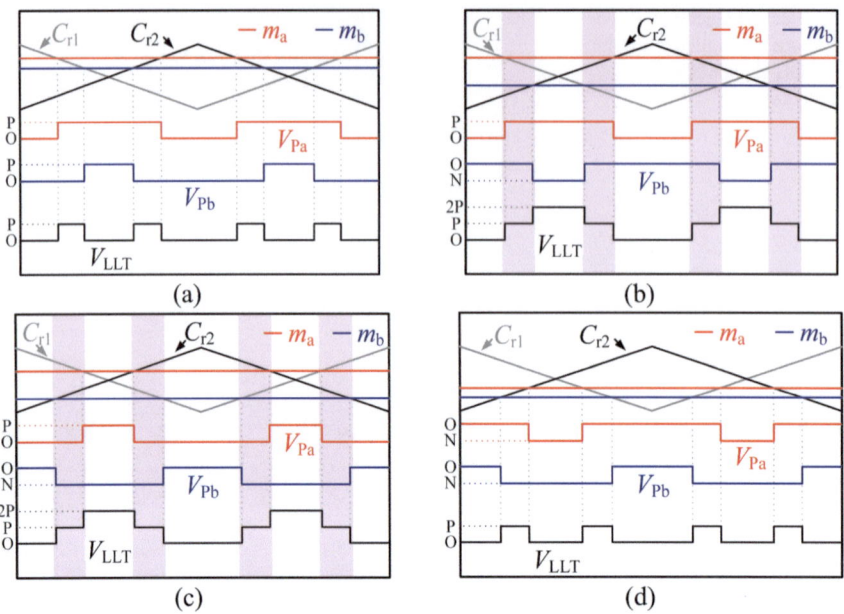

Fig. 5.25 The V_{LLT} and V_{Px} when PS-PWM is applied: **a** $m_a > 0$, $m_b > 0$; **b** $m_a > 0$, $m_b < 0$, $|m_a| > |m_b|$; **c** $m_a > 0$, $m_b < 0$, $|m_a| < |m_b|$; **d** $m_a > 0$, $m_b > 0$

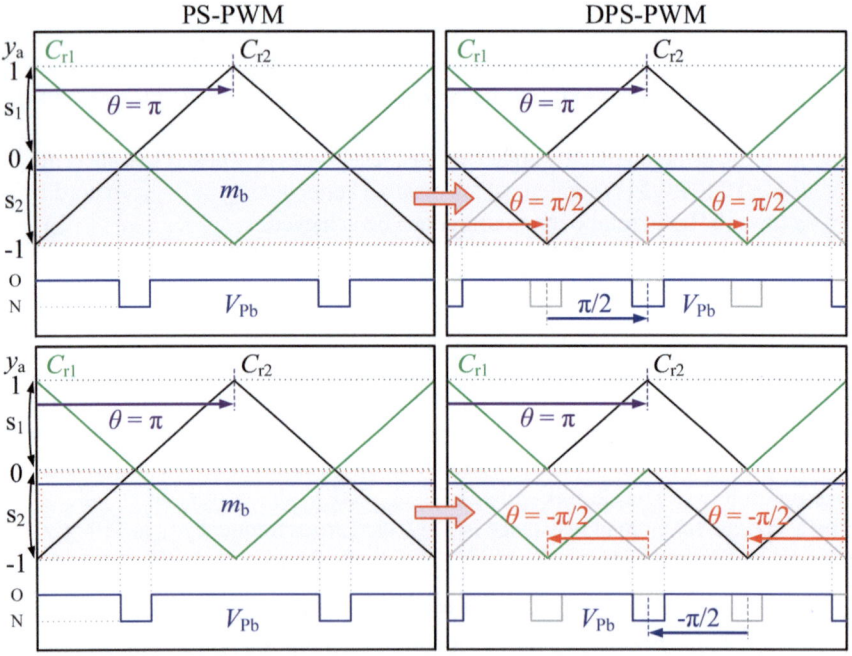

Fig. 5.26 The carrier diagram of the DPS-PWM in two parallel inverters

5.2 Control Technology for AC-Side Current Quality Improvement

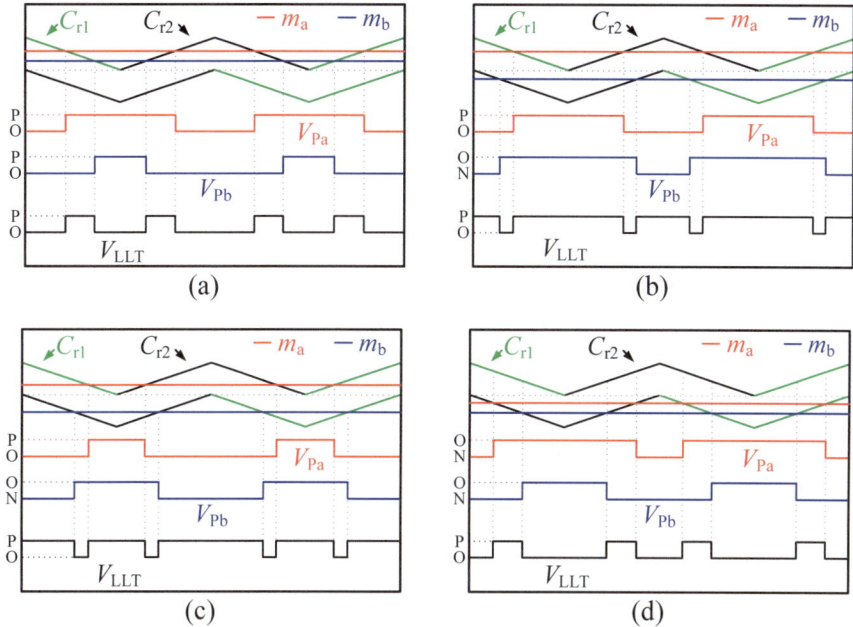

Fig. 5.27 The V_{LLT} and V_{Px}, when DPS-PWM is applied: **a** $m_a > 0$, $m_b > 0$; **b** $m_a > 0$, $m_b < 0$, $|m_a| > |m_b|$; **c** $m_a > 0$, $m_b < 0$, $|m_a| < |m_b|$; **d** $m_a > 0$, $m_b > 0$

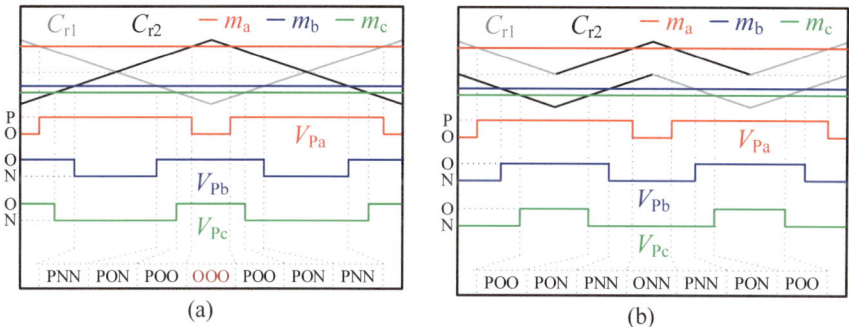

Fig. 5.28 The voltage vector sequence in one carrier period: **a** PS-PWM; **b** DPS-PWM

Figure 5.32 shows the experimental results for two parallel two-level inverters when the modulation index is 0.9. Where u_{ab} represents the sum of line voltage between phase A and phase B in each inverter. The i_c represents the total output current of phase c in parallel inverters. As shown in Fig. 5.32a, when Method-1 is applied, the u_{ab} contains a large number of harmonics, which results in a high current ripple in i_c. When Method-2 is applied, odd carrier multiple groups (harmonics near frequency f_c, $3f_c$, $5f_c$, ...) in u_{ab} are effectively reduced, which

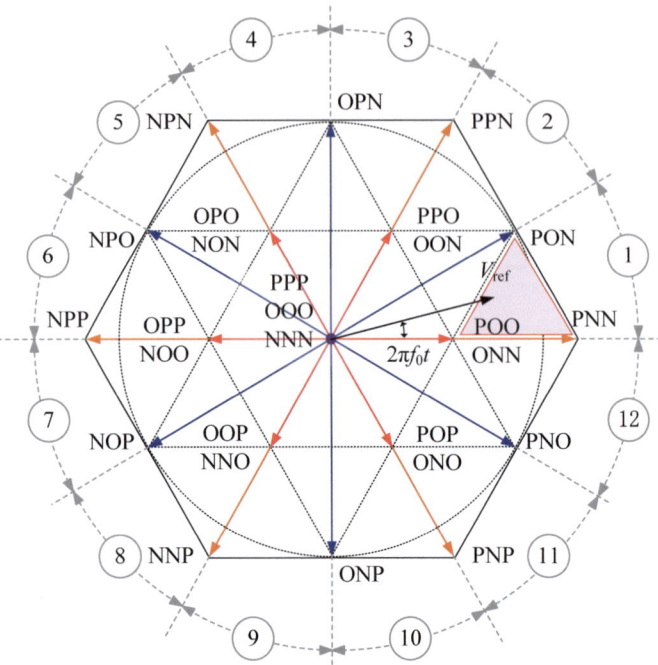

Fig. 5.29 Three-level voltage vector diagram

leads to a lower current ripple and THD compared to Method-1. In Fig. 5.32c, it is worth noticing that the equivalent total output line voltage u_{ab} could provide a satisfactory five voltage levels waveform like a common three-level inverter, due to the space voltage vectors diagram for the three-level inverter being applied in the two parallel two-level inverters. Moreover, the voltage harmonics amplitude of even carrier multiple groups (harmonics near frequency $2f_c$, $4f_c$, $6f_c$, …) in Fig. 5.32b are further minimized, and the current THD is 68% of that in Fig. 5.32b, which proves the effectiveness of the proposed ICBM.

The same conclusion can be obtained when the proposed method is extended to three-level inverters. As shown in Fig. 5.33, it is worth noticing that Method-2 could not significantly improve the output current quality compared with Method-1, due to the voltage harmonics amplitude of odd carrier multiple groups in u_{ab} of the three-level inverter already being low. By contrast, the proposed method could provide a better equivalent total output line voltage quality. The voltage weighted total harmonic distortion (WTHD) of the proposed method is 0.31%, which is 57% and 62% of that in Method-1 and Method-2. That results in a decreased current THD by 48.3% and 42.5% respectively compared with Method 1 and Method-2.

Figure 5.34 shows the comparison results of output currents between proposed Method-3 and Method-2. Where i_{c1} and i_{c2} represent the current of

Fig. 5.30 The carriers of the DPS-PWM in multiple-parallel inverters: **a** $\theta_{p2\_j} < 0$; **b** $\theta_{p2\_j} > 0$

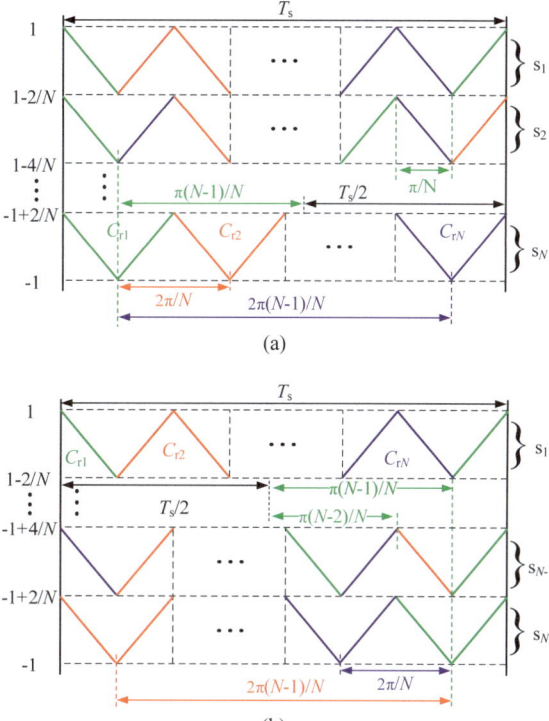

inverter 1 and inverter 2 in phase c, the modulation index is 0.9. Both Method-2 and Method-3 could provide a stable sinusoidal phase current in each inverter, and the total output current quality is much better than that of the phase current of each inverter. Moreover, compared to Method-2, the output current in each inverter with the proposed Method-3 will contain more current ripple to maintain a lower total output current ripple. This will result in an increase of THD in the phase current. For example, in parallel two-level inverters, the THD of i_{c1} in Method-3 is 104.8% of that in Method-2 while the THD of i_c in Method-3 is 68.0% of that in Method-2. The similar experimental results are also drawn in parallel three-level inverters and shown in Fig. 5.34c, d.

To further prove the advantages of the proposed method in output current quality improvement, the experimental comparisons of total current THD for various methods are shown in Fig. 5.35, where the modulation index varied from 0.1 to 1.15. As shown in Fig. 5.35a, when the proposed method is applied in parallel two-level inverters, the THDs of the proposed scheme are significantly lower than those of Method-1 and Method-2 in both high and low modulation indices. Furthermore, a similar conclusion can be derived in parallel three-level inverters, as shown in Fig. 5.35b. This further demonstrates the correctness and

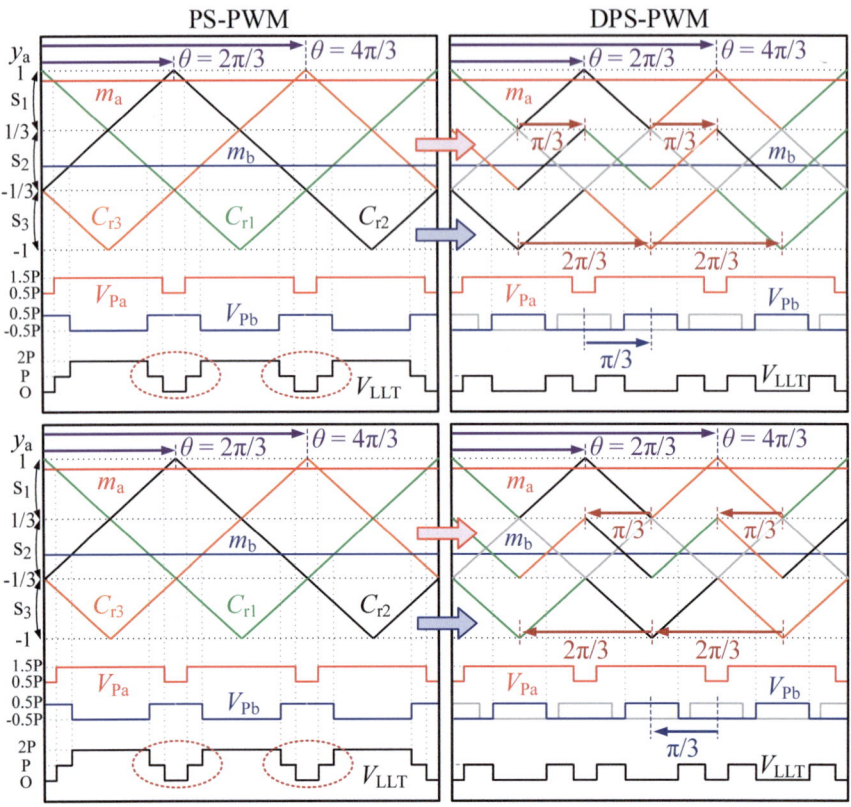

Fig. 5.31 The carrier diagram of the DPS-PWM in three parallel inverters

effectiveness of the proposed ICBM and the extended scheme of the proposed ICBM.

(2) **Experimental results for DPS-PWM method based on phase-shifted angles**

To verify the effectiveness and superiority of the proposed DPS-PWM, the experimental results are given in this section. Figure 5.36a, b show the total line voltage waveform (V_{LLT}) and total output current in phase c (i_c) of both DPS-PWM and the PS-PWM where the number of parallel inverters is 2, and the modulation index is 0.9. The corresponding Fast Fourier Transform (FFT) results are also given. It can be noted that the case that V_{LLT} varies between three voltage levels in one carrier period is avoided in Fig. 5.36b, which means the proposed method could provide a better voltage waveform than that of PS-PWM. The FFT results also prove the point. In Fig. 5.36a, the voltage and current harmonics at around 2 times the switching frequency have a high amplitude, which leads to a higher WTHD of line voltage and higher THD of output current compared with DPS-PWM. When the proposed DPS-PWM is applied, the harmonics at 2 times the switching frequency are effectively reduced. The

WTHD and THD in Fig. 5.36b are 0.27% and 4.43%, which are only 58.7% and 55.4% of those in Fig. 5.36a.

A similar conclusion can be drawn under different modulation indices. When the modulation index is 0.5, the V_{LLT} and i_c contain higher harmonics compared to those in Fig. 5.37. This will also lead to an increased WTHD in V_{LLT} and THD in i_c. In Fig. 5.37b, when the DPS-PWM is applied, the voltage harmonics of V_{LLT} around 2 times the switching frequency can be significantly reduced compared to Fig. 5.37a. This reduced the THD of i_c from 17.23% to 7.71%.

Figures 5.38 and 5.39 show the corresponding experimental results when the number of parallel inverters is 3.

It can be seen that the output waveforms are obviously better than those when $N = 2$. This proves that both proposed method and PS-PWM can further improve the quality of total output waveform when the number of parallel inverters is increased. Also, the DPS-PWM could effectively reduce the voltage harmonics in V_{LLT}, which results in a better total output current quality. In Fig. 5.38b, the THD of i_c in DPS-PWM is 3.25%, which is 40.6% lower than that in PS-PWM. When the modulation index decreases, the proposed DPS-PWM is still valid. In Fig. 5.39b, the THD of i_c in DPS-PWM is 6.04%, which is 35.3% lower than that in PS-PWM.

To further demonstrate the advantages of the proposed method, the experimental results comparison for total line voltage and total output current is given in Figs. 5.40 and 5.41, where the number of parallel inverters is 2 and 3, the modulation index is varied from 0.1 to 1. In Fig. 5.40, when the modulation index changes, the WTHDs of V_{LLT} in the DPS-PWM are almost lower than that in PS-PWM at both $N = 2$ and $N = 3$. It is worth noticing that the proposed method is similar to PS-PWM when the number of parallel inverters (N) is an odd number and the modulation index is less than $1/N$. That results in a similar total output waveform quality when the modulation index is less than $1/N$. The similar conclusion can also be drawn from Fig. 5.41, and the proposed method could provide a satisfactory total output current in both high and low modulation indices.

To summarize, to meet the high current quality requirement in high-power rating photovoltaic applications, the ICBM and DPS-PWM methods are proposed for total current ripple suppression. In the ICBM method, a proper zero-sequence modulation wave is injected into the three-phase modulation waves to achieve a low total current ripple based on the root mean square value analysis of the output current ripple. According to the model of the current ripple, the proper zero-sequence component modulation wave is determined. In the DPS-PWM method, by adding a modified angle to the initial angle of PS-PWM, the case that the total line voltage waveform varies between three voltage levels in one carrier period is avoided, and a better total output current quality can be obtained.

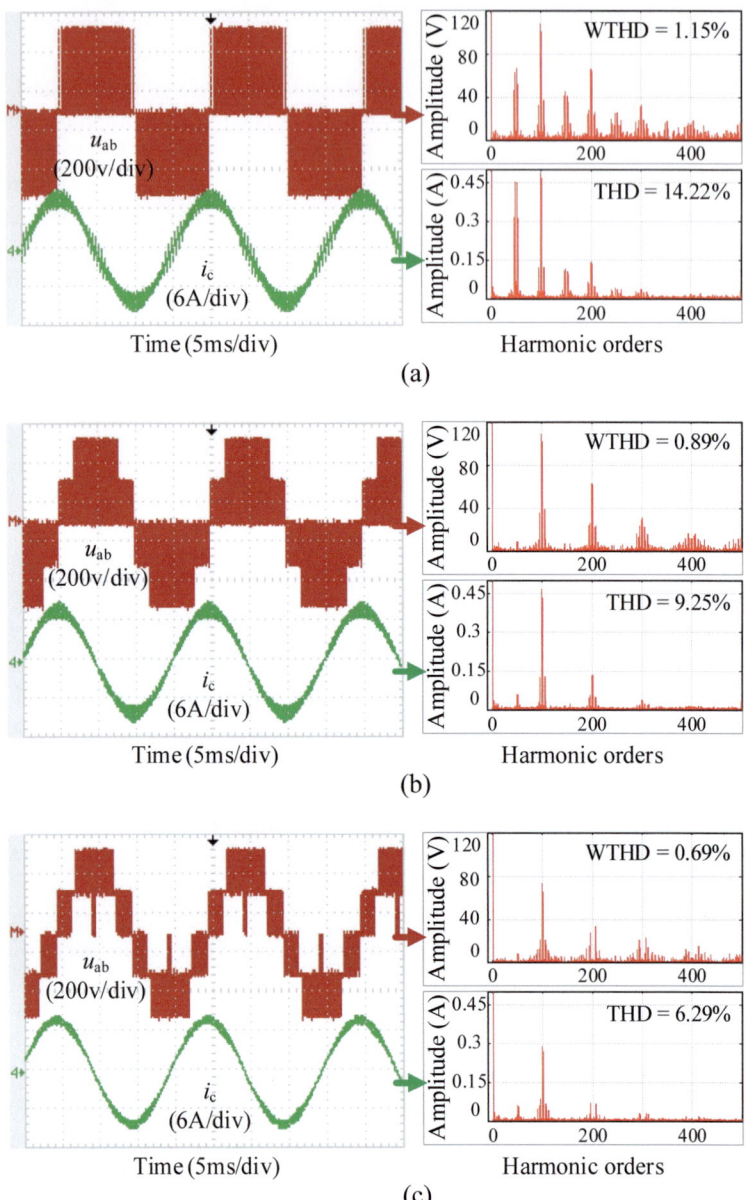

Fig. 5.32 The experimental results for two-level inverters when $m = 0.9$: **a** Methos-1; **b** Method-2; **c** Method-3

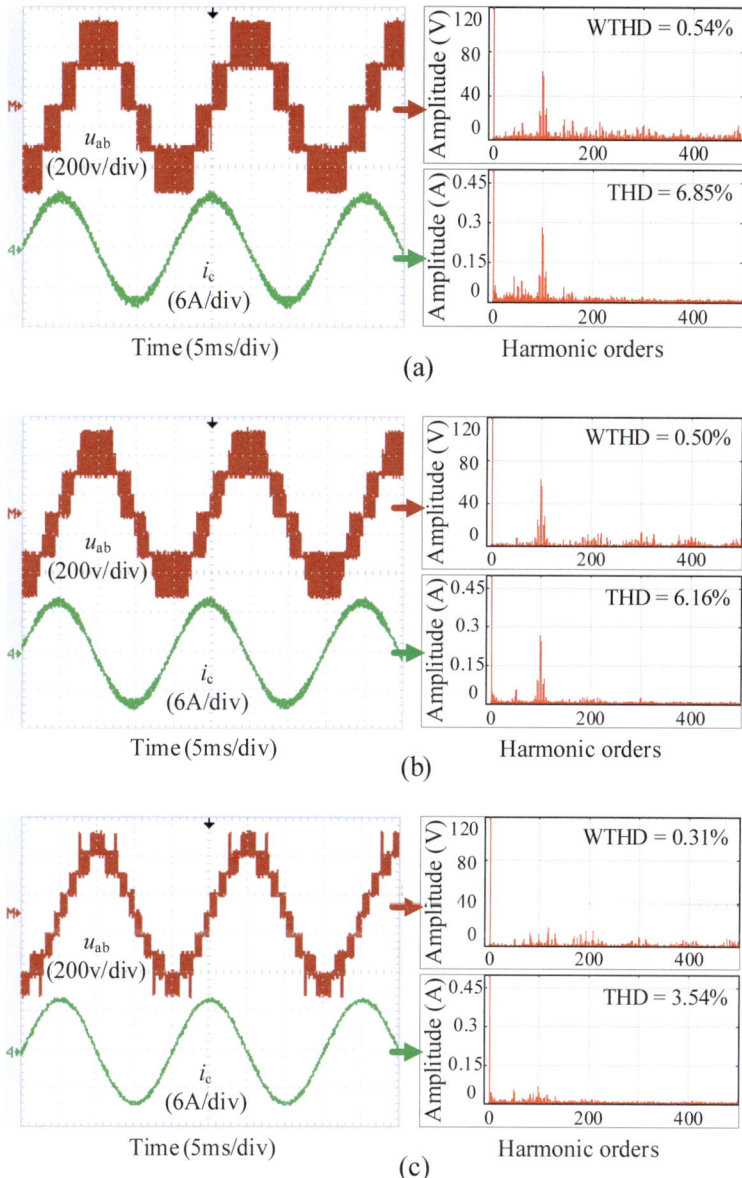

Fig. 5.33 The experimental results for three-level inverters when $m = 0.9$: **a** Methos-1; **b** Method-2; **c** Method-3

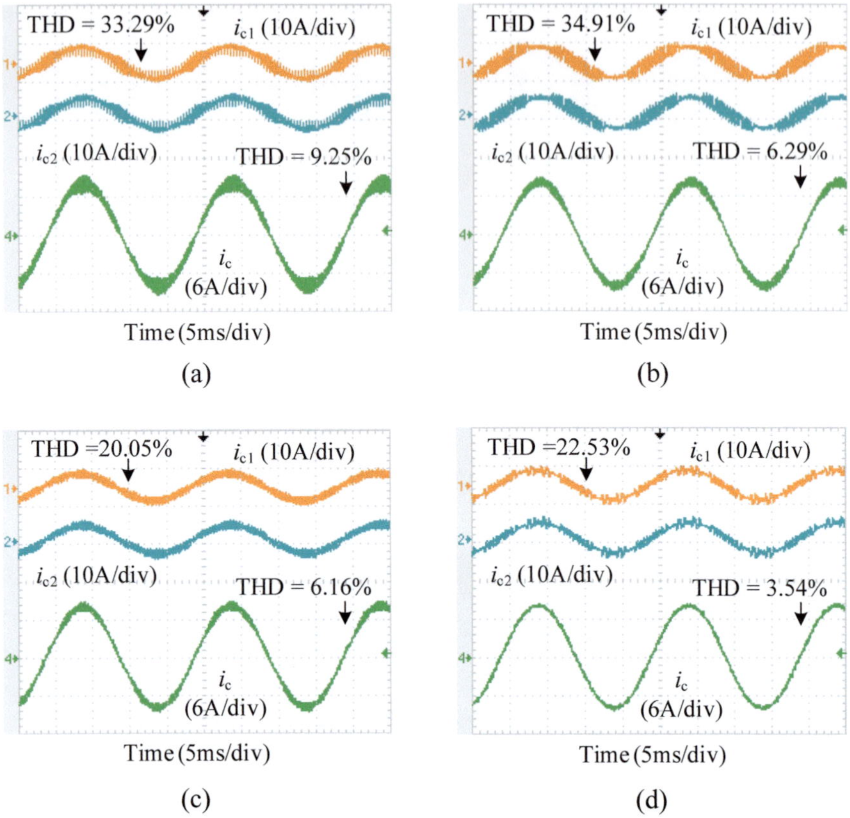

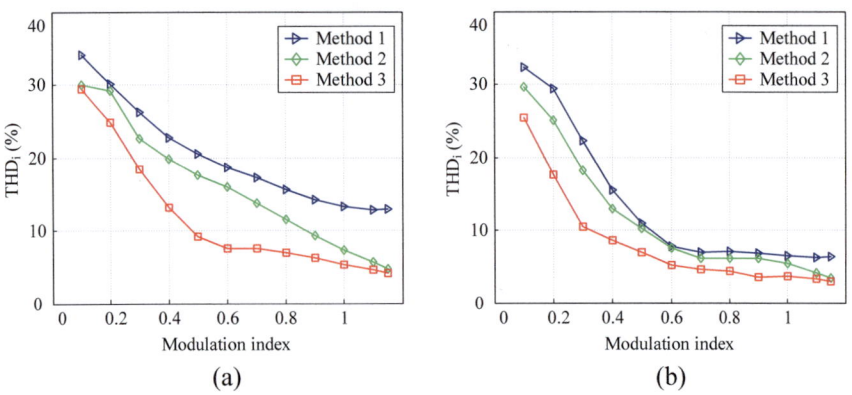

Fig. 5.34 The comparison results of output currents: **a** Method-2 in two-level inverters; **b** Method-3 in two-level inverters; **c** Method-2 in three-level inverters; **d** Method-3 in three-level inverters

Fig. 5.35 The experimental comparison results of total output current: **a** parallel two-level inverters; **b** parallel three-level inverters

5.3 Control Technology for DC-Side Neutral-Point Voltage Oscillation ...

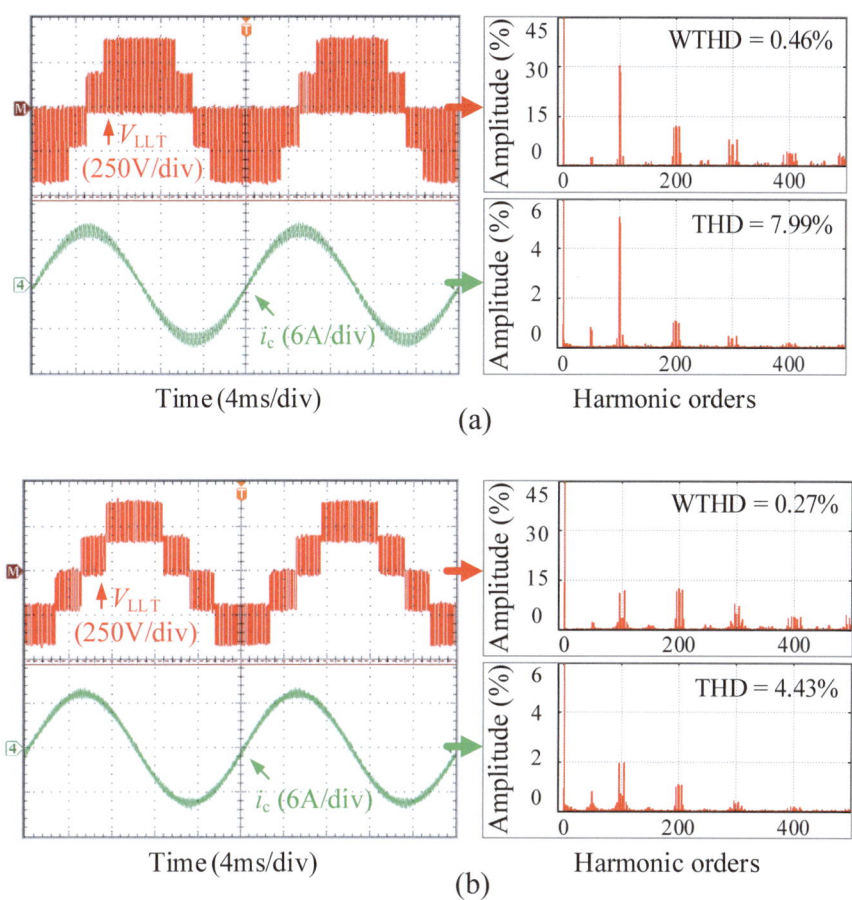

Fig. 5.36 The experimental results when $m = 0.9$, $N = 2$: **a** PS-PWM; **b** DPS-PWM

5.3 Control Technology for DC-Side Neutral-Point Voltage Oscillation Suppression

In 3LT^{2}I, two capacitors are connected in series in dc-link to generate three output voltage levels. However, the structure of two dc-link capacitors connected in series easily causes the voltages of capacitors to be unequal and leads to NPV unbalance. The NPV unbalance is divided into dc unbalance and low-frequency oscillations. The dc unbalance problem means that two dc-link voltages do not have the same voltage, and the voltage difference between these voltages is not changed with time. The dc unbalance problem will cause the distortion of output currents, which is a serious problem that must be solved. Lots of researches have been proposed to solve the dc unbalance problem. The low-frequency oscillations problem, which means the ac ripple of the voltages across top- and bottom-side capacitors, changes

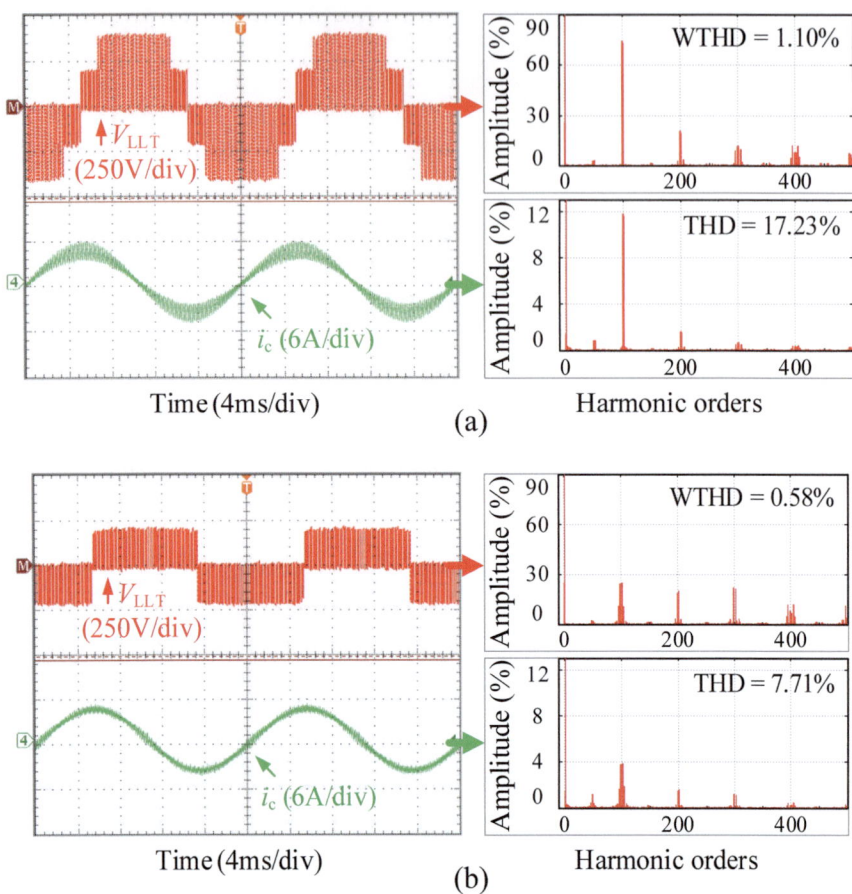

Fig. 5.37 The experimental results when $m = 0.5$, $N = 2$: **a** PS-PWM; **b** DPS-PWM

with time. The frequency of the low-frequency oscillations is three times the output fundamental frequency, and the magnitude varies depending on the transferred power, the power factor, and the dc-link capacitor value. A large low-frequency oscillation causes output currents to contain the ac ripple's frequency. Furthermore, the large low-frequency oscillation imposes stress on converter devices. To mitigate the low-frequency oscillations, the large dc-link capacitor value is required, which increases the volume and the cost of the $3LT^2I$. For a reliable and cost-effective system, small capacitors are preferably used for the dc-link capacitors. Thus, the low-frequency oscillations problem has to be solved.

In this section, the technology to suppress the low-frequency oscillation of NPV in $3LT^2I$ is given. The power flowing from DC side to AC side is taken as the example to illustrate the principle of the technology. The technology includes two groups: the first group is based on space vector modulation (SVM), and the second group is

5.3 Control Technology for DC-Side Neutral-Point Voltage Oscillation …

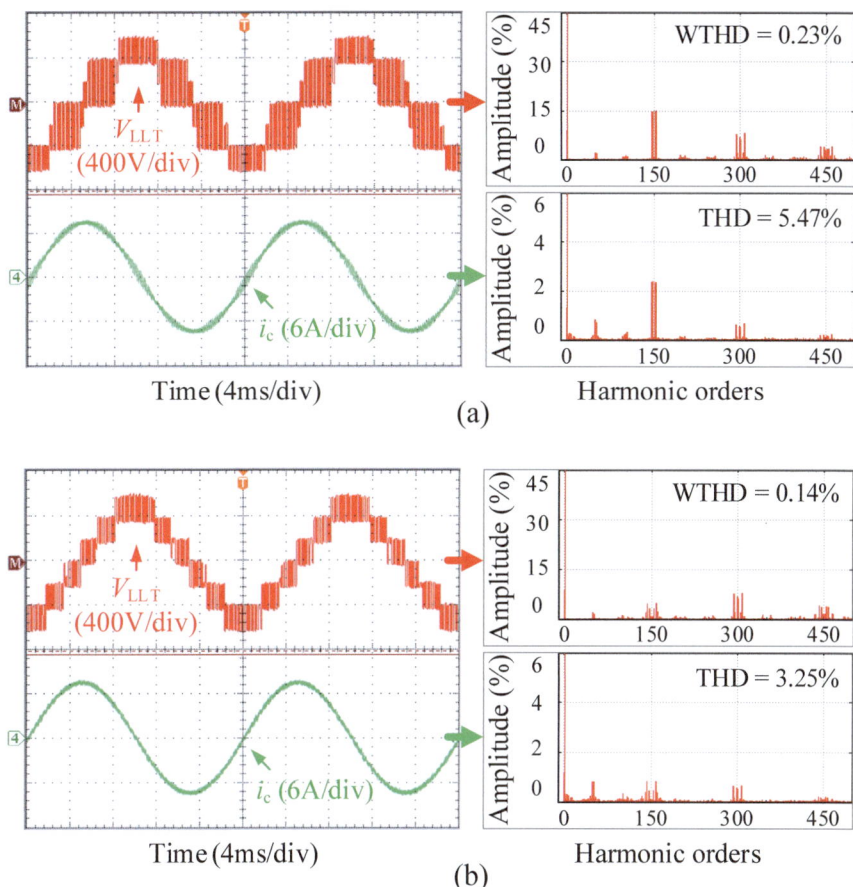

Fig. 5.38 The experimental results when $m = 0.9$, $N = 3$: **a** PS-PWM; **b** DPS-PWM

based on carrier-based pulse width modulation (CB-PWM). The first group is further classified into two types of methods according to the magnitudes of the CMV. The first type method is characterized with high CMV magnitude, and the second type method is characterized with low CMV magnitude. This section is organized as follows. First, the generation mechanism of NPV oscillation is given. Then two groups of methods are illustrated step by step.

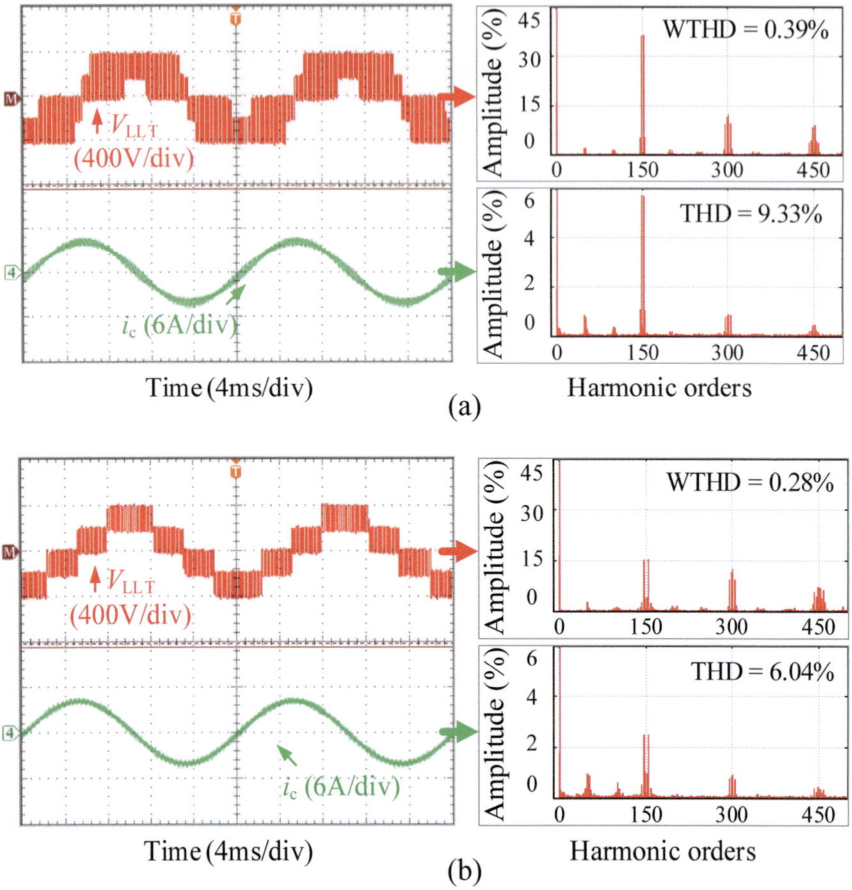

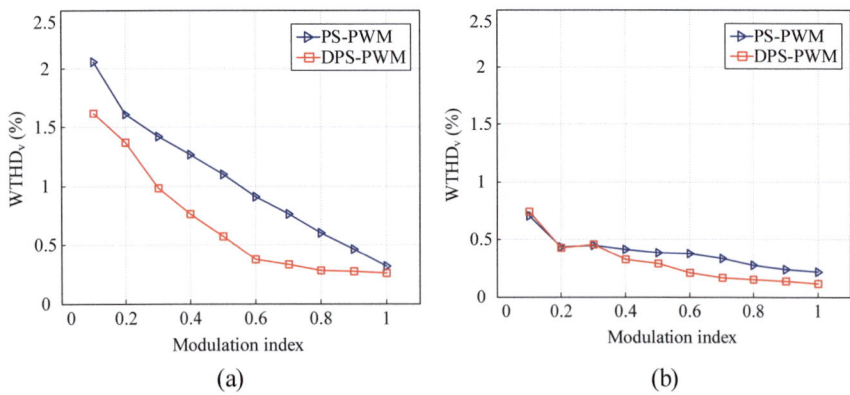

Fig. 5.39 The experimental results when $m = 0.5$, $N = 2$: **a** PS-PWM; **b** DPS-PWM

Fig. 5.40 The experimental results comparison of WTHD in total line voltage: **a** $N = 2$; **b** $N = 3$

5.3 Control Technology for DC-Side Neutral-Point Voltage Oscillation ... 269

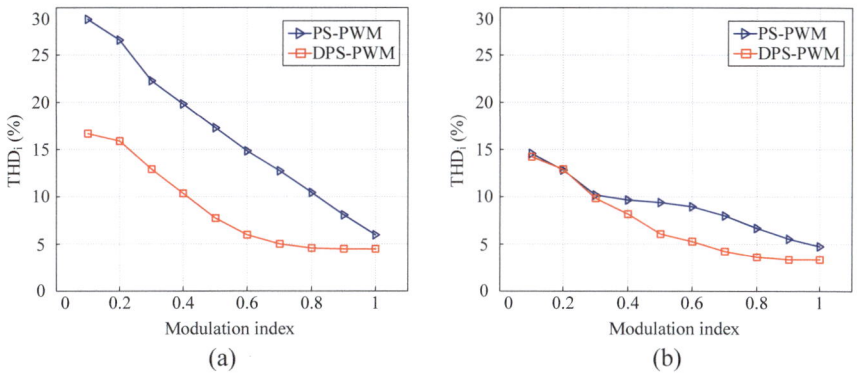

Fig. 5.41 The experimental results comparison of THD in total output current: **a** $N = 2$; **b** $N = 3$

5.3.1 The Generation Mechanism of Neutral-Point Voltage Oscillation

To suppress the low-frequency oscillation of NPV, the generation mechanism and corresponding characteristics of oscillation in NPV will be given in this section. For ease of analysis, the space vector diagram of 3LT^{2}I is re-depicted in Fig. 5.42. The space vector diagram (SVD) contains 27 basic voltage vectors in total. The basic voltage vectors can be categorized into zero vector, small vector, medium vector, and large vector based on their magnitudes.

The NPV unbalance is essentially caused by the fact that the current injected into neutral point O does not equal to zero in one switching control period. Different basic voltage vector have different neutral point current. Table 5.3 shows the relationship of basic vectors type, switching states, and neutral point current of 3LT^{2}I. It can be seen that the neutral point currents generated by large vectors and zero vectors are equal to zero, and are unequal to zero for small and medium vectors.

The effects of basic vectors on NPV are shown in Fig. 5.43. Although the zero vector [OOO] is connected to the neutral point O, the total sum of the currents is zero. Thus, it does not affect the NPV balancing. Large vector [PNN] does not affect the NPV balancing because the neutral point O is disconnected. Because the P-type switching states are connected between the positive dc-link and neutral point O, the neutral point current flows into the neutral point O. In this condition, C_1 is discharged and C_2 is charged, and ΔU_{np} ($\Delta U_{np} = U_p - U_n$) is decreased. On the other hand, the N-type small vector [OON] increases the ΔU_{np} since the three phases are connected between the neutral point and the negative dc-link. The medium vector [PON] also affects the NPV. Depending on the current direction of the phase connected to the neutral point, the ΔU_{np} increases or decreases with the medium vector. When current direction is positive, the ΔU_{np} increases. While the ΔU_{np} decreases, when current direction is negative. Among the four types of basic vectors, only the medium and small vectors influence the NPV unbalance. The medium vector

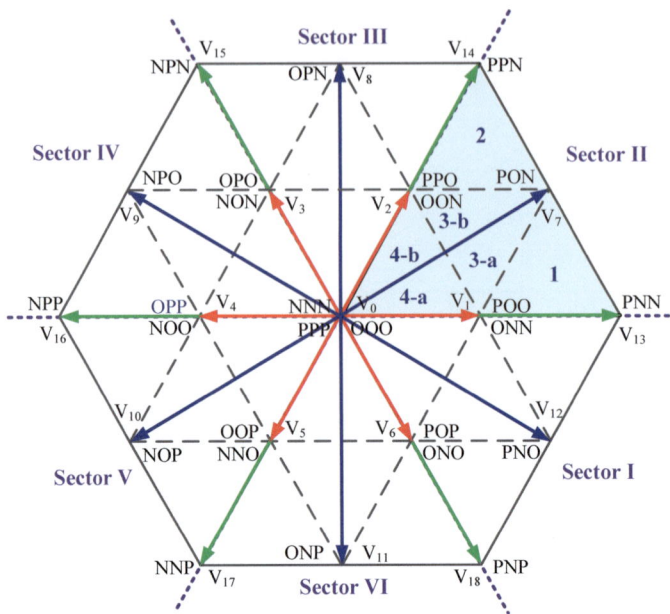

Fig. 5.42 Space vectors diagram of $3LT^2I$ containing 27 basic voltage vectors

charges or discharges the two dc-link capacitors with different current direction and causes the NPV oscillation.

When conventional SVM method is applied, the oscillations of NPV are induced. The frequency of NPV oscillations is three times of fundamental frequency. This oscillation can be eliminated by using the small vector.

5.3.2 Neutral-Point Voltage Oscillation Suppression Methods

In this section, two groups of NPV oscillation methods based on SVM and CB-PWM are introduced.

Based on Space Vector Pulse Width Modulation

To reduce the low-frequency voltage oscillation, the methods based on the symmetric SVM are illustrated in this section. These methods are classified into two types according to the magnitudes of the CMV. The first type method features a high CMV magnitude, and the second type method features a low CMV magnitude.

The first type method is derived from the symmetric SVM method, in which a compensation time is added into the three phase turn-on times to minimize the

5.3 Control Technology for DC-Side Neutral-Point Voltage Oscillation …

Table 5.3 The relationship of basic vectors type, switching states, and neutral point current

Vector type	Switching state		Neutral point current	
Zero vector	[OOO]		0	
Small vector	P-type	N-type	P-type	N-type
	[POO]	[ONN]	$i_b + i_c$	i_a
	[PPO]	[OON]	i_c	$i_a + i_b$
	[OPO]	[NON]	$i_a + i_c$	i_b
	[OPP]	[NOO]	i_a	$i_b + i_c$
	[POP]	[ONO]	i_b	$i_a + i_c$
	[OOP]	[NNO]	$i_a + i_b$	i_c
Medium vector	[PON]		i_b	
	[OPN]		i_a	
	[NPO]		i_c	
	[NOP]		i_b	
	[ONP]		i_a	
	[PNO]		i_c	
Large vector	[PNN]		0	
	[PPN]		0	
	[NPN]		0	
	[NPP]		0	
	[NNP]		0	
	[PNP]		0	

low-frequency voltage oscillation of NPV. This type of method has many variations depending on the different applications. The method proposed by South Korean scholar Ui-Min Choi is the most typical representative [21]. The second type method is proposed by the author's team [22], which is evolved from the first method. The significant features of the second type method are that the CMV magnitude is only one-half of that in the first method. The above two methods are elaborated as the typical NPV oscillation suppression methods in this section.

(1) **NPV Oscillation Suppression Method Featuring with High CMV Magnitude**

This method is proposed by South Korean scholar Ui-Min Choi. By adding a compensation time to the three phase turn-on times, the low-frequency voltage oscillation is suppressed. The proper compensation time is simply calculated by considering the phase currents and dwell time of small and medium basic vectors in a switching sequence.

The small vectors and the medium vectors oscillate the two capacitor voltages. To minimize the oscillation, the average neutral current should be zero during the switching sequence. This can be obtained by adding a proper compensation time (T_{comp}) to the three-phase turn-on times (T_a, T_b, T_c). Sector I of

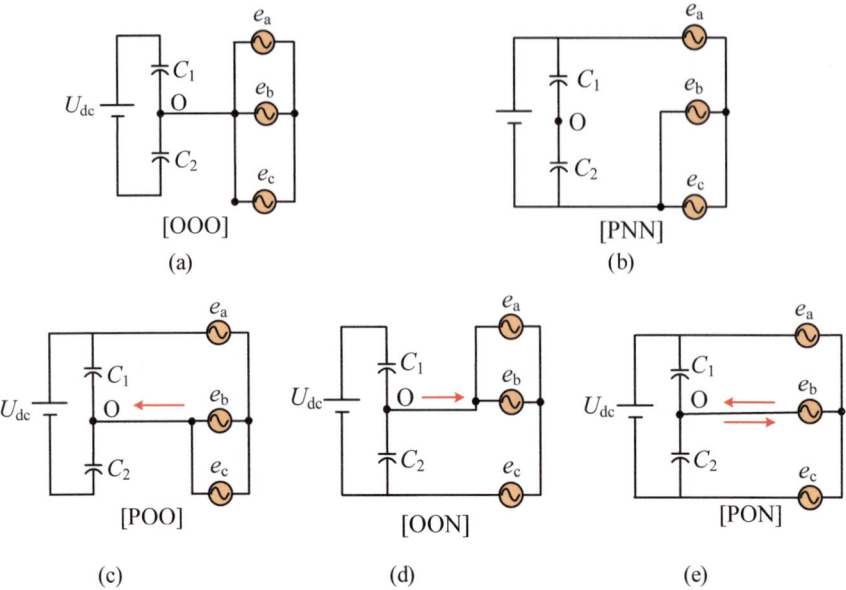

Fig. 5.43 The effects of basic vectors on NPV: **a** zero vector [OOO]; **b** large vector [PNN]; **c** P-type small vector [POO]; **d** N-type small vector [OON]; **e** medium vector [PON]

SVD is considered as an example to explain the proposed method. Sector I is further divided into six regions, which consist of 1, 2, 3-a, 3-b, 4-a, and 4-b. Thus, six cases are elaborated as follows.

Case 1, when the reference voltage V_{ref} is in region 1 of Sector I, it is synthesized by four vectors [ONN], [PNN], [PON], and [POO]. These four vectors are symmetric distribution to form the switching sequence as ONN-PNN-PON-POO-PON-PNN-ONN. In conventional modulation method, the dwell time of redundant N-type small vector [ONN] and P-type small vector [POO] is equally divided to reduce the effect of small vectors on NPV. However, in this sequence, the medium vector [PON] makes the average of the neutral point current not zero during the switching sequence. The ΔU_{np} is increased or decreased depending on the current direction of phase B.

Since the neutral point currents corresponding to zero vector and large vector are equal to zero, the average neutral point current in one switching period is only determined by the current and dwell times of small and medium vector. By analyzing the switching sequence, the T_{comp} is simultaneously added into three phase turn-on times, the dwell times of redundant P-type and N-type small vectors can be adjusted, while the dwell time of medium vector remains unchanged. Thus, adding T_{comp} can realize zero average neutral point current, and NPV oscillation can be suppressed. The switching sequences before and after T_{comp} are added as shown in Fig. 5.44.

5.3 Control Technology for DC-Side Neutral-Point Voltage Oscillation ...

The dwell times of [PON], [ONN] and [POO] before T_{comp} added can be expressed as

$$\begin{cases} T_{PON} = 2(T_c - T_b) \\ T_{ONN} = 2T_a \\ T_{POO} = 2(0.5T_s - T_c) = 2T_a \end{cases} \quad (5.88)$$

Assuming $C_1 = C_2 = C$ The deviation of NPV is proportional to the integration of i_{np}. Thus, the deviation of NPV caused by [PON] $\Delta U_{np\text{-}PON}$ can be expressed as

$$\Delta U_{np-PON} = \frac{1}{C} \int_{t=0}^{t=T_{PON}} i_{np} dt \quad (5.89)$$

Since the switching time is small enough, there is no change in the current. So, the $\Delta U_{np\text{-}PON}$ can be further expressed as

$$\Delta U_{np-PON} = \frac{1}{C} \int_{t=0}^{t=T_{PON}} i_{np} dt = \frac{1}{C} 2(T_c - T_b) I_b \quad (5.90)$$

After T_{comp} added, the $\Delta U_{np\text{-}PON}$ remained unchanged because the T_{PON} is no changed. While the deviation of NPV caused by small vectors of [ONN] and [POO] $\Delta U_{np\text{-}ONN+POO}$ is changed as

$$\left| \Delta U_{np-ONN+POO} \right| = \left| \frac{1}{C} \int_{t=0}^{t=T_{ONN}+T_{POO}} i_{np} dt \right|$$
$$= \left| \frac{1}{C}(2T_a + T_{comp})I_a + \frac{1}{C}(T_s - 2T_c - T_{comp})(-I_a) \right|$$
$$= \left| \frac{1}{C} 2T_{comp} I_a \right| \quad (5.91)$$

In order to suppress the NPV oscillation, the magnitude of $\Delta U_{np\text{-}PON}$ should be equal to $\Delta U_{np\text{-}ONN+POO}$. Based on this theory, the magnitude of T_{comp} can be derived as

$$\left| T_{comp} \right| = \left| \frac{I_b}{I_a} \right| (T_c - T_b) \quad (5.92)$$

The sign of T_{comp} is determined by the phase current direction of [O] state in medium vector. For example, in the medium vector [PON], if $i_b < 0$, C_1 is discharged and U_p is decreased, while C_2 is charged and U_n is increased. In this case, [PON] decreases ΔU_{np}. To eliminate this decrement of ΔU_{np}, the T_{comp} should be a positive value to increase the dwell time of N-type small vector and to decrease the dwell time of P-type small vector. Conversely, if $i_b > 0$, the T_{comp} should be a negative value.

When 3LT²I operates in high power factor, [PON] decreases ΔU_{np} because $i_b < 0$. Therefore, the dwell time of P-type should be reduced by T_{comp}, and the dwell time of N-type should be increased by T_{comp}. The turn-on times of three phase are redefined as

$$\begin{cases} T_{a\text{-comp}} = T_a + 0.5 \left| \frac{I_b}{I_a} \right| (T_c - T_b) \\ T_{b\text{-comp}} = T_b + 0.5 \left| \frac{I_b}{I_a} \right| (T_c - T_b) \\ T_{c\text{-comp}} = T_c + 0.5 \left| \frac{I_b}{I_a} \right| (T_c - T_b) \end{cases} \quad (5.93)$$

After T_{comp} is added, the dwell times of [ONN] and [POO] are changed as

$$\begin{cases} T_{ONN} = 2T_a + \left| \frac{I_b}{I_a} \right| (T_c - T_b) \\ T_{POO} = 2T_a - \left| \frac{I_b}{I_a} \right| (T_c - T_b) \end{cases} \quad (5.94)$$

Case 2, when the reference voltage V_{ref} is in region 2 of Sector I, it is synthesized by four vectors [OON], [PNN], [PON], and [PPO]. The switching sequence is OON-PON-PPN-PPO-PPN-PON-OON. The dwell times of [PON], [OON], and [PPO] should be considered and are expressed as

$$\begin{cases} T_{PON} = 2(T_b - T_a) \\ T_{OON} = 2T_a \\ T_{PPO} = 2(0.5T_s - T_c) = 2T_a \end{cases} \quad (5.95)$$

The principle for calculating the precise T_{comp} is the same as that in case 1. T_{comp} is added to adjust the dwell time of redundant N-type and P-type small vectors so that the $\Delta U_{np\text{-PON}}$ caused by medium vector can be offset. The T_{comp} is calculated as

$$\left| T_{comp} \right| = \left| \frac{I_b}{I_c} \right| (T_b - T_a) \quad (5.96)$$

In this region, $i_b > 0$, and [PON] increases ΔU_{np}. Therefore, the T_{comp} should be a negative value to decrease the dwell time of N-type small vector and increase the dwell time of P-type small vector. Thus, the turn-on times of the three phase are changed as

$$T_{x\text{-comp},(x=a,b,c)} = T_x - 0.5T_{x\text{-comp}} = T_x - 0.5 \left| \frac{I_b}{I_c} \right| (T_b - T_a) \quad (5.97)$$

5.3 Control Technology for DC-Side Neutral-Point Voltage Oscillation ...

Case 3, when the reference voltage V_{ref} is in region 3-a of Sector I, four vectors [OON], [PON], [ONN], and [POO] are adopted to synthesize V_{ref}. The switching sequence is ONN-OON-PON-POO-PON-OON-ONN. Unlike the above two cases, the ΔU_{np} is not only induced by the medium vector [PON] but also induced by the non-redundant N-type small vector [OON]. Both of the ΔU_{np} should be offset by adjusting the dwell times of the redundant small vectors of [ONN] and [POO]. The dwell times of four vectors are expressed as

$$\begin{cases} T_{PON} = 2(T_c - T_a) \\ T_{OON} = 2(T_a - T_b) \\ T_{ONN} = 2T_b \\ T_{OOP} = 2(0.5T_s - T_c) = 2T_b \end{cases} \quad (5.98)$$

The $T_{comp\text{-}OON}$ is added to adjust the dwell times of redundant N-type and P-type small vectors [ONN] and [POO], thereby eliminating the $\Delta U_{np\text{-}OON}$ induced by the N-type small vector [OON]. $T_{comp\text{-}OON}$ is calculated as

$$\left|T_{comp\text{-}ONN}\right| = \left|\frac{I_c}{I_a}\right|(T_a - T_b) \quad (5.99)$$

Similarly, $T_{comp\text{-}PON}$ is used to eliminate the $\Delta U_{np\text{-}PON}$ induced by the medium vector [PON], which is derived as

$$\left|T_{comp\text{-}PON}\right| = \left|\frac{I_b}{I_a}\right|(T_c - T_a) \quad (5.100)$$

In this region, since $i_b < 0$, the medium vector [PON] decreases ΔU_{np}, and the deviation is $\Delta U_{np\text{-}PON}$. To eliminate the $\Delta U_{np\text{-}PON}$, $T_{comp\text{-}PON}$ should be a positive value to increase the dwell time of the N-type small vector [ONN]. While the non-redundant N-type small vector [OON] increases ΔU_{np}, and the deviation is $\Delta U_{np\text{-}OON}$. To eliminate the $\Delta U_{np\text{-}OON}$, $T_{comp\text{-}OON}$ should be a negative value to reduce the dwell time of the N-type small vector [ONN]. Thus, the total time compensation is expressed as

$$T_{comp} = \left|T_{comp\text{-}PON}\right| - \left|T_{comp\text{-}OON}\right| \quad (5.101)$$

Using the calculated T_{comp}, the turn-on times of three phases are revised as

$$T_{x\text{-}comp} = T_x + 0.5\left(\left|T_{comp\text{-}PON}\right| - \left|T_{comp\text{-}OON}\right|\right) \quad (5.102)$$

Case 4, when the reference voltage V_{ref} is in region 3-b of Sector I, the vectors of [POO], [PON], [OON], and [PPO] are adopted, and the switching sequence is arranged as OON-PON-POO-PPO-POO-PON-OON. Similar to case 3, the ΔU_{np} includes $\Delta U_{np\text{-}PON}$ caused by the medium vector [PON] and

$\Delta U_{\text{np-POO}}$ caused by the non-redundant P-type small vector [POO]. $T_{\text{comp-PON}}$ and $T_{\text{comp-POO}}$ are used to adjust the dwell times of redundant small vectors of [OON] and [PPO] and thus to offset $\Delta U_{\text{np-PON}}$ and $\Delta U_{\text{np-POO}}$, respectively. These compensation times are calculated as

$$\begin{cases} |T_{\text{comp-PON}}| = \left|\dfrac{I_b}{I_c}\right|(T_c - T_a) \\ |T_{\text{comp-POO}}| = \left|\dfrac{I_a}{I_c}\right|(T_b - T_c) \end{cases} \tag{5.103}$$

In this region, the medium vector [PON] increases ΔU_{np}, since $i_b > 0$, and the P-type small vector [POO] decreases ΔU_{np}. Thus, the total time compensation is expressed as

$$T_{\text{comp}} = |T_{\text{comp-POO}}| - |T_{\text{comp-PON}}| \tag{5.104}$$

Then the turn-on times of three phase are revised as

$$T_{x\text{-comp}} = T_x + 0.5(|T_{\text{comp-POO}}| - |T_{\text{comp-PON}}|) \tag{5.105}$$

Case 5, when the reference voltage V_{ref} is in region 4-a of Sector I, the vectors of [OON], [OOO], [ONN], and [POO] are adopted. The switching sequence is arranged as ONN-OON-OOO-POO-OOO-OON-ONN. In this case, only non-redundant N-type small vector [OON] induces the ΔU_{np}, which can be offset by the redundant small vectors of [ONN] and [POO] by adding the proper T_{comp}. Similar to the above analyzing method, T_{comp} is derived as

$$|T_{\text{comp}}| = \left|\dfrac{I_c}{I_a}\right|(T_c - T_b) \tag{5.106}$$

Since [OON] increase ΔU_{np}, the T_{comp} should be a negative value. Thus, the turn-on times of three phase are revised as

$$T_{x-\text{comp}} = T_x - 0.5 T_{\text{comp}} = T_x - 0.5\left|\dfrac{I_c}{I_a}\right|(T_c - T_b) \tag{5.107}$$

Case 6, the reference voltage V_{ref} is in region 4–6 of Sector I, and the switching sequence is arranged as OON-OOO-POO-PPO-POO-OOO-OON. Only the $\Delta U_{\text{np-POO}}$ induced by [POO] should be offset by adding compensation time to adjust the dwell times of redundant small vectors [OON] and [PPO]. The compensation time T_{comp} is calculated as

$$|T_{\text{comp}}| = \left|\dfrac{I_a}{I_c}\right|(T_b - T_a) \tag{5.108}$$

5.3 Control Technology for DC-Side Neutral-Point Voltage Oscillation ...

Because [POO] decreases ΔU_{np}, the T_{comp} should be a positive value to increase the dwell time of N-type small vector [OON]. Thus, the turn-on times of the three phase are changed by T_{comp}, and are expressed as

$$T_{x\text{-comp}} = T_x + 0.5T_{comp} = T_x + 0.5\left|\frac{I_a}{I_c}\right|(T_b - T_a) \quad (5.109)$$

T_{min} is defined as $T_{min} = \min(T_a, T_b, T_c)$. It should be noted that if the magnitude of T_{comp} is larger than T_{min}, distortion will appear in the output voltages. Therefore, the magnitude of T_{comp} should be limited to T_{min}. Table 5.4 gives the final compensation time and revised turn-on times of three phases when the reference voltage V_{ref} is in Sector I.

According to the same principles of calculation and compensation as mentioned above, the compensation time in other sectors can be obtained.

This compensation time adding method only changes the dwell times of the redundant N-type and P-type small vectors to reduce the NPV oscillations. The N-type and P-type switching states produce the same output voltages while only affecting the NPV. The total dwell times of the redundant N-type and P-type small vectors are not changed by the proposed method, and the dwell times of other voltage vectors are also not changed. Therefore, the outputs are not distorted by the proposed method.

(2) **NPV Oscillation Suppression Method Featuring with Low CMV Magnitude**

Since all small vectors are adopted in the first type method, the magnitude of CMV is high, up to one-third of the dc-link voltage. As described in the previous chapter, high CMV will induce high leakage current, which will bring security risks to users. The CMV reduction is significant in practice, but conventional NPV oscillation suppression methods have not taken this issue into account. To solve the above issues, an improved space vector modulation (ISVM) method is proposed by the book's author team, so as to reduce the CMV and NPV imbalance simultaneously.

Different from the traditional idea of three-vector modulation, the idea of four-vector modulation is proposed in ISVM for the first time. The proposed ISVM adopts a large vector, a medium vector, a small vector and a zero vector to synthesize the V_{ref} in each switching period. The CMV magnitudes of the adopted small and zero vectors are one-sixth of dc-link voltage and zero, respectively. Thus, the CMV magnitude of the 3LT^{2}I is restricted to one-sixth of the dc-link voltage. The NPV oscillation is suppressed by adjusting the dwell times of the four vectors. The main advantages of this method are listed as follows.

(1) The control objectives of CMV reduction and NPV oscillation suppression are coupled in conventional modulation methods due to redundant small vectors being adopted. The proposed four-vector modulation idea easily decouples the two control objectives.

(2) By selecting small vector with low CMV magnitude and adjusting the dwell times of large, medium, small, zero vectors, the NPV oscillation and CMV reduction can be realized simultaneously. Therefore, the reliability and safety of PV generation system are improved.

(3) Since the NPV oscillation is suppressed, the smaller capacitors can be preferably adopted for dc-link capacitors. Thus, the cost and volume of the 3LT^{2}I are reduced.

(4) A novel seven-segment switching sequence is proposed in ISVM, which achieves the minimal number of switching between two adjacent states in one switching cycle. Besides, the proposed switching sequence starts and ends with the same vector [OOO] within cycles, which can eliminate the extra switching transition between cycles. Hence, the switching losses are reduced, and the energy conversion efficiency is improved.

The detailed implementation process is illustrated as follows.

In order to decrease the magnitude of CMV, partial small and zero voltage vectors have lower CMV magnitude are adopted. Hence, this modulation uses the large, medium, partial small and zero vectors. In which the adopted small vectors are [POO], [OON], [OPO], [ONO], [OOP], [NOO], and the zero vector is [OOO]. The medium and small vectors have effects on NPV for the non-zero i_{np}. Table 5.5 gives the CMV magnitude and i_{np} of basic vector adopted in the proposed ISVM method. It can be seen that all the CMV magnitudes are restricted to one-sixth of dc-link voltage, which is only one-half of conventional modulation method. Then the small vectors with lower CMV magnitude are selected to suppress the NPV oscillation caused by medium vector. Thus, the CMV reduction and NPV oscillation suppression are decoupled.

The proposed ISVM method adopts large, medium, small and zero vectors, which add a degree of freedom to achieve the NPV balance control. The proposed ISVM method should satisfy the following conditions.

Firstly, the relationship between the dwell times and large, medium, small and zero vectors with reduced CMV should satisfy

$$\begin{cases} V_{\text{ref}}T_s = V_Z T_Z + V_S T_s + V_M T_M + V_L T_L \\ T_s = T_L + T_M + T_S + T_Z \end{cases} \quad (5.110)$$

where the dwell times of zero, small, medium and large vectors V_Z, V_S, V_M, V_L, are denoted by T_Z, T_S, T_M and T_L, respectively. Moreover, the dwell times should be a positive value.

Secondly, the NPV oscillation caused by the medium vector should be eliminated by the small vector. Hence, the magnitude of NPV changed by the medium vector should be equal to that of the small vector, which is expressed as

$$\Delta U_{np} = \frac{1}{C} \left| \int_{t=0}^{t=T_S} I_S(t) dt \right| = \frac{1}{C} \left| \int_{t=0}^{t=T_M} I_M(t) dt \right| \quad (5.111)$$

5.3 Control Technology for DC-Side Neutral-Point Voltage Oscillation ...

where the currents I_S and I_M are the neutral point current corresponding to the small vector and medium vector, respectively.

To simplify the complexity of the algorithm, the SVD of 3LT²I is divided into 12 sectors as shown in Fig. 5.45, in which the vectors with high CMV magnitude are excluded. The polarity of output currents in each sector is also labeled in SVD.

The reference voltage V_{ref} in SVD has phase difference with the output current I. If the filter impedance is large, the phase difference increases. When the polarities of output currents and voltage are different, the same medium vector has different effects on the NPV. In order to mitigate the NPV oscillation preferably, the ISVM method is divided into two parts depending on the location of V_{ref} and I.

Part 1: V_{ref} and I Are in the Same Sector

As described previously, the small vectors and medium vectors oscillate the two capacitor voltages. The NPV oscillation can be eliminated by adjusting the dwell time of large, medium, small and zero vectors. Sector 1 is considered as a representative example to explain the operating principles of reducing the NPV oscillation. In Sector 1, medium vector V_7 [PON] decreases ΔU_{np} because i_b is negative. It is indicated that the N-type small vector should be selected to reduce the NPV oscillation. Therefore, the N-type small vector V_2[OON] is selected, and the duty cycle of the V_2[OON] (denoted by y_1) is used in each switching cycle. Based on the volt–second balance

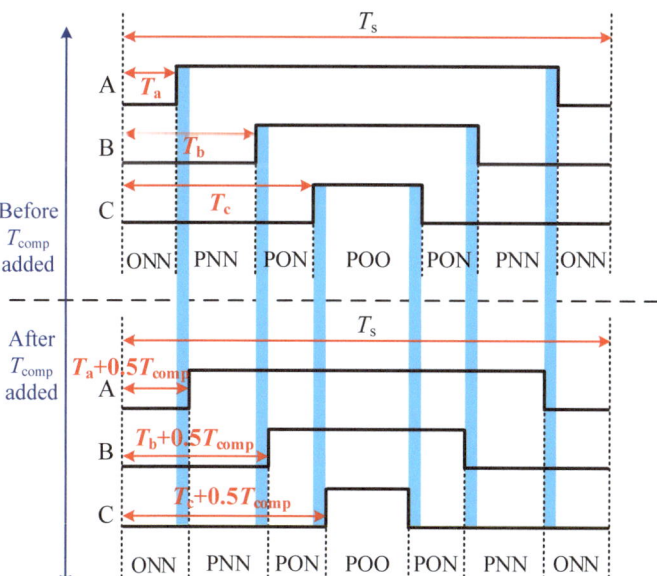

Fig. 5.44 The switching sequences before and after T_{comp} added in region 1 of Sector I

Table 5.4 Final compensation time and revised turn-on times of three phases for Sector I

Case	Region	$\|T_{comp}\|$	$T_{x\text{-}comp}, (x = a, b, c)$
1	1	$\|T_{comp}\| = \min\left(T_{min}, \left\|\frac{I_b}{I_a}\right\|(T_c - T_b)\right)$	$T_{x\text{-}comp} = T_x + 0.5\|T_{comp}\|$
2	2	$\|T_{comp}\| = \min\left(T_{min}, \left\|\frac{I_b}{I_c}\right\|(T_b - T_a)\right)$	$T_{x\text{-}comp} = T_x - 0.5\|T_{comp}\|$
3	3-a	$\|T_{comp}\| = \min(T_{min}, \|T_{comp\text{-}PON}\| - \|T_{comp\text{-}OON}\|)$ $\|T_{comp\text{-}PON}\| = \left\|\frac{I_b}{I_a}\right\|(T_c - T_a), \|T_{comp\text{-}ONN}\| = \left\|\frac{I_c}{I_a}\right\|(T_a - T_b)$	$T_{x\text{-}comp} = T_x + 0.5\|T_{comp}\|$
4	3-b	$\|T_{comp}\| = \min(T_{min}, \|T_{comp\text{-}PON}\| - \|T_{comp\text{-}POO}\|)$ $\|T_{comp\text{-}PON}\| = \left\|\frac{I_b}{I_c}\right\|(T_c - T_a), \|T_{comp\text{-}POO}\| = \left\|\frac{I_a}{I_c}\right\|(T_b - T_c)$	$T_{x\text{-}comp} = T_x + 0.5\|T_{comp}\|$
5	4-a	$\|T_{comp}\| = \min\left(T_{min}, \left\|\frac{I_c}{I_a}\right\|(T_c - T_b)\right)$	$T_{x\text{-}comp} = T_x - 0.5\|T_{comp}\|$
6	4-b	$\|T_{comp}\| = \min\left(T_{min}, \left\|\frac{I_a}{I_c}\right\|(T_b - T_a)\right)$	$T_{x\text{-}comp} = T_x + 0.5\|T_{comp}\|$

Table 5.5 The basic vectors corresponding to switching states, i_{np}, and CMV in ISVM

Vector type	Switching state	i_{np}	CMV	Switching state	i_{np}	CMV
Zero vector	[OOO]	0	0			
Small vector	[POO]	$i_b + i_c$	$1/6U_{dc}$	[OON]	$i_a + i_b$	$-1/6U_{dc}$
	[OPO]	$i_a + i_c$	$1/6U_{dc}$	[NOO]	$i_b + i_c$	$-1/6U_{dc}$
	[OOP]	$i_a + i_b$	$1/6U_{dc}$	[ONO]	$i_a + i_c$	$-1/6U_{dc}$
Medium vector	[PON]	i_b	0	[OPN]	i_a	0
	[NPO]	i_c	0	[NOP]	i_b	0
	[ONP]	i_a	0	[PNO]	i_c	0
Large vector	[PNN]	0	$-1/6U_{dc}$	[PPN]	0	$1/6U_{dc}$
	[NPN]	0	$-1/6U_{dc}$	[NPP]	0	$1/6U_{dc}$
	[NNP]	0	$-1/6U_{dc}$	[PNP]	0	$1/6U_{dc}$

principle, the relationship between the dwell times and voltage vectors V_0, V_7, V_{13}, and V_2 are obtained as

$$\begin{cases} V_{ref}T_s = V_0 T_0 + V_7 T_7 + V_{13} T_{13} + V_2 y_1 T_s \\ T_s = T_0 + T_7 + T_{13} + y_1 T_s \end{cases} \quad (5.112)$$

where the dwell times of vectors V_0, V_7, V_{13} are denoted by T_0, T_7 and T_{13}, respectively. The voltage vectors V_0, V_7, V_{13}, and V_2 are expressed as

5.3 Control Technology for DC-Side Neutral-Point Voltage Oscillation ...

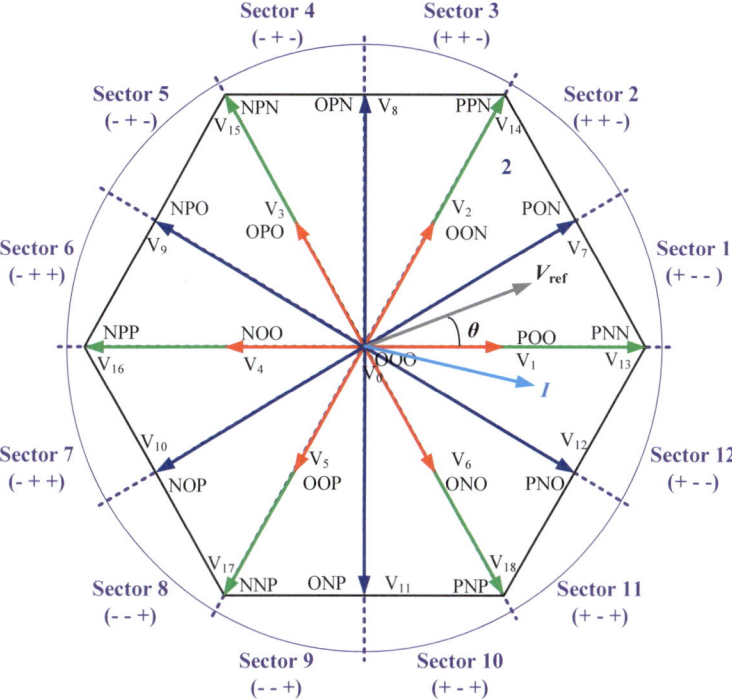

Fig. 5.45 Space vectors diagram of 3LT^{2}I with 12 sectors and excluding the voltage vectors with high CMV magnitude

$$\begin{cases} V_0 = 0, \quad V_2 = \frac{1}{3}U_{dc}e^{j\frac{\pi}{3}} \\ V_7 = \frac{\sqrt{3}}{3}U_{dc}e^{j\frac{\pi}{6}}, \quad V_{13} = \frac{2}{3} \cdot U_{dc} \\ V_{ref} = \frac{mU_{dc}}{\sqrt{3}} \cdot e^{j\theta} \end{cases} \quad (5.113)$$

Thus, the dwell times of four vectors are calculated as

$$\begin{cases} T_0 = T_s - \frac{\sqrt{3}}{2}T_s m \sin\theta - \frac{1}{2}y_1 T_s \\ T_2 = y_1 T_s \\ T_7 = (2m\sin\theta - y_1)T_s \\ T_{13} = \left(\frac{\sqrt{3}}{2}m\cos\theta - \frac{3}{2}m\sin\theta\right)T_s + \frac{y_1}{2}T_s \end{cases} \quad (5.114)$$

In order to maintain the dwell times are positive value. The variable y_1 satisfies

$$0 \leq y_1 \leq \min\left[1 - \frac{\sqrt{3}}{2}m\sin\theta, 2m\sin\theta\right] \quad (5.115)$$

In Sector 1, assuming that there is no change in the current during the control period, the deviations of $\Delta U_{\text{np-PON}}$ caused by $V_7[\text{PON}]$ and $\Delta U_{\text{np-OON}}$ caused by the N-type small vector [OON] are expressed as

$$\begin{cases} \Delta U_{\text{np-PON}} = \frac{1}{C} \int_{t=0}^{t=T_7} I_b(t)\mathrm{d}t = \frac{1}{C} T_7 \cdot I_b \\ \Delta U_{\text{np-OON}} = \frac{1}{C} \int_{t=0}^{t=T_2} I_b(t)\mathrm{d}t = \frac{1}{C} y_1 T_s \cdot I_c \end{cases} \quad (5.116)$$

In order to eliminate the NPV oscillation, the above deviations should be equal. Therefore, the duty cycle of the small vector $V_2[\text{OON}]$ is calculated as

$$y_1 = \frac{I_b}{I_c + I_b} \cdot 2m \sin \theta \quad (5.117)$$

Considering the limitation of positive value, the duty cycle y_1 is obtained as

$$y_1 = \min\left\{ \frac{I_b}{I_c + I_b} \cdot 2m \sin \theta, 2 - \sqrt{3} m \sin \theta \right\} \quad (5.118)$$

In order to reduce the total switching loss, the overall requirements for switching sequence design in the $3LT^2I$ are as follows. Firstly, only one times switching exchange in one switching cycle. Secondly, when V_{ref} moves from one sector to the next, the transition time requires no or minimum number of switching. If the switching sequence ends and starts with the same vector within two cycles, the extra switching events between two cycles can be eliminated. Therefore, the switching sequence is arranged as OOO-OON-PON-PNN-PON-OON-OOO. Besides, the CMV is defined as $u_{\text{cm}} = (u_{AO} + u_{BO} + u_{CO})/3$, and the waveform is given in Fig. 5.46 with the designed switching sequence. It can be seen that the CMV magnitude is restricted within one-sixth of dc-link voltage using the proposed ISVM method, which is half that of conventional SVM.

In the other sectors, a similar procedure is used for calculating the dwell times of the selected voltage vectors to suppress NPV oscillation. The sectors, the duty cycle of small vector and the switching sequence are shown in Table 5.6 when V_{ref} and I are in the same sectors.

Part 2: V_{ref} and I Are in the Different Sector

In practice, the filter impedance will cause current and voltage to be in different sectors. Assuming V_{ref} is located in Sector 2 and I is located in Sector 1, medium vector $V_7[\text{PON}]$ decreases ΔU_{np} because i_b is negative. Thus, the N-type small vector $V_2[\text{OON}]$ is selected to decrease the NPV oscillation. In this case, the vectors V_0, V_2, V_7, V_{14} are used to synthesize voltage vector V_{ref}. In order to achieve the minimal number of switches changing, the switching sequence is formed as [OOO] – [OON] – [PON] – [PPN] – [PON] – [OON] – [OOO]. The switching sequence and CMV waveform are shown in Fig. 5.47. The duty cycle of the $V_2[\text{OON}]$ (denoted as y_2) is used in each switching cycle. Although the small vector $V_2[\text{OON}]$ is selected, the CMV magnitude is still restricted within one-sixth of dc-link voltage.

5.3 Control Technology for DC-Side Neutral-Point Voltage Oscillation ...

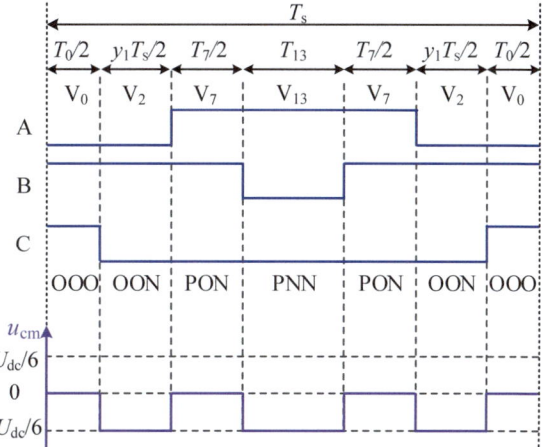

Fig. 5.46 Switching sequence and CMV of ISVM method when the V_{ref} and I are all in Sector 1

Based on the volt-second balance principle, the dwell times T_0, T_2, T_7 and T_{14} are calculated as

$$\begin{cases} T_0 = \left[1 - 2m\sin\left(\frac{\pi}{3} - \theta\right) - \sqrt{3}m\sin\left(\theta - \frac{\pi}{6}\right)\right]T_s - \frac{1}{2}y_2 T_s \\ T_2 = y_2 T_s \\ T_7 = 2m\sin\left(\frac{\pi}{3} - \theta\right)T_s \\ T_{14} = \sqrt{3}m\sin\left(\theta - \frac{\pi}{6}\right)T_s - \frac{1}{2}y_2 T_s \end{cases} \quad (5.119)$$

In order to mitigate the NPV oscillation, the magnitude of $\Delta U_{np\text{-}PON}$ caused by V_7[PON] and $\Delta U_{np\text{-}OON}$ caused by N-type small vector V_2[OON] should be the same. Based on this principle, the duty cycle of the small vector V_2[OON] is obtained as

$$y_2 = \frac{I_b}{I_c} \cdot 2m\sin\left(\frac{\pi}{3} - \theta\right) \quad (5.120)$$

Meanwhile, to ensure that the dwell times are positive during the switching period, the duty cycle y_2 is equal as

$$y_2 = \min\left\{ \begin{array}{l} \frac{I_b}{I_c} \cdot 2m\sin\left(\frac{\pi}{3} - \theta\right), 2\sqrt{3}m\sin\left(\theta - \frac{\pi}{6}\right), \\ \left[2 - 4m\sin\left(\frac{\pi}{3} - \theta\right) - 2\sqrt{3}m\sin\left(\theta - \frac{\pi}{6}\right)\right] \end{array} \right\} \quad (5.121)$$

A similar procedure is used for calculating the duty cycle of the selected small vector in all the other sectors. The sectors, duty cycle of small vector and switching sequence are shown in Table 5.7 when V_{ref} and I are in different sectors.

Besides the suppression of NPV oscillations, the NPV dc unbalance control principle is also given in the proposed ISVM. The implementation process is illustrated as follows.

Table 5.6 Sectors, duty cycle of small vector and switching sequence when V_{ref} and I are in same sectors

Sector	Duty cycle of small vector y_1	Switching sequence
1	$\min\left\{\frac{I_b}{I_c+I_b} \cdot 2m\sin\theta, \ 2-\sqrt{3}m\sin\theta\right\}$	OOO-OON-PON-PNN-PON-OON-OOO
2	$\min\left\{\frac{I_b}{I_a+I_b} \cdot 2m\sin\left(\frac{\pi}{3}-\theta\right), \ 2-4m\sin\left(\frac{\pi}{3}-\theta\right)-2\sqrt{3}m\sin\left(\theta-\frac{\pi}{6}\right)\right\}$	OOO-POO-PON-PPN-PON-POO-OOO
3	$\min\left\{\frac{I_a}{I_a+I_b} \cdot 2m\sin\left(\theta-\frac{\pi}{3}\right), \ 2-4m\sin\left(\theta-\frac{\pi}{3}\right)-2\sqrt{3}m\sin\left(\frac{\pi}{2}-\theta\right)\right\}$	OOO-OPO-OPN-PPN-OPN-OPO-OOO
4	$\min\left\{\frac{I_a}{I_a+I_c} \cdot 2m\sin\left(\frac{2\pi}{3}-\theta\right), \ 2-4m\sin\left(\frac{2\pi}{3}-\theta\right)-2\sqrt{3}m\sin\left(\theta-\frac{3\pi}{6}\right)\right\}$	OOO-OON-OPN-NPN-OPN-OON-OOO
5	$\min\left\{\frac{I_c}{I_a+I_c} \cdot 2m\sin\left(\theta-\frac{4\pi}{6}\right), \ 2-4m\sin\left(\theta-\frac{4\pi}{6}\right)-2\sqrt{3}m\sin\left(\frac{5\pi}{6}-\theta\right)\right\}$	OOO-NOO-NPO-NPN-NPO-NOO-OOO
6	$\min\left\{\frac{I_c}{I_b+I_c} \cdot 2m\sin(\pi-\theta), \ 2-4m\sin(\pi-\theta)-2\sqrt{3}m\sin\left(\theta-\frac{5\pi}{6}\right)\right\}$	OOO-OPO-NPO-NPP-NPO-OPO-OOO
7	$\min\left\{\frac{I_c}{I_b+I_c} \cdot 2m\sin(\theta-\pi), \ 2-4m\sin(\theta-\pi)-2\sqrt{3}m\sin\left(\frac{7\pi}{6}-\theta\right)\right\}$	OOO-OOP-NOP-NPP-NOP-OOP-OOO
8	$\min\left\{\frac{I_b}{I_a+I_b} \cdot 2m\sin\left(\frac{8\pi}{6}-\theta\right), \ 2-4m\sin\left(\frac{8\pi}{6}-\theta\right)-2\sqrt{3}m\sin\left(\theta-\frac{7\pi}{6}\right)\right\}$	OOO-NOO-NOP-NNP-NOP-NOO-OOO
9	$\min\left\{\frac{I_a}{I_a+I_b} \cdot 2m\sin\left(\theta-\frac{8\pi}{6}\right), \ 2-4m\sin\left(\theta-\frac{8\pi}{6}\right)-2\sqrt{3}m\sin\left(\frac{9\pi}{6}-\theta\right)\right\}$	OOO-ONO-ONP-NNP-ONP-ONO-OOO

(continued)

5.3 Control Technology for DC-Side Neutral-Point Voltage Oscillation … 285

Table 5.6 (continued)

Sector	Duty cycle of small vector y_1	Switching sequence
10	$\min\left\{\begin{array}{l}\frac{I_a}{I_a+I_c}\cdot 2m\sin\left(\frac{10\pi}{6}-\theta\right),\\ 2-4m\sin\left(\frac{10\pi}{6}-\theta\right)-2\sqrt{3}m\sin\left(\theta-\frac{9\pi}{6}\right)\end{array}\right\}$	OOO-OOP-ONP-PNP-ONP-OOP-OOO
11	$\min\left\{\begin{array}{l}\frac{I_c}{I_a+I_c}\cdot 2m\sin\left(\theta-\frac{10\pi}{6}\right),\\ 2-4m\sin\left(\theta-\frac{10\pi}{6}\right)-2\sqrt{3}m\sin\left(\frac{11\pi}{6}-\theta\right)\end{array}\right\}$	OOO-POO-PNO-PNP-PNO-POO-OOO
12	$\min\left\{\begin{array}{l}\frac{I_c}{I_c+I_b}\cdot 2m\sin(2\pi-\theta),\\ 2-4m\sin(2\pi-\theta)-2\sqrt{3}m\sin\left(\theta-\frac{11\pi}{6}\right)\end{array}\right\}$	OOO-ONO-PNO-PNN-PNO-ONO-OOO

Fig. 5.47 Switching sequence and CMV of ISVM method when the V_{ref} and I are all in Sector 1

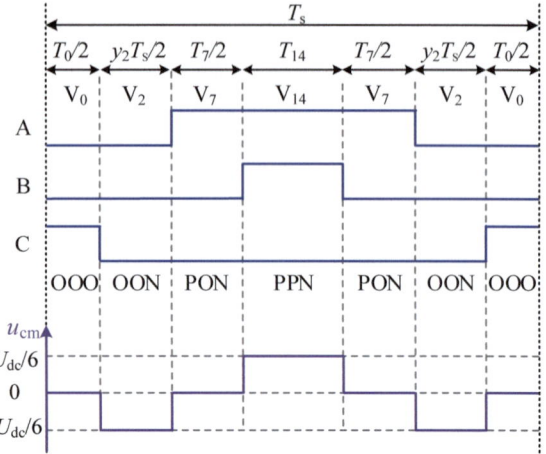

When the NPV dc unbalance occurs, the PI controller is adopted for NPV recovery. The selected small vector type is determined by the sign of ΔU_{np}. It should be noted that the duty cycle of the selected small vector is not only determined by the output of the PI controller, but also restricted by the volt-second balance principle and NPV oscillation suppression rule simultaneously. The control diagram for NPV dc unbalance suppression is shown in Fig. 5.48. In order to control the NPV dc unbalance to zero, the reference value is set to zero. The error between ΔU_{np} and the reference value is fed into the PI controller to generate the control variable y_i.

When the dc unbalance occurs, the average of the neutral current should be changed by selecting the P-type small vector or N-type small vector. The proposed method is divided into two cases and Sector 1 is considered as an example to explain the implementation process.

Case 1: $\Delta U_{np} > k$

In this case, to balance the NPV, the average of the neutral current should be a negative value; the P-type vector should be selected. Assuming the reference voltage is located in Sector 1, the adopted four vectors are V_0, V_1, V_7, and V_{13}, respectively. The switching sequence and CMV waveform with the small vector V_1[POO] are shown in Fig. 5.49. The magnitude of CMV is still maintained lower than one-sixth of the dc-link voltage.

The duty cycle of the V_1 [POO] (denoted by y_3) is used in each switching cycle, which is expressed as

$$y_3 = |Y_3| = \left| \frac{k_p(\tau_i s + 1)}{s} \left(\Delta U_{np}^* - \Delta U_{np} \right) \right| \tag{5.122}$$

Therefore, the dwell times of vector V_0, V_1, V_7, V_{13} are expressed as

5.3 Control Technology for DC-Side Neutral-Point Voltage Oscillation … 287

Table 5.7 Sectors, duty cycle of small vector and switching sequence when V_{ref} and I are in different sectors

Sector	Duty cycle of small vector y_2	Switching sequence
V_{ref} in 2 I in 1	$\min\left\{\begin{array}{l}\frac{I_b}{I_c}\cdot 2m\sin\left(\frac{\pi}{3}-\theta\right),\ 2\sqrt{3}m\sin\left(\theta-\frac{\pi}{6}\right),\\ 2-4m\sin\left(\frac{\pi}{3}-\theta\right)-2\sqrt{3}m\sin\left(\theta-\frac{\pi}{6}\right)\end{array}\right\}$	OOO-OON-PON-PPN-PON-OON-OOO
V_{ref} in 3 I in 1	$\min\left\{\begin{array}{l}\frac{I_b}{I_a+I_b}\cdot 2m\sin\left(\theta-\frac{\pi}{3}\right),\\ 2-4m\sin\left(\theta-\frac{\pi}{3}\right)-2\sqrt{3}m\sin\left(\frac{\pi}{2}-\theta\right)\end{array}\right\}$	OOO-OPO-OPN-PPN-OPN-OPO-OOO
V_{ref} in 4 I in 1	$\min\left\{\begin{array}{l}\frac{I_a}{I_b}\cdot 2m\sin\left(\frac{2\pi}{3}-\theta\right),\ 2\sqrt{3}m\sin\left(\theta-\frac{3\pi}{6}\right),\\ 2-4m\sin\left(\frac{2\pi}{3}-\theta\right)-2\sqrt{3}m\sin\left(\theta-\frac{3\pi}{6}\right)\end{array}\right\}$	OOO-OPO-OPN-NPN-OPN-OPO-OOO
V_{ref} in 5 I in 1	$\min\left\{\begin{array}{l}\frac{I_a}{I_a+I_c}\cdot 2m\sin\left(\theta-\frac{4\pi}{6}\right),\\ 2-4m\sin\left(\theta-\frac{4\pi}{6}\right)-2\sqrt{3}m\sin\left(\frac{5\pi}{6}-\theta\right)\end{array}\right\}$	OOO-NOO-NPO-NPN-NPO-NOO-OOO
V_{ref} in 6 I in 1	$\min\left\{\begin{array}{l}\frac{I_c}{I_a}\cdot 2m\sin(\pi-\theta),\ 2\sqrt{3}m\sin\left(\theta-\frac{5\pi}{6}\right),\\ 2-4m\sin(\pi-\theta)-2\sqrt{3}m\sin\left(\theta-\frac{5\pi}{6}\right)\end{array}\right\}$	OOO-NOO-NPO-NPP-NPO-NOO-OOO
V_{ref} in 7 I in 1	$\min\left\{\begin{array}{l}\frac{I_b}{I_b+I_c}\cdot 2m\sin(\theta-\pi),\\ 2-4m\sin(\theta-\pi)-2\sqrt{3}m\sin\left(\frac{7\pi}{6}-\theta\right)\end{array}\right\}$	OOO-OOP-NOP-NPP-NOP-OOP-OOO
V_{ref} in 8 I in 1	$\min\left\{\begin{array}{l}\frac{I_b}{I_c}\cdot 2m\sin\left(\frac{8\pi}{6}-\theta\right),\ 2\sqrt{3}m\sin\left(\theta-\frac{7\pi}{6}\right),\\ 2-4m\sin\left(\frac{8\pi}{6}-\theta\right)-2\sqrt{3}m\sin\left(\theta-\frac{7\pi}{6}\right)\end{array}\right\}$	OOO-OOP-NOP-NNP-NOP-OOP-OOO
V_{ref} in 9 I in 1	$\min\left\{\begin{array}{l}\frac{I_b}{I_a+I_b}\cdot 2m\sin\left(\theta-\frac{8\pi}{6}\right),\\ 2-4m\sin\left(\theta-\frac{8\pi}{6}\right)-2\sqrt{3}m\sin\left(\frac{9\pi}{6}-\theta\right)\end{array}\right\}$	OOO-ONO-ONP-NNP-ONP-ONO-OOO]

(continued)

Table 5.7 (continued)

Sector	Duty cycle of small vector y_2	Switching sequence
V_{ref} in 10 I in 1	$\min\left\{\frac{I_a}{I_b} \cdot 2m\sin\left(\frac{10\pi}{6} - \theta\right), 2\sqrt{3}m\sin\left(\theta - \frac{9\pi}{6}\right), 2 - 4m\sin\left(\frac{10\pi}{6} - \theta\right) - 2\sqrt{3}m\sin\left(\theta - \frac{9\pi}{6}\right)\right\}$	OOO-ONO-ONP-PNP-ONP-ONO-OOO
V_{ref} in 11 I in 1	$\min\left\{\frac{I_a}{I_a+I_c} \cdot 2m\sin\left(\theta - \frac{10\pi}{6}\right), 2 - 4m\sin\left(\theta - \frac{10\pi}{6}\right) - 2\sqrt{3}m\sin\left(\frac{11\pi}{6} - \theta\right)\right\}$	OOO-POO-PNO-PNP-PNO-POO-OOO
V_{ref} in 12 I in 1	$\min\left\{\frac{I_c}{I_a} \cdot 2m\sin(2\pi - \theta), 2\sqrt{3}m\sin\left(\theta - \frac{11\pi}{6}\right), 2 - 4m\sin(2\pi - \theta) - 2\sqrt{3}m\sin\left(\theta - \frac{11\pi}{6}\right)\right\}$	OOO-POO-PNO-PNN-PNO-POO-OOO
V_{ref} in 1 I in 12	$\min\left\{\frac{I_c}{I_c+I_b} \cdot 2m\sin\theta, 2 - \sqrt{3}m\sin\theta\right\}$	OOO-OON-PON-PNN-PON-OON-OOO

5.3 Control Technology for DC-Side Neutral-Point Voltage Oscillation ...

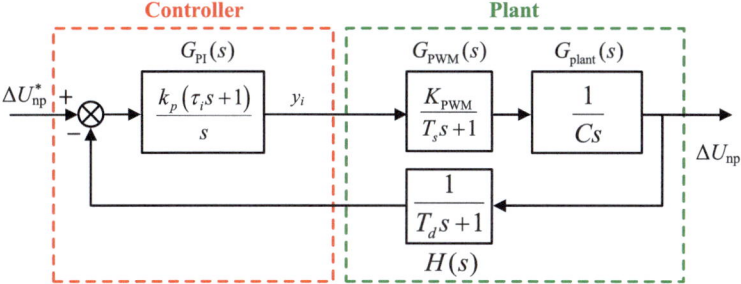

Fig. 5.48 The control diagram for NPV dc suppression

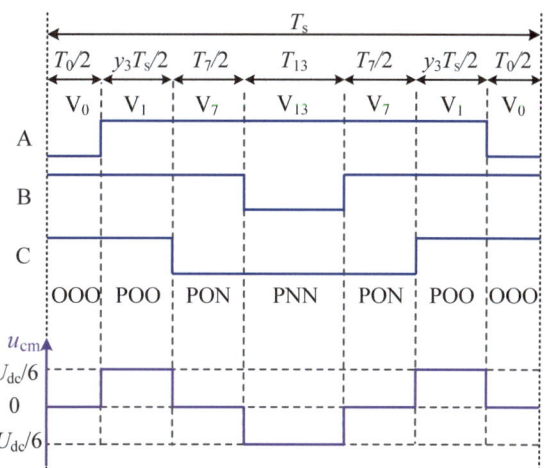

Fig. 5.49 Switching sequence and CMV of ISVM method for dc unbalance suppression when V_{ref} is in Sector 1 and $\Delta U_{np} > k$

$$\begin{cases} T_0 = \left(1 - \frac{1}{2}m\sin\theta - \frac{\sqrt{3}}{2}m\cos\theta\right)T_s - \frac{1}{2}y_3 T_s \\ T_2 = y_3 T_s \\ T_7 = 2m\sin\theta T_s \\ T_{13} = \left(\frac{\sqrt{3}}{2}m\cos\theta - \frac{3}{2}m\sin\theta\right)T_s - \frac{1}{2}y_3 T_s \end{cases} \quad (5.123)$$

To guarantee the dwell times being positive, the duty cycle y_3 should satisfy

$$0 \leq y_3 \leq \min\left[\begin{array}{l} 2 - m\sin\theta - \sqrt{3}m\cos\theta, \\ \sqrt{3}m\cos\theta - 3m\sin\theta \end{array}\right] \quad (5.124)$$

In the other sectors, the duty cycle of small voltage vector can be calculated by this method. The relationship between the sectors, the duty cycle of P-type small vector and switching sequence are shown in Table 5.8 when ΔU_{np} is larger than k.

Table 5.8 Sectors, duty cycle of small vector and switching sequence when $\Delta U_{np} > k$

Sector	Duty cycle of small vector y_3	Switching sequence
1	$0 \leq y_3 \leq \min\left\{2\sqrt{3}m\sin\left(\frac{\pi}{6} - \theta\right), 2 - \sqrt{3}m\sin\theta\right\}$	OOO-POO-PON-PNN-PON-POO-OOO
2	$0 \leq y_3 \leq \min\left\{\begin{array}{l}2m\sin\left(\frac{\pi}{3} - \theta\right), \\ 2 - 4m\sin\left(\frac{\pi}{3} - \theta\right) - 2\sqrt{3}m\sin\left(\theta - \frac{\pi}{6}\right)\end{array}\right\}$	OOO-POO-PON-PPN-PON-POO-OOO
3	$0 \leq y_3 \leq \min\left\{\begin{array}{l}2m\sin\left(\theta - \frac{\pi}{3}\right), \\ 2 - 4m\sin\left(\theta - \frac{\pi}{3}\right) - 2\sqrt{3}m\sin\left(\frac{\pi}{2} - \theta\right)\end{array}\right\}$	OOO-OPO-OPN-PPN-OPN-OPO-OOO
4	$0 \leq y_3 \leq \min\left\{\begin{array}{l}2\sqrt{3}m\sin\left(\theta - \frac{3\pi}{6}\right), \\ 2 - 4m\sin\left(\frac{2\pi}{3} - \theta\right) - 2\sqrt{3}m\sin\left(\theta - \frac{3\pi}{6}\right)\end{array}\right\}$	OOO-OPO-OPN-NPN-OPN-OPO-OOO
5	$0 \leq y_3 \leq \min\left\{\begin{array}{l}2\sqrt{3}m\sin\left(\frac{5\pi}{6} - \theta\right), \\ 2 - 4m\sin\left(\theta - \frac{4\pi}{6}\right) - 2\sqrt{3}m\sin\left(\frac{5\pi}{6} - \theta\right)\end{array}\right\}$	OOO-OPO-NPO-NPN-NPO-OPO-OOO
6	$0 \leq y_3 \leq \min\left\{\begin{array}{l}2m\sin(\pi - \theta), \\ 2 - 4m\sin(\pi - \theta) - 2\sqrt{3}m\sin\left(\theta - \frac{5\pi}{6}\right)\end{array}\right\}$	OOO-OPO-NPO-NPP-NPO-OPO-OOO
7	$0 \leq y_3 \leq \min\left\{\begin{array}{l}2m\sin(\theta - \pi), \\ 2 - 4m\sin(\theta - \pi) - 2\sqrt{3}m\sin\left(\frac{7\pi}{6} - \theta\right)\end{array}\right\}$	OOO-OOP-NOP-NPP-NOP-OOP-OOO
8	$0 \leq y_3 \leq \min\left\{\begin{array}{l}2\sqrt{3}m\sin\left(\theta - \frac{7\pi}{6}\right), \\ 2 - 4m\sin\left(\frac{8\pi}{6} - \theta\right) - 2\sqrt{3}m\sin\left(\theta - \frac{7\pi}{6}\right)\end{array}\right\}$	OOO-OOP-NOP-NNP-NOP-OOP-OOO
9	$0 \leq y \leq \min\left\{\begin{array}{l}2\sqrt{3}m\sin\left(\theta - \frac{9\pi}{6}\right), \\ 2 - 4m\sin\left(\frac{8\pi}{6} - \theta\right) - 2\sqrt{3}m\sin\left(\theta - \frac{9\pi}{6}\right)\end{array}\right\}$	OOO-OOP-ONP-NNP-ONP-OOP-OOO

(continued)

5.3 Control Technology for DC-Side Neutral-Point Voltage Oscillation …

Table 5.8 (continued)

Sector	Duty cycle of small vector y_3	Switching sequence
10	$0 \le y_3 \le \min\left\{\begin{array}{l}2m\sin\left(\frac{10\pi}{6}-\theta\right),\\ 2-4m\sin\left(\frac{10\pi}{6}-\theta\right)-2\sqrt{3}m\sin\left(\theta-\frac{9\pi}{6}\right)\end{array}\right\}$	OOO-OOP-ONP-PNP-ONP-OOP-OOO
11	$0 \le y_3 \le \min\left\{\begin{array}{l}2m\sin\left(\theta-\frac{10\pi}{6}\right),\\ 2-4m\sin\left(\theta-\frac{10\pi}{6}\right)-2\sqrt{3}m\sin\left(\frac{11\pi}{6}-\theta\right)\end{array}\right\}$	OOO-POO-PNO-PNP-PNO-POO-OOO
12	$0 \le y_3 \le \min\left\{\begin{array}{l}2m\sin(2\pi-\theta),\\ 2-4m\sin(2\pi-\theta)-2\sqrt{3}m\sin\left(\theta-\frac{11\pi}{6}\right)\end{array}\right\}$	OOO-POO-PNO-PNN-PNO-POO-OOO

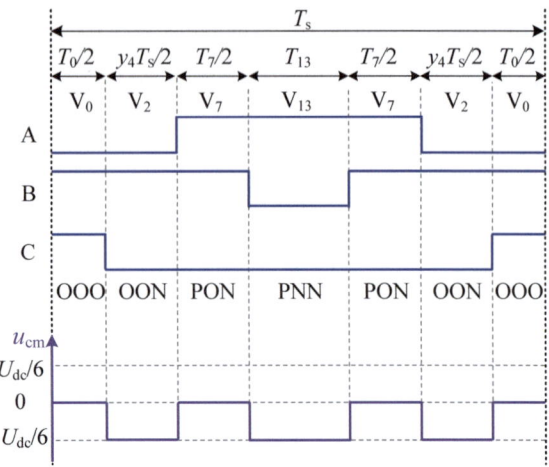

Fig. 5.50 Switching sequence and CMV of ISVM method for dc unbalance suppression when V_{ref} is in Sector 1 and $\Delta U_{np} > -k$

Case 2: $\Delta U_{np} < -k$

In opposition to the above case, the N-type small vector should be selected in Sector 1 where the average of the neutral current is positive. The modified switching sequence and CMV waveform with the small vector V_2[OON] are shown in Fig. 5.50. The magnitude of CMV is equal to one-sixth of the dc-link voltage.

In order to mitigate the NPV dc unbalance, the duty cycle of the V_2[OON] (denoted by y_4) is used in each switching cycle, which is expressed as

$$y_4 = |Y_4| = \left| \frac{k_p(\tau_i s + 1)}{s} \left(\Delta U_{np}^* - \Delta U_{np} \right) \right| \tag{5.125}$$

Therefore, the dwell times of vector V_0, V_1, V_7, and V_{13} are expressed as

$$\begin{cases} T_0 = \left(1 - \frac{\sqrt{3}}{2} m \sin\theta\right) T_s - \frac{1}{2} y_4 T_s \\ T_2 = y_4 T_s \\ T_7 = (2m \sin\theta - y_4) T_s \\ T_{13} = \left(\frac{\sqrt{3}}{2} m \cos\theta - \frac{3}{2} m \sin\theta\right) T_s + \frac{y_4}{2} T_s \end{cases} \tag{5.126}$$

The duty cycle y_4 satisfies as

$$0 \leq y_4 \leq \min\left(2 - \sqrt{3} m \sin\theta, 2m \sin\theta\right) \tag{5.127}$$

The duty cycle of corresponding small vector in all sectors can be calculated by this method. The voltage sectors, switching sequence and the duty cycle of N-type small vector are shown in Table 5.9 when ΔU_{np} is larger than $-k$.

All in all, in the proposed ISVM, the large, medium, small and zero vectors corresponding to the lower CMV amplitude are selected. By selecting the small vector and

5.3 Control Technology for DC-Side Neutral-Point Voltage Oscillation … 293

Table 5.9 Sectors, duty cycle of small vector and switching sequence when when $\Delta U_{np} < -k$

Sector	Duty cycle of small vector y_4	Switching sequence
1	$0 \leq y_4 \leq \min\left\{2m\sin\theta, 2 - \sqrt{3}m\sin\theta\right\}$	OOO-OON-PON-PNN-PON-OON-OOO
2	$0 \leq y_4 \leq \min\left\{\begin{array}{l}2\sqrt{3}m\sin\left(\theta - \frac{\pi}{6}\right), \\ 2 - 4m\sin\left(\frac{\pi}{3} - \theta\right) - 2\sqrt{3}m\sin\left(\theta - \frac{\pi}{6}\right)\end{array}\right\}$	OOO-OON-PON-PPN-PON-OON-OOO
3	$0 \leq y_4 \leq \min\left\{\begin{array}{l}2\sqrt{3}m\sin\left(\frac{\pi}{2} - \theta\right), \\ 2 - 4m\sin\left(\theta - \frac{\pi}{3}\right) - 2\sqrt{3}m\sin\left(\frac{\pi}{2} - \theta\right)\end{array}\right\}$	OOO-OON-OPN-PPN-OPN-OON-OOO
4	$0 \leq y_4 \leq \min\left\{\begin{array}{l}2m\sin\left(\frac{2\pi}{3} - \theta\right), \\ 2 - 4m\sin\left(\frac{2\pi}{3} - \theta\right) - 2\sqrt{3}m\sin\left(\theta - \frac{3\pi}{6}\right)\end{array}\right\}$	OOO-OON-OPN-NPN-OPN-OON-OOO
5	$0 \leq y_4 \leq \min\left\{\begin{array}{l}2m\sin\left(\theta - \frac{4\pi}{6}\right), \\ 2 - 4m\sin\left(\theta - \frac{4\pi}{6}\right) - 2\sqrt{3}m\sin\left(\frac{5\pi}{6} - \theta\right)\end{array}\right\}$	OOO-NOO-NPO-NPN-NPO-NOO-OOO
6	$0 \leq y_4 \leq \min\left\{\begin{array}{l}2\sqrt{3}m\sin\left(\theta - \frac{5\pi}{6}\right), \\ 2 - 4m\sin(\pi - \theta) - 2\sqrt{3}m\sin\left(\theta - \frac{5\pi}{6}\right)\end{array}\right\}$	OOO-NOO-NPO-NPP-NPO-NOO-OOO
7	$0 \leq y_4 \leq \min\left\{\begin{array}{l}2\sqrt{3}m\sin\left(\frac{7\pi}{6} - \theta\right), \\ 2 - 4m\sin(\theta - \pi) - 2\sqrt{3}m\sin\left(\frac{7\pi}{6} - \theta\right)\end{array}\right\}$	OOO-NOO-NOP-NPP-NOP-NOO-OOO
8	$0 \leq y_4 \leq \min\left\{\begin{array}{l}2m\sin\left(\frac{8\pi}{6} - \theta\right), \\ 2 - 4m\sin\left(\frac{8\pi}{6} - \theta\right) - 2\sqrt{3}m\sin\left(\theta - \frac{7\pi}{6}\right)\end{array}\right\}$	OOO-NOO-NOP-NNP-NOP-NOO-OOO
9	$0 \leq y_4 \leq \min\left\{\begin{array}{l}2m\sin\left(\theta - \frac{8\pi}{6}\right), \\ 2 - 4m\sin\left(\theta - \frac{8\pi}{6}\right) - 2\sqrt{3}m\sin\left(\frac{9\pi}{6} - \theta\right)\end{array}\right\}$	OOO-ONO-ONP-NNP-ONP-ONO-OOO

(continued)

Table 5.9 (continued)

Sector	Duty cycle of small vector y_4	Switching sequence
10	$0 \le y_4 \le \min\left\{\begin{array}{l} 2\sqrt{3}m\sin(\theta - \frac{9\pi}{6}), \\ 2 - 4m\sin(\frac{10\pi}{6} - \theta) - 2\sqrt{3}m\sin(\theta - \frac{9\pi}{6}) \end{array}\right\}$	OOO-ONO-ONP-PNP-ONP-ONO-OOO
11	$0 \le y_4 \le \min\left\{\begin{array}{l} 2\sqrt{3}m\sin(\frac{11\pi}{6} - \theta), \\ 2 - 4m\sin(\theta - \frac{10\pi}{6}) - 2\sqrt{3}m\sin(\frac{11\pi}{6} - \theta) \end{array}\right\}$	OOO-ONO-PNO-PNP-PNO-ONO-OOO
12	$0 \le y_4 \le \min\left\{\begin{array}{l} 2m\sin(2\pi - \theta), \\ 2 - 4m\sin(2\pi - \theta) - 2\sqrt{3}m\sin(\theta - \frac{11\pi}{6}) \end{array}\right\}$	OOO-ONO-PNO-PNN-PNO-ONO-OOO

5.3 Control Technology for DC-Side Neutral-Point Voltage Oscillation ...

adjusting the dwell times of large, medium, small, zero vectors, the oscillation and dc unbalance on the NPV and CMV reduction can be realized simultaneously. In addition, a novel switching sequence arrangement method with the minimal number of switches transition in one switching cycle and between switching cycles is proposed to reduce the total switching loss.

Based on Carrier-Based Pulse Width Modulation

In this section, the duty signal compensation (DSC) method is presented, which is derived from the theory of carrier-based pulse width modulation. This method is proposed by a South Korean scholar June-Seok Lee in [23]. The main advantage of DSC is that the complex trigonometric operations in SVM are avoided.

In this method, the NPV oscillation is suppressed by adding an optimal duty signal compensation value to the three-phase reference duty signals. This compensation value is calculated simply by the currents and the reference duty signals. The theory and implementation process are illustrated as follows.

The three-phase sinusoidal and normalized reference voltages are re-expressed as

$$\begin{cases} u_{a,\text{ref}} = m\cos(2\pi f_1 t) \\ u_{b,\text{ref}} = m\cos\left(2\pi f_1 t - \frac{2\pi}{3}\right) \\ u_{c,\text{ref}} = m\cos\left(2\pi f_1 t + \frac{2\pi}{3}\right) \end{cases} \quad (5.128)$$

To improve the dc-link voltage utilization ratio, u_o is injected into three-phase reference voltages to obtain the redefined reference voltage $u_{x,\text{ref,off}}$, which is expressed as

$$\begin{cases} u_{x,\text{ref,off}} = u_{x,\text{ref}} + u_o \\ u_o = -\frac{1}{2}\left(u_{\text{ref,max}} + u_{\text{ref,min}}\right) \\ u_{\text{ref,max}} = \max\left(u_{a,\text{ref}}, u_{b,\text{ref}}, u_{c,\text{ref}}\right) \\ u_{\text{ref,min}} = \min\left(u_{a,\text{ref}}, u_{b,\text{ref}}, u_{c,\text{ref}}\right) \end{cases} \quad (5.129)$$

The waveforms of zero-sequence voltage and redefined reference voltage are shown in Fig. 5.51.

The relationship between reference voltages and reference duty signals are shown in Fig. 5.52. In which $u_{\max}$, u_{mid} and $u_{\min}$ are the maximum, minimum and medium of $u_{x,\text{ref,off}}$. $u_{\max}$, u_{mid} and $u_{\min}$ are compared with two vertically disposed in-phase carriers to generate three switching states represented by [P], [O], and [N], respectively.

According to the above geometric relationship, the reference duty signals can be calculated as

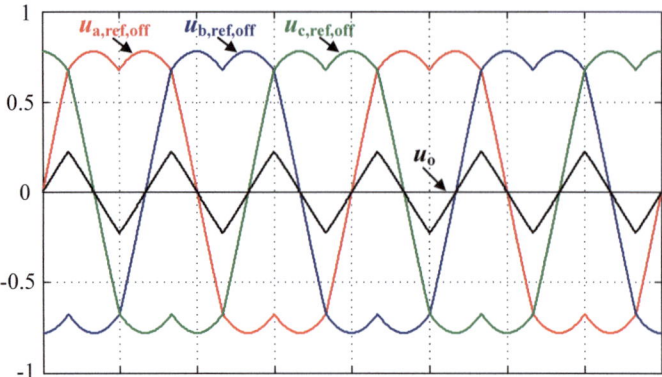

Fig. 5.51 The waveforms of zero-sequence voltage and redefined three phase reference voltage

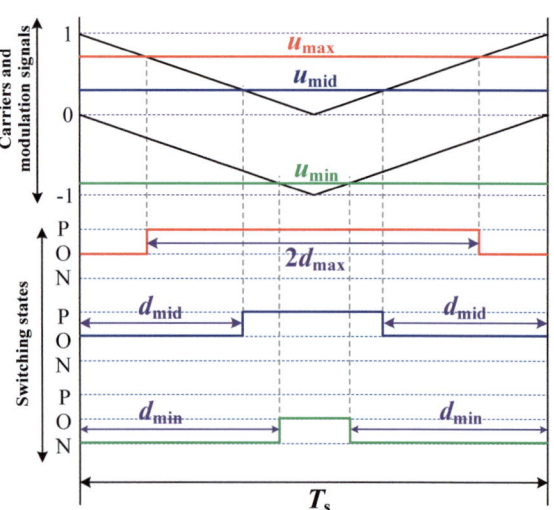

Fig. 5.52 The relationship of reference voltages and reference duty signals

$$\begin{cases} d_{\max} = 2u_{\max} \\ d_{\mathrm{mid}} = 2u_{\mathrm{mid}} \\ d_{\min} = 2u_{\min} \end{cases} \quad (5.130)$$

As the above description, $d_{\max}$ always a positive value, and $d_{\min}$ always a negative value. Meanwhile, the magnitudes of them satisfy $|d_{\max}| = |d_{\min}|$ all the time. $d_{\max}$ always causes the output switching state to be [P] during $2d_{\max}T_s$, and $d_{\min}$ always causes the output switching state to be [N] during $2|d_{\min}|T_s$. However, the output switching state made by d_{mid} can be [P] or [N] depending on the sign of d_{mid}. If $d_{\mathrm{mid}} > 0$, the output switching state is [P] during $2d_{\mathrm{mid}}T_s$, while the output switching state is [N] during $2|d_{\mathrm{mid}}|T_s$. Thus, the average neutral-point current (I_{np}) for T_s can

5.3 Control Technology for DC-Side Neutral-Point Voltage Oscillation ...

be calculated using reference duty signals and expressed as

$$I_{np} = -(|d_{max}|I_{max} + |d_{mid}|I_{mid} + |d_{min}|I_{min}) \tag{5.131}$$

where I_{max}, I_{mid}, and I_{min} are the currents of the maximum, medium and minimum reference voltages. I_{np} is calculated by the reference duty signals and the three phase currents, which directly causes the NPV oscillation. I_{np} can be zero by adding the optimal duty signal compensation value (d_{comp}) to the reference duty signals, thus to suppress the NPV oscillation. After d_{comp} is added, the I_{np} can be revised as $I_{np,comp}$, and expressed as

$$I_{np,comp} = -(|d_{max} + d_{comp}|I_{max} + |d_{mid} + d_{comp}|I_{mid} + |d_{min} + d_{comp}|I_{min}) \tag{5.132}$$

It should be noted that $|d_{comp}|$ should be smaller than $|d_{max}|$ (or $|d_{min}|$) to prevent over-modulation. The optimal d_{comp} is obtained when $I_{np,comp} = 0$. Four cases are included according to the sign of d_{comp} and d_{mid}.

Case 1: $d_{comp} > 0$ and $d_{mid} > 0$
In this case, the sign of $d_{comp} + d_{mid}$ is not changed. $I_{np,comp}$ can be represented as

$$I_{np,comp,case1} = -(|d_{max}|I_{max} + |d_{mid}|I_{mid} + |d_{min}|I_{min}) \\ - (|d_{comp}|I_{max} + |d_{comp}|I_{mid} - |d_{comp}|I_{min}) \tag{5.133}$$

By using I_{np} before adding d_{comp} to the reference duty signals, the optimal d_{comp} can be obtained, which can realize $I_{np,comp,case1}$ equal to zero. The optimal d_{comp} is expressed as

$$|d_{comp}| = \frac{I_{np}}{I_{max} - I_{min} + I_{mid}}, \quad I_{np,comp,case1} = 0 \tag{5.134}$$

Since $(I_{max} - I_{min} + I_{mid}) > 0$, the optimal d_{comp} is finally expressed as

$$d_{comp} = \frac{I_{np}}{I_{max} - I_{min} + I_{mid}} \text{ for } I_{np} > 0 \tag{5.135}$$

Case 2: $d_{comp} > 0$ and $d_{mid} < 0$
Depending on the magnitude of d_{comp} and d_{mid}, Case 2 is further divided into two cases (Cases 2-1 and 2-2). For Case 2-1, in which $|d_{comp}|$ is larger than $|d_{mid}|$, $I_{np,comp}$ can be represented as

$$I_{np,comp,case2-1} = I_{np} - \left[|d_{comp}|I_{max} - |d_{comp}|I_{min} + (|d_{comp}| - 2|d_{mid}|)I_{mid}\right] \tag{5.136}$$

To make $I_{np,comp,case2-1}$ equal to zero, d_{comp} can be calculated as

$$|d_{comp}| = \frac{I_{np} + 2|d_{mid}|I_{mid}}{I_{max} - I_{min} + I_{mid}}, \quad I_{np,comp,case2-1} = 0 \quad (5.137)$$

Because $(I_{max} - I_{min} + I_{mid}) > 0$, the optimal d_{comp} is calculated as

$$d_{comp} = \frac{I_{np} + 2|d_{mid}|I_{mid}}{I_{max} - I_{min} + I_{mid}}, \quad \text{for } I_{np} + 2|d_{mid}|I_{mid} > 0 \quad (5.138)$$

For Case 2-2, in which $|d_{comp}|$ is smaller than $|d_{mid}|$, $I_{np,comp}$ can be expressed as

$$I_{np,comp,case2-2} = I_{np} - \left(|d_{comp}|I_{max} - |d_{comp}|I_{min} - |d_{comp}|I_{mid}\right) \quad (5.139)$$

Similar to Case 1, when $I_{np,comp,case2-2}$ equal to zero, d_{comp} can be obtained as

$$d_{comp} = \frac{I_{np}}{I_{max} - I_{min} - I_{mid}} \text{ for } I_{np} > 0 \quad (5.140)$$

Additionally, d_{comp} of Case 3 ($d_{comp} < 0$ and $d_{mid} > 0$) and Case 4 ($d_{comp} < 0$ and $d_{mid} < 0$) can be calculated in a similar manner as Case 1 and Case 2. The conclusions are given as follows.

Case 3: $d_{comp} < 0$ and $d_{mid} > 0$

Case 3 is further divided into two cases (Cases 3-1 and 3-2). For Case 3-1, $|d_{comp}|$ is larger than $|d_{mid}|$, $I_{np,comp}$, the d_{comp} is calculated as

$$d_{comp} = \frac{I_{np} + 2|d_{mid}|I_{mid}}{I_{max} - I_{min} - I_{mid}}, \quad \text{for } I_{np} + 2|d_{mid}|I_{mid} < 0 \quad (5.141)$$

For Case 3-2, $|d_{comp}|$ is smaller than $|d_{mid}|$, $I_{np,comp}$, the d_{comp} is calculated as

$$d_{comp} = \frac{I_{np}}{I_{max} - I_{min} + I_{mid}} \text{ for } I_{np} < 0 \quad (5.142)$$

Case 4: $d_{comp} < 0$ and $d_{mid} < 0$

To make the average neutral-point current equal to zero, the d_{comp} is calculated as

$$d_{comp} = \frac{I_{np}}{I_{max} - I_{min} - I_{mid}} \text{ for } I_{np} < 0 \quad (5.143)$$

It should be noted that the requirements of Case2-1 and Case 3-1 are not obvious, so the intuitive conditions should be given from the theory. Because the magnitudes of $|d_{max}|$ and $|d_{min}|$ are equal all the time, the I_{np} before d_{comp} is added can be re-expressed and simplified as

$$I_{np} = -(KI_{max} + |d_{mid}|I_{mid} + KI_{min}) \quad (5.144)$$

5.3 Control Technology for DC-Side Neutral-Point Voltage Oscillation …

where $K = |d_{\max}| = |d_{\min}|$. Besides, for symmetrical three-phase output current, $I_{\max} + I_{\mathrm{mid}} + I_{\mathrm{mid}} = 0$. The I_{np} can be further simplified as

$$I_{\mathrm{np}} = (K - |d_{\mathrm{mid}}|)I_{\mathrm{mid}} \tag{5.145}$$

Since the sign of $(K - |d_{\mathrm{mid}}|)$ is positive all the time, the sign of I_{np} is always the same as I_{mid}. Consequently, the d_{comp} and the requirements of Case 2-1 and Case 3-1 are simplified as

$$\begin{cases} \text{Case 2-1}: d_{\mathrm{comp}} = \frac{I_{\mathrm{np}}+2|d_{\mathrm{mid}}|I_{\mathrm{mid}}}{I_{\max}-I_{\min}+I_{\mathrm{mid}}}, & \text{for } I_{\mathrm{np}} > 0 \\ \text{Case 3-1}: d_{\mathrm{comp}} = \frac{I_{\mathrm{np}}+2|d_{\mathrm{mid}}|I_{\mathrm{mid}}}{I_{\max}-I_{\min}-I_{\mathrm{mid}}}, & \text{for } I_{\mathrm{np}} < 0 \end{cases} \tag{5.146}$$

As the above expression of d_{comp}, we know that the sign of d_{comp} is always the same as the sign of I_{np} in all cases. Thus, two requirements (I_{np} and d_{comp}) are needed to calculate d_{comp} and to add it to the reference duty signals. The duty signal compensation d_{comp} for the NPV oscillation suppression are summarized in Table 5.10.

In Case 2 and Case 3, first, Case x-1 ($x = 2, 3$) is calculated, and then Case x-2 ($x = 2, 3$) is calculated if d_{comp} calculated from Case x-1 does not satisfy its requirement ($|d_{\mathrm{comp}}| > |d_{\mathrm{mid}}|$).

The DSC technology based on carrier-based pulse width modulation is presented for NPV oscillation suppression. This method avoids complex trigonometric operations in SVM. The optimal duty signal compensation value is calculated simply by the currents and the reference duty signals.

In conclusion, to suppress the low-frequency oscillation of NPV of 3LT^{2}I, two groups of technology are presented. The first group is based on SVM, and the second group is a method based on CB-PWM. According to the magnitudes of the CMV, the first group is further classified into two types: the method featuring high CMV

Table 5.10 Duty signal compensations d_{comp} for the NPV oscillation suppression

Requirement	$d_{\mathrm{mid}} > 0$	$d_{\mathrm{mid}} < 0$
$I_{\mathrm{np}} > 0$	Case 1	Case 2-1: $|d_{\mathrm{comp}}| > |d_{\mathrm{mid}}|$
	$d_{\mathrm{comp}} = \frac{I_{\mathrm{np}}}{I_{\max}-I_{\min}+I_{\mathrm{mid}}}$	$d_{\mathrm{comp}} = \frac{I_{\mathrm{np}}+2|d_{\mathrm{mid}}|I_{\mathrm{mid}}}{I_{\max}-I_{\min}+I_{\mathrm{mid}}}$
		Case 2-2: $|d_{\mathrm{comp}}| < |d_{\mathrm{mid}}|$
		$d_{\mathrm{comp}} = \frac{I_{\mathrm{np}}}{I_{\max}-I_{\min}-I_{\mathrm{mid}}}$
$I_{\mathrm{np}} < 0$	Case 3-1: $|d_{\mathrm{comp}}| > |d_{\mathrm{mid}}|$	Case 4
	$d_{\mathrm{comp}} = \frac{I_{\mathrm{np}}+2|d_{\mathrm{mid}}|I_{\mathrm{mid}}}{I_{\max}-I_{\min}-I_{\mathrm{mid}}}$	$d_{\mathrm{comp}} = \frac{I_{\mathrm{np}}}{I_{\max}-I_{\min}-I_{\mathrm{mid}}}$
	Case 3-2: $|d_{\mathrm{comp}}| < |d_{\mathrm{mid}}|$	
	$d_{\mathrm{comp}} = \frac{I_{\mathrm{np}}}{I_{\max}-I_{\min}+I_{\mathrm{mid}}}$	

magnitude, and the method featuring low CMV magnitude. The deriving and implementation process of each method is illustrated step by step. Using the above technology, the NPV oscillation is suppressed, which is beneficial for increasing system reliability and economy.

5.4 Summary

With the constantly increasing of power generation capacity and installed capacity of PV stations, the high robustness and high performance of PV inverter are gradually becoming important indicators for PV equipment suppliers and users.

To satisfy these newly emerging requirements, the advanced control technology is proposed in this chapter. For high robustness, the nonlinear control technology is proposed, which enhances the dynamic and steady-state response of PV inverter under load disturbance and parameter uncertainties. For AC-side current quality improvement, the current harmonic elimination and current ripple suppression technologies are proposed, which improve current quality with low switching frequency. For DC-side NPV performance improvement, the NPV oscillation suppression technology is given, which reduces the fluctuations of NPV with small dc-link capacitors. The nonlinear control technology is integrated into middle loop and inner loop. The control technologies for AC-side current harmonic and ripple suppression and DC-side NPV oscillation suppression are integrated into PWM modulator and auxiliary loop.

To facilitate readers' understanding of the technologies presented in this chapter, the theoretical analyses and implementation process of each control method are illustrated step by step by. Besides, the corresponding results of each method are given for intuitive understanding.

References

1. C. Fu, C. Zhang, G. Zhang, Z. Zhang, High-performance finite-time adaptive control strategy for three-level t-type converters. ISA Trans. **150**, 404–411 (2024)
2. B. Chen, C. Lin, Finite-time stabilization-based adaptive fuzzy control design. IEEE Trans. Fuzzy Syst. **29**(8), 2438–2443 (2021)
3. Y. Liu, Q. Zhu, Adaptive fuzzy finite-time control for nonstrict-feedback nonlinear systems. IEEE Trans. Cybern. **52**(10), 10420–10429 (2022)
4. C. Fu, C. Zhang, G. Zhang, J. Song, C. Zhang, B. Duan, Disturbance observer-based finite-time control for three-phase AC/DC converter. IEEE Trans. Ind. Electron. **69**(6), 5637–5647 (2022)
5. J. Hu, L. Shang, Y. He, Z.Q. Zhu, Direct active and reactive power regulation of grid-connected dc/ac converters using sliding mode control approach. IEEE Trans. Power Electron. **26**(1), 210–222 (2011)
6. Y. Gui, M. Li, J. Lu, S. Golestan, J.M. Guerrero, J.C. Vasquez, A voltage modulated DPC approach for three-phase PWM rectifier. IEEE Trans. Ind. Electron. **65**(10), 7612–7619 (2018)
7. IEC Standard Photovoltaic (PV), Systems—characteristics of the utility Interface. IEC Standard **61727** (2009)

8. R.N. Ray, D. Chatterjee, S.K. Goswami, Harmonics elimination in a multilevel inverter using the particle swarm optimisation technique. IET Power Electron. **2**(6), 646–652 (2009)
9. K.P. Panda, G. Panda, Application of swarm optimisation-based modified algorithm for selective harmonic elimination in reduced switch count multilevel inverter. IET Power Electron. **11**(8), 1472–1482 (2018)
10. A. Routray, R.K. Singh, R. Mahanty, Modified particle swarm optimization based harmonic minimization in hybrid cascaded multilevel inverter, in *2018 IEEE Energy Conversion Congress and Exposition* (2018), pp. 5573–5578
11. Y. Jiang, X. Li, C. Qin, X. Xing, Z. Chen, Improved particle swarm optimization based selective harmonic elimination and neutral point balance control for three-level inverter in low-voltage ride-through operation. IEEE Trans. Ind. Informat. **18**(1), 642–652 (2022)
12. J. Kennedy, R. Eberhart, Particle swarm optimization, in *Proceedings of ICNN'95—International Conference on Neural Networks*, (1995), pp. 1942–1948
13. Y. Shi, R. Eberhart, A modified particle swarm optimizer, in *1998 IEEE International Conference on Evolutionary Computation Proceedings. IEEE World Congress on Computational Intelligence*, (1998), pp. 69–73
14. S. Bouckaert et al., Net zero by 2050: a roadmap for the global energy sector. International Energy Agency. https://www.iea.org/reports/net-zero-by-2050. Accessed May 2021
15. S. Angadi, U.R. Yaragatti, Y. Suresh, A.B. Raju, Comprehensive review on solar, wind and hybrid wind-PV water pumping systems-an electrical engineering perspective. CPSS Trans. Power Electron. Appl. **6**(1), 1–19 (2021)
16. L. Zhang, K. Sun, Y. Xing, J. Zhao, Parallel operation of modular single-phase transformerless grid-tied PV inverters with common DC bus and AC bus. IEEE J. Emerg. Sel. Top Power Electron. **3**(4), 858–869 (2015)
17. IEEE standard for interconnecting distributed resources with electric power systems, IEEE Standard 1547 (2003)
18. X. Liu, C. Zhang, X. Xing, R. Zhang, T. Liu, Q. Ren, A carrier-based low total current ripple modulation strategy for parallel converters with leakage current attenuation. IEEE Trans. Ind. Electron. **70**(4), 3751–3761 (2023)
19. X. Liu, C. Zhang, X. Xing, X. Li, A double PS-PWM strategy for improving total output current quality in parallel converters. IEEE Trans. Power Electron. **39**(3), 3277–3288 (2024)
20. D. Holmes, T. Lipo, *Pulse Width Modulation for Power Converters: Principles and Practice* (Wiley, New York, NY, 2003)
21. U.M. Choi, F. Blaabjerg, K.B. Lee, Method to minimize the low frequency neutral-point voltage oscillations with time-offset injection for neutral-point-clamped inverters. IEEE Trans. Ind. Appl. **51**(2), 1678–1691 (2015)
22. X. Xing, X. Li, F. Gao et al., Improved space vector modulation technique for neutral-point voltage oscillation and common-mode voltage reduction in three-level inverter. IEEE Trans. Power Electron. **34**(9), 8697–8714 (2018)
23. J.S. Lee, K.B. Lee, Time-offset injection method for neutral-point AC ripple voltage reduction in a three-level inverter. IEEE Trans. Power Electron. **31**(3), 1931–1941 (2016)

Open Access This chapter is licensed under the terms of the Creative Commons Attribution 4.0 International License (http://creativecommons.org/licenses/by/4.0/), which permits use, sharing, adaptation, distribution and reproduction in any medium or format, as long as you give appropriate credit to the original author(s) and the source, provide a link to the Creative Commons license and indicate if changes were made.

The images or other third party material in this chapter are included in the chapter's Creative Commons license, unless indicated otherwise in a credit line to the material. If material is not included in the chapter's Creative Commons license and your intended use is not permitted by statutory regulation or exceeds the permitted use, you will need to obtain permission directly from the copyright holder.

Chapter 6
Advanced Control Technology of Photovoltaic Inverters Under Grid Faults

With the integration of large-scale renewable power generation system and renewable power consumption system into the power grid, the power system has been evolved from traditional power system to the modern power system. The modern power system has significant "weak inertia" characteristics, which frequently leads to grid fault. When the grid fault occurs, the traditional control technologies are no longer applicable, which leads to the DC voltage instability and grid current distortion in the PV inverter. More seriously, these issues lead to fault protection, and even burn out the PV inverter equipment.

To solve the above issues, the characteristics of grid voltages and grid standard requirements for PV inverter are analyzed and illustrated in the first part of this chapter. To ensure PV inverter stable operation and satisfy the grid standard requirements under grid fault condition, a series of advanced control technologies are subsequently presented. First, the advanced control technology for DC voltage and grid current control is given, which maintain the PV inverter transmitting high-quality power energy to power grid even during a grid fault. The advanced control technology is integrated into middle loop and inner loop. According to types of control variables in the inner loop, the advanced control technology has two control schemes: dual-loop voltage and current control scheme and dual-loop voltage and power control scheme. Each control scheme has its own advantages and meets different control requirements in practical applications. Then the advanced control technology for neutral-point voltage (NPV) balance under grid fault is presented, which maintains the NPV of the three-level T-type inverter ($3LT^2I$) in balance. Thus, the inverter realizes its own safe and stable operation. The NPV balance control technology is integrated into PWM modulator and auxiliary loop. Both the current direction and NPV are fed into the auxiliary loop to obtain the output control variable. The output control variables of inner loop and auxiliary loop are added up to drive the modulator. The modulator adjusts the switching state of power semiconductor switches in real time according to the input variable. The modules related to the above control technologies are depicted in Fig. 6.1 with highlight color.

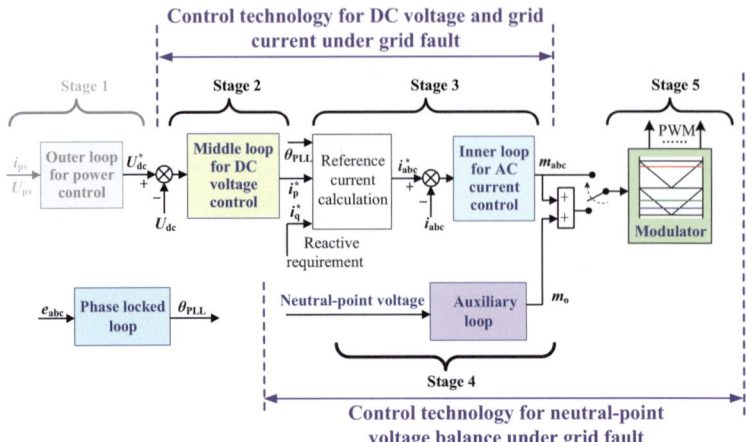

Fig. 6.1 The modules associated with high robustness and performance in overall control diagram

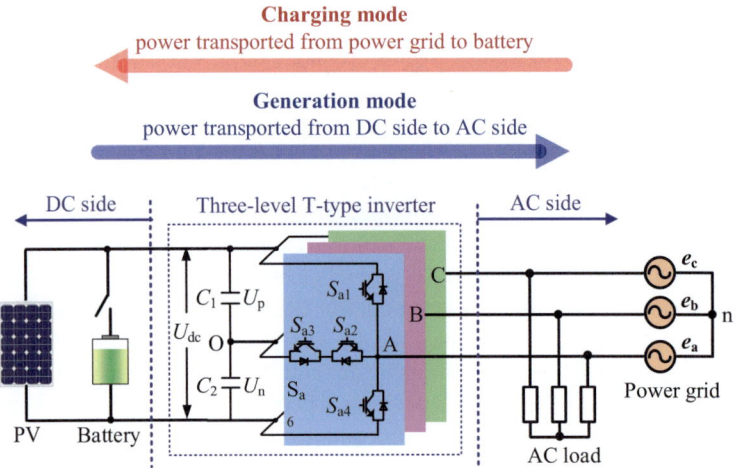

Fig. 6.2 PV generation system using the 3LT^{2}I with both power flowing directions

It should be emphasized that the topology and power flowing direction affect the system performance. Thus, the generation mode and charging mode (shown in Fig. 6.2) of 3LT^{2}I are adopted to elaborate corresponding control technology.

This chapter is organized as follows: the characteristics of grid voltages and grid standard requirements for PV inverter are illustrated since they determine the system control objectives. To satisfy the standard requirements, advanced control technologies for different control objectives under different operating modes are elaborated step by step. Meanwhile the output results are also provided to help readers better understand the control algorithm intuitively.

6.1 Overview of Grid Faults

With the integration of large-scale renewable power generation system and renewable power consumption system into the power grid, the power system has been evolved from the traditional power system to the modern power system. The modern power system has the distinctive features of "dual high": high proportion of renewable energy and high proportion of power converters. Unlike the traditional power system, the modern power system can flexibly adjust the power generation equipment and power-consuming equipment according to the demand and operation status by connecting or disconnecting the power connected converter.

However, the flexibility feature leads to the modern power system having "weak inertia" characteristic, which is significantly different from the traditional power systems that has the "rigid" characteristic. Additionally, in the traditional power system, users are power consumers and the loads are predictable and planned, which are relatively stable. While in the modern power system, the users change to "producers + consumers" and loads become strong uncertainty. All the above features make the power grid weak. In the weak power grid environment, unbalance, sag, swell, and harmonics of power grid voltages occur frequently [1]. Among them, voltage imbalance and harmonics are critical power quality issues since it directly determines the stability of power system on the generation side and consume side.

The harmonics are caused by converter switching operation and local non-linear loads. The problems caused by harmonics and the corresponding control technologies will be elaborated in Chap. 7. While the voltage imbalance is caused by the short circuit fault, the uneven penetration of single-phase renewable energy, and unequal distribution of single-phase loads, as shown in Fig. 6.3. The advanced control technology for addressing the problems caused by grid voltage imbalance will be discussed in detail in this chapter.

When a grid fault occurs, the grid voltages usually dips. The voltages dip refers to a sudden reduction in the voltage magnitude, usually between 0% and 90% of the rated value, and its duration is generally from half a fundamental cycle to a few seconds. According to the different types of power grid fault and transformer wiring methods, the voltage dips are divided into seven types: A ~ G. These seven types are further classified into three-phase symmetrical voltage dip fault and asymmetrical voltage dip fault. The symmetrical voltage dip means that the magnitudes of all three-phase voltages are balanced dipped, while the asymmetric voltage dip directly causes the three-phase grid voltage imbalance [2].

For PV power generation system, the main types of grid voltage dip include Type A, Type B and Type E [3]. The magnitude and waveforms of three-phase grid voltages corresponding to Type A, Type B and Type E dips are shown in Fig. 6.4. When Type A dip occurs, the magnitudes of three-phase voltages are dipped with the same amplitude. Compared to the pre-fault condition, the magnitudes of three-phase voltages remain balanced while the magnitude value is reduced. When Type B and Type E dips occur, the magnitudes of all three-phase voltages are imbalanced.

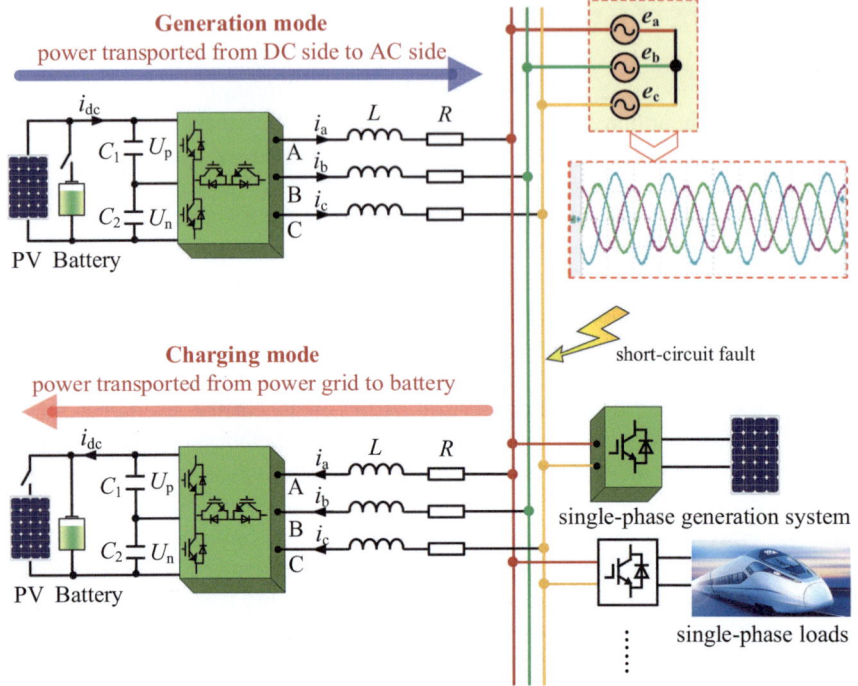

Fig. 6.3 The configuration of PV generation system with grid faults

Specifically, only one phase magnitude is reduced in Type B dip, while two phase magnitudes are reduced in Type E dip.

When the grid voltages are unbalanced, the output current of PV inverter with traditional control technology will produce non-characteristic subharmonics and unbalanced components. This issue leads to an increase in system losses, and triggers fault protection, and may even burn out the inverter equipment.

To mitigate the above problems, grid operators worldwide have issued relevant connection guidelines, setting clear requirements for the output characteristics of renewable energy systems to ensure the safety and stability of modern power system. One of the core requirements outlined in these guidelines is low voltage ride-through (LVRT), which is a key challenge in the integration of renewable energy generation technologies. The LVRT requirement means that when the grid voltage dip appears, the PV inverter system must remain connected to the grid and deliver a certain amount of reactive power to the grid to support grid voltage recovery until the grid returns to normal.

Chinese grid code of GB/T 19964-2012 stipulated the dynamic and static performance of PV inverter in LVRT operation [4]. Especially in terms of dynamic performance, the PV inverter should meet the requirements of access time and the amount of reactive power injected into the power grid. Figure 6.5 shows the access time requirement for PV inverter. When the grid voltage drops to 0 V, the PV inverter should

6.1 Overview of Grid Faults

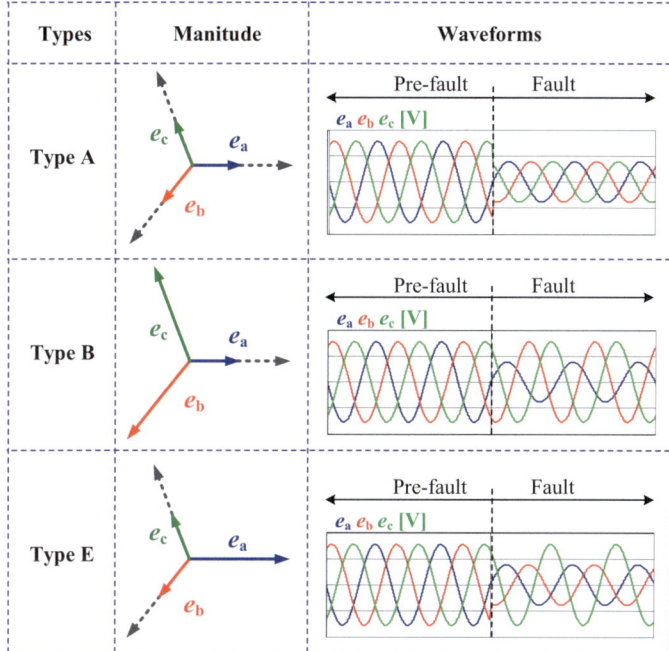

Fig. 6.4 The magnitude and waveforms of three-phase grid voltages corresponding to Type A, Type B and Type E dips

be kept connected to the power grid for at least 150 ms. When the grid voltage falls below the curve, the PV inverter can be disconnected from the power grid. For the response time aspects, the dynamic reactive current response time of the inverter is not more than 30 ms from the time of grid voltage dip. After the fault is cleared, the grid-connected PV inverter should be restored to the normal power generation state with a power change rate of at least 30% rated power per second.

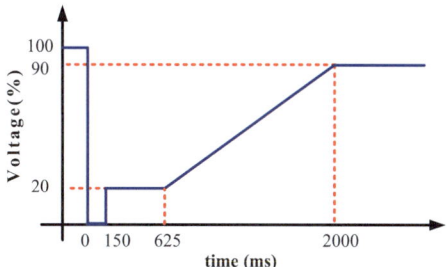

Fig. 6.5 The Chinese grid code requirements of access time of PV inverters connecting into the power grid

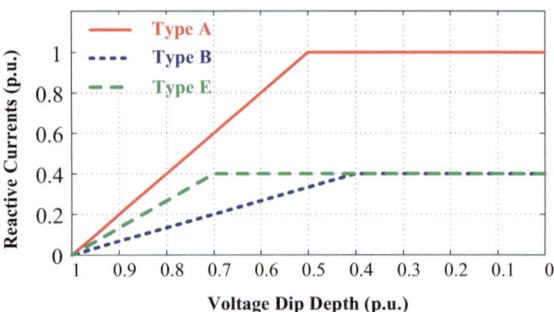

Fig. 6.6 Voltage dip types and required reactive current

In addition, the dynamic reactive current injected into the grid by the PV inverter should track the grid voltage change in real time and meet certain requirements.

The German E.ON grid code provides the detailed requirements of injected reactive current for different voltage dip types [5]. Figure 6.6 gives the relationship between the reactive current and the percentage of grid voltage reduction during the dip. It can be seen that when the three-phase power grid voltage dips symmetrically, the reactive current compensation amplitude does not exceed 100%. When the three-phase power grid voltage dips asymmetrically, the reactive current compensation amplitude does not exceed 40%. The ratio of the output reactive current to the rated current of the system is at least twice the voltage dip depth (this ratio is taken as twice the voltage dip depth).

It should be noted that the amplitude of the positive-sequence component of the grid voltage is selected to measure the voltage dip depth. The reactive current reference injected into the power grid is determined by the depth of grid voltage dip and the system rated current. The active current reference i_d^* can be calculated as

$$i_d^{*2} + i_q^{*2} = I_m^2 \tag{6.1}$$

where I_m is the rated current of the system.

According to the above requirement, the PV inverter should provide the reactive current to the grid during LVRT operation to help grid voltage recovery. Additionally, the inverter should also maintain its own safe and stable operation under grid fault condition. Thus, in a weak power grid, to ensure the normal operation of the PV inverter, and meet the standard requirements, advanced control technologies are presented in the following section.

6.2 Advanced Control Technology for DC Voltage and Grid Current of PV Inverters Under Grid Faults

As described above, the LVRT operation capability is a key function for the PV inverter to successfully cope with the power grid fault. To put it more precisely, the most important factors for LVRT function are that the PV inverter maintains its own safe and stable operation, and transmits high-quality power to the grid. Specifically, for the $3LT^2I$ topology, which is commonly used in PV generation systems, fast and accurate current or power tracking control is the key factor determining the success or failure of high-quality power transmission, the effective control of NPV balance guarantees the reliable operation of the inverter system.

In this section, the control technologies for DC voltage, grid current or transmitted power are elaborated. According to control variable types in the inner loop, the advanced control technology includes two control schemes: dual-loop voltage and current control scheme and dual-loop voltage and power control scheme. Each control scheme has its own advantages and meets different control requirements in practical applications. The dual-loop voltage and current control scheme is convenient for overcurrent protection, which can be achieved by limiting the current reference in the current control loop. The dual-loop voltage and power control scheme aims to regulate the transmitted active and reactive power directly, in which the current control loop and phase-locked loop (PLL) are no needed any more. In each control scheme, multiple control methods are illustrated to satisfy different application requirements.

6.2.1 Dual-Loop Voltage and Current Control Schemes

In the dual-loop voltage and current control scheme, the voltage loop is used to control dc-link voltage maintaining in the dc reference value, and the current loop is used to realize current tracking the ac or dc reference value.

For the voltage loop, the control variable is a dc value. The control methods are relative mature and not introduced further. Generally, the proportional-integral (PI) control method is adopted for the voltage loop in this section. For the current loop, the control variable is a dc or ac value. In particular, when a grid fault occurs, the both dc and ac components appear in the current, which increases control difficulties. Thus, control methods for the current loop are the research focus in industry and academia, and it is the topic in this section. Three methods are mainly introduced in this section, which consisting the proportional integral quasi-resonant (PIQR) control method [6], the hybrid passivity-based control (HPBC) method [7] and the generalized model predictive control (GMPC) method [8]. The theoretical analysis and implementation process of each method are elaborated as follows.

Proportional Integral Quasi-Resonant Control Method

In this part, the PIQR control method is presented for the 3LT^{2}I in generation mode under unbalanced grid voltage conditions.

Since negative-sequence components are generated by unbalanced grid voltages, the active and reactive currents have ac ripples at twice fundamental frequency. When PI controllers are employed, the active and reactive currents cannot track the reference values without steady-state error, and three-phase symmetrical output currents cannot be maintained.

The PIQR controller is proposed to achieve harmonic compensation. The main control objective is to maintain three-phase symmetrical output currents. Therefore, the negative-sequence current should be eliminated. In the positive-sequence synchronous reference frame, the frequency of negative-sequence components is 100 Hz. Therefore, the quasi-resonant term is added to a conventional PI controller to eliminate the negative-sequence components. The control bandwidth around resonant frequency can be increased due to the adoption of a quasi-resonant controller rather than a resonant controller.

The transfer function of the PIQR controller in the d-axis and q-axis can be expressed as

$$G_c(s) = k_p + \frac{k_i}{s} + \frac{2k_r \omega_c s}{s^2 + 2\omega_c s + \omega_0^2} \tag{6.2}$$

where ω_0 should be set as twice the angular frequency of grid voltage (ω_1), so as to eliminate the negative-sequence components in the active and reactive currents. ω_c decides the bandwidth around resonant frequency, and its value can be set according to

$$\omega_c = 2\pi f_1 \cdot \Delta f \tag{6.3}$$

where f_1 is the fundamental frequency of the ac grid voltage, and Δf is the percentage of the frequency fluctuation of the ac grid voltage.

Considering that the percentage of frequency fluctuation of the ac grid voltage is ±2%, ω_c should be set as $\omega_c = 2\pi \times 50 \times 2\% = 6.28$ rad/s.

Figures 6.7 and 6.8 depict the simulated results of the 3LT^{2}I with the traditional PI controller and the PIQR controller, respectively. As shown in Fig. 6.7a, when the currents in d-axis and q-axis are controlled by PI controllers, there exist ac ripples in waveforms of i_d and i_q. The frequency of ac ripples is 100 Hz. These ac ripples are negative-sequence components caused by unbalanced grid voltage. Therefore, the currents in d-axis and q-axis cannot track their reference values without steady-state error. The three-phase output current becomes unbalanced, as displayed in Fig. 6.7b. As shown in Fig. 6.8, when proportional integral resonant controllers are adopted for the d-axis and q-axis, the ac ripples in i_d and i_q are eliminated, and tracking performance without steady-state error can be achieved. Thus, balanced three-phase output currents are obtained.

6.2 Advanced Control Technology for DC Voltage and Grid Current of PV … 311

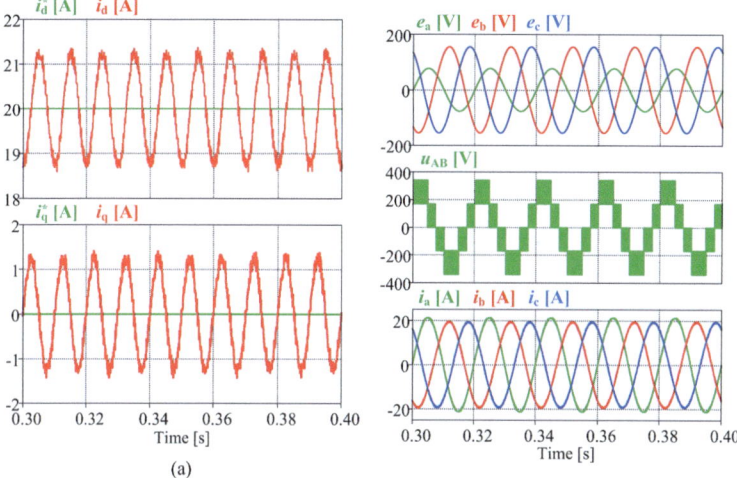

Fig. 6.7 Simulation results of the $3LT^2I$ with the traditional PI controller **a** d-axis and q-axis current components. **b** Unbalanced grid voltage, line voltage, and three-phase output currents

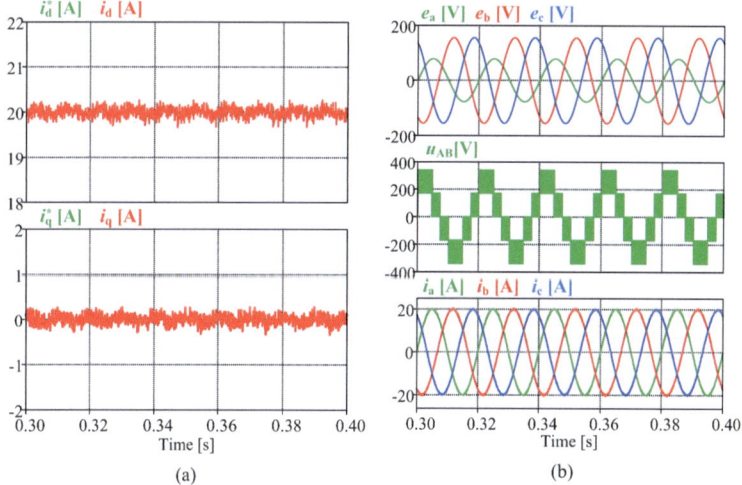

Fig. 6.8 Simulation results of the $3LT^2I$ with the PIQR controller **a** d-axis and q-axis current components. **b** Unbalanced grid voltage, line voltage, and three-phase output currents

Hybrid Passivity-Based Control Method

Fast dynamic performance of PV inverter is one of the most important issues in LVRT operation, because the voltage dip period is short. Nevertheless, it is difficult to control the grid currents with conventional methods, especially during transients or under low power factor operation. In order to satisfy the LVRT requirements and improve the performances of the 3LT^{2}I in PV system, the HPBC strategy with fast transient response and strong robustness to system parameter changes and external disturbances is proposed in this section.

The 3LT^{2}I operating in generation mode is taken as an example to illustrate the HPBC principle, which is embedded into the inner loop for current control. Three parts are included in this section: modeling and analysis of the 3LT^{2}I under unbalanced grid voltages, the design process of the HPBC controller and demonstration of the 3LT^{2}I's performance with HPBC.

Part 1: Modeling and Analysis of 3LT^{2}I Under Unbalanced Grid Voltages
On the basis of Kirchhoff's voltage law, the grid side model can be described as

$$\begin{cases} L\frac{di_a}{dt} + Ri_a + e_a + u_{nO} = u_{AO} \\ L\frac{di_b}{dt} + Ri_b + e_b + u_{nO} = u_{BO} \\ L\frac{di_c}{dt} + Ri_c + e_c + u_{nO} = u_{CO} \end{cases} \quad (6.4)$$

where L and R denote the inductor and equivalent series resistor of the filter, e_a, e_b, and e_c are the three-phase grid voltages, u_{nO} is the voltage between the ac common point n and dc neutral point O. When unbalanced grid fault occurs, the unbalanced grid voltages are expressed as

$$\begin{bmatrix} e_a \\ e_b \\ e_c \end{bmatrix} = E_P \begin{bmatrix} \cos \omega t \\ \cos(\omega t - \frac{2\pi}{3}) \\ \cos(\omega t + \frac{2\pi}{3}) \end{bmatrix} + E_N \begin{bmatrix} \cos \omega t \\ \cos(\omega t + \frac{2\pi}{3}) \\ \cos(\omega t - \frac{2\pi}{3}) \end{bmatrix} \quad (6.5)$$

where ω is the fundamental angular frequency of the grid, and E_P and E_N are the amplitudes of positive-sequence and negative-sequence voltages, respectively. Since the three-phase three-wire system does not have a zero-sequence current loop, the zero-sequence component is ignored here.

Therefore, combining the above equations, the grid side model is expressed as

$$L\frac{d}{dt}\begin{bmatrix} i_a \\ i_b \\ i_c \end{bmatrix} = -E_P \begin{bmatrix} \cos \omega t \\ \cos(\omega t - \frac{2\pi}{3}) \\ \cos(\omega t + \frac{2\pi}{3}) \end{bmatrix} - E_N \begin{bmatrix} \cos \omega t \\ \cos(\omega t + \frac{2\pi}{3}) \\ \cos(\omega t - \frac{2\pi}{3}) \end{bmatrix}$$

$$+ \begin{bmatrix} u_{AO} \\ u_{BO} \\ u_{CO} \end{bmatrix} - u_{nO}\begin{bmatrix} 1 \\ 1 \\ 1 \end{bmatrix} - R\begin{bmatrix} i_a \\ i_b \\ i_c \end{bmatrix} \quad (6.6)$$

In order to simplify the control of active and reactive currents, the above mathematical model is transformed from the stationary reference frame to the positive-sequence synchronous reference frame. The transformation matrix is expressed as

$$T = \frac{2}{3}\begin{bmatrix} \cos \omega t & \cos(\omega t - \frac{2\pi}{3}) & \cos(\omega t + \frac{2\pi}{3}) \\ -\sin \omega t & -\sin(\omega t - \frac{2\pi}{3}) & -\sin(\omega t + \frac{2\pi}{3}) \end{bmatrix} \quad (6.7)$$

6.2 Advanced Control Technology for DC Voltage and Grid Current of PV ...

Then the mathematical model in the positive-sequence synchronous reference frame can be written as

$$\begin{cases} L\frac{di_d}{dt} - \omega L i_q + R i_d = u_d - e_d \\ L\frac{di_q}{dt} + \omega L i_d + R i_q = u_q + e_q \end{cases} \quad (6.8)$$

where $e_d = E_P + E_N \cos 2\omega t$ and $e_q = E_N \sin 2\omega t$.

Thus, the Euler-Lagrange model of 3LT²I under unbalanced grid voltage conditions can be expressed as

$$M\dot{x} + Jx + Rx = u \quad (6.9)$$

where M and R are positive definite matrices, and J is an antisymmetric matrix and satisfies $J = -J^T$, and

$$x = \begin{bmatrix} i_d \\ i_q \end{bmatrix}, M = \begin{bmatrix} L & 0 \\ 0 & L \end{bmatrix}, J = \begin{bmatrix} 0 & -\omega L \\ \omega L & 0 \end{bmatrix}$$
$$R = \begin{bmatrix} R & 0 \\ 0 & R \end{bmatrix}, u = \begin{bmatrix} u_d - E_P - E_N \cos 2\omega t \\ u_q + E_N \sin 2\omega t \end{bmatrix} \quad (6.10)$$

It can be seen from the above model that the output current is closely related to the passive device filter inductor L and the equivalent series resistor R. u is the control input and contains the positive- and negative-sequence components of the grid voltages, which is different from the models under normal grid conditions [9] for conventional passivity-based control (PBC). Therefore, the control strategy needs to be redesigned to achieve good current output waveforms and meet LVRT requirements, which will be elaborated in the following sections.

Part 2: Design Process of HPBC Controller

In order to better study the current control performance of the proposed HPBC strategy under LVRT conditions, the dc-link voltage is assumed to be controlled by the maximum power point tracking (MPPT), and remains constant.

In LVRT operation, the active and reactive current references of the 3LT²I should be modified to satisfy grid requirements, which has been given in above section. The premise of passivity-based controller design is to ensure the strict passivity of the system. In the Euler-Lagrange model of the 3LT²I, the filter inductor L and the equivalent series resistor R are passive devices, so the 3LT²I system is passive. From the mathematical point of view, the definition of strictly passive and the detailed proof of the passivity for 3LT²I system are given first, and then the specific design details of the hybrid passivity-based controller are proposed.

A system is said to be strictly passive, if there is a non-negative storage function $H(x)$ and a positive definite function $Q(x)$, such that

$$H[x(T)] - H[x(0)] \leq \int_0^T u^T y d\tau - \int_0^T Q(x) d\tau, \ T > 0 \qquad (6.11)$$

or

$$\dot{H}(x) \leq u^T y - Q(x) \qquad (6.12)$$

where $x(t)$, $u(t)$, and $y(t)$ are the state, input, and output vectors with the appropriate dimensions, respectively.

The term $\int_0^T u^T y d\tau$ indicates the energy supplied to the system, and $H[x(T)] - H[x(0)]$ and $\int_0^T Q(x) d\tau$ denote the stored energy and dissipated energy of the system, respectively. Since the dissipated energy always propels the state $x(t)$ back to the desired equilibrium point, the passive system is inherently stable. That is, as long as the system is passive, it must be internally stable. The detailed proof of strict passivity for the 3LT^{2}I system is given mathematically below.

For the mathematical model of the 3LT^{2}I under LVRT conditions, multiplying both sides of the two equations in the mathematical model in the positive-sequence synchronous reference frame by i_d and i_q, respectively, and then adding the two equations together, a power equation is obtained as

$$Li_d \frac{di_d}{dt} + Li_q \frac{di_q}{dt} + Ri_d^2 + Ri_q^2$$
$$= u_d i_d + u_q i_q - i_d (E_P + E_N \cos 2\omega t) + i_q E_N \sin 2\omega t \qquad (6.13)$$

which can be further simplified to

$$\frac{d}{dt} \left(\frac{Li_d^2}{2} + \frac{Li_d^2}{2} \right)$$
$$= u_d i_d + u_q i_q + i_q E_N \sin 2\omega t - i_d (E_P + E_N \cos 2\omega t) - R(i_d^2 + i_q^2) \qquad (6.14)$$

Letting

$$H(x) = \frac{1}{2} x^T M x = \frac{L}{2} (i_d^2 + i_q^2) > 0 \qquad (6.15)$$

$$u^T y = u_d i_d + u_q i_q + i_q E_N \sin 2\omega t - i_d (E_P + E_N \cos 2\omega t) \qquad (6.16)$$

$$Q(x) = x^T R x = R(i_d^2 + i_q^2) > 0 \qquad (6.17)$$

With the above equations, it is easy to conclude that the 3LT^{2}I system is strictly passive. Therefore, a passivity-based controller design can be implemented for the 3LT^{2}I.

6.2 Advanced Control Technology for DC Voltage and Grid Current of PV ...

The passivity-based controller of the 3LT²I is designed to achieve $i_d \to i_d^*$ and $i_q \to i_q^*$, where i_d^* and i_q^* represent the desired equilibrium points of i_d and i_q, respectively.

Let error $x_e = x - x^*$, and $H(x_e)$ represent the error storage function as

$$H(x_e) = \frac{1}{2} x_e^T M x_e > 0 \tag{6.18}$$

where $x = [i_d \ i_q]^T$. Thus,

$$\dot{H}(x_e) = x_e^T M \dot{x}_e \tag{6.19}$$

The error Euler-Lagrange model of the 3LT²I can be expressed as

$$M\dot{x}_e + R x_e = M\dot{x} + R x - \left(M\dot{x}^* + R x^*\right) = u - \left(M\dot{x}^* + J x + R x^*\right) \tag{6.20}$$

In order to achieve $x \to x^*$, it is desirable that $H(x_e)$ rapidly converges to the desired point of 0, which requires damping injection R_a. The damping injection dissipation is

$$R_e x_e = (R + R_a) x_e \tag{6.21}$$

where the positive definite diagonal damping matrix R_a is

$$R_a = \begin{bmatrix} r_a & 0 \\ 0 & r_a \end{bmatrix}, r_a > 0. \tag{6.22}$$

Thus, the error Euler-Lagrange model is rewritten as

$$M\dot{x}_e + R_e x_e = u - \left(M\dot{x}^* + J x + R x^* - R_a x_e\right) \tag{6.23}$$

The passivity-based controller can be designed as

$$u = M\dot{x}^* + J x + R x^* - R_a x_e \tag{6.24}$$

For error storage function $H(x_e)$, we get

$$\dot{H}(x_e) = x_e^T M \dot{x}_e = -x_e^T R_e x_e < 0 \tag{6.25}$$

Owing to the positive definiteness of R_e. It follows from (6.20) and (6.25) that $x \to x^*$ based on the Lyapunov stability criterion, which means that the control objective is achieved with the passivity-based controller (6.26). The speed of $H(x_e) \to 0$ mainly depends on $R_a \gg R$ while the speed of $H(x_e) \to 0$ is mainly dependent on R_a, which principally determines the robustness of the 3LT²I to system parameter changes and external disturbances.

From Euler-Lagrange model of 3LT²I under unbalanced grid voltage conditions, the passivity-based controller becomes

$$\begin{bmatrix} u_d - E_P - E_N \cos 2\omega t \\ u_q + E_N \sin 2\omega t \end{bmatrix} = \begin{bmatrix} L & 0 \\ 0 & L \end{bmatrix} \begin{bmatrix} i_d^* \\ i_q^* \end{bmatrix} + \begin{bmatrix} 0 & -\omega L \\ \omega L & 0 \end{bmatrix} \begin{bmatrix} i_d \\ i_q \end{bmatrix} + \begin{bmatrix} R & 0 \\ 0 & R \end{bmatrix} \begin{bmatrix} i_d^* \\ i_q^* \end{bmatrix} - \begin{bmatrix} r_a & 0 \\ 0 & r_a \end{bmatrix} \begin{bmatrix} i_d - i_d^* \\ i_q - i_q^* \end{bmatrix} \quad (6.26)$$

So, the output variables of the controller are derived as

$$\begin{cases} u_d = e_d + L\frac{di_d^*}{dt} - \omega L i_q + R i_d^* - r_a(i_d - i_d^*) \\ u_q = -e_q + L\frac{di_q^*}{dt} + \omega L i_d + R i_q^* - r_a(i_q - i_q^*) \end{cases} \quad (6.27)$$

where $e_d = E_P + E_N \cos 2\omega t$ and $e_q = E_N \sin 2\omega t$ in theory. It should be noted that the control variables in the above expression contain the grid voltage negative-sequence components with a frequency of 100 Hz in the positive-sequence synchronous reference frame. Consequently, the current negative-sequence components are eliminated by using the proposed HPBC strategy and three-phase symmetrical output currents are obtained to meet the LVRT requirements.

The specific control block diagram is shown in Fig. 6.9, where p is the differential operator ($p = \frac{d}{dt}$). It can be seen that the proposed method avoids the positive- and negative-sequence separation of the grid voltage and thus is simpler in calculation. Furthermore, it is noted that the proposed HPBC strategy is derived based on the system mathematical model, which has accurate control performance. In addition, it is noteworthy that the transient response of the proposed HPBC strategy is much faster than that of the PI controller since the introduction of integrator is avoided.

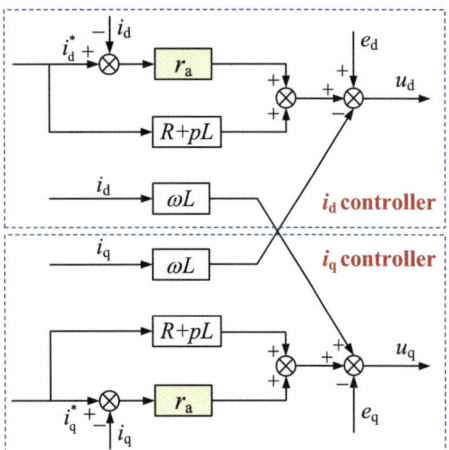

Fig. 6.9 Control block diagram of the proposed HPBC strategy

6.2 Advanced Control Technology for DC Voltage and Grid Current of PV ...

It can be seen from the above derivation of the HPBC that the inverter system is stable. The following is a vivid convergence analysis from a mathematical perspective.

Combining the mathematical model in the positive-sequence synchronous reference frame and output variables of the HPBC, it is easy to obtain

$$\begin{cases} L\frac{d}{dt}(i_d - i_d^*) = -(R + r_a)(i_d - i_d^*) \\ L\frac{d}{dt}(i_q - i_q^*) = -(R + r_a)(i_d - i_d^*) \end{cases} \quad (6.28)$$

which means that the dynamic decoupling of the currents is achieved by using the hybrid passivity-based controller. The current response is derived as

$$\begin{cases} i_d = i_d^* + C_{01}e^{-\frac{R+r_a}{L}t} \\ i_q = i_q^* + C_{02}e^{-\frac{R+r_a}{L}t} \end{cases} \quad (6.29)$$

which implies that, for appropriate r_a, i_d and i_q can quickly converge to i_d^* and i_q^*, respectively. It should be noted that C_{01} and C_{02} are finite constants depending on circuit parameters.

From the expression of current response, the transfer function of the current is

$$G(s) = \frac{1}{C_o s + 1}, \quad \text{where} \quad C_o = \frac{L}{R + r_a} \quad (6.30)$$

When delay τ is considered, the current transfer function is rewritten as

$$G_d(s) = \frac{1}{C_o s + 1} e^{-\tau s} \quad (6.31)$$

By Padé approximation, $G_d(s)$ can be expressed as

$$G_d(s) = \frac{1}{C_o s + 1} \cdot \frac{1 - \frac{\tau s}{2}}{1 + \frac{\tau s}{2}} \quad (6.32)$$

It is pointed out that if the transfer function $G(s)$ of a system satisfies $Re[G(j\omega)] > 0$, the system is passive and internally stable. Therefore, the damping term r_a satisfies

$$r_a > \frac{4L\omega^2 \tau}{4 - \omega^2 \tau^2} - R \quad (6.33)$$

The symmetry of the three-phase currents can be achieved under unbalanced grid voltage conditions by utilizing the proposed HPBC strategy. Consequently, the NPV balance can be achieved by adjusting the dwell time of the redundant small vectors, which will be elaborated in the next section and will not be repeated in this section.

The whole control block diagram of the 3LT²I system is shown in Fig. 6.10. The dual second-order generalized integrator PLL [10] is used to lock the phase θ of the grid voltage. The dual second-order generalized integrator PLL provides a high-performance detection system that can quickly and accurately characterize positive-sequence voltage components even under extremely unbalanced and distorted grid operating conditions, which is perfectly applicable to this part.

The current reference is calculated based on LVRT requirements. The grid-side currents are controlled by the proposed HPBC strategy. The P controller is adopted to guarantee NPV balance. SVM is applied as the modulation strategy. The outputs of the current controller and NPV the balance controller are used as the input to the modulator.

Part 3: Demonstration of 3LT²I Performance with HPBC

In order to validate the effectiveness of the proposed HPBC strategy, a simulation test is carried out utilizing MATLAB/Simulink. The parameters for simulations are summarized in Table 6.1.

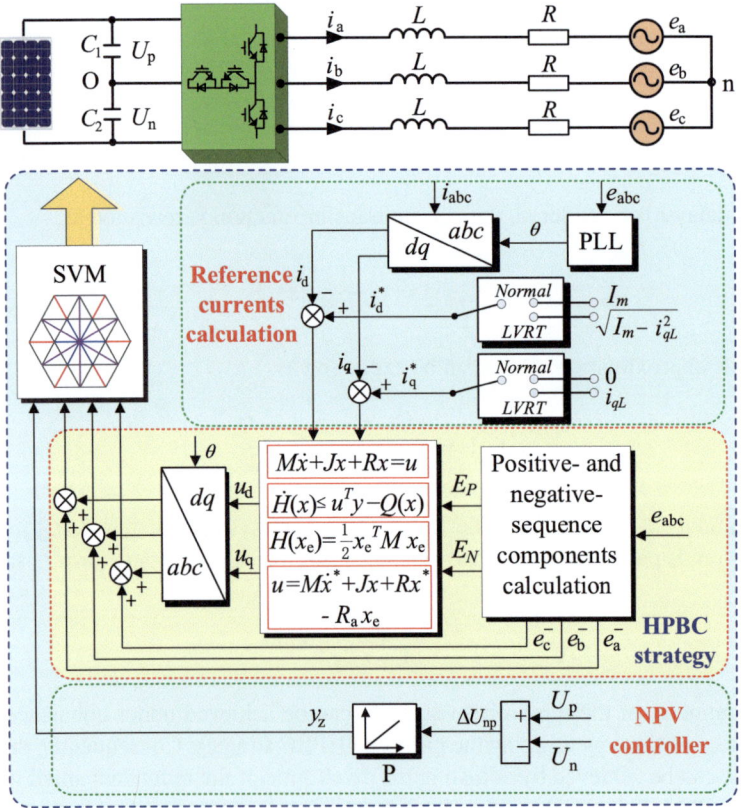

Fig. 6.10 Control block diagram of the 3LT²I system using HPBC method

6.2 Advanced Control Technology for DC Voltage and Grid Current of PV …

Table 6.1 Parameters for simulations of the $3LT^2I$ with HPBC control method

Parameters	Values
DC bus voltage	320 V
DC-link capacitors	3300 µF
AC grid voltage (RMS)	110 V
Grid frequency	50 Hz
Switching frequency	10 kHz
Sampling frequency	10 kHz
LC filter	$L = 1$ mH, $C_f = 10$ µF

Consider three different grid voltage dip types, namely Type A, Type B, and Type E. The performance of the proposed HPBC strategy under these three cases are simulated and compared with the conventional PBC method in both steady-state and transient-state conditions.

Case 1: Type A Operation

Figure 6.11 shows the simulated dynamic tracking performance of active and reactive currents during normal ($i_d{}^* = 20$ A and $i_q{}^* = 0$ A) and Type A ($i_d{}^* = 0$ A and $i_q{}^* = -20$ A) operations for different control strategies.

Simulation waveforms with the conventional PBC are presented in Fig. 6.11a, whereas the ones for the proposed HPBC strategy are shown in Fig. 6.11b. It can be seen that the transient response time is about 0.015 s for the conventional PBC strategy, while only 0.012 s for the proposed HPBC strategy. The comparison of transient response time confirms that the response speed is dramatically improved with the proposed HPBC strategy, which coincides with the theoretical analysis. The simulated waveforms of active and reactive power during normal and Type A

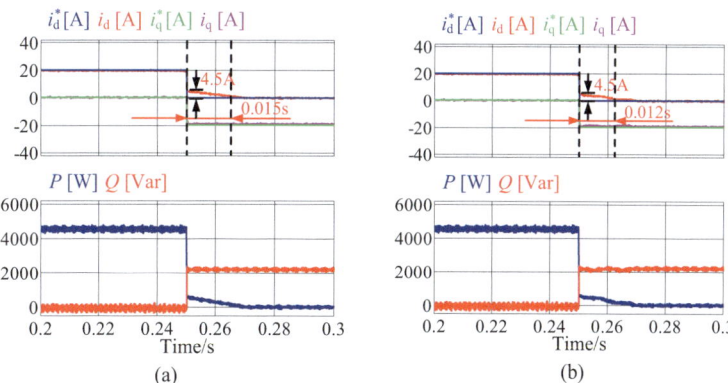

Fig. 6.11 Dynamic tracking performances of active and reactive currents before and after LVRT for Type A. **a** Conventional PBC method. **b** Proposed HPBC strategy

operations for two different control methods show that both methods can achieve fast tracking of active and reactive power under a three-phase balanced dip.

The steady-state waveforms for both the conventional PBC method and the proposed HPBC strategy during normal and Type A operations are shown in Fig. 6.12. A three-phase grid voltage dip to 50% of the rated voltage occurs at $t = 0.25$ s. It can be observed that the grid-side currents of both methods exhibit good quality. This is because Type A is a three-phase balanced dip that does not affect the symmetry of the three-phase currents.

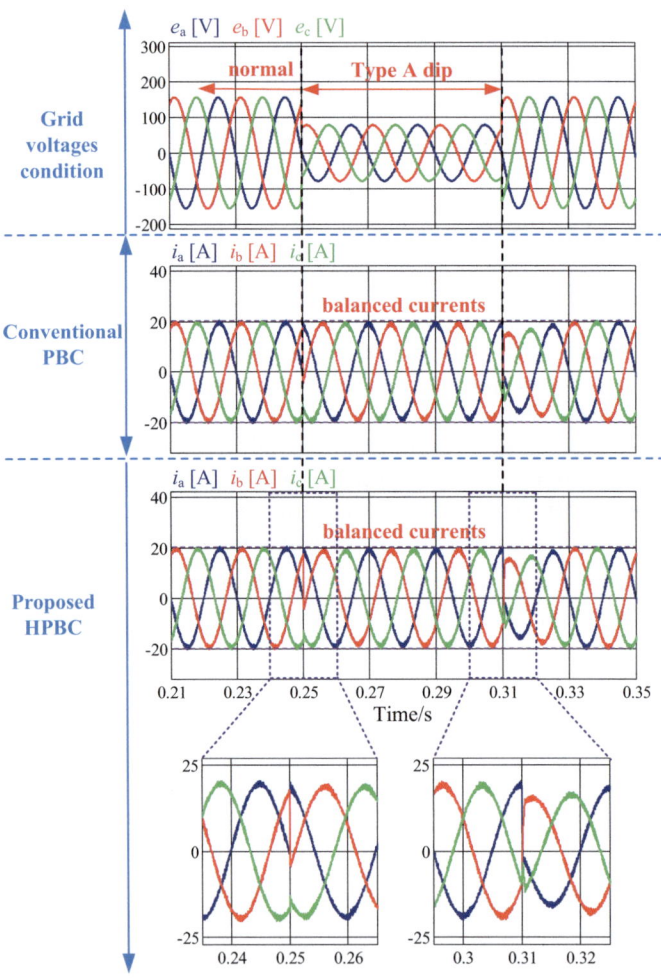

Fig. 6.12 Steady-state results of grid voltages and currents during normal and Type A operations with conventional PBC and proposed HPBC strategy

6.2 Advanced Control Technology for DC Voltage and Grid Current of PV ...

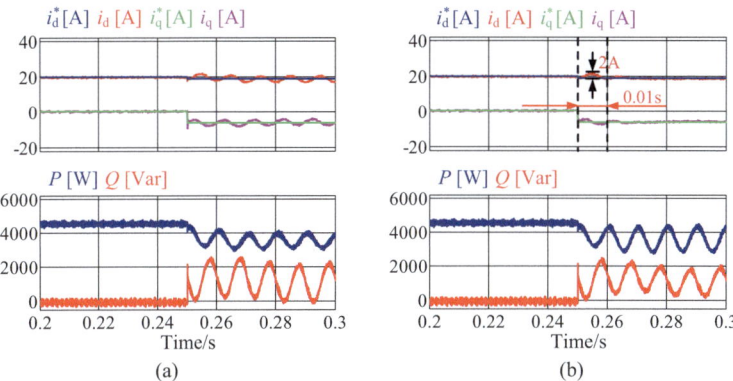

Fig. 6.13 Dynamic tracking performances of active and reactive currents before and after LVRT for Type B **a** Conventional PBC method. **b** Proposed HPBC strategy

Case 2: Type B Operation

When a Type B fault occurs, the reference active and reactive current are $i_d^* = 18.9$ A and $i_q^* = -6.7$ A. The simulated dynamic tracking performance with different control strategies are shown in Fig. 6.13. When the conventional PBC method is adopted, ac ripples appear in the waveforms of active current i_d and reactive current i_q. These ac ripples, with a frequency of 100 Hz, are negative-sequence components resulting from the unbalanced grid voltage. Therefore, the reference currents i_d^* and i_q^* cannot be tracked without a steady-state error. Fortunately, when the proposed HPBC strategy is employed, reference tracking without a steady-state error can be realized without ac ripples. The simulated waveforms of active and reactive power during normal and Type B operations for three different control methods are also given. It can be clearly seen that there are double-frequency oscillations in both active and reactive power waveforms with three different control methods. Even if the grid-side currents are balanced, there will be oscillations in the active and reactive power due to the negative-sequence components in grid voltages.

Figure 6.14 presents the steady-state performance during normal and Type B operations. An a-phase grid voltage dip to 50% of the rated voltage occurs at $t = 0.25$ s. The grid-side currents of the proposed HPBC strategy are compared with those of the conventional PBC method. It can be clearly seen that with the conventional PBC method, the current difference between the highest phase and the lowest phase is 3 A, which indicates that the conventional PBC method is no longer applicable under unbalanced grid conditions. When adopting the proposed HPBC strategy, grid-side currents can be quickly balanced because the negative-sequence currents are significantly suppressed.

Case 3: Type E Operation

When a Type E fault occurs, the reference active and reactive current are $i_d^* = 18.3$ A and $i_q^* = -8$ A. The simulated dynamic tracking performance with different control strategies are shown in Fig. 6.15. There exist ac ripples in the current waveforms

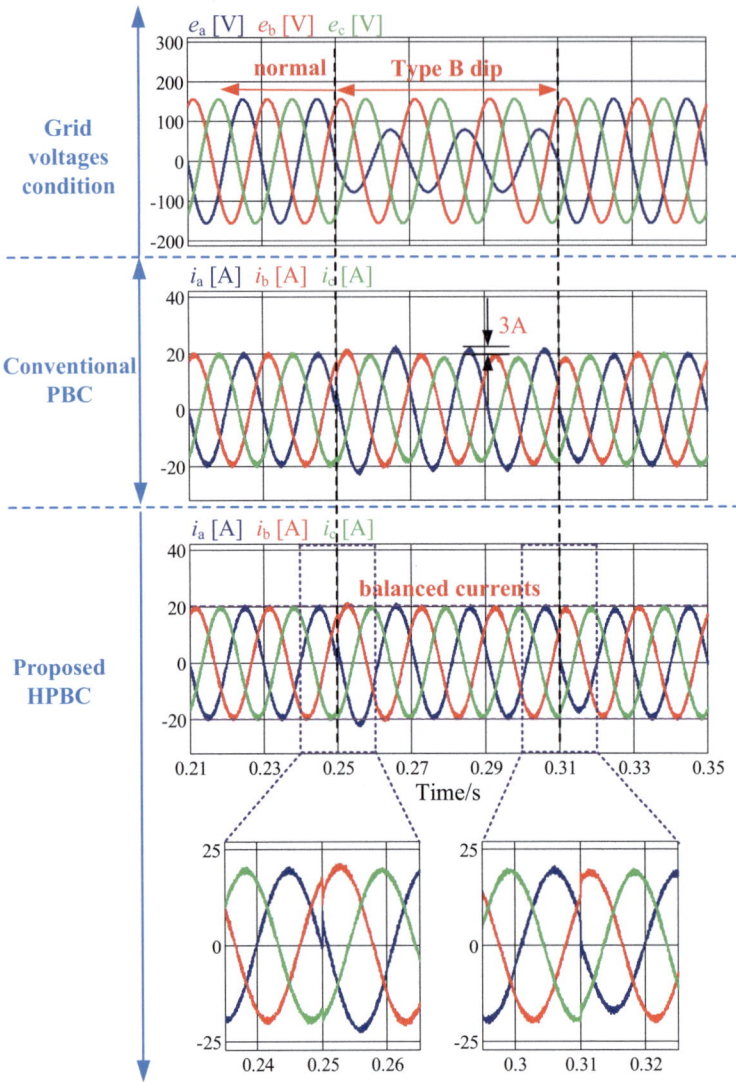

Fig. 6.14 Steady-state results of grid voltages and currents during normal and Type B operations with conventional PBC and proposed HPBC strategy

when the conventional PBC method is applied. When adopting the proposed HPBC strategy, these high-frequency ripples in i_d and i_q are completely eliminated, and reference tracking without a steady-state error can be implemented with a significantly faster response. It can also be seen that there are double-frequency oscillations in the simulated waveforms of active and reactive power under two-phase grid voltage

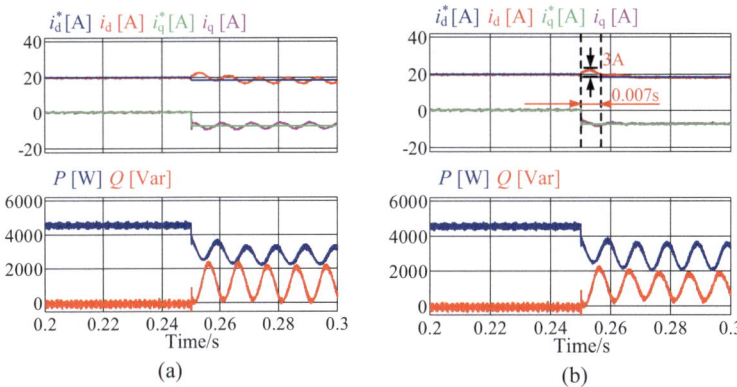

Fig. 6.15 Dynamic tracking performances of active and reactive currents before and after LVRT for Type E **a** Conventional PBC method. **b** Proposed HPBC strategy

dip. Even if the grid-side currents are balanced, the negative-sequence components in the grid voltages inevitably cause oscillations in both active and reactive power.

Figure 6.16 shows the simulated waveforms of grid voltages and grid-side currents during normal and Type E operations. A two-phase (phases a and b) grid voltage dip to 50% of the rated voltage occurs at $t = 0.25$ s. Grid-side currents are simulated using the conventional PBC method to compare with the proposed HPBC strategy. It can be seen that with the conventional PBC method, the grid-side currents are no longer balanced, while three-phase symmetrical currents with high quality can be quickly obtained by the proposed HPBC strategy.

In order to obtain three-phase symmetrical currents with high quality as well as fast dynamic performance when the voltage dips, a HPBC strategy to control the grid side currents for a $3LT^2I$ is presented to satisfy LVRT requirements. The detailed Euler-Lagrange model of the $3LT^2I$ under LVRT conditions is firstly developed. Then, the HPBC strategy including a negative-sequence grid voltage feedforward control term is proposed. Thus, the grid current negative-sequence component is effectively suppressed and the output waveforms quality is improved significantly. According to LVRT requirements, reactive power is injected into the grid to keep voltage and frequency stable. With the proposed strategy, the superior performance, such as fast response, non-overshoot, and accurate current tracking, is achieved in both transient and steady-state conditions compared to the other two conventional control methods when the grid voltage dip is asymmetrical. Simulation results confirm the correctness of the theoretical analysis and the effectiveness of the proposed strategy. The proposed strategy has a simple structure and does not require additional hardware investment, so it can be easily implemented in engineering applications. The proposed scheme has great potential to be extended to the PV generation system, active power filter and static var generator.

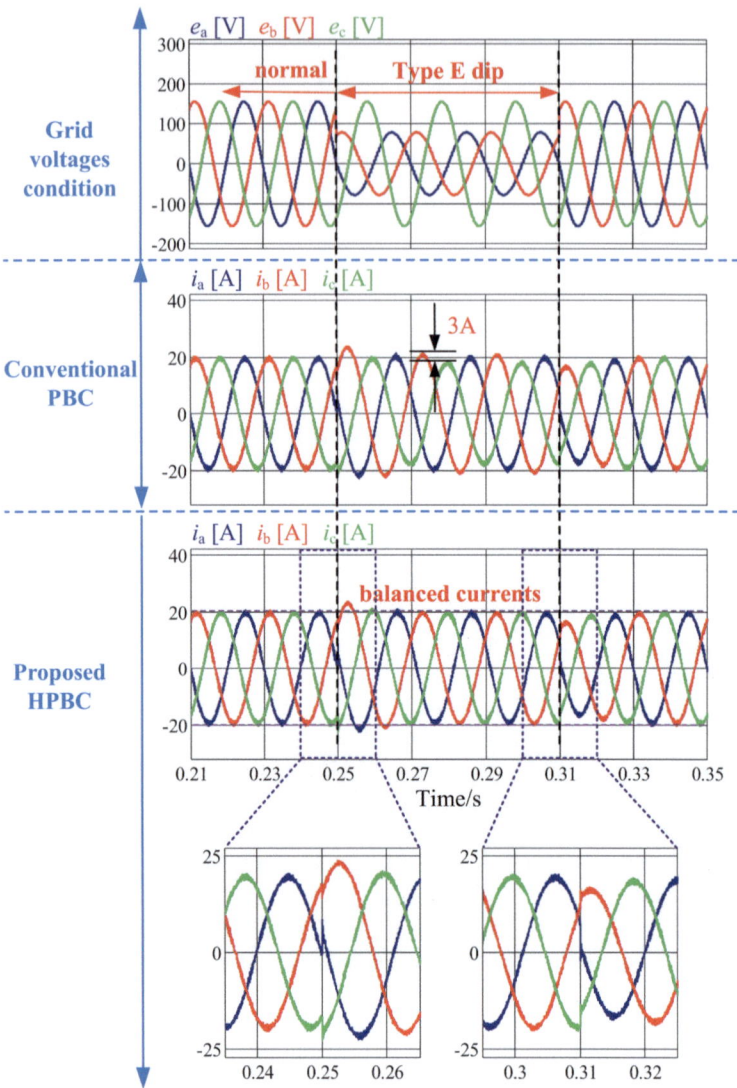

Fig. 6.16 Steady-state results of grid voltages and currents during normal and Type E operations with conventional PBC and proposed HPBC strategy

Model Predictive Control Method

With the rapid development of advanced microprocessors, the model predictive control (MPC) method has been proposed as a powerful alternative method for three-phase inverter, since it has good steady-state and dynamic performance [11]. Using MPC, the optimal voltage vector minimizing the pre-defined cost function

6.2 Advanced Control Technology for DC Voltage and Grid Current of PV …

is selected and directly applied to the inverter [12]. With this method, the positive- and negative-sequence separation, PLL and PWM modulators are no longer needed which provides a great simplification and flexibility. Under unbalanced grid voltage or load conditions, the MPC method is used to regulate the three-phase currents to be sinusoidal with low harmonics for two-level inverter. However, the MPC methods for two-level inverters are not suitable for three-level inverter due to its inherent NPV balance problem. For three-level inverter, a cost function with weighting factors is adopted in the MPC method to achieve the minimum load current tracking error and NPV balance [13]. Nevertheless, the weighting factors in the MPC scheme are usually tuned by trial and error. In [14], an improved MPC method without weighting factors is proposed. However, all the above MPC methods for NPV balance mainly focus on normal grid conditions. According to the grid code, power inverters should provide a certain amount of reactive power under unbalanced grid voltage, and the NPV unbalance of three-level inverters is increased as the power factor decreases.

Thus, for the 3LT²I, to eliminate current distortion and balance NPV with full power factor operation under unbalanced grid voltage condition, a GMPC scheme is proposed by the authors' team. Three parts are included in this section: sinusoidal reference current derivation with different power control objectives, implementation process of GMPC and the control performance of GMPC.

Part 1: Sinusoidal Reference Current Derivation with Different Power Control Objectives

It should be noted that the 3LT²I operating in charging mode is taken as an example to illustrate the GMPC. In this mode, system power is transported from the power grid to the battery, and the positive direction of current is defined as flowing from AC side to DC side. The topology of 3LT²I and the defined positive current direction are depicted in Fig. 6.1. The unbalanced grid voltage and currents in phase j (j = a, b, c) are decomposed into positive-, negative- and zero-sequence components, and expressed as

$$\begin{cases} e_j = e_j^p + e_j^n + e_j^0 \\ i_j = i_j^p + i_j^n + i_j^0 \end{cases} \tag{6.34}$$

and i_j^p, i_j^n and i_j^0 are the positive-, negative-, and zero-sequence components of the grid current, respectively. The zero-sequence component is ignored in a three-phase three-wire system without a neutral wire. Using Clarke and Park transformations, the unbalanced grid voltage and current are expressed as

$$\begin{cases} e_{\alpha\beta} = e_{\alpha\beta}^p + e_{\alpha\beta}^n = e_{dq}^p + e_{dq}^n \\ i_{\alpha\beta} = i_{\alpha\beta}^p + i_{\alpha\beta}^n = i_{dq}^p + i_{dq}^n \end{cases} \tag{6.35}$$

where $e_{\alpha\beta}^p$, $e_{\alpha\beta}^n$ and $i_{\alpha\beta}^p$, $i_{\alpha\beta}^n$ are the positive- and negative-sequence components of the grid voltage and grid current in the $\alpha\beta$ stationary frame. e_{dq}^p, e_{dq}^n and i_{dq}^p, i_{dq}^n are the positive- and negative-sequence components of grid voltage and grid current in the dq rotating frame.

The relationship between voltage and current in the $\alpha\beta$ stationary frame and the dq rotating frame is satisfied as

$$\begin{cases} e_{dq}^p = e_{\alpha\beta}^p e^{j\omega t} \\ e_{dq}^n = e_{\alpha\beta}^n e^{-j\omega t} \\ i_{dq}^p = i_{\alpha\beta}^p e^{j\omega t} \\ i_{dq}^n = i_{\alpha\beta}^n e^{-j\omega t} \end{cases} \tag{6.36}$$

The instantaneous complex power is calculated as

$$s = \frac{2}{3} e \times i^* = p + jq \tag{6.37}$$

where i^* is the conjugate of current, and p and q are the instantaneous active power and reactive power, respectively. Substituting the voltage and current in the dq rotating frame, the instantaneous complex power is rewritten as

$$s = \left[p_0 + p_{s2} \sin(2\omega t) + p_{c2} \cos(2\omega t) \right] \\ + j \left[q_0 + q_{s2} \sin(2\omega t) + q_{c2} \cos(2\omega t) \right] \tag{6.38}$$

where $p_0, p_{s2}, p_{c2}, q_0, q_{s2}$ and q_{c2} are satisfied as

$$\begin{bmatrix} p_0 \\ p_{s2} \\ p_{c2} \\ q_0 \\ q_{s2} \\ q_{c2} \end{bmatrix} = \begin{bmatrix} e_d^p & e_q^p & e_d^n & e_q^n \\ e_q^n & -e_d^n & -e_q^p & e_d^p \\ e_d^n & e_q^n & e_d^p & e_q^p \\ e_q^p & -e_d^p & e_q^n & -e_d^n \\ -e_d^n & -e_q^n & e_d^p & e_q^p \\ e_q^n & -e_d^n & e_q^p & -e_d^p \end{bmatrix} \begin{bmatrix} i_d^p \\ i_q^p \\ i_d^n \\ i_q^n \end{bmatrix} \tag{6.39}$$

To reduce the complexity and computational burden of the 3LT²I in the dq rotating frame, the powers are expressed with the original grid voltage and currents and their 1/4 delayed values. The 1/4 delayed values of the grid voltage and currents are defined as

$$\begin{cases} e_{\alpha\beta}^m = e_{dq}^p e^{j(\omega t - \frac{\pi}{2})} + e_{dq}^n e^{-j(\omega t - \frac{\pi}{2})} \\ i_{\alpha\beta}^m = i_{dq}^p e^{j(\omega t - \frac{\pi}{2})} + i_{dq}^n e^{-j(\omega t - \frac{\pi}{2})} \end{cases} \tag{6.40}$$

Thus, the positive- and negative-sequence components of the grid voltage and grid current in the dq rotating frame can be derived from the original grid voltage and currents and its 1/4 delayed values, which are expressed as

6.2 Advanced Control Technology for DC Voltage and Grid Current of PV ...

$$\begin{bmatrix} x_{dq}^p \\ x_{dq}^n \end{bmatrix} = \frac{1}{2} \begin{bmatrix} e^{-j\omega t} & je^{-j\omega t} \\ e^{j\omega t} & -je^{j\omega t} \end{bmatrix} \begin{bmatrix} x \\ x^m \end{bmatrix} \quad (6.41)$$

where $x = e$ or i presents the grid voltage or currents. By expanding the above formula, it can be further expressed as

$$\begin{cases} x_d^p = \left(x_\alpha - x_\beta^m\right)\cos\omega t + \left(x_\alpha^m + x_\beta\right)\sin\omega t \\ x_q^p = \left(x_\beta + x_\alpha^m\right)\cos\omega t + \left(x_\beta^m - x_\alpha\right)\sin\omega t \\ x_d^n = \left(x_\alpha + x_\beta^m\right)\cos\omega t + \left(x_\alpha^m - x_\beta\right)\sin\omega t \\ x_q^n = \left(x_\beta - x_\alpha^m\right)\cos\omega t + \left(x_\beta^m + x_\alpha\right)\sin\omega t \end{cases} \quad (6.42)$$

Thus, the instantaneous active and reactive powers are re-derived as

$$\begin{cases} p_0 = \frac{3}{4}\left(i_\alpha e_\alpha + i_\beta e_\beta + i_\alpha^m e_\alpha^m + i_\beta^m e_\beta^m\right) \\ p_{s2} = \frac{3}{4}[-A_1\cos(2\omega t) + A_2\sin(2\omega t)] \\ p_{c2} = \frac{3}{4}[A_2\cos(2\omega t) + A_1\sin(2\omega t)] \\ q_0 = \frac{3}{4}\left(i_\alpha e_\beta - i_\beta e_\alpha + i_\alpha^m e_\beta^m - i_\beta^m e_\alpha^m\right) \\ q_{s2} = \frac{3}{4}[-B_1\cos(2\omega t) + B_2\sin(2\omega t)] \\ q_{c2} = \frac{3}{4}[B_2\cos(2\omega t) + B_1\sin(2\omega t)] \end{cases} \quad (6.43)$$

where the variables of A_1, A_2, B_1 and B_2 are expressed as

$$\begin{cases} A_1 = i_\alpha e_\alpha^m + i_\beta e_\beta^m + i_\alpha^m e_\alpha + i_\beta^m e_\beta \\ A_2 = i_\alpha e_\alpha + i_\beta e_\beta - i_\alpha^m e_\alpha^m - i_\beta^m e_\beta^m \\ B_1 = i_\alpha e_\beta^m - i_\beta e_\alpha^m + i_\alpha^m e_\beta - i_\beta^m e_\alpha \\ B_2 = i_\alpha e_\beta - i_\beta e_\alpha - i_\alpha^m e_\beta^m + i_\beta^m e_\alpha^m \end{cases} \quad (6.44)$$

It can be observed from the above expressions that the six power variables cannot be simultaneously controlled using only four positive- and negative-sequence active and reactive current components. Therefore, in the actual system, the controlled power should be flexibly selected according to the requirements of working conditions. In this section, two control objectives of active power oscillation elimination and reactive power oscillation elimination are selected, and the corresponding reference current calculation methods are given. In addition, when the grid voltages are balanced, the corresponding reference currents are also calculated.

Case 1: Active Power Oscillation Elimination

When the system requires high control performance for active power or dc-link bus voltage, the active power oscillations should be eliminated. At the same time, the 3LT²I needs to meet the requirement oscillations of reactive power in LVRT operation. In this case, the power variables must meet the following conditions:

$$\begin{cases} p_0 = \frac{3}{4}\left(i_\alpha e_\alpha + i_\beta e_\beta + i_\alpha^m e_\alpha^m + i_\beta^m e_\beta^m\right) = p_{ref} \\ p_{s2} = \frac{3}{4}[-A_1 \cos(2\omega t) + A_2 \sin(2\omega t)] = 0 \\ p_{c2} = \frac{3}{4}[A_2 \cos(2\omega t) + A_1 \sin(2\omega t)] = 0 \\ q_0 = \frac{3}{4}\left(i_\alpha e_\beta - i_\beta e_\alpha + i_\alpha^m e_\beta^m - i_\beta^m e_\alpha^m\right) = q_{ref} \end{cases} \quad (6.45)$$

The above equations are further expressed as

$$\begin{bmatrix} e_\alpha & e_\beta & e_\alpha^m & e_\beta^m \\ e_\alpha^m & e_\beta^m & e_\alpha & e_\beta \\ e_\alpha & e_\beta & -e_\alpha^m & -e_\beta^m \\ e_\beta & -e_\alpha & e_\beta^m & -e_\alpha^m \end{bmatrix} \begin{bmatrix} i_\alpha \\ i_\beta \\ i_\alpha^m \\ i_\beta^m \end{bmatrix} = \frac{4}{3} \begin{bmatrix} p_{ref} \\ 0 \\ 0 \\ q_{ref} \end{bmatrix} \quad (6.46)$$

Thus, the sinusoidal reference currents that satisfy the active power requirements are obtained as

$$\begin{bmatrix} i_{\alpha ref\_A} \\ i_{\beta ref\_A} \end{bmatrix} = \frac{2}{3} \begin{bmatrix} \dfrac{2e_\beta}{e_\alpha^2 + (e_\alpha^m)^2 + e_\beta^2 + (e_\beta^m)^2} & \dfrac{e_\beta^m}{e_\alpha e_\beta^m - e_\alpha^m e_\beta} \\ \dfrac{-2e_\alpha}{e_\alpha^2 + (e_\alpha^m)^2 + e_\beta^2 + (e_\beta^m)^2} & \dfrac{-e_\alpha^m}{e_\alpha e_\beta^m - e_\alpha^m e_\beta} \end{bmatrix} \begin{bmatrix} q_{ref} \\ p_{ref} \end{bmatrix} \quad (6.47)$$

Case 2: Reactive Power Oscillation Elimination

When the system requires high performance of reactive power, it is necessary to suppress reactive power oscillations. At the same time, the average transmitted active power should satisfy the standard requirements. In this case, the power variables need to meet the following conditions:

$$\begin{cases} p_0 = \frac{3}{4}\left(i_\alpha e_\alpha + i_\beta e_\beta + i_\alpha^m e_\alpha^m + i_\beta^m e_\beta^m\right) = p_{ref} \\ q_0 = \frac{3}{4}\left(i_\alpha e_\beta - i_\beta e_\alpha + i_\alpha^m e_\beta^m - i_\beta^m e_\alpha^m\right) = q_{ref} \\ q_{s2} = \frac{3}{4}[-B_1 \cos(2\omega t) + B_2 \sin(2\omega t)] = 0 \\ q_{c2} = \frac{3}{4}[B_2 \cos(2\omega t) + B_1 \sin(2\omega t)] = 0 \end{cases} \quad (6.48)$$

Similarly, the above equations are further expressed as

$$\begin{bmatrix} e_\alpha & e_\beta & e_\alpha^m & e_\beta^m \\ e_\beta^m & -e_\alpha^m & e_\beta & -e_\alpha \\ e_\beta & -e_\alpha & -e_\beta^m & e_\alpha^m \\ e_\beta & -e_\alpha & e_\beta^m & -e_\alpha^m \end{bmatrix} \begin{bmatrix} i_\alpha \\ i_\beta \\ i_\alpha^m \\ i_\beta^m \end{bmatrix} = \frac{4}{3} \begin{bmatrix} p_{ref} \\ 0 \\ 0 \\ q_{ref} \end{bmatrix} \quad (6.49)$$

Thus, the sinusoidal reference currents that satisfy the reactive power requirements are obtained as

6.2 Advanced Control Technology for DC Voltage and Grid Current of PV ...

$$\begin{bmatrix} i_{\alpha\text{ref\_R}} \\ i_{\beta\text{ref\_R}} \end{bmatrix} = \frac{2}{3} \begin{bmatrix} \dfrac{-e_\alpha^m}{e_\alpha{}^2+(e_\alpha^m)^2+e_\beta{}^2+(e_\beta^m)^2} & \dfrac{2e_\alpha}{e_\alpha e_\beta^m - e_\alpha^m e_\beta} \\ \dfrac{-e_\beta^m}{e_\alpha{}^2+(e_\alpha^m)^2+e_\beta{}^2+(e_\beta^m)^2} & \dfrac{2e_\beta}{e_\alpha e_\beta^m - e_\alpha^m e_\beta} \end{bmatrix} \begin{bmatrix} q_{\text{ref}} \\ p_{\text{ref}} \end{bmatrix} \quad (6.50)$$

It can be seen that the calculation process avoids positive- and negative-sequence separation and PLL calculation. Thus, the control complexity and calculation burden are effectively reduced. When the actual grid current tracks the reference currents, the active power oscillations or reactive power oscillations are eliminated and sinusoidal grid currents are simultaneously achieved.

Case 3: Grid Voltages Are Balanced

When the grid voltages are balanced, the reference currents are calculated as

$$\begin{bmatrix} i_{\alpha\text{ref\_B}} \\ i_{\beta\text{ref\_B}} \end{bmatrix} = \frac{2}{3(e_\alpha{}^2 + e_\beta{}^2)} \begin{bmatrix} e_\alpha & e_\beta \\ e_\beta & -e_\alpha \end{bmatrix} \begin{bmatrix} q_{\text{ref}} \\ p_{\text{ref}} \end{bmatrix} \quad (6.51)$$

If a proper current controller is used to make the current track the above calculated reference current according to the requirements of working conditions.

Part 2: Implementation Process of GMPC

For the $3\text{LT}^2\text{I}$, the switching function S_j ($j = \text{a, b, c}$) is defined as

$$S_j = \begin{cases} 1 & S_{j1} = S_{j2} = 1, S_{j3} = S_{j4} = 0 \\ 0 & S_{j2} = S_{j3} = 1, S_{j1} = S_{j4} = 0 \\ -1 & S_{j3} = S_{j4} = 1, S_{j1} = S_{j2} = 0 \end{cases} \quad (6.52)$$

The output states of each phase corresponding to $S_j = 1, 0, -1$ are [P], [O], and [N].

To apply the proposed GMPC scheme, the model of $3\text{LT}^2\text{I}$ is constructed firstly. The average model of $3\text{LT}^2\text{I}$ in the stationary frame is obtained based on Kirchhoff's voltage law.

$$\begin{cases} e_\text{a} = L\dfrac{di_\text{a}}{dt} + Ri_\text{a} + u_{\text{AO}} + u_{\text{on}} \\ e_\text{b} = L\dfrac{di_\text{b}}{dt} + Ri_\text{b} + u_{\text{BO}} + u_{\text{on}} \\ e_\text{c} = L\dfrac{di_\text{c}}{dt} + Ri_\text{c} + u_{\text{CO}} + u_{\text{on}} \end{cases} \quad (6.53)$$

The model of the $3\text{LT}^2\text{I}$ in the $\alpha\beta$ orthogonal coordinates frame is derived by using the Clarke transformation, and it is expressed as

$$\begin{cases} e_\alpha = L\dfrac{di_\alpha}{dt} + Ri_\alpha + u_\alpha \\ e_\beta = L\dfrac{di_\beta}{dt} + Ri_\beta + u_\beta \end{cases} \quad (6.54)$$

The MPC control algorithm is designed in the discrete-time domain. Thus, the discrete mathematical model is derived based on the forward Euler approximation, which is expressed as

$$\begin{bmatrix} i_\alpha(k+1) \\ i_\beta(k+1) \end{bmatrix} = \frac{T_s}{L} \begin{bmatrix} e_\alpha(k) - u_\alpha(k) \\ e_\beta(k) - u_\beta(k) \end{bmatrix} + \frac{L - T_s R}{L} \begin{bmatrix} i_\alpha(k) \\ i_\beta(k) \end{bmatrix} \quad (6.55)$$

where T_s is the sampling time, $i_\alpha(k), i_\beta(k), u_\alpha(k), u_\beta(k), e_\alpha(k), e_\beta(k)$ are $\alpha\beta$ components of output currents, voltages and grid voltage at kth instant. $i_\alpha(k+1), i_\beta(k+1)$ are $\alpha\beta$ components of predictive current value at $(k+1)$th instant. As the above equations show that the currents at the time $(k+1)$th instant can be controlled to track the reference current by controlling the output voltage of $3LT^2I$ at time kth instant.

However, in real-time implementation, the time delay caused by measurements, analogue-to-digital conversion and algorithm calculations is unavoidable. This time delay will not only increase the control error, but also cause the PWM pulse confusion and reduce the system stability. In order to improve the accuracy of current control and reduce the risk of system instability, the time delay of digital controller must be compensated.

Figure 6.17 gives the execution process of the MPC control algorithm. The ideal MPC considers that measurements, analogue-to-digital conversion, algorithm calculations and output of switching control signals are implemented instantaneously at the execution instant. Thus, there is no time delay in idea MPC. However, in practice, the delay of the digital controller is unavoidable, and the execution process of actual MPC includes control variable calculation time and action time. The control variable calculation time is defined as t_{cal}, in which the digital controller completes the process of the measurements, analogue-to-digital conversion and algorithm calculations. Besides, during t_{cal} period, the control variable in use remains the control variable of the previous cycle. For example, in the kth control period, the time consumed for measurements, analogue-to-digital conversion and algorithm calculations is t_{cal}. After that, the control variable y_k is obtained. During t_{cal} period, the control variable remains the control variable of y_{k-1}, which results in the action time of y_k is only $T_s - t_{cal}$. Obviously, the time delay will produce control errors and reduce the control accuracy. In order to mitigate the adverse effects of time delay, the one-step prediction method is adopted to compensate the time delay. In MPC with time delay compensation method, the y_k is obtained in $(k-1)$th control period, and maintained unchanged until it is output at kth instant. After the delay compensation algorithm, the action time of y_k is unchanged, and the action time remains T_s.

The discrete mathematical model with shifted one-step forward is written as

$$\begin{bmatrix} i_\alpha(k+2) \\ i_\beta(k+2) \end{bmatrix} = \frac{T_s}{L} \begin{bmatrix} e_\alpha(k+1) - u_\alpha(k+1) \\ e_\beta(k+1) - u_\beta(k+1) \end{bmatrix} + \frac{L - T_s R}{L} \begin{bmatrix} i_\alpha(k+1) \\ i_\beta(k+1) \end{bmatrix} \quad (6.56)$$

where the $e_{\alpha\beta}(k+1)$ can be obtained from Lagrange extrapolation formula

$$e_{\alpha\beta}(k+1) = 3e_{\alpha\beta}(k) - 3e_{\alpha\beta}(k-1) + e_{\alpha\beta}(k-2) \quad (6.57)$$

The optimal voltage vector $u_{\alpha\beta}(k+1)$ is selected to minimize errors between the two-step future reference $i_{\alpha\beta\text{-ref}}(k+2)$ and $i_{\alpha\beta}(k+2)$ in one sampling period. Meanwhile,

6.2 Advanced Control Technology for DC Voltage and Grid Current of PV ...

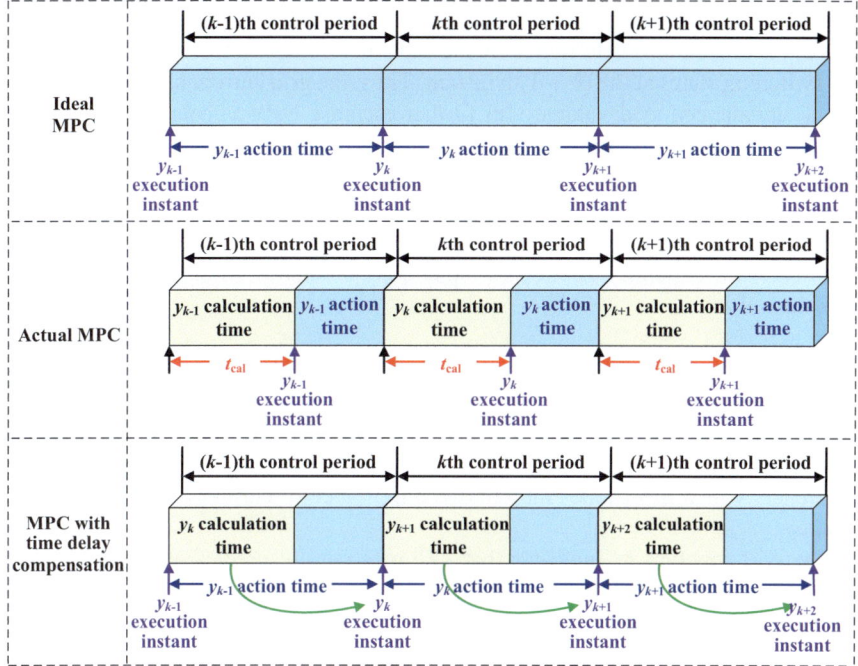

Fig. 6.17 Execution process of the MPC control algorithm

the discrete model of dc-link capacitor voltages with time delay compensation is derived as

$$\begin{bmatrix} U_p(k+2) \\ U_n(k+2) \end{bmatrix} = \frac{T_s}{C} \begin{bmatrix} i_p(k+1) \\ i_n(k+1) \end{bmatrix} - \begin{bmatrix} U_p(k+1) \\ U_n(k+1) \end{bmatrix} \quad (6.58)$$

where $i_p(k+1)$ and $i_n(k+1)$ are currents through upper and lower dc-link capacitors at $(k+1)$th instant. Consequently, the predictive voltage difference $\Delta U_{np}(k+2) = U_p(k+2) - U_n(k+2)$ is derived as

$$\Delta U_{np}(k+2) = \Delta U_{np}(k+1) + \frac{T_s}{C} i_{np}(k+1) \quad (6.59)$$

where $i_{np}(k+1) = i_p(k+1) - i_n(k+1)$ is the current flowing into NP. It is expressed as

$$i_{np}(k+1) = \sum_{j=a,b,c}^{c} [1 - S_j(k+1)][1 + S_j(k+1)] i_j(k+1) \quad (6.60)$$

According to the above expressions, the predictive capacitor voltage difference at the $(k + 2)$th instant is determined by the capacitor voltage difference, grid currents and switching states at the $(k + 1)$th instant. Thus, the grid current harmonics caused by the unbalanced grid voltage will further increase NPV unbalance. In order to eliminate grid currents harmonics under unbalanced grid voltage, the actual current should be controlled to track the sinusoidal reference currents derived in Part 1.

Multiple objectives are incorporated into MPC controller using the pre-defined cost function to achieve high-quality output performance. For the 3LT^{2}I, the control objectives are reference current tracking and NPV balance. Thus, the cost function is expressed as

$$g_1(k) = \left| i_{\alpha\beta}(k+2) - i_{\alpha\beta\text{-ref}}(k+2) \right| + \lambda \left| \Delta U_{np}(k+2) \right| \tag{6.61}$$

The reference currents $i_{\alpha\beta\text{-ref}}(k + 2)$ are calculated from Part 1 using the two-step Lagrange extrapolation formula. λ is the weighting factor associate with NPV difference, which determines the relative importance of current tracking and NPV balance.

The cost function g_1 indicates that 27 times current predictive model calculations, 27 times capacitor voltage predictive model calculations and 27 times cost function evaluations are required to select the optimal voltage vector. The process has cumbersome calculations and is extremely time-consuming. Thus, the real-time implementation is limited in practice.

To reduce the computation complexity and execution time, the deadbeat-like dynamic model prediction is used to achieve grid currents tracking. The redundant small vectors are used to balance NPV. Thus, the cost function is written as

$$g_2(k) = \left| u_{\alpha\beta}(k+2) - u_{\alpha\beta\text{-ref}}(k+2) \right| \tag{6.62}$$

where $u_{\alpha\beta\text{-ref}}(k + 2)$ are the reference voltage vectors. They are obtained by the deadbeat-like method with $i_{\alpha\beta}(k + 2) = i_{\alpha\beta\text{-ref}}(k + 2)$, and are expressed as

$$\begin{bmatrix} u_{\alpha-\text{ref}}(k+1) \\ u_{\beta-\text{ref}}(k+1) \end{bmatrix} = \begin{bmatrix} e_\alpha(k+1) \\ e_\beta(k+1) \end{bmatrix} - \begin{bmatrix} i_\alpha(k+1) \\ i_\beta(k+1) \end{bmatrix} - \frac{T_s}{L} \begin{bmatrix} i_{\alpha-\text{ref}}(k+2) - i_\alpha(k+1) \\ i_{\beta-\text{ref}}(k+2) - i_\beta(k+1) \end{bmatrix} \tag{6.63}$$

This equation shows that the grid currents tracking is realized if the selected optimal voltage vector follows the reference voltage vector. With this MPC method, the computational burden of predictive model is reduced.

When the inverter operates at high power factor, and the effects of small vectors on NPV are determinate. In this case, the redundant small vectors are selected based on the voltage difference of dc-link capacitors. However, the NPV unbalance increases as the power factor decreases. Especially for low power factor operation, the large NPV unbalance cannot be mitigated using the conventional MPC method. To overcome this issue, under an unbalanced grid voltage, both the voltage difference of

6.2 Advanced Control Technology for DC Voltage and Grid Current of PV ...   333

dc-link capacitors and the grid current directions are considered when selecting the appropriate small vector in the proposed GMPC scheme. The process of selecting the appropriate small vector for NPV balance control will be elaborated in following section, and will not be introduced any more in here.

In the proposed GMPC, to further reduce cost function evaluation times, the triangle candidate region, which is the subset of 27 space voltage vectors, is presented to select the optimal voltage vector. The triangle candidate regions are determined by large and medium voltage vectors as shown in Fig. 6.18. Consequently, only five space voltage vectors are included in one triangle candidate region. Moreover, when one of appropriate redundant small vectors is selected to control NPV balance, only four vectors are included in cost function evaluation. So, the computational burden is further reduced.

In order to elaborate the process of GMPC, the flow chart of the proposed algorithm is shown in Fig. 6.19. The reference of active power and reactive power are obtained from the outer dc-link voltage loop. Unlike the conventional MPC method, both the dc-link capacitor voltage difference and grid current direction are considered to select the small vector, so as to control NPV balance.

The procedure is presented in detail as follows:

1. The sinusoidal reference currents are calculated under unbalanced grid voltage according to the requirements of working conditions.
2. The reference voltage vector is obtained based on the discrete-time dynamic model using the deadbeat-like method.

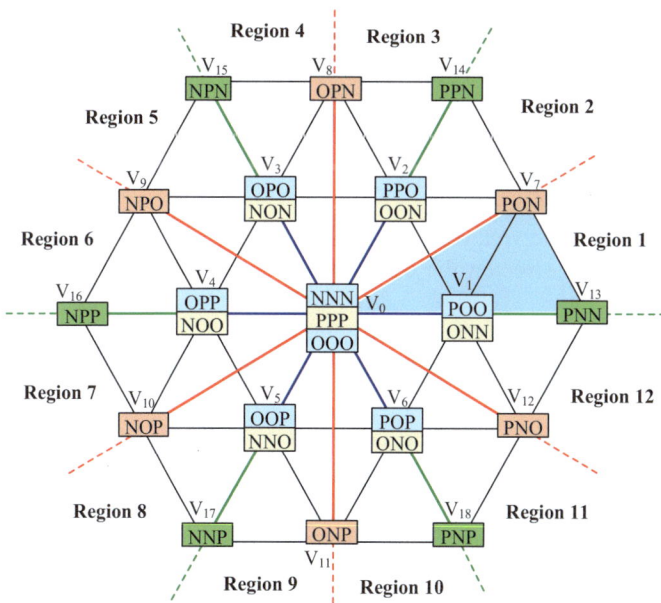

Fig. 6.18 The triangle candidate regions in SVD

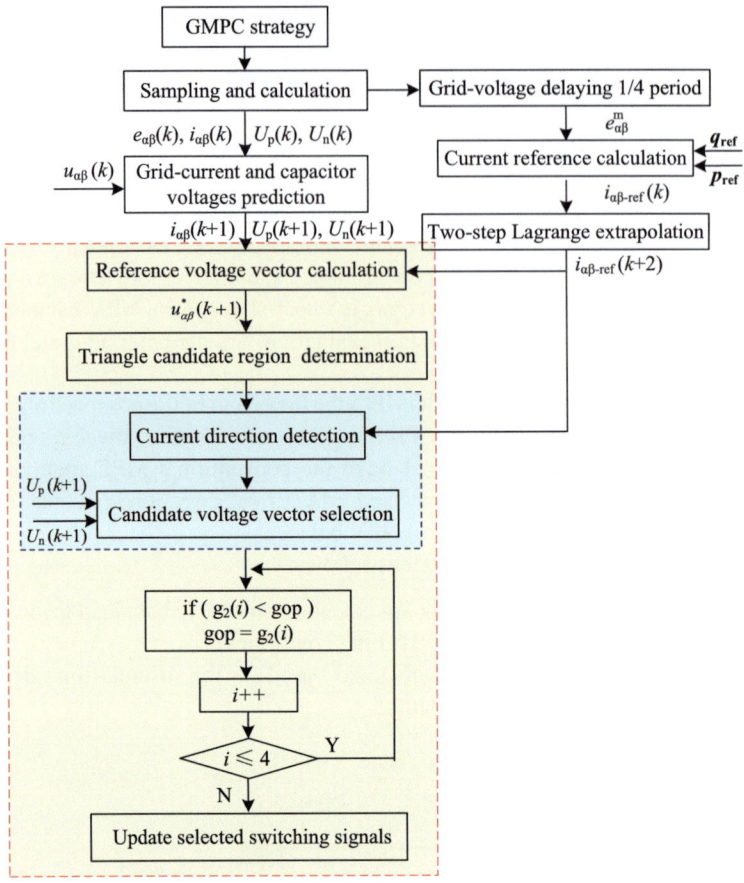

Fig. 6.19 Flow chart of the proposed GMPC

3. The four candidate voltage vectors are selected according to the location of the reference voltage vector, the dc-link capacitor voltage difference and the grid current direction.
4. The optimal switching vector, which minimizes the pre-defined cost function is selected and applied to achieve the optimal control objectives.

Part 3: Control Performance of GMPC

To verify the effectiveness of the proposed GMPC scheme, experimental tests have been conducted. Two control methods are compared: Method-1 is the proposed GMPC and Method-2 is the reference derived based on the balanced grid voltage. Two cases are tested in experimental. The results are depicted as follows.

Case 1: Power Factor Angle is 0°

When the power factor angle is 0°, the dc-link load resistor is 25 Ω, and the reference reactive power is set to 0 kvar. The results are shown in Fig. 6.20. The grid voltages are unbalanced, and the root mean square (RMS) values of three phase grid voltages are 77 V, 110 V, 110 V, respectively. The RMS voltage of phase A dips by 30%, which is different from that of phases B and C. Since Method-1 is designed based on the balanced grid voltage, it cannot handle the unbalanced grid voltage conditions. As a result, the grid currents are highly distorted, and the THD reaches 12.56%. After adopting Method-3, the grid current distortion is effectively suppressed, and the grid current THD is reduced to 3.25%. In this condition, the negative-sequence currents appear, resulting in different peak values in the grid currents. When the inverter operates at a 0° power factor angle, the grid current tracks the reference current and remains in phase with the grid voltage.

Case 2: Power Factor Angle is 90°

When the inverter operates at a 90° power factor angle, the load resistor in the dc-link is removed, and the reference reactive power is set to 5 kvar. The experimental results

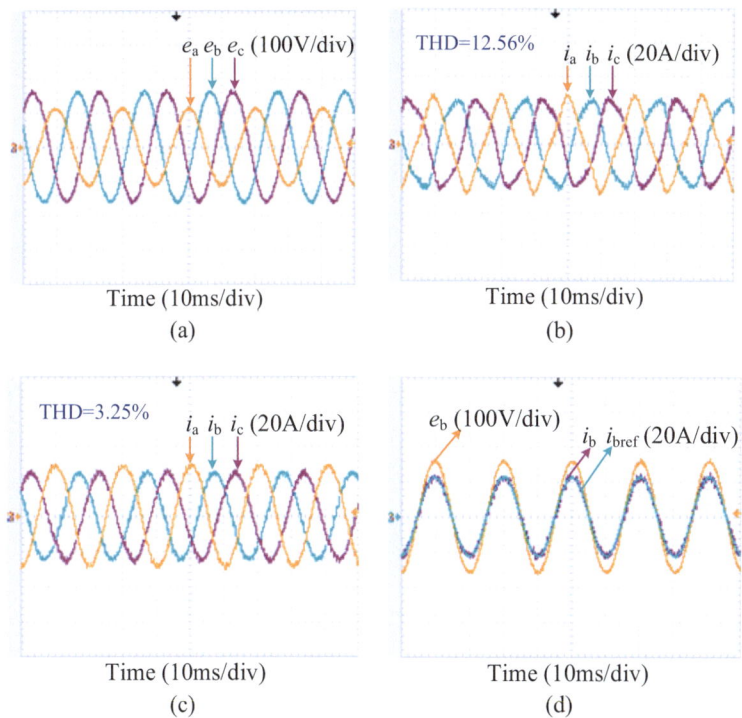

Fig. 6.20 Experimental waveforms when the power factor angle is 0° **a** Unbalanced grid voltage. **b** Grid currents with Method-1. **c** Grid currents with the proposed Method-2. **d** Grid voltage, grid current and reference current

are shown in Fig. 6.21. When Method-1 is used, the harmonic and negative-sequence components are generated in grid currents, and the current THD reaches 12.41%. After Method-2 is adopted, since the NPV balance is maintained and the reference current is sinusoidal, the THD is reduced to 3.25%. Moreover, the grid current tracks the reference current, and the phase difference with the grid voltage is approximately 90°, which confirm the effectiveness of GMPC in current tracking performance.

To summarize, according to the above-mentioned analysis, the proposed GMPC scheme achieves accurate current tracking and NPV balance simultaneously during full power factor operation. In addition, only one reference voltage vector calculation and four cost function evaluation are required, which greatly simplifies the computational burden and reduces the execution time.

In this section, the dual-loop voltage and current control scheme to control dc-link voltage and current tracking are presented. Regarding the voltage loop, the control methods are relatively mature and thus are not introduced. As for the current loop, three methods are mainly elaborated.

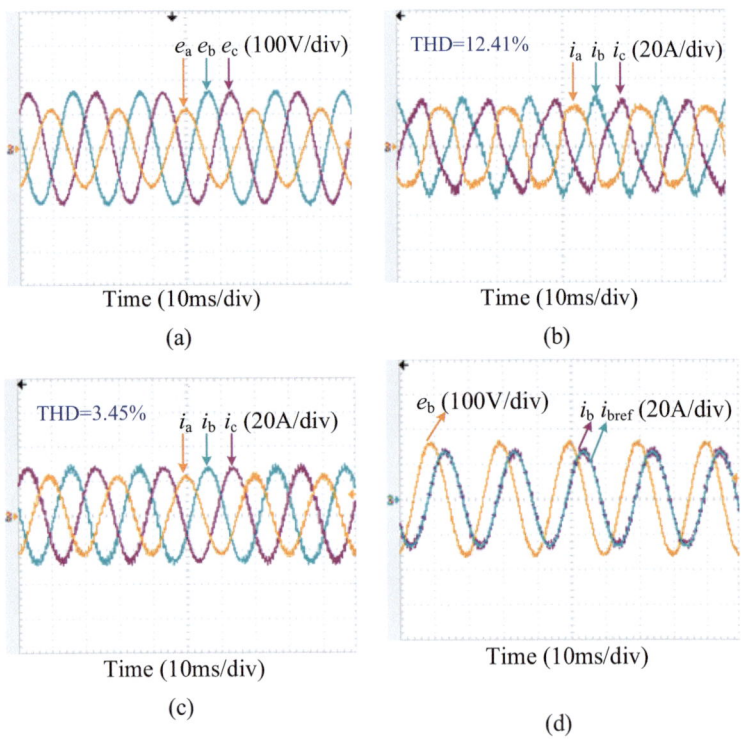

Fig. 6.21 Experimental waveforms when the power factor angle is 90° **a** Unbalanced grid voltage. **b** Grid currents with Method-1. **c** Grid currents with the proposed Method-2. **d** Grid voltage, grid current and reference current

6.2.2 Dual-Loop Voltage and Power Control Schemes

Although dual-loop voltage and current control scheme is convenient for engineering implementation. It should be noted that the control performance relies on the quality of the current loop and the bandwidth of the PLL. Thus, the dual-loop voltage and power control scheme is presented. In this scheme, the voltage loop is used to maintain the dc-link voltage stable, and the power loop is used to control transmitted power to reach its reference value. Unlike the dual-loop voltage and current control scheme, the active and reactive powers are independently regulated without the current control loop and the PLL in this approach. The author has conducted extensive researches in this area. Among these researches, two advanced control methods are presented as typical example to illustrate the theoretical analysis and implementation process of this control scheme. The two advanced control methods consist of the finite-time command filtered control (FTCFC) method [15] and the improved passivity-based direct power control (IPBDPC) method [16].

Finite-Time Command Filtered Control Method

For a three-phase inverter, unbalanced grid voltages cause grid current distortions and dc-link voltage fluctuations, threatening the safe operation of the power grid or loads. To address this issue, the control objectives of the inverter under unbalanced grid conditions are transformed into a unified power tracking problem by introducing modified power references. Then, the FTCFC is proposed in this part.

The 3LT^{2}I operating in charging mode is taken as an example to illustrate the principle of FTCFC. Three parts are included in this section: ac side and dc side modeling and analysis, design process of FTCFC and stability analysis.

Part 1: AC Side and DC Side Modeling and Analysis
The following system model is obtained under several assumptions, such as ignoring switching losses and grid resonances. In order to facilitate the theoretical analysis.

1. **AC side modeling**

    By using the Kirchhoff voltage law (KVL) and the Clark transformation, the following equations hold:

$$\begin{cases} L\dot{i}_\alpha = e_\alpha - Ri_\alpha - u_\alpha \\ L\dot{i}_\beta = e_\beta - Ri_\beta - u_\beta \end{cases} \tag{6.64}$$

    where $e_{\alpha,\beta}$, $i_{\alpha,\beta}$ and $u_{\alpha,\beta}$ represent three-phase grid voltages, grid currents, and the output voltage of a inverter in the $\alpha\beta$ frame. Considering unbalanced grid conditions, the grid voltages and currents consist of both positive-sequence and negative-sequence components. For the uniform notation, the superscripts $+$ and $-$ represent positive-sequence and negative-sequence components, respectively. Then it is analyzed and obtained that

$$e_\alpha = e_\alpha^+ + e_\alpha^-, \quad e_\beta = e_\beta^+ + e_\beta^- \qquad (6.65)$$

The positive-sequence and negative-sequence voltages can be are expressed as

$$\begin{cases} e_\alpha^+ = |E^+|\cos(\omega_1 t), \ e_\beta^+ = |E^+|\sin(\omega_1 t) \\ e_\alpha^- = |E^-|\cos(\omega_1 t), \ e_\beta^- = |E^-|\sin(-\omega_1 t) \end{cases} \qquad (6.66)$$

where ω_1 is the grid angle frequency, and $|E^+|$ and $|E^-|$ are the amplitudes of the positive-sequence and negative-sequence grid voltages, respectively. As described in the above expression, $e_{\alpha,\beta}^+$ and $e_{\alpha,\beta}^-$ satisfy the following dynamic characteristics

$$\begin{cases} \dot{e}_\alpha^+ = -\omega_1 e_\beta^+, \ \dot{e}_\beta^+ = \omega_1 e_\alpha^+ \\ \dot{e}_\alpha^- = \omega_1 e_\beta^-, \ \dot{e}_\beta^- = -\omega_1 e_\alpha^- \end{cases} \qquad (6.67)$$

Based on the instantaneous power theory, the active power P and reactive power Q are calculated by

$$\begin{cases} P = \frac{3}{2} e_\alpha i_\alpha + \frac{3}{2} e_\beta i_\beta \\ Q = \frac{3}{2} e_\beta i_\alpha - \frac{3}{2} e_\alpha i_\beta \end{cases} \qquad (6.68)$$

Based on the above analysis and considering the effects of disturbances, the model of the AC-side subsystem is built as

$$\dot{P} = -\frac{3}{2L} u_P - \frac{1}{L} RP + \frac{3}{2L}(e_\alpha^2 + e_\beta^2) - \omega_1 Q + 3\omega_1 \left(e_\beta^- i_\alpha - e_\alpha^- i_\beta\right) + d_1 \quad (6.69)$$

$$\dot{Q} = \frac{3}{2L} u_Q - \frac{1}{L} RQ + \omega_1 P - 3\omega_s \left(e_\alpha^- i_\alpha + e_\beta^- i_\beta\right) + d_2 \qquad (6.70)$$

where d_1 and d_2 are considered as the lumped disturbances caused by system delays, model uncertainties, etc., which are bounded in practice. The new control inputs are

$$\begin{bmatrix} u_P \\ u_Q \end{bmatrix} = \begin{bmatrix} e_\alpha & e_\beta \\ -e_\beta & e_\alpha \end{bmatrix} \begin{bmatrix} u_\alpha \\ u_\beta \end{bmatrix} \qquad (6.71)$$

The input signals in the above expression are designed to decouple the original control inputs u_α and u_β, as well as to design independent control laws for active and reactive power regulation.

2. **DC side modeling**

   Assuming the resistor R_L is connected on the DC side to simulate the consumed power in charging mode. Additionally, to simplify the analysis, the conditions of $C_1 = C_2 = C$ and NPV balance are satisfied.

6.2 Advanced Control Technology for DC Voltage and Grid Current of PV …

According to the power conservation law, by neglecting losses, the dc-link capacitor power variation is expressed as

$$CU_{dc}\dot{U}_{dc} = P - U_{dc}\frac{U_{dc}}{R_L} \tag{6.72}$$

Then the model of dc-link subsystem is described as

$$\begin{cases} \dot{U}_{dc} = f_1(U_{dc}) + \frac{P}{CU_{dc}} \\ f_1(U_{dc}) = -\frac{U_{dc}}{CR_L} \end{cases} \tag{6.73}$$

where $f_1(U_{dc})$ is taken as the unknown disturbance caused by load variations.

Combining the AC-side model and DC-side model, the resulting dynamic model is a typical nonlinear system with external disturbances. This model accounts for unbalanced grid conditions and external disturbances, making it more suitable for practical applications.

Part 2: Design Process of FTCFC

In this section, the FTCFC method is designed for both the power tracking loop in AC side and dc-link voltage tracking loop in DC side.

1. **Power tracking loop in AC side**

   The control objective is to regulate active and reactive power to track modified power references P_{new}^{ref} and Q_{new}^{ref} respectively, i.e.,

   $$P \to P_{new}^{ref}, \quad Q \to Q_{new}^{ref} \tag{6.74}$$

   in which

   $$\begin{cases} P_{new}^{ref} = P_0^{ref} + \frac{3}{2}\varsigma\left(e_\alpha^- i_\alpha^+ + e_\beta^- i_\beta^+\right) \\ Q_{new}^{ref} = Q_0^{ref} + \frac{3}{2}(2-\varsigma)\left(e_\beta^- i_\alpha^+ - e_\alpha^- i_\beta^+\right) \end{cases} \tag{6.75}$$

   where the original power references P_0^{ref} and Q_0^{ref} are dc signals. The selection of the coordination coefficient ς depends practical requirements.

   The following three control modes are discussed:

   Mode 1: $\varsigma = 0$, the objective is to obtain sinusoidal grid currents, maintain a constant dc-link voltage, and remove active power ripples. However, there exist reactive power ripples.

   Mode 2: $\varsigma = 1$, the objective is to obtain sinusoidal and balanced grid currents at the cost of active power ripples, reactive power ripples, and dc-link voltage ripples.

   Mode 3: $\varsigma = 2$, the objective is to obtain sinusoidal grid currents and eliminate reactive power ripples. However, there exist active power ripples and dc-link voltage ripples.

Modes 1 and Mode 3 achieve sinusoidal currents under unbalanced grid conditions after the realization of power tracking. The difference from Mode 2 is that the grid currents in Mode 1 and Mode 3 are unbalanced. It should be noted that the modified active and reactive power references area time-varying signals with twice the fundamental frequency.

Figure 6.22 shows the control diagram for the power tracking loop. The objective is to ensure $P \to P_{new}^{ref}$ and $Q \to Q_{new}^{ref}$, where the modified references are calculated based on above equations. Specifically, the positive- and negative-sequence grid voltages and currents are obtained using the $T/4$ delay technique. The command filter is used to estimate the derivative of power reference. Meanwhile, the finite-time control laws are constructed to achieve fast and accurate power tracking performance.

Define the power tracking error as

$$\begin{cases} z_P = P - P_{new}^{ref} \\ z_Q = Q - Q_{new}^{ref} \end{cases} \quad (6.76)$$

Then, the control objective is transformed to ensure $z_P \to 0$ and $z_Q \to 0$. Choose the Lyapunov function V_1 for the power tracking loop as

$$V_1 = \frac{1}{2}z_P^2 + \frac{1}{2}z_Q^2 \quad (6.77)$$

Based on the dynamic model and error variables, taking the derivative of V_1 yields

$$\begin{aligned}
\dot{V}_1 &= z_P \dot{z}_P + z_Q \dot{z}_Q \\
&= z_P\left(\dot{P} - \dot{P}_{new}^{ref}\right) + z_Q\left(\dot{Q} - \dot{Q}_{new}^{ref}\right) \\
&= z_P\left[-\frac{3}{2L}u_P + f_P + 3\omega_1\left(e_\beta^- i_\alpha - e_\alpha^- i_\beta\right) + d_1 - \dot{P}_{new}^{ref}\right] \\
&\quad + z_Q\left[\frac{3}{2L}u_Q + f_Q - 3\omega_1\left(e_\alpha^- i_\alpha + e_\beta^- i_\beta\right) + d_2 - \dot{Q}_{new}^{ref}\right]
\end{aligned} \quad (6.78)$$

where $f_P = -\frac{1}{L}RP + \frac{3}{2L}\left(e_\alpha^2 + e_\beta^2\right) - \omega_1 Q$, $f_Q = -\frac{1}{L}RQ + \omega_1 P$.

Based on above equation, the derivatives of active and reactive power references are required for constructing the control laws. The direct numerical difference method is sensitive to noise, and the analytical calculation method is excessively complicated. Thus, we introduce the following command filter to estimate and smooth the derivatives of power references.

6.2 Advanced Control Technology for DC Voltage and Grid Current of PV ...

$$\begin{cases} \dot{z}_0 = \upsilon \\ \dot{\upsilon} = -\tau_1 \left[\text{sig}(\sigma_0)^{\frac{1}{2}} + \text{sig}(\sigma_0)^{\frac{3}{2}} \right] + z_1 \\ \dot{z}_1 = -\tau_2 \left[\frac{1}{2|\sigma_0|} \text{sig}(\sigma_0) + 2\sigma_0 + \frac{3}{2} \text{sig}(\sigma_0)^2 \right] \end{cases} \quad (6.79)$$

where $\sigma_0 = z_0 - \rho_0$, $\text{sig}(\sigma_0)^x = |\sigma_0|^x \, \text{sgn}(\sigma_0)$, ρ_0 is the command filter's input signal, and τ_1 and τ_2 are filter parameters.

Based on [17], the filter errors

$$|z_0 - \rho_0| \leq \mathcal{O}_1, \quad |\upsilon - \dot{\rho}_0| \leq \mathcal{O}_2 \quad (6.80)$$

within a fixed-time transient process t_f, where $\mathcal{O}_{1,2}$ are bounded and arbitrarily small by selecting appropriate filter parameters. Apparently, the command filter is able to precisely estimate ρ_0 and $\dot{\rho}_0$. If we set the filter input $\rho_0 = \left[P_{\text{new}}^{\text{ref}}, Q_{\text{new}}^{\text{ref}} \right]^T$, $\left[\hat{P}_{\text{new}}^{\text{ref}}, \hat{Q}_{\text{new}}^{\text{ref}} \right]^T = \upsilon$.

In order to satisfy the fast finite-time practical stability criterion in [18], we construct the finite-time control laws as

$$\begin{cases} u_P = -\frac{2L}{3} \left[-k_{11} z_P - k_{12} z_P^{\gamma} - f_P - 3\omega_1 \left(e_\beta^- i_\alpha - e_\alpha^- i_\beta \right) + \hat{P}_{\text{new}}^{\text{ref}} \right] \\ u_Q = \frac{2L}{3} \left[-k_{21} z_Q - k_{22} z_Q^{\gamma} - f_Q + 3\omega_1 \left(e_\alpha^- i_\alpha + e_\beta^- i_\beta \right) + \hat{Q}_{\text{new}}^{\text{ref}} \right] \end{cases} \quad (6.81)$$

where k_{ij} are control parameters with $i, j = 1, 2$ and $\gamma = 3/5$. The selection guidelines of k_{ij} are discussed in the following part. According to the above expressions, $\dot{V}_1$ is rewritten as

$$\dot{V}_1 = -k_{11} z_P^2 - k_{21} z_Q^2 - k_{12} z_P^{\gamma+1} - k_{22} z_Q^{\gamma+1} \\ + z_P \mathcal{O}_2 + z_Q \mathcal{O}_2 + z_P d_1 + z_Q d_2 \quad (6.82)$$

The proposed finite-time control law guarantees fast and accurate power tracking control.

2. **DC-link voltage tracking loop in DC side**

   In mode 1, it is expected to maintain the dc-link voltage at its reference value. Note that, the effects of load disturbances cannot be ignored and may cause significant fluctuations in the dc-link voltage.

   In conventional dc-link voltage control methods, dc-link currents are measured and then fed forward compensation to achieve good control performance. However, these additional sensors are undesirable from the perspectives of reliability and cost. Thus, the author constructs a fixed-time disturbance observer to estimate the unknown load disturbances. The estimated results are then incorporated into the finite-time control law, ensuring dc-link voltage control performance while avoiding the use of dc-link current sensors. The control structure of the dc-link voltage tracking loop is shown in Fig. 6.23.

First, a disturbance estimation process is given in this section. Construct a nominal dc-link model as

$$\dot{\hat{U}}_{dc} = \frac{P}{CU_{dc}} \tag{6.83}$$

According to dc-link model, it is naturally obtained that

$$f_1 = \dot{U}_{dc} - \dot{\hat{U}}_{dc} = \dot{\tilde{U}}_{dc} \tag{6.84}$$

where $\tilde{U}_{dc} = U_{dc} - \hat{U}_{dc}$, the problem of disturbance estimation is transformed into the calculation of $\dot{\tilde{U}}_{dc}$. According to the command filter, the output signal $\dot{\hat{\tilde{U}}}_{dc}$ is generated with $\tilde{U}_{dc}$ as input signal. The estimation error is expressed as

$$\left|\hat{f}_1 - f_1\right| = \left|\dot{\hat{\tilde{U}}}_{dc} - \dot{\tilde{U}}_{dc}\right| \leq \mathcal{O}_{dc} \tag{6.85}$$

within a fixed time t_f. The low-pass filter in dc-link voltage tracking loop helps to attenuate high-frequency noise. Next, the finite-time control law is constructed using the disturbance estimation $\hat{f}_1$. Define the dc-link voltage tracking error as

$$z_{dc} = U_{dc} - U_{dc}^{ref} \tag{6.86}$$

where U_{dc}^{ref} is the dc-link voltage reference. Similar to the power tracking loop, the Lyapunov function is chosen as

$$V_2 = \frac{1}{2}z_{dc}^2 \tag{6.87}$$

For a three-phase inverter with power flowing from AC side to DC side, the power dynamics are typically configured to be faster than the dc-link voltage dynamics. Based on the singular perturbation theory, the dc-link model in mode 1 can be expressed as

$$\dot{U}_{dc} = f_1(U_{dc}) + \frac{P_0^{ref}}{CU_{dc}} \tag{6.88}$$

Taking the derivative of V_2 yields

$$\dot{V}_2 = z_{dc}\dot{z}_{dc} = z_{dc}\left[f_1(U_{dc}) + \frac{1}{CU_{dc}}P_0^{ref}\right] \tag{6.89}$$

Similarly, the finite-time control law for the dc-link voltage tracking loop is construct as

6.2 Advanced Control Technology for DC Voltage and Grid Current of PV …

$$P_0^{\text{ref}} = CU_{\text{dc}}\left(-k_{31}z_{\text{dc}} - k_{32}z_{\text{dc}}^{\gamma} - \hat{f}_1\right) \tag{6.90}$$

where k_{31} and k_{32} are the control parameters. Substituting the above finite-time control law into $\dot{V}_2$, derivative of V_2 can be simplified as

$$\dot{V}_2 = -k_{31}z_{\text{dc}}^2 - k_{32}z_{\text{dc}}^{\gamma+1} + z_{\text{dc}}O_{\text{dc}} \tag{6.91}$$

The finite-time control law P_0^{ref} maintains the dc-link voltage at its reference value even under load disturbances.

Part 3: Stability Analysis

To demonstrate the effectiveness of finite-time control laws for the power tracking loop on the AC side and dc-link voltage tracking loop in DC side, the Lyapunov function of the closed-loop control system is chosen as

$$V = V_1 + V_2 = \frac{1}{2}z_P^2 + \frac{1}{2}z_Q^2 + \frac{1}{2}z_{\text{dc}}^2 \tag{6.92}$$

Using the expression of $\dot{V}_1$ and $\dot{V}_2$, $\dot{V}$ is given by

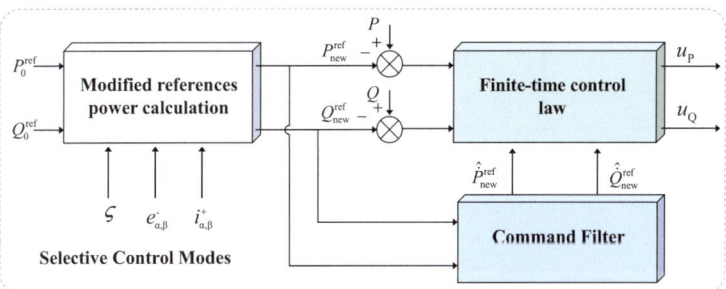

Fig. 6.22 The proposed FTCFC diagram for the power tracking control

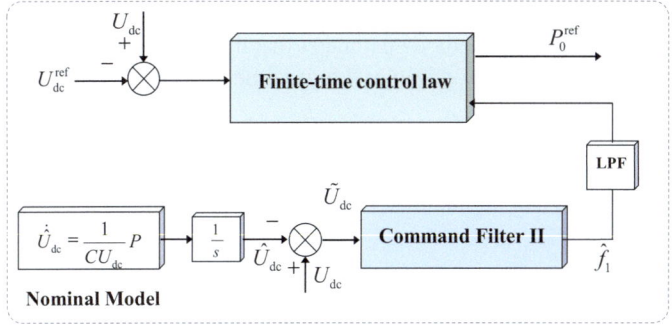

Fig. 6.23 The proposed FTCFC diagram for the dc-link voltage control

$$\begin{aligned}\dot{V} &= \dot{V}_1 + \dot{V} \\ &= -k_{11}z_P^2 - k_{21}z_Q^2 - k_{31}z_{dc}^2 - k_{12}z_P^{\gamma+1} - k_{22}z_Q^{\gamma+1} \\ &\quad - k_{32}z_{dc}^{\gamma+1} + z_P O_2 + z_Q O_2 + z_{dc} O_{dc} + z_P d_1 + z_Q d_2\end{aligned} \quad (6.93)$$

According to Young's inequality, the following holds

$$\begin{cases} z_P O_2 \leq 0.5z_P^2 + 0.5O_2^2, \ z_Q O_2 \leq 0.5z_Q^2 + 0.5O_2^2 \\ z_{dc} O_{dc} \leq 0.5z_{dc}^2 + 0.5O_{dc}^2, \ z_P d_1 \leq 0.5z_p^2 + 0.5d_1^2 \\ z_Q d_2 \leq 0.5z_Q^2 + 0.5d_2^2 \end{cases} \quad (6.94)$$

Using the inequality to the above equation, $\dot{V}$ is rewritten as

$$\begin{aligned}\dot{V} &\leq -(k_{11}-1)z_P^2 - (k_{21}-1)z_Q^2 - (k_{31}-0.5)z_{dc}^2 - k_{12}z_P^{\gamma+1} \\ &\quad -k_{22}z_Q^{\gamma+1} - k_{32}z_{dc}^{\gamma+1} + O_2^2 + 0.5O_{dc}^2 + 0.5d_1^2 + 0.5d_2^2 \\ &\leq aV - bV^h + c\end{aligned} \quad (6.95)$$

where a, b, c, h, and r satisfy the following expression

$$\begin{cases} a = \min\{2k_{11}-2, \ 2k_{21}-2\} \\ b = \min\{2^h k_{i2}\}, \ (i=1,2,3) \\ c = O_2^2 + 0.5O_{dc}^2 + 0.5d_1^2 + 0.5d_2^2 \\ h = (\gamma+1)/2 \end{cases} \quad (6.96)$$

Selecting $a > 0$ and $b > 0$, the fast finite-time practical stability criterion is satisfied, one has

$$|z_P| \leq \sqrt{\frac{2c}{a-\tau}}, \ |z_Q| \leq \sqrt{\frac{2c}{a-\tau}} \quad (6.97)$$

within a finite time $T_s = t_0 + t_f + \frac{1}{\tau(1-h)} \ln\left[\frac{\frac{b}{\tau}+V(t_0)^{1-h}}{\frac{b}{\tau}+\left(\frac{c}{a-\tau}\right)^{1-h}}\right]$, and $0 < \tau < a$.

Considering the dynamic models of active power, reactive power and dc-link voltage model, the FTCFC method guarantees that the tracking errors converge to an arbitrarily small region of the origin in a finite time and ensures robustness against disturbances.

Figure 6.24 shows the control diagram of the proposed FTCFC method for the 3LT²I under unbalanced grid conditions. The FTCFC method is a typical direct power control structure, directly regulating the power without relying on the current control loop and the PLL link.

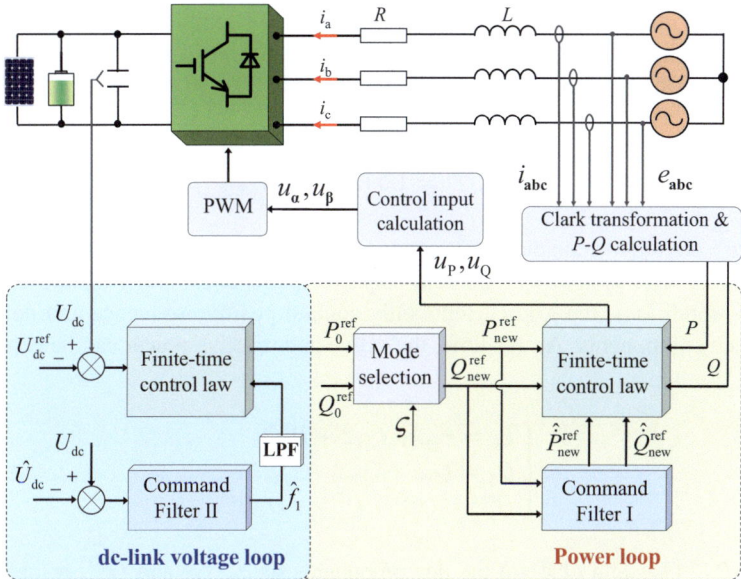

Fig. 6.24 The control diagram of the FTCFC method under unbalanced grid conditions

Improved Passivity-Based Direct Power Control Method

This part illustrates an IPBDPC strategy based on extended reactive power theory in the $\alpha\beta$ reference frame, which can effectively suppress dc-link voltage fluctuations and current harmonics.

The charging mode of 3LT²I is also taken as an example to illustrate the IPBDPC principle. This section consists of three parts: modeling and analysis, the design process of IPBDPC, and the robustness analysis.

Part 1: Modeling and Analysis

According to the KVL, in the $\alpha\beta$ reference frame, the currents and voltages of the 3LT²I satisfy the following relationship:

$$L\frac{di_{\alpha\beta}}{dt} = e_{\alpha\beta} - Ri_{\alpha\beta} - u_{\alpha\beta} \tag{6.98}$$

Using instantaneous power theory, the active power and reactive power on the grid side are re-calculated as follows. To distinguish these power variables from those in the previous control model, the active and reactive power are donates as P_g, Q_g, respectively.

$$\begin{cases} P_g = 1.5 \operatorname{Re}\left(i^*_{\alpha\beta} e_{\alpha\beta}\right) \\ Q_g = 1.5 \operatorname{Im}\left(i^*_{\alpha\beta} e_{\alpha\beta}\right) \end{cases} \tag{6.99}$$

It should be noted that '*' indicates the conjugate of a complex vector.

Under unbalanced grid voltage conditions, the extended reactive power theory can better represent the instantaneous reactive power change than the Q_g [19], which is expressed as

$$Q_g^{ext} = 1.5\,\mathrm{Re}\!\left(i_{\alpha\beta}^* e_{\alpha\beta}^m\right) \tag{6.100}$$

where $e_{\alpha\beta}^m$ denotes the quadrature components that lag $e_{\alpha\beta}$ by 90°. The 3LT²I is a three-phase three-wire system with no zero-sequence path. Thus, under unbalanced grid voltage conditions, the grid currents only contain positive-sequence and negative-sequence components. At this time, the active and reactive power can be expressed as

$$\begin{cases} P_g = P_{g0} + P_{gc2} + P_{gs2} \\ Q_g = Q_{g0} + Q_{gc2} + Q_{gs2} \\ Q_g^{ext} = Q_{g0}^{ext} + Q_{gc2}^{ext} + Q_{gs2}^{ext} \end{cases} \tag{6.101}$$

where P_{g0}, Q_{g0}, and Q_{g0}^{ext} are the dc components of P_g, Q_g and Q_g^{ext}, respectively. The remaining terms correspond to the second-order harmonic components caused by negative-sequence voltage and current, which are expressed as

$$\begin{cases} P_{gc2} = 1.5\left(e_\alpha^+ i_\alpha^- + e_\beta^+ i_\beta^-\right) = 1.5\left|U_g^+\right|\left|I_g^-\right|\cos(2\omega_1 t + \theta_u^+ - \theta_i^-) \\ P_{gs2} = 1.5\left(e_\alpha^- i_\alpha^+ + e_\beta^- i_\beta^+\right) = 1.5\left|U_g^-\right|\left|I_g^+\right|\cos(2\omega_1 t + \theta_i^+ - \theta_u^-) \end{cases} \tag{6.102}$$

$$\begin{cases} Q_{gc2} = 1.5\left(e_\beta^+ i_\alpha^- - e_\alpha^+ i_\beta^-\right) = 1.5\left|U_g^+\right|\left|I_g^-\right|\sin(2\omega_1 t + \theta_u^+ - \theta_i^-) \\ Q_{gs2} = 1.5\left(e_\beta^- i_\alpha^+ - e_\alpha^- i_\beta^+\right) = -1.5\left|U_g^-\right|\left|I_g^+\right|\sin(2\omega_1 t + \theta_i^+ - \theta_u^-) \end{cases} \tag{6.103}$$

$$\begin{cases} Q_{gc2}^{ext} = 1.5\left(e_\alpha^{m+} i_\alpha^- + u_\beta^{m+} i_\beta^-\right) = 1.5\left|U_g^+\right|\left|I_g^-\right|\sin(2\omega_1 t + \theta_u^+ - \theta_i^-) \\ Q_{gs2}^{ext} = 1.5\left(e_\alpha^{m-} i_\alpha^+ + e_\beta^{m-} i_\beta^+\right) = 1.5\left|U_g^-\right|\left|I_g^+\right|\sin(2\omega_1 t + \theta_i^+ - \theta_u^-) \end{cases} \tag{6.104}$$

where $e_{\alpha\beta}^+$, $e_{\alpha\beta}^-$, $i_{\alpha\beta}^+$ and $i_{\alpha\beta}^-$ are positive- and negative-sequence components of the grid voltage and current, respectively, $|U_g^+|$, $|I_g^+|$, $|U_g^-|$ and $|I_g^-|$ are the amplitudes of the positive- and negative-sequence components of the grid voltage and current, respectively, θ_i^+, θ_u^+, θ_i^- and θ_u^- are the initial phase of the positive- and negative-sequence component of the grid voltage and current, respectively. ω_1 denotes the fundamental frequency.

To achieve stable active power and sinusoidal current, the second-order harmonic components of active and reactive power needs to satisfy the following expression.

$$\begin{cases} P_{gc2} + P_{gs2} = 0 \\ Q_{gc2}^{ext} + Q_{gs2}^{ext} = 0 \end{cases} \tag{6.105}$$

6.2 Advanced Control Technology for DC Voltage and Grid Current of PV ...

Therefore, if the fluctuations in active power P_g can be eliminated, the fluctuations in Q_{gc2}^{ext} can also be suppressed simultaneously. Furthermore, under balanced grid voltage conditions, the grid voltages and those quadrature components satisfy the following relationship:

$$\begin{cases} e'_\alpha = e_\beta \\ e'_\beta = -e_\alpha \end{cases} \quad (6.106)$$

At this time, $Q_g^{ext} = Q_g$ can be obtained, indicating that the novel power theory combination is also applicable under balanced grid voltage conditions.

Then, the dynamic equation of P_g and Q_g^{ext} can be derived as

$$\begin{cases} \frac{dP_g}{dt} = 1.5\left(e_\alpha \frac{di_\alpha}{dt} + i_\alpha \frac{de_\alpha}{dt} + e_\beta \frac{di_\beta}{dt} + i_\beta \frac{de_\beta}{dt}\right) \\ \frac{dQ_g^{ext}}{dt} = 1.5\left(e'_\alpha \frac{di_\alpha}{dt} + i_\alpha \frac{de'_\alpha}{dt} + e'_\beta \frac{di_\beta}{dt} + i_\beta \frac{de'_\beta}{dt}\right) \end{cases} \quad (6.107)$$

And the dynamic equation of e_α, e_β, e_α^m and e_β^m can satisfy the following formula

$$\begin{cases} \frac{de_\alpha}{dt} = -\omega_1 e_\alpha^m \\ \frac{de_\beta}{dt} = -\omega_1 e_\beta^m \\ \frac{de_\alpha^m}{dt} = -\omega_1 e_\alpha \\ \frac{de_\beta^m}{dt} = \omega_1 e_\beta \end{cases} \quad (6.108)$$

According to the above equations, the dynamic model of the active and reactive power can be given by

$$\begin{cases} \frac{2}{3}L\frac{dP_g}{dt} - e_\alpha^2 + e_\beta^2 - (e_\alpha u_\alpha + e_\beta u_\beta) - \frac{2}{3}RP_g - \frac{2}{3}\omega_1 LQ_g^{ext} \\ \frac{2}{3}L\frac{dQ_g^{ext}}{dt} = e_\alpha^m e_\alpha + e_\beta^m e_\beta - (e_\alpha^m u_\alpha + e_\beta^m u_\beta) - \frac{2}{3}RQ_g^{ext} + \frac{2}{3}\omega_1 LP_g \end{cases} \quad (6.109)$$

Thus, the EL model of the 3LT²I can be obtained as

$$M\dot{x} + Jx + Rx = u \quad (6.110)$$

where M is a positive definite symmetric coefficient matrix, and R is the system dissipation term. J is an anti-symmetric coefficient matrix and $J^T = -J^T$. x represents the state variable vector, and u is the matrix of converter voltage vector. They are expressed as

$$\begin{cases} \mathbf{x} = \begin{bmatrix} P_g \\ Q_g^{\text{ext}} \end{bmatrix}, \mathbf{M} = \begin{bmatrix} \frac{2}{3}L & 0 \\ 0 & \frac{2}{3}L \end{bmatrix} \\ \mathbf{J} = \begin{bmatrix} 0 & \frac{2}{3}\omega_1 L \\ -\frac{2}{3}\omega_1 L & 0 \end{bmatrix}, \mathbf{R} = \begin{bmatrix} \frac{2}{3}R & 0 \\ 0 & \frac{2}{3}R \end{bmatrix} \\ \mathbf{u} = \begin{bmatrix} e_\alpha^2 + e_\beta^2 - (e_\alpha u_\alpha + e_\beta u_\beta) \\ e_\alpha^m e_\alpha + e_\beta^m e_\beta - (e_\alpha^m u_\alpha + e_\beta^m u_\beta) \end{bmatrix} \end{cases} \quad (6.111)$$

Part 2: Design Process of IPBDPC

To track the power references, let error $\mathbf{x}_e = \mathbf{x}_{ref} - \mathbf{x}$, and the error storage function is given by

$$H_e(\mathbf{x}) = \frac{1}{2} \mathbf{x}_e^T \mathbf{M} \mathbf{x}_e \quad (6.112)$$

Injecting damping $\mathbf{R}_a$ into the dissipation term can achieve $H_e(\mathbf{x}) \to 0$ rapidly, thereby increases the convergence rate of $\mathbf{x} \to \mathbf{x}_{ref}$. Therefore, the new dissipation term $\mathbf{R}_d$ is

$$\mathbf{R}_d \mathbf{x}_e = (\mathbf{R} + \mathbf{R}_a) \mathbf{x}_e \quad (6.113)$$

where the additional damping matrix $\mathbf{R}_a$ is a positive definite symmetric matrix, as well as $\mathbf{R}$.

$$\mathbf{R}_a = \begin{bmatrix} r_a & 0 \\ 0 & r_a \end{bmatrix}, r_a > 0 \quad (6.114)$$

Then the error EL model is defined as follows

$$\mathbf{M}\dot{\mathbf{x}}_e + \mathbf{R}_d \mathbf{x}_e = -\mathbf{u} + \mathbf{M}\dot{\mathbf{x}}_{ref} + \mathbf{J}(\mathbf{x}_{ref} - \mathbf{x}_e) + \mathbf{R}\mathbf{x}_{ref} + \mathbf{R}_a \mathbf{x}_e \quad (6.115)$$

Therefore, the PBC can be designed as

$$\mathbf{u} = \mathbf{M}\dot{\mathbf{x}}_{ref} + \mathbf{J}\mathbf{x} + \mathbf{R}\mathbf{x}_{ref} + \mathbf{R}_a \mathbf{x}_e \quad (6.116)$$

When the power tracking is sufficiently accurate, the error $\mathbf{x}_e$ can fluctuate around zero. The system rapidly converges to the desired equilibrium point and achieves the global stability, when

$$\dot{H}_e(\mathbf{x}) = \mathbf{x}_e^T \mathbf{M} \dot{\mathbf{x}}_e = -\mathbf{x}_e^T \mathbf{R}_d \mathbf{x}_e = -\frac{H_e(\mathbf{x})}{t_{cr}} < 0 \quad (6.117)$$

where $t_{cr} = L_g/(2R_g + 3r_a)$ means the time constant of convergence. Because of $R_a \gg R$, the convergence speed of $H_e(\mathbf{x})$ mainly depends on the R_a, which determining the robustness of the 3LT^{2}I to against system parameter changes and external

6.2 Advanced Control Technology for DC Voltage and Grid Current of PV ...

disturbances. The PBC control principle can be extended as

$$\begin{cases} E_P = e_\alpha u_\alpha + e_\beta u_\beta \\ \quad = -\frac{2}{3}L\dot{P}_{gref} - \frac{2}{3}\omega_1 LQ_g^{ext} - \frac{2}{3}RP_{gref} - r_a\left(P_{gref} - P_g\right) + e_\alpha^2 + e_\beta^2 \\ E_Q = e_\beta^m u_\alpha + e_\alpha^m u_\beta \\ \quad = -\frac{2}{3}L\dot{Q}_{gref}^{ext} + \frac{2}{3}\omega_1 LP_g - \frac{2}{3}RQ_{gref}^{ext} - r_a\left(Q_{gref}^{ext} - Q_g^{ext}\right) + e_\alpha^m e_\alpha + e_\beta^m e_\beta \end{cases}$$

(6.118)

To simplify the above dynamics, E_P and E_Q are defined as the new control inputs. Therefore, the original control inputs u_α and u_β can be calculated by

$$\begin{cases} u_\alpha = \frac{e_\beta^m E_P - e_\beta E_Q}{e_\alpha e_\beta^m - e_\beta e_\alpha^m} \\ u_\beta = \frac{e_\alpha E_Q - e_\alpha^m E_P}{e_\alpha e_\beta^m - e_\beta e_\alpha^m} \end{cases}$$

(6.119)

The above control method is applied under non-distorted grid voltage conditions. However, when unbalanced and distorted grid voltage coexist, the grid currents become distorted again. Consequently, if only the fundamental frequency components of the voltages $e_{f\alpha}$ and $e_{f\beta}$ at PCC are considered, the grid currents will reduce the content of low-order harmonics (e.g., fifth, seventh, ...). The fundamental voltages and their quadrature components are extracted using the second order generalized integral-quadrature signal generator (SOGI-QSG). The transfer function of SOGI-QSG is expressed as

$$\begin{cases} G_I(s) = \frac{e_{f\alpha\beta}(s)}{e_{\alpha\beta}(s)} = \frac{k_s \omega_g s}{s^2 + k_s \omega_g s + \omega_g^2} \\ G_{qI}(s) = \frac{e_{f\alpha\beta}^m(s)}{e_{\alpha\beta}} = \frac{k_s \omega_g^2}{s^2 + k_s \omega_g s + \omega_g^2} \end{cases}$$

(6.120)

where $\omega_g = \omega_1 = 100\pi$ rad/s is considered, and k_s represents the damping ratio. The signal $e_{f\alpha\beta}^m$ is 90° phase-shifted relative to the input signal $e_{f\alpha\beta}$. Thus, the final control laws of $u_{f\alpha}$ and $u_{f\beta}$ with fundamental voltage injection can be derived as

$$\begin{cases} E_{Pf} = e_{f\alpha} u_\alpha + e_{f\beta} u_\beta \\ \quad = -\frac{2}{3}L\dot{P}_{gref} - \frac{2}{3}\omega_1 LQ_{fg}^{ext} - \frac{2}{3}RP_{gref} - r_a\left(P_{gref} - P_{fg}\right) + e_{f\alpha}^2 + e_{f\beta}^2 \\ E_{Qf} = e_{f\alpha}^m u_\alpha + e_{f\beta}^m u_\beta \\ \quad = -\frac{2}{3}L_g \dot{Q}_{gref}^{ext} + \frac{2}{3}\omega_1 LP_{fg} - \frac{2}{3}RQ_{gref}^{ext} - r_a\left(Q_{gref}^{ext} - Q_{fg}^{ext}\right) + e_{f\alpha}^m e_{f\alpha} + e_{f\beta}^m e_{f\beta} \end{cases}$$

(6.121)

$$\begin{cases} u_{f\alpha} = \frac{e_{f\beta}^m E_{Pf} - e_{f\beta} E_{Qf}}{e_{f\alpha} e_{f\beta}^m - e_{f\beta} e_{f\alpha}^m} \\ u_{f\beta} = \frac{e_{f\alpha} E_{Qf} - e_{f\alpha}^m E_{Pf}}{e_{f\alpha} e_{f\beta}^m - e_{f\beta} e_{f\alpha}^m} \end{cases}$$

(6.122)

where E_{Pf} and E_{Qf} are the fundamental components of E_P and E_Q, $P_{fg} = 1.5\,\mathrm{Re}\left(i^*_{\alpha\beta}e_{f\alpha f\beta}\right)$, $Q^{ext}_{fg} = 1.5\,\mathrm{Re}\left(i^*_{\alpha\beta}e^m_{\alpha\beta}\right)$.

The control diagram is shown in Fig. 6.25. In order to enhance the dc voltage control capability and robustness, a PI-based dc-link voltage outer loop control is designed. The NPV balance is achieved by injecting appropriate zero-sequence components into modulated waves, as discussed in the previous section and therefore not repeated here. More notably, for the inner power loop, the proposed IPBDPC method employs a compound control that includes a feedback term, a grid voltage feedforward term, and a model-based inverse feedforward term. The feedback term is a damping control. The grid voltage feedforward term can improve the system's response speed and stability under grid voltage variations, while the model-based inverse input is used as a feedforward to enhance the output tracking performance. Thus, the 3LT^{2}I system will have a better steady-state and dynamic performance under the proposed method.

Part 3: Robustness Analysis

To analyze the grid-connected stability and robustness of the 3LT^{2}I, the sequence impedance model of the IPBDPC is developed. Since the proposed IPBDPC is a direct power control method designed in the $\alpha\beta$ frame, the fixed dc operating point of voltage and current are absent. Therefore, a linearization method based on harmonic interference is designed to linearize the nonlinear terms in the model. The proposed model indicates the closed-loop relation between e_{abc} and i_{abc}. Subsequently, the stability criteria of impedance model under parameter perturbations and external

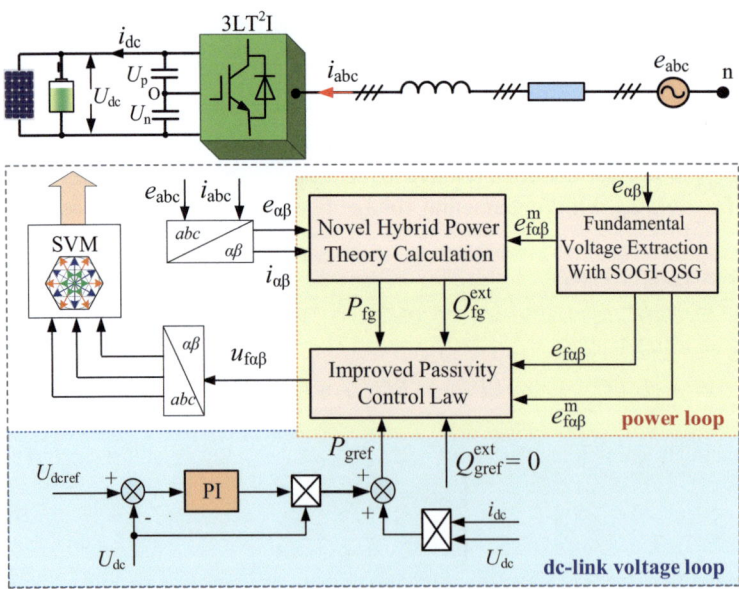

Fig. 6.25 Control block diagram of the proposed IPBDPC strategy

6.2 Advanced Control Technology for DC Voltage and Grid Current of PV …

disturbances is carried out to analyze the robustness of the proposed control system. To establish the impedance model, the deduction is carried out from the final control law as

$$\begin{cases} u_{f\alpha} = \frac{2}{3}L(A_\alpha + B_\alpha) + \frac{2}{3}RC_\alpha + r_aD_\alpha + e_\alpha \\ u_{f\beta} = \frac{2}{3}L(A_\beta + B_\beta) + \frac{2}{3}RC_\beta + r_aD_\beta + e_\beta \end{cases} \quad (6.123)$$

where A_α, A_β, B_α, B_β, C_α, and C_β are four different coupling terms, D_α, and D_β are the coupling terms of injection damping, which can be written as

$$\begin{cases} A_\alpha = \frac{-\omega_1 Q_{fg}^{ext} e_{f\alpha} + \omega_f P_{fg} e_{f\beta}}{U_{fg}^2}, & A_\beta = \frac{-\omega_1 P_{fg} e_{f\alpha} - \omega_1 Q_{fg}^{ext} e_{f\beta}}{U_{fg}^2} \\ B_\alpha = \frac{-\dot{P}_{ref} e_{f\alpha} - \dot{Q}_{ref}^{ext} e_{f\beta}}{U_{fg}^2}, & B_\beta = \frac{-\dot{P}_{ref} e_{f\beta} + \dot{Q}_{ref}^{ext} e_{f\alpha}}{U_{fg}^2} \end{cases} \quad (6.124)$$

$$\begin{cases} C_\alpha = \frac{-P_{ref} e_{f\alpha} - Q_{ref}^{ext} e_{f\beta}}{U_{fg}^2}, \ C_\beta = \frac{-P_{ref} e_{f\beta} + Q_{ref}^{ext} e_{f\alpha}}{U_{fg}^2} \\ D_\alpha = -\frac{e_{f\alpha}}{U_{fg}^2}\left(P_{ref} - P_{fg}\right) - \frac{e_{f\beta}}{U_{fg}^2}\left(Q_{ref} - Q_{fg}^{ext}\right) \\ D_\beta = -\frac{e_{f\beta}}{U_{fg}^2}\left(P_{ref} - P_{fg}\right) + \frac{e_{f\alpha}}{U_{fg}^2}\left(Q_{ref} - Q_{fg}^{ext}\right) \end{cases} \quad (6.125)$$

where $U_{fg}^2 = e_{f\alpha}^2 + e_{f\beta}^2$.

To further simplify A_α and A_β, substitute P_g, Q_g and Q_g^{ext} into the expression of $u_{f\alpha}$ and $u_{f\beta}$. In addition, P_{ref} and Q_{ref}^{ext} are constant under small signal steady-state condition, hence B_α and B_β can be written as follows

$$\begin{cases} A_\alpha = 1.5\omega_f i_\beta, & A_\beta = -1.5\omega_f i_\alpha \\ B_\alpha = 0, & B_\beta = 0 \end{cases} \quad (6.126)$$

However, C_α, C_β, D_α, and D_β are non-linear terms that cannot be directly linearized. The grid voltage $e_{\alpha\beta}$ and current $i_{\alpha\beta}$ after harmonic injection contain both fundamental components $e_{f\alpha\beta}$, $i_{f\alpha\beta}$ and harmonic components $e_{h\alpha\beta}$, $i_{h\alpha\beta}$, respectively. For D_α and D_β, after SOGI filtering, the proportion of the harmonic component $e_{h\alpha\beta}$ in $e_{\alpha\beta}$ is very small and can be neglected. The U_{fg}^2 in C_α, C_β, D_α and D_β can be approximated as a dc value. Therefore, the C_α and C_β can be linearized, the active and reactive power components P_{fhg} and Q_{fhg}^{ext} after harmonic injection can be approximated as

$$\begin{cases} P_{fhg} \approx \underbrace{\left(\frac{3}{2}i_{f\alpha\beta} \odot e_{f\alpha\beta}\right)}_{P_{g0}} + \underbrace{\left(\frac{3}{2}i_{h\alpha\beta} \odot e_{f\alpha\beta}\right)}_{P_{ac}} \\ Q_{fhg}^{ext} \approx \underbrace{\left(\frac{3}{2}i_{f\alpha\beta} \otimes u_{f\alpha\beta}\right)}_{Q_{g0}^{ext}} + \underbrace{\left(\frac{3}{2}i_{h\alpha\beta} \otimes u_{f\alpha\beta}\right)}_{Q_{ac}} \end{cases} \quad (6.127)$$

where $\odot$ and $\otimes$ denote the dot product and cross product of two complex vectors, respectively. P_{g0} and Q_{g0}^{ext} are approximately equal to their reference P_{ref} and Q_{ref} in steady state, respectively, which has no impact on harmonic frequency domain analysis. By substituting (6.127) into (6.125), D_α and D_β can be simplified as

$$\begin{cases} D_\alpha \approx \dfrac{e_{f\alpha}P_{ac}+e_{f\beta}Q_{ac}}{U_{fg}^2} \approx \dfrac{3}{2}\sum_{h=1}^{N} i_{h\alpha} \\ D_\beta \approx \dfrac{e_{f\beta}P_{ac}-e_{f\alpha}Q_{ac}}{U_{fg}^2} \approx \dfrac{3}{2}\sum_{h=1}^{N} i_{h\beta} \end{cases} \quad (6.128)$$

According to above equations, the harmonic reference voltage vectors $u_{h\alpha\beta}$ can be converted to

$$\begin{bmatrix} u_{h\alpha} \\ u_{h\beta} \end{bmatrix} = \underbrace{\begin{bmatrix} \frac{3}{2}r_a & L_g\omega_1 \\ -L_g\omega_1 & \frac{3}{2}r_a \end{bmatrix}}_{Z_{PB}}\begin{bmatrix} i_{h\alpha} \\ i_{h\beta} \end{bmatrix} + \underbrace{\begin{bmatrix} -\frac{2R_g P_{ref}G_I(s)}{3U_{fg}^2}+G_I(s) & -\frac{2RQ_{ref}G_I(s)}{3U_{fg}^2} \\ \frac{2R_g Q_{ref}G_I(s)}{3U_{fg}^2} & -\frac{2RP_{ref}G_I(s)}{3U_{fg}^2}+G_I(s) \end{bmatrix}}_{T_{cv}}\begin{bmatrix} e_{h\alpha} \\ e_{h\beta} \end{bmatrix}$$

$$(6.129)$$

where Z_{PB} and T_{cv} are the transfer matrix of i_h and e_h to u_h, respectively. The harmonic impedance of the 3LT^{2}I is expressed by Laplace transform from current model as

$$\begin{bmatrix} e_{h\alpha} \\ e_{h\beta} \end{bmatrix} = \underbrace{\begin{bmatrix} R+sL & 0 \\ 0 & R+sL \end{bmatrix}}_{Z_{LR}}\begin{bmatrix} i_{h\alpha} \\ i_{h\beta} \end{bmatrix} + \begin{bmatrix} u_{h\alpha} \\ u_{h\beta} \end{bmatrix} \quad (6.130)$$

Thus, the impedance model considering harmonic components can be calculated as

$$\begin{bmatrix} e_{h\alpha} \\ e_{h\beta} \end{bmatrix} = (\mathbf{I}-T_{cv})^{-1}(Z_{LR}+Z_{PB})\begin{bmatrix} i_{h\alpha} \\ i_{h\beta} \end{bmatrix} = \underbrace{\begin{bmatrix} Z_{\alpha\alpha}(s) & Z_{\alpha\beta}(s) \\ Z_{\beta\alpha}(s) & Z_{\beta\beta}(s) \end{bmatrix}}_{Z_{T^2C}}\begin{bmatrix} i_{h\alpha} \\ i_{h\beta} \end{bmatrix} \quad (6.131)$$

where the Z_{T^2C} represents the final impedance matrix. The equivalent sequence-impedance transfer function can be expressed as

$$\begin{cases} Z_{T^2C,p}(s) = \dfrac{Z_{\alpha\alpha}(s)+Z_{\beta\beta}(s)}{2}+j\dfrac{Z_{\beta\alpha}(s)-Z_{\alpha\beta}(s)}{2} \\ Z_{T^2C,n}(s) = \dfrac{Z_{\alpha\alpha}(s)-Z_{\beta\beta}(s)}{2}+j\dfrac{Z_{\beta\alpha}(s)+Z_{\alpha\beta}(s)}{2} \end{cases} \quad (6.132)$$

Since the diagonal elements of the two-dimensional impedance matrix have symmetric characteristics, $Z_{\alpha\alpha}=Z_{\beta\beta}$, $Z_{\alpha\beta}=-Z_{\beta\alpha}$, it can be inferred that $Z_{T^2C,n}(s)=0$. Then, the impedance of the power grid is assumed as

6.2 Advanced Control Technology for DC Voltage and Grid Current of PV ...

$$Z_{\text{grid}}(s) = \frac{L_n s + R_n}{(L_n s + R_n)C_n s + 1} \quad (6.133)$$

where the R_n and L_n are grid resistance and inductance, respectively, C_n is the parallel capacitor. In this case, the stability of the system can be identified by the impedance ratio of $IR_{\text{recp}}(s) = Z_{\text{grid}}(s)/Z_{T^2C,p}(s)$ and its closed loop transfer function is $G_Z(s) = 1/[1 + IR_{\text{recp}}(s)]$.

When r_a changes, the Nyquist diagrams of $IR_{\text{recp}}(s)$ and Pole-zero map of G_Z are shown in Fig. 6.26. The characteristic loci will not surround the stable point $(-1, j0)$ under five different control parameters, demonstrating the passivity and stability of the system as shown in Fig. 6.26a. From the overall view, it can be observed that the larger the damping value, the smaller the intersection value between the Nyquist diagram and the imaginary axis, which means that control degree is increasing. For larger/smaller values of damping ($r_a = 1, 10, \ldots 500$), the dominant poles move closer to the imaginary axis, potentially leading to an undesired stability margin as shown in Fig. 6.26b. Thus, the values of r_a can be manually fine-tuned from small to large until a satisfactory response is achieved within the given constraint conditions.

As shown in Fig. 6.27, with the change of the filter inductance value, the Nyquist diagram never surrounds $(-1, j0)$ points, further confirming the stability of the system. In general, the proposed IPBDPC method is robust to parameters perturbations in both the controller and the hardware circuit.

To summarize, in this section, the dual-loop voltage and power control scheme is presented. In this scheme, the voltage loop is used to maintain the dc-link voltage stable, and the power loop is used to control transmitted power to reach its reference value. In the power loop, the active and reactive power are independently regulated without relying on a current control loop or a PLL. Two methods for power control are introduced.

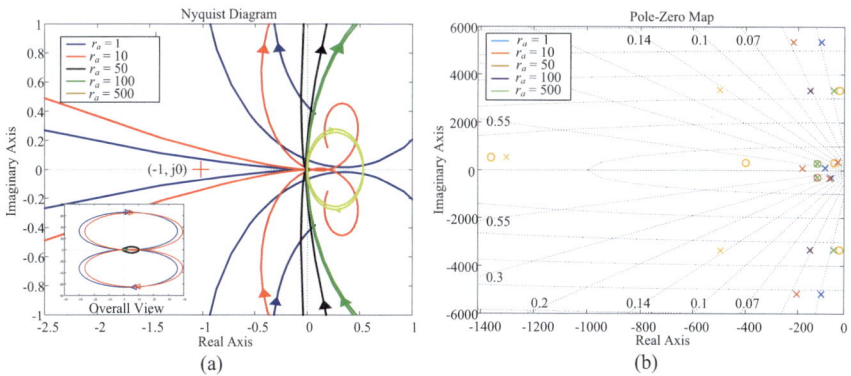

Fig. 6.26 The stability analysis with different control parameters r_a. **a** Nyquist diagram of IR_{recp}. **b** Pole-zero map of G_Z

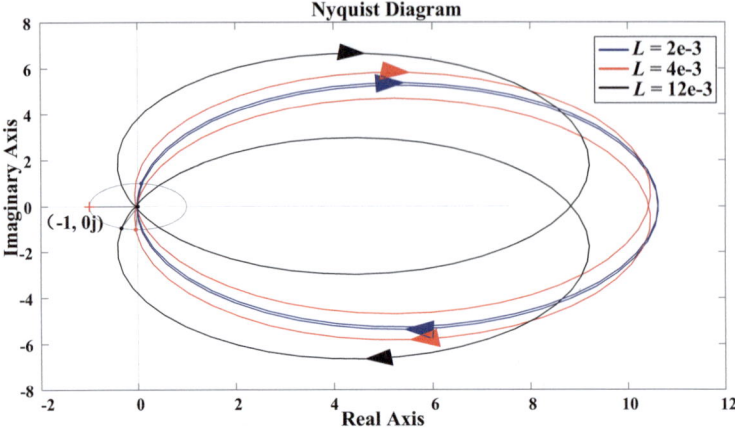

Fig. 6.27 The Nyquist diagram of IR_{recp} with filtering inductance changes

6.3 Advanced Control Technology for Neutral-Point Voltage Balance Under Grid Faults

As is well-know that the effective suppression of NPV unbalance is the guarantee of reliable operation of $3LT^2I$. In the previous research, numerous NPV balance algorithms have been proposed. In Chap. 5, several typically control methods have been introduced in detail. These methods mainly divided into two groups according to the types of NPV unbalance. One group is dc NPV unbalance control methods. Another group is the NPV oscillation mitigation methods. These methods have been proven to be effective in normal operation, which are easy to implement and have the strong NPV balance ability at high power factors.

Under power grid fault conditions, the grid standards require that that the inverter must have LVRT function. When a short-circuit fault of power grid is detected, the PV inverter is switched into LVRT mode. In this case, the PV inverters are required not only to maintain their connection to the grid but also to inject an amount of reactive power to support the recovery of grid voltage.

As the introduction of overview of grid fault, the reactive current that injected into grid is determined by the percentage of grid voltage reduction during the dip. Then, the active current is reduced and obtained by the rated current and reactive current. With the increase of reactive current, the power factor of PV inverter system decreases. In this condition, the NPV balance control methods for normal power grid condition are not applicable any more. The NPV will easily fluctuate to unacceptable levels for the effects of grid voltage dips and low power factor. How to suppress NPV unbalance effectively and ensure the reliable operation of $3LT^2I$ under grid fault condition is an important issue.

To address this issue, the author has conducted extensive researches and put forward various effective methods for NPV balance under grid fault. This section

6.3 Advanced Control Technology for Neutral-Point Voltage Balance Under ...

mainly elaborates three method for NPV balance under grid fault. These methods are integrated into PWM modulator and auxiliary loop. The three methods include based on model predictive control method, based on selective harmonics elimination pulse width modulation and based on common-mode voltage reduction modulation method.

6.3.1 Based on Model Predictive Control

For the $3LT^2I$, the GMPC is proposed to realize two control objectives: current tracking and NPV balance. The implementation process current tracking has been presented in previous section. Thus, the GMPC for NPV balance is presented in this section.

When unbalanced power grid voltages occur, the grid code requires that the PV inverter should provide a certain reactive power to power grid. In this case, the NPV unbalance of $3LT^2I$ is increased as the power factor decreases. Especially, in severe grid fault condition, the NPV unbalance cannot be effectively suppressed, as the effects of small voltage vectors on NPV will be reversed with the change of power factor. To deal with this problem, both current direction and dc-link capacitors voltage difference are used to select the appropriate small vector. By doing so, the NPV balance can be achieved under different power factors.

First, the effects of small vectors on NPV are analyzed. Taking redundant small vectors [POO] and [ONN] as an example, the effects of redundant small vectors on NPV under different grid currents directions are shown in Fig. 6.28. It can be observed that, the small vector [POO] is connected between the positive dc-link and O. When $i_a > 0$, the neutral current flows out the NP, and the upper capacitor is charged depicted in Fig. 6.28a. Conversely, the upper capacitor is discharged when $i_a < 0$, as shown in Fig. 6.28b. On the other hand, the small vector [ONN] is connected between the O and negative dc-link. As shown in Fig. 6.28c, since the neutral current flows into the NP when $i_a > 0$, the small vector [ONN] charges the lower capacitor. Conversely, the small vector [ONN] discharges the lower capacitor when $i_a < 0$, as shown in Fig. 6.28d. To summarize, the effects of P-type and N-type small vectors on capacitor voltages are determined by the grid current direction.

Since the current directions change with the power factor, the effects of small vectors on NPV are uncertain under different power factor operation. Figure 6.29 shows the change in grid current direction with changes in power factor angle φ ($-90° \sim 90°$) and electrical angle ωt. It should be noted that the power factor is defined as $\cos\varphi$. Figure 6.29a shows the current direction of i_a associated with the redundant voltage vectors [POO] and [ONN] for region 1. It is indicated that as the power factor and electrical angle change, the polarity of the grid current is also changed. Consequently, the effects of [POO] and [ONN] associated with i_a on NPV are changed. Figure 6.29b, c are the current direction of i_b and i_c associated with the redundant voltage vectors of [OPO], [NON] for region 4 and [PPO], [OON] for region 2, respectively. Similarly, the directions of grid currents are changed with

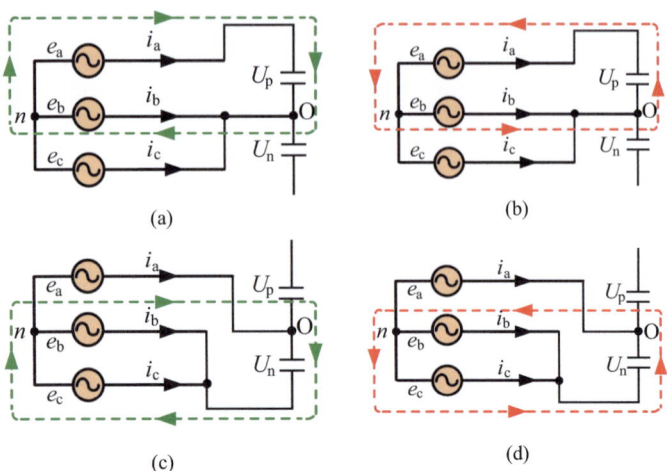

Fig. 6.28 Effects of small vectors on capacitor voltage. **a** [POO] with $i_a > 0$. **b** [POO] with $i_a < 0$. **c** [ONN] with $i_a > 0$. **d** [ONN] with $i_a < 0$

power factor changes. This indicates that the effects of small vectors on NPV are determined by the corresponding grid currents directions under different power factor operation.

According to the above analysis, in order to effectively control the NPV balance during the full power factor range, both the voltage difference of dc-link capacitors and the grid current directions are considered when selecting the appropriate small vector. It should be noted that to reduce cost function evaluation times, the triangle candidate region is adopted. Each triangle candidate region has five space voltage vectors, including both P-type and N-type redundant small vectors. Once an appropriate redundant small vector is selected for NPV balancing according to above theory, only four vectors are included in cost function evaluation, thereby further reducing the computational burden. The four candidate voltage vectors selection process is elaborated as follows.

Take region 1 as an example to illustrate this process. When the reference voltage vector locates in region 1, the small vectors of [POO] and [ONN] are used to balance the NPV. The effects of small vectors on capacitor voltage are different with different current directions. Therefore, the proposed scheme in region 1 consists of four cases.

Case 1: If $U_p < U_n$ and $i_a > 0$, [POO] is selected to increase U_p, so the candidate voltage vectors are [OOO], [POO], [PON] and [PNN].
Case 2: If $U_p < U_n$ and $i_a < 0$, [ONN] is selected to increase U_p, so the candidate voltage vectors are [OOO], [ONN], [PON] and [PNN].
Case 3: If $U_p > U_n$ and $i_a > 0$, [ONN] is selected to increase U_n, so the candidate voltage vectors are [OOO], [ONN], [PON] and [PNN].
Case 4: If $U_p > U_n$ and $i_a < 0$, [POO] is selected to increase U_n, so the candidate voltage vectors are [OOO], [POO], [PON] and [PNN].

6.3 Advanced Control Technology for Neutral-Point Voltage Balance Under …

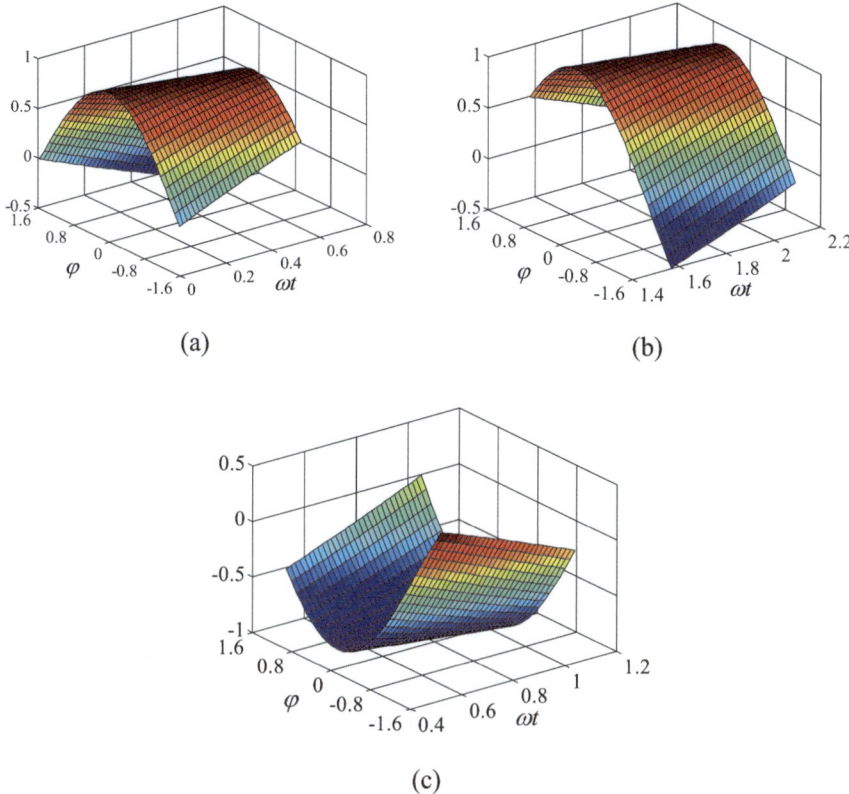

Fig. 6.29 Change in grid current direction with power factor **a** i_a associated with the [POO] and [ONN] for region 1 **b** i_b associated with the [OPO] and [NON] for region 4 **c** i_c associated with the [PPO] and [OON] for region 2

According to the symmetry of the space voltage vector diagram, a similar procedure of small vector selection can be extended to all the other triangle candidate regions. Based on the location of the reference voltage vector, the capacitor voltage difference, and the grid current direction, the candidate vectors participating in the cost function evaluation are determined and summarized in Table 6.2. The small vectors highlighted in bold are selected to control NPV balance during the full power factor range.

To verify the effectiveness of the proposed GMPC scheme for NPV balance, experimental tests have been conducted. Two control methods are compared: Method-1 is the conventional MPC without considering of grid currents directions and Mthod-2 is the proposed method considering current directions of small vectors. The results are presented as follows.

When the inverter operates at high power factor, the voltages of dc-link capacitors are depicted in Fig. 6.30. The voltage difference of dc-link capacitors is effectively eliminated using both Method-1 and Method-2. Since the effects of small vectors

Table 6.2 Summarisation of candidate voltage vectors with the proposed scheme in all regions

Region	Criteria	Candidate vectors	Criteria	Candidate vectors
1	$U_p < U_n, i_a > 0$	[OOO] [POO] [PON] [PNN]	$U_p < U_n, i_a < 0$	[OOO] [ONN] [PON] [PNN]
	$U_p > U_n, i_a > 0$	[OOO] [ONN] [PON] [PNN]	$U_p > U_n, i_a < 0$	[OOO] [POO] [PON] [PNN]
2	$U_p < U_n, i_c > 0$	[OOO] [OON] [PON] [PPN]	$U_p < U_n, i_c < 0$	[OOO] [PPO] [PON] [PPN]
	$U_p > U_n, i_c > 0$	[OOO] [PPO] [PON] [PPN]	$U_p > U_n, i_c < 0$	[OOO] [OON] [PON] [PPN]
3	$U_p < U_n, i_c > 0$	[OOO] [OON] [OPN] [PPN]	$U_p < U_n, i_c < 0$	[OOO] [PPO] [OPN] [PPN]
	$U_p > U_n, i_c > 0$	[OOO] [PPO] [OPN] [PPN]	$U_p > U_n, i_c < 0$	[OOO] [OON] [OPN] [PPN]
4	$U_p < U_n, i_b > 0$	[OOO] [OPO] [OPN] [NPN]	$U_p < U_n, i_b < 0$	[OOO] [NON] [OPN] [NPN]
	$U_p > U_n, i_b > 0$	[OOO] [NON] [OPN] [NPN]	$U_p > U_n, i_b < 0$	[OOO] [OPO] [OPN] [NPN]
5	$U_p < U_n, i_b > 0$	[OOO] [OPO] [NOP] [NPN]	$U_p < U_n, i_b < 0$	[OOO] [NON] [NOP] [NPN]
	$U_p > U_n, i_b > 0$	[OOO] [NON] [NOP] [NPN]	$U_p > U_n, i_b < 0$	[OOO] [OPO] [NOP] [NPN]
6	$U_p < U_n, i_a > 0$	[OOO] [NOO] [NPO] [NPP]	$U_p < U_n, i_a < 0$	[OOO] [OPP] [NPO] [NPP]
	$U_p > U_n, i_a > 0$	[OOO] [OPP] [NPO] [NPP]	$U_p > U_n, i_a < 0$	[OOO] [NOO] [NPO] [NPP]
7	$U_p < U_n, i_a > 0$	[OOO] [NOO] [NOP] [NPP]	$U_p < U_n, i_a < 0$	[OOO] [OPP] [NOP] [NPP]
	$U_p > U_n, i_a > 0$	[OOO] [OPP] [NOP] [NPP]	$U_p > U_n, i_a < 0$	[OOO] [NOO] [NOP] [NPP]
8	$U_p < U_n, i_c > 0$	[OOO] [OOP] [NOP] [NNP]	$U_p < U_n, i_c < 0$	[OOO] [NNO] [NOP] [NNP]
	$U_p > U_n, i_c > 0$	[OOO] [NNO] [NOP] [NNP]	$U_p > U_n, i_c < 0$	[OOO] [OOP] [NOP] [NNP]
9	$U_p < U_n, i_c > 0$	[OOO] [OOP] [ONP] [NNP]	$U_p < U_n, i_c < 0$	[OOO] [NNO] [ONP] [NNP]
	$U_p > U_n, i_c > 0$	[OOO] [NNO] [ONP] [NNP]	$U_p > U_n, i_c < 0$	[OOO] [OOP] [ONP] [NNP]
10	$U_p < U_n, i_b > 0$	[OOO] [ONO] [ONP] [PNP]	$U_p < U_n, i_b < 0$	[OOO] [POP] [ONP] [PNP]
	$U_p > U_n, i_b > 0$	[OOO] [POP] [ONP] [PNP]	$U_p > U_n, i_b < 0$	[OOO] [ONO] [ONP] [PNP]
11	$U_p < U_n, i_b > 0$	[OOO] [ONO] [PNO] [PNP]	$U_p < U_n, i_b < 0$	[OOO] [POP] [PNO] [PNP]

(continued)

6.3 Advanced Control Technology for Neutral-Point Voltage Balance Under ...

Table 6.2 (continued)

Region	Criteria	Candidate vectors	Criteria	Candidate vectors
	$U_p > U_n, i_b > 0$	[OOO] [POP] [PNO] [PNP]	$U_p > U_n, i_b < 0$	[OOO] [ONO] [PNO] [PNP]
12	$U_p < U_n, i_a > 0$	[OOO] [POO] [PNO] [PNN]	$U_p < U_n, i_a < 0$	[OOO] [ONN] [PNO] [PNN]
	$U_p > U_n, i_a > 0$	[OOO] [ONN] [PNO] [PNN]	$U_p > U_n, i_a < 0$	[OOO] [POO] [PNO] [PNN]

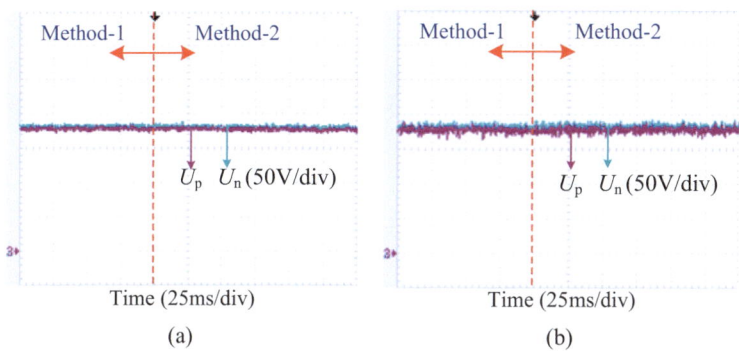

Fig. 6.30 Experimental waveforms of capacitor voltages with Method-1 and Method-2 when power factor angle is 0°. **a** Balanced grid voltage. **b** Unbalanced grid voltage

on NPV are certain at high power factor operation, both Method-1 and Method-2 can control the NPV balance with similar performance. When the grid voltages are unbalanced, the NPV difference is slightly increased compared to the case of balanced grid voltage.

When the power factor angle is 60°, the voltages across the dc-link capacitors are shown in Fig. 6.31. It can be seen that the NPV difference is increased with large power factor angle. Compared to Method-1, the NPV difference of Method-2 can be reduced under both balanced and unbalanced grid voltages, as the grid current directions are considered in Method-3.

The experimental waveforms with 90° power factor angle operation are shown in Fig. 6.32. It is clear that the NPV difference of Method-1 exceeds 30 V under balanced and unbalanced grid voltages. The reason is that only the voltages across the dc-link capacitors are adopted to control the NPV balance in Method-1, and the change in grid current directions results in opposite effects of small vector. Unlike Method-1, Method-2 considers both the dc-link capacitor voltages and the grid current directions. As a result, the voltage difference is effectively suppressed, thereby improving the system performance.

In conclusion, the relationship between the influence of small vectors on the NPV balance and the grid current directions is analyzed for the first time, which indicates that the effects of small vectors on NPV vary across different power factors, and the

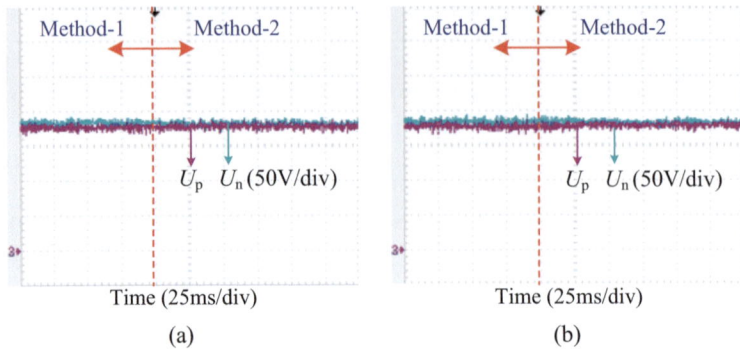

Fig. 6.31 Experimental waveforms of capacitor voltages with Method-1 and Method-2 when power factor angle is 60°. **a** Balanced grid voltage. **b** Unbalanced grid voltage

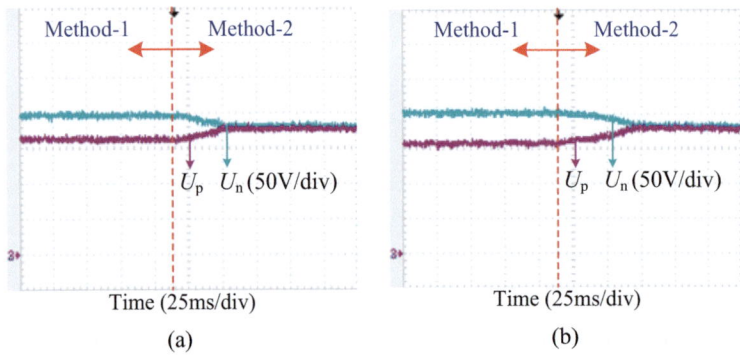

Fig. 6.32 Experimental waveforms of capacitor voltages with Method-1 and Method-2 when power factor angle is 90°. **a** Balanced grid voltage. **b** Unbalanced grid voltage

grid current direction plays a crucial role in determining their impact. The proposed GMPC scheme simultaneously considers the dc-link capacitors voltage difference and grid current directions to select the appropriate small vector. By doing so, the NPV balance can be achieved in different power factors.

6.3.2 Based on Selective Harmonics Elimination Pulse Width Modulation

The NPV balance can be achieved with the MPC method in above section. However, the switching frequency of MPC is not fixed, and the filter parameters are difficult to design. Moreover, the MPC method performs well at high switching frequency,

6.3 Advanced Control Technology for Neutral-Point Voltage Balance Under ...

but the effect is not so good at low switching frequency. Fortunately, the selective harmonic elimination pulse width modulation (SHEPWM) method can effectively eliminate low-order harmonics and generate high-quality output waveforms at low switching frequency. SHEPWM methods with novel NPV balance control strategies have been presented for both active NP clamped three-level inverters and 3LT^{2}I, achieving low-order harmonics elimination and NPV balance. Nevertheless, the above literature using SHEPWM primarily focus based on normal grid voltage conditions. When a grid fault occurs, the grid code requires the inverter to inject reactive power to the grid during LVRT operation to support grid voltage recovery. However, the compensated reactive power increases the unbalance of NPV, which is difficult to control with the conventional SHEPWM method.

Therefore, it is quite urgent to study a SHEPWM method with enhanced NPV balance control capability for LVRT applications. In order to solve these problems, an improved SHEPWM (IM-SHEPWM) method based on IPSO algorithm is presented by the author's team [20]. An improved particle swarm optimization (IPSO) algorithm has been illustrated in Chap. 5, in which the harmonic elimination equations are constructed and the switching angles are solved using IPSO algorithm. In this section, the IM-SHEPWM mainly elaborated on NPV balance control for LVRT applications. This section contains two parts: theoretical analysis of IM-SHEPWM and experimental verifications.

Part 1: Theoretical Analysis of IM-SHEPWM

Grid fault may lead to the grid voltage dip from Type A to G, among which Types A, B, and E attract great interest in PV generation systems and related test organizations. Under power grid fault condition, grid codes require that the PV inverter inject an amount of reactive power into the grid to help grid voltage restoration. The reactive current reference i_q^* satisfying

$$i_q^* = \sqrt{I_m^2 - i_d^{*2}} \tag{6.134}$$

where i_d^* is the reference active current, which is determined by the power controller or DC voltage controller. I_m is the system rated current.

Under normal conditions, i_q^* is usually set to 0 in order to generate as much active power as possible. When $i_q^* = 0$, P-type small vectors will decrease upper capacitor voltage, while N-type small vectors decrease the bottom capacitor voltage. When the PV inverter operates under LVRT conditions, the i_q^* is no longer zero, making the effects of each small vector on NPV uncertain. The effects of small vector on NPV are not only determined by the type of small vector but also the current direction flows into NP. Thus, the IM-SHEPWM increases the degree of freedom of the output current direction and combines with the NPV deviation to achieve the NPV balance control.

Taking the small vector [POO] as an example, it has a redundant small vector [ONN], which is used to control the NPV. For ease of analysis, the three-phase output voltages of 3LT^{2}I using the SHEPWM method are re-drawn in Fig. 6.33.

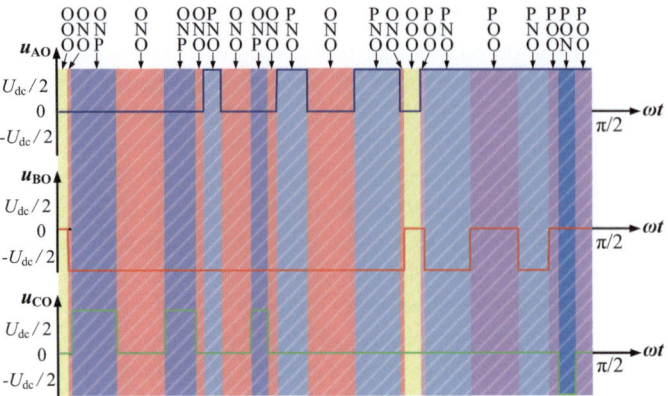

Fig. 6.33 Three-phase output voltages of the 3LT^{2}I using SHEPWM method

Conventional methods are mostly used for unity power factor or high-power factor applications, and the effects of P- and N-type small vectors on NPV is certain. Increasing dwell time of P-type small vector can reduce U_p, while increasing dwell time of N-type small vector can decrease U_n, and their effects on NPV are opposite. However, after detailed research, it is found that under LVRT conditions, the power factor is significantly low, and the effect of P- and N-type small vectors on NPV is related to the direction of output current. Therefore, the capacitor voltage deviation and current direction should be considered simultaneously in NPV balance control during LVRT operation. Based on this concept, an enhanced NP balance control method is proposed. The effects of [POO] and [ONN] on NPV are analysed in detail as follows.

When $U_p - U_n < 0$, if the small vector [POO] appears in the output vectors, as shown in Fig. 6.34, its corresponding NP current is $i_o = i_b + i_c = -i_a$. If the direction of i_a is positive, U_p will be further reduced, as depicted in Fig. 6.34a. For the purpose of effectively controlling the NPV balance, [POO] should be replaced by [ONN] without changing the output line voltage. If the direction of i_a is negative, it is beneficial for NPV balance; therefore, no replacement will be made to maintain the NPV balance, as depicted in Fig. 6.34b. When $U_p - U_n > 0$, similar analysis can be performed.

Table 6.3 summarizes the criteria and the replacing small vectors in the proposed IM-SHEPWM method. This control scheme can quickly achieve NPV balance during LVRT operation.

Figure 6.35 depicts the overall control block diagram of the 3LT^{2}I. The grid voltages are sampled, and the grid phase θ is obtained by PLL. The active and reactive currents i_d and i_q are derived from the sampled ac-side output currents through *abc/dq* transform. On the basis of LVRT requirements, the active and reactive current references can be obtained. Subsequently, current error between reference and actual values is fed into the hybrid passivity-based controller, and particularly

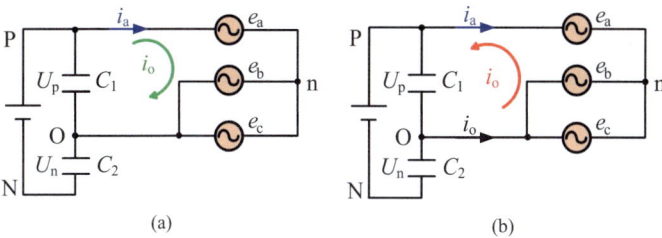

Fig. 6.34 Effect of POO NPV and output current directions of the 3LT^{2}I in full power factor range when $U_p < U_n$ (**a**) POO with $i_a > 0$ (**b**) POO with $i_a < 0$

Table 6.3 Summarisation of the criteria and the replacing small vectors in the proposed IM-SHEPWM method

Criteria	Replacing small vectors
$U_p < U_n$, [POO] with $i_a > 0$	[POO] is replaced by [ONN]
$U_p < U_n$, [PPO] with $i_c < 0$	[PPO] is replaced by [OON]
$U_p < U_n$, [OPO] with $i_b > 0$	[OPO] is replaced by [NON]
$U_p < U_n$, [OPP] with $i_a < 0$	[OPP] is replaced by [NOO]
$U_p < U_n$, [OOP] with $i_c > 0$	[OOP] is replaced by [NNO]
$U_p < U_n$, [POP] with $i_b < 0$	[POP] is replaced by [ONO]
$U_p < U_n$, [ONN] with $i_a < 0$	[ONN] is replaced by [POO]
$U_p < U_n$, [OON] with $i_c > 0$	[OON] is replaced by [PPO]
$U_p < U_n$, [NON] with $i_b < 0$	[NON] is replaced by [OPO]
$U_p < U_n$, [OPP] with $i_a > 0$	[OPP] is replaced by [NOO]
$U_p < U_n$, [NNO] with $i_c < 0$	[NNO] is replaced by [OOP]
$U_p < U_n$, [ONO] with $i_b > 0$	[ONO] is replaced by [POP]
$U_p > U_n$, [POO] with $i_a < 0$	[POO] is replaced by [ONN]
$U_p > U_n$, [PPO] with $i_c > 0$	[PPO] is replaced by [OON]
$U_p > U_n$, [OPO] with $i_b < 0$	[OPO] is replaced by [NON]
$U_p > U_n$, [OPP] with $i_a > 0$	[OPP] is replaced by [NOO]
$U_p > U_n$, [OOP] with $i_c < 0$	[OOP] is replaced by [NNO]
$U_p > U_n$, [POP] with $i_b > 0$	[POP] is replaced by [ONO]
$U_p > U_n$, [ONN] with $i_a > 0$	[ONN] is replaced by [POO]
$U_p > U_n$, [OON] with $i_c < 0$	[OON] is replaced by [PPO]
$U_p > U_n$, [NON] with $i_b > 0$	[NON] is replaced by [OPO]
$U_p > U_n$, [OPP] with $i_a < 0$	[OPP] is replaced by [NOO]
$U_p > U_n$, [NNO] with $i_c > 0$	[NNO] is replaced by [OOP]
$U_p > U_n$, [ONO] with $i_b < 0$	[ONO] is replaced by [POP]

suitable for LVRT conditions, to generate output voltages V_d and V_q as input variables of modulation scheme.

In the IM-SHEPWM scheme, the proposed IPSO algorithm is used to solve the harmonic elimination equations, then the switching angles are obtained and stored in the switching angle lookup table. Combined with the modulation index and phase obtained from the current loop, the switching angles are searched online for SHEPWM generation. According to the capacitor voltage deviation and output current direction under different power factors, the proposed enhanced NP balance control method is applied to achieve high-quality grid-connected currents and NPV balance simultaneously in LVRT operation.

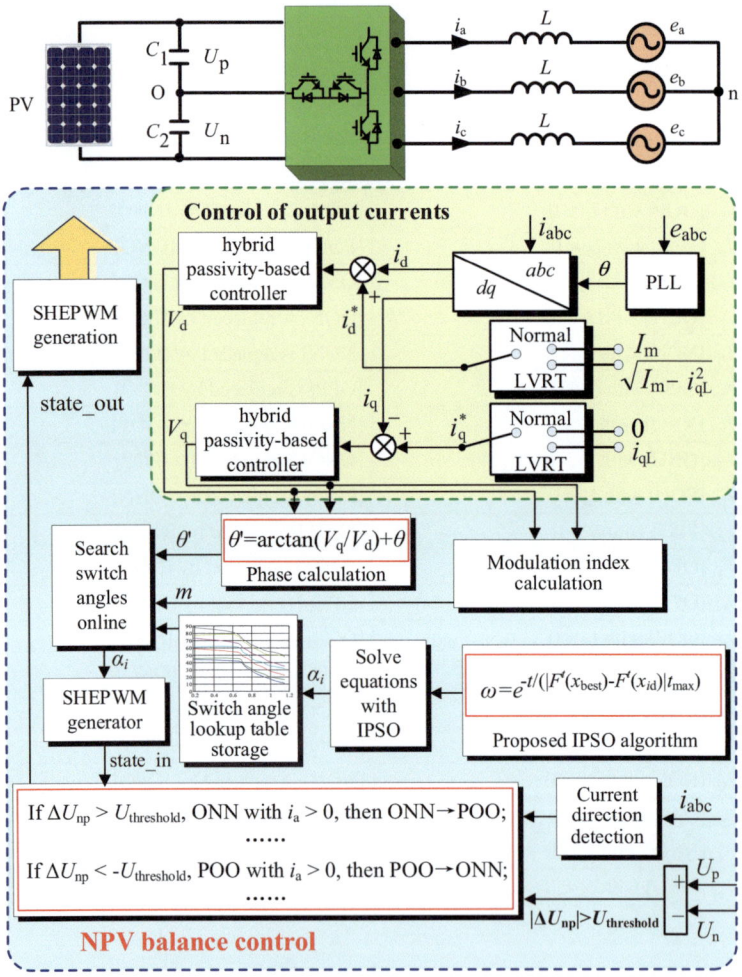

Fig. 6.35 Control diagram of IM-SHEPWM

Part 2: Experimental Verification for IM-SHEPWM

In order to verify the effectiveness of the proposed scheme, experimental tests are conducted. The dc-link voltage U_{dc} is set to 340 V, the RMS value of power grid voltage set 110 V in normal conditions. The parameter L and C are set to 10 mH and 940 μF.

The reference currents are $i_d^* = 20$ A, $i_q^* = 0$ A when the grid is normal. In the experimental verification, three cases are taken as the verification conditions. In the three cases, the dip depths of Types A, B, and E are 50% for three-phase dip, 40% for single-phase dip, and 70% for two-phase dip respectively, which are the conditions with maximum reactive power injection. Three methods are used for comparison. Method-1 is the conventional SHEPWM method without NPV balance control. Method-2 is the conventional SHEPWM method with NPV balance control, which does not consider the output current directions and is suitable for normal grid. Method-3 is the proposed IM-SHEPWM method, in which the output current directions and NPV deviation are considered simultaneously under different power factors to realize the NPV balance control.

Case 1: Types A with Dip Depths is 50% for Three-Phase

Figure 6.36 illustrates experimental waveforms of grid voltages and inverter output currents during LVRT operation for Type A dip. Since three-phase grid voltages drop symmetrically 50% by nominal value shown in Fig. 6.36a, the currents delivered to the grid are $i_d^* = 0$ A, $i_q^* = -20$ A. The power factor in this case is very low. The total harmonic distortion (THD) of output currents for Method-3 is 3.3% after reactive power injection. It can also be seen from Fig. 6.36b that the phase difference between grid voltage and output current is close to π/2, indicating that the power factor in this case is very small.

Figure 6.37 depicts the line voltage frequency spectrum of the proposed SHEPWM in LVRT operation for Type A dip. The aimed 5th, 7th, 11th, 13th, 17th, and 19th harmonic contents have been effectively eliminated (the percentage of each harmonic

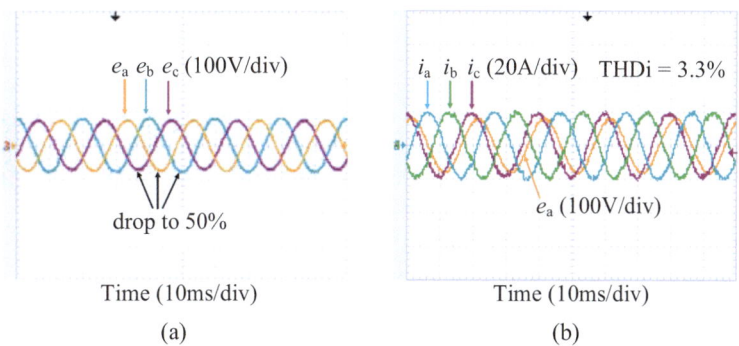

Fig. 6.36 Experimental waveforms of grid voltages and output currents of Method-3 in LVRT operation for Type A dip. **a** Grid voltages. **b**. Output currents

content in the fundamental amplitude is less than 0.20%) which shows that the proposed SHEPWM technology has achieved good results in closed-loop control.

Line voltage and capacitor voltages experimental waveforms are presented in Fig. 6.38 during method switching for Type A dip. When Method-1 without NPV control is used, the maximum value of NPV deviation reaches 35 V. After switching to Method-2, this value still reaches 35 V because the effect of output current direction on NPV is not considered at low power factor as shown in Fig. 6.38a. When Method-1 is switched to Method-3, the NPV can be effectively balanced since both the output current direction and the NPV deviation are considered in Method-3 to select an appropriate small vector shown in Fig. 6.38b.

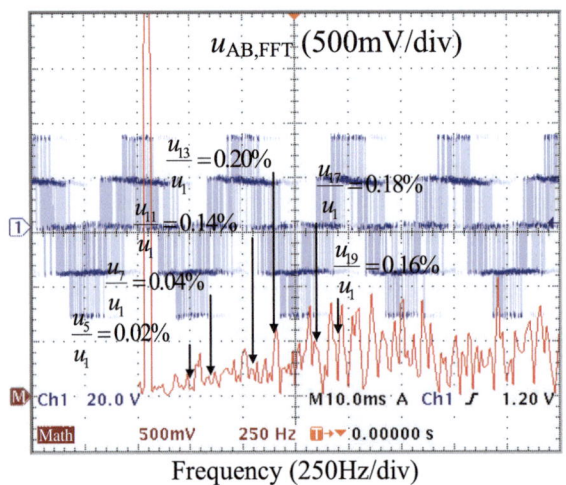

Fig. 6.37 Experimental waveforms of line voltage frequency spectrum of Method-3 in LVRT operation for Type A dip

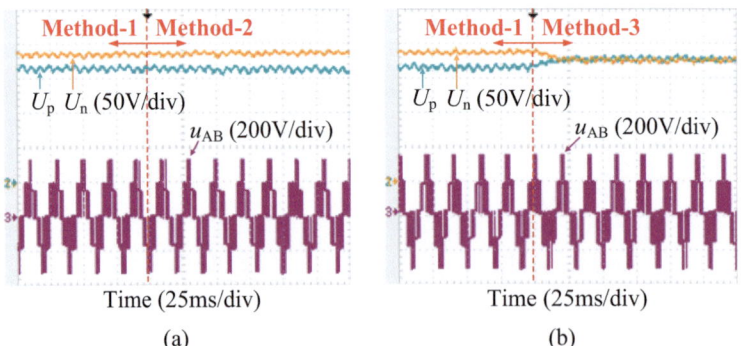

Fig. 6.38 Experimental results of line voltage and capacitor voltages during methods switching for Type A dip. **a** Method-1 switches to Method-2. **b** Method-1 switches to Method-3

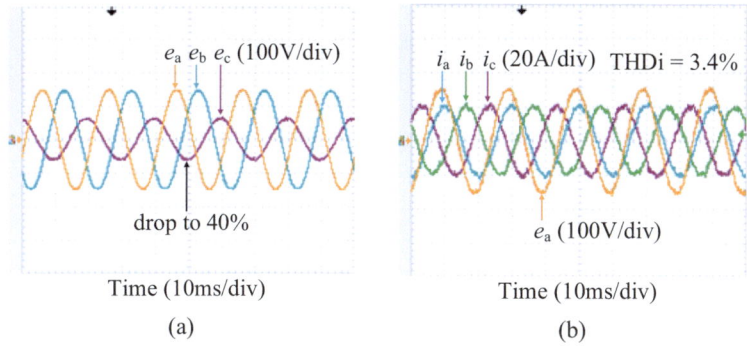

Fig. 6.39 Experimental waveforms of grid voltages and output currents of Method-3 in LVRT operation for Type B dip. **a** Grid voltages. **b** Output currents

Case 2: Types A with Dip Depths is 40% for Single-Phase

Figure 6.39 shows the grid voltage waveforms for Type B dip, where C-phase grid voltage drops to 40% of the nominal value. In accordance with LVRT requirements, the reactive power injection is $i_d^* = 18$ A, $i_q^* = -8$ A. The proposed method can achieve high-quality output current with THD of 3.4%. Additionally, it can also be observed that the phase difference between grid voltage and output current is small, indicating that the power factor in this case is very high.

The line voltage frequency spectrum of Method-3 for Type B dip is presented in Fig. 6.40. It can be seen that among the selected 5th, 7th, 11th, 13th, 17th, and 19th harmonics, the largest percentage of the fundamental amplitude is 0.10%, which corresponds to the 17th harmonic. It can be seen that Method-3 can effectively eliminate the selected low-order harmonics when Type B dip occurs, which is consistent with theoretical analysis.

Figure 6.41 illustrates the line voltage and capacitor voltage waveforms during method switching for Type B dip. Because the NPV balance control is not applied in LVRT operation, the NPV deviation of Method-1 exceeds 60 V. After switching methods, the NPV deviations of both Method-2 and Method-3 are significantly reduced as the influence of small vectors on NPV is certain at high power factor.

Case 3: Types A with Dip Depths Is 70% for Two-Phase

A- and b-phase grid voltages drop to 70% of the nominal value for Type E, as demonstrated in Fig. 6.42a. The active and reactive current references are 18 A and 8 A, respectively, to meet LVRT requirements. The THD of output currents using Method-3 is 3.7%, which satisfies the requirements of IEEE standard. It can be seen from Fig. 6.42b that the phase difference between grid voltage and output current is extremely small, which demonstrates that the power factor in this case is very close to 1.

Figure 6.43 depicts the line voltage frequency spectrum of Method-3 under LVRT conditions for Type E dip. The selected harmonic components are the 5th, 7th, 11th, 13th, 17th, and 19th harmonics, and their percentages relative to the fundamental

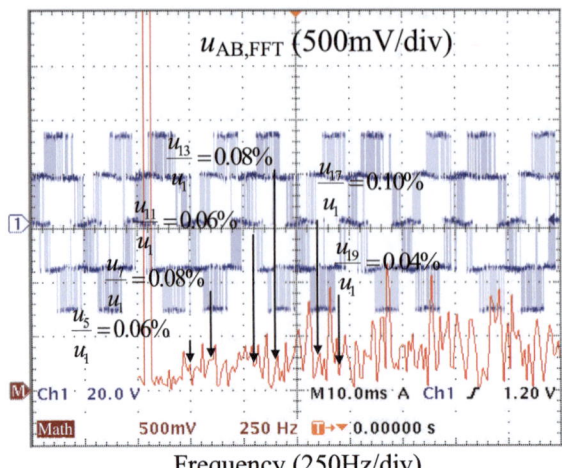

Fig. 6.40 Experimental waveforms of line voltage frequency spectrum of Method-3 in LVRT operation for Type B dip

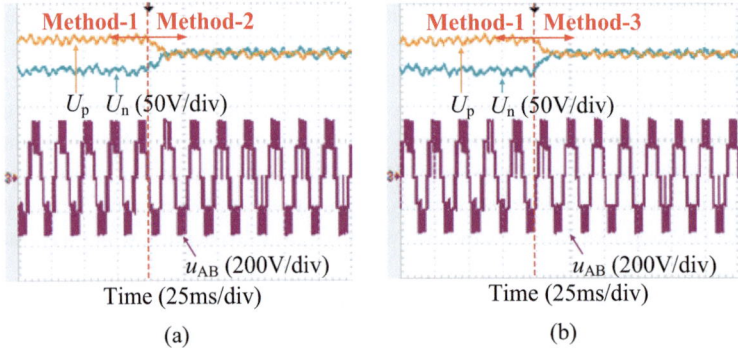

Fig. 6.41 Experimental results of line voltage and capacitor voltages during methods switching for Type B dip. **a** Method-1 switches to Method-2. **b** Method-1 switches to Method-3

amplitude are 0.01%, 0.08%, 0.20%, 0.27%, 0.13%, and 0.12%, respectively. The experimental results show that the switching angles solved by the proposed method can effectively eliminate the selected low-order harmonics in LVRT operation.

Figure 6.44 presents line voltage and capacitor voltages experimental waveforms during method switching for Type E dip. The NPV deviation of Method-1 exceeds 60 V due to the absence of NPV balance control. After switching methods, both Method-2 and Method-3 can decrease the capacitor voltage difference to within 10 V. Since the influence of small vectors on the NPV is certain when the power factor is close to 1, Method-2 and Method-3 can successfully control the NPV balance with similar performance.

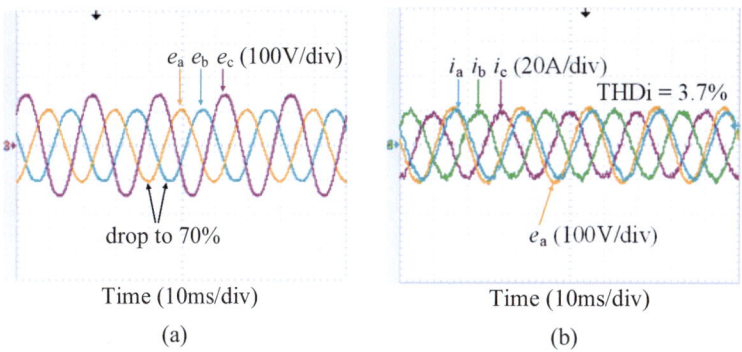

Fig. 6.42 Experimental waveforms of grid voltages and output currents of Method-3 in LVRT operation for Type C dip. **a** Grid voltages. **b** Output currents

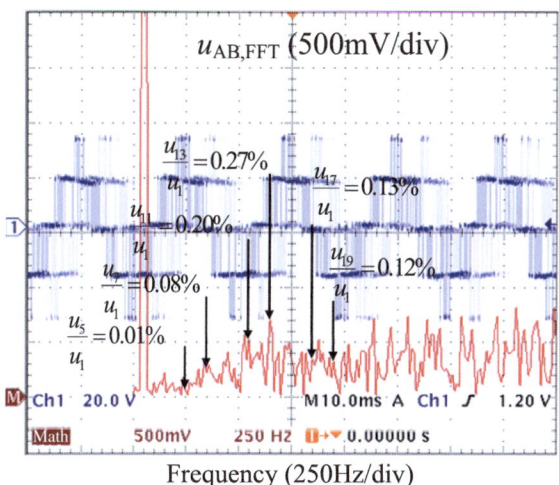

Fig. 6.43 Experimental waveforms of line voltage frequency spectrum of Method-3 in LVRT operation for Type E dip

Conclusion: Reactive power injection increases the NPV unbalance during LVRT operation, which makes the NPV balance control tremendously difficult. Through detailed research, it is found that in LVRT operation, the power factor is very low, and the effect of P- and N-type small vectors on NPV depends on the output current direction. To address this issue, an IM-SHEPWM for LVRT operation is proposed, which increases the degree of freedom of the output current direction and combines with the NPV deviation to better realize the NP balance control. It overcomes the problem that the conventional NP balance control method of SHEPWM, which is designed for normal power grid, only considers the NPV deviation and does not

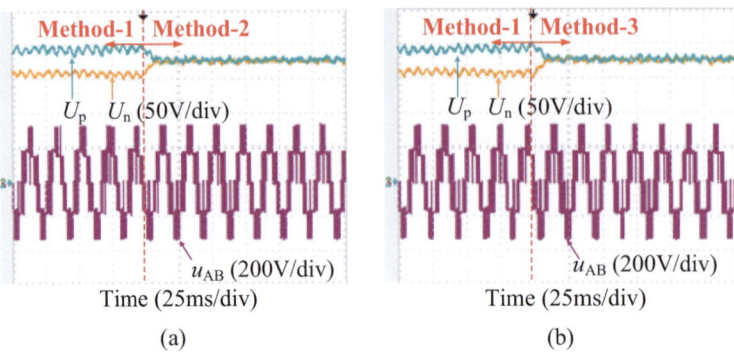

Fig. 6.44 Experimental results of line voltage and capacitor voltages during methods switching for Type C dip. **a** Method-1 switches to Method-2. **b** Method-1 switches to Method-3

consider the output current direction in LVRT operation. Therefore, it cannot balance the NPV well.

6.3.3 Based on Common-Mode Voltage Reduction Modulation

For transformerless PV generation systems, another problem that must be effectively addressed is the leakage current, which associated with common-mode voltage (CMV). In the previous section, the NPV balance control methods have high CMV, which will easily increase the leakage current [21]. Thus, the author proposes a selective SVM (SSVM) method in this section [22], which reduces NPV oscillation and CMV for $3LT^2I$ operating in generation mode under grid fault conditions. The SSVM includes four steps, and they are presented as follows.

Step 1: Analysis of the Reasons of Uncontrollable NPV Balance

For ease of analysis, the space voltage vector diagram is re-depicted in Fig. 6.45, which includes 27 basic voltage vectors in total.

When the neutral-point O of the dc-link voltage is selected as the reference point, the CMV is defined as the average value of three phase output voltage.

$$u_{cm} = \frac{u_{AO} + u_{BO} + u_{CO}}{3} \quad (6.135)$$

The space vectors and their corresponding CMVs are listed in Table 6.4. It is clear that the CMV magnitudes of the bolded voltage vectors are relatively high. The CMV magnitude is reduced by avoiding the utilization of these vectors. For example, in large medium small zero vector modulation (LMSZVM) method, the large, medium, selected small and zero vectors are used to reduce the CMV. The CMV value is $U_{dc}/$

6.3 Advanced Control Technology for Neutral-Point Voltage Balance Under …

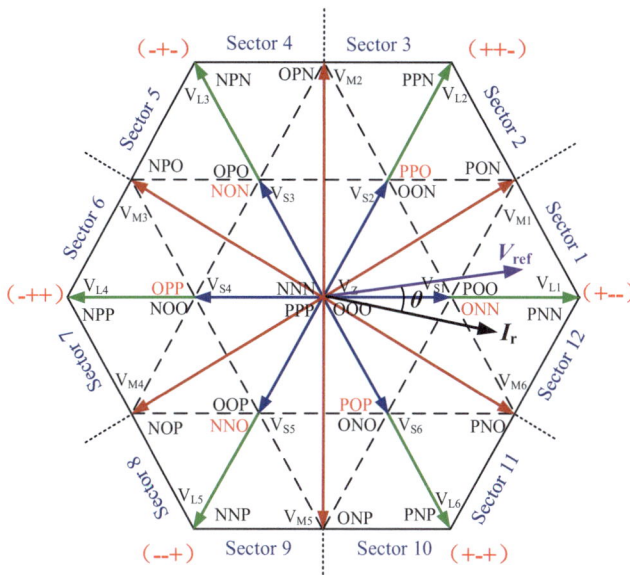

Fig. 6.45 Space vector diagram of the 3LT^{2}I

6 and $-U_{dc}/6$ generated by large and partial small ([POO], [OON], [OPO], [NOO], [OOP], [ONO]) vectors, while the CMV magnitude is 0 produced by medium vectors and zero vector [OOO]. Thus, the ground current can be suppressed effectively.

On the other hand, the medium vector charges or discharges the dc-link capacitors with different current direction, leading to the NPV oscillation. This oscillation can only be compensated by using the P-type or N-type partial small vector based on the location of reference output voltage V_{ref}. However, this method is only suitable for normal grid conditions and high-power-factor operation. In this case, the power factor angle θ between output voltage V_{ref} and output current I_r is less than $\pi/6$. For example, in the space voltage vector diagram, when V_{ref} locates in Sector 1 and θ is $\pi/15$, small vector [POO] and [OON] are selected to balance the NPV.

Nevertheless, in LVRT operation, the grid code requires the inverter to deliver a certain amount of reactive current to the utility grid. Thus, the power factor angle easily exceeds $\pi/6$. Under such conditions, the conventional LMSZVM method struggles to maintain NPV balance while reducing the CMV.

For example, when the V_{ref} locates in Sector 1 and the power factor angle θ is equal to $\pi/4$ as shown in Fig. 6.46a, the effects of small vector [POO] and [OON] on NPV are identical to those of medium vector [PON]. Consequently, the NPV oscillation and dc unbalance cannot be controlled any more. A similar analysis is used for all other sectors when θ is $\pi/4$, and the NPV uncontrolled area with patterns is presented in Fig. 6.46b. Unfortunately, this NPV uncontrolled area further expands when power factor angle θ is equal to $5\pi/12$. Meanwhile, compared to normal conditions, the unbalance grid voltages will worsen NPV balance.

Table 6.4 The space vectors and CMV of 3LT^{2}I

Vectors	Names	u_{cm}	Names	u_{cm}
Zero	**V$_{Zn}$[NNN]**	**$-3U_{dc}/6$**	V$_{Z0}$[OOO]	0
	V$_{Zp}$[PPP]	**$3U_{dc}/6$**		
Small	V$_{S1n}$[ONN]	$-2U_{dc}/6$	V$_{S1p}$[POO]	$U_{dc}/6$
	V$_{S2p}$[PPO]	$2U_{dc}/6$	V$_{S2n}$[OON]	$-U_{dc}/6$
	V$_{S3n}$[NON]	$-2U_{dc}/6$	V$_{S3p}$[OPO]	$U_{dc}/6$
	V$_{S4p}$[OPP]	$2U_{dc}/6$	V$_{S4n}$[NOO]	$-U_{dc}/6$
	V$_{S5n}$[NNO]	$-2U_{dc}/6$	V$_{S5p}$[OOP]	$U_{dc}/6$
	V$_{S6p}$[POP]	$2U_{dc}/6$	V$_{S6n}$[ONO]	$-U_{dc}/6$
Medium	V$_{M1}$[PON]	0	V$_{M4}$[NOP]	0
	V$_{M2}$[OPN]	0	V$_{M5}$[ONP]	0
	V$_{M3}$[NPO]	0	V$_{M6}$[PNO]	0
Large	V$_{L1}$[PNN]	$-U_{dc}/6$	V$_{L4}$[NPP]	$U_{dc}/6$
	V$_{L2}$[PPN]	$U_{dc}/6$	V$_{L5}$[NNP]	$-U_{dc}/6$
	V$_{L3}$[NPN]	$-U_{dc}/6$	V$_{L6}$[PNP]	$U_{dc}/6$

Bold: The CMV magnitudes of the bolded voltage vectors are relatively high.

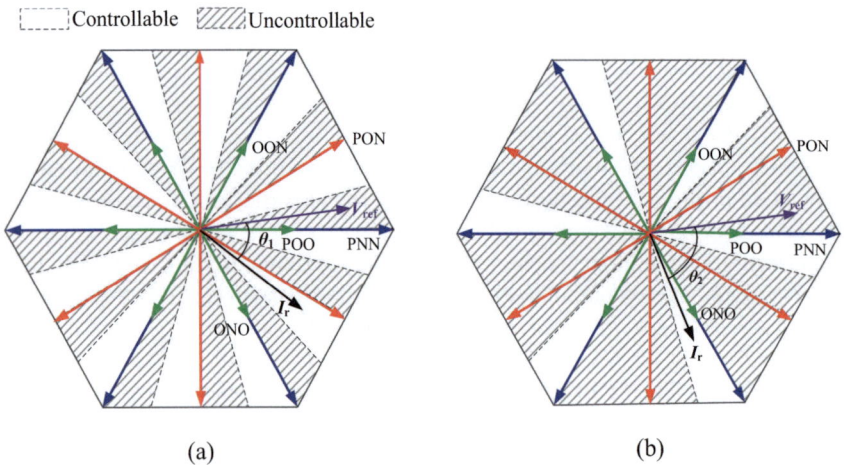

Fig. 6.46 Uncontrollable area with LMSZVM method under different power factor. **a** θ is $\pi/4$. **b** θ is $5\pi/12$

Based on the above analysis, we can obtain that the NPV has uncontrollable area using LMSZVM in LVRT operation. The reasons are analyzed and given as follows.

The effects of each small vector on NPV are not only determined by the type of small vector but also the current direction flows into NP. In the conventional LMSZVM method, only the type of small vector is considered, which leads to the

6.3 Advanced Control Technology for Neutral-Point Voltage Balance Under ...

effect of small vector on NPV is same as that of medium vector when power factor is increased. Thus, the uncontrollable area is produced in LVRT operation.

Under LVRT operating conditions, the relative position of V_{ref} and I_r varies with the injected reactive current. Therefore, the effects of small vector on NPV are uncertain, which is different from normal conditions. In LVRT operation, to reduce the NPV oscillation in all areas, the correct small vectors which have opposite effect on NPV with medium vector should be selected.

Step 2: Correct Small Vector Selection
The effects of each small vector on NPV are determined by the current direction. The current directions depending on regions are shown in Table 6.5 and space vector diagram of the $3LT^2I$.

Sector 2 is taken as a representative example to explain the principle of the correct small vector selection. It is easily obtained that the range of θ is from 0 to $\pi/2$ under LVRT condition. Thus, three current directions of $(+--)$, $(++-)$ and $(+-+)$ are included when V_{ref} locates in Sector 2. In this case, the voltage vectors to synthesize the V_{ref} includes medium vector [PON], small vectors with low CMV [POO] [ONO] and [OON], large vector [PPN] and zero vector [OOO]. The current paths of the medium and small vector are shown in Fig. 6.47. It can be observed that the NPV is determined by the type of vectors and its current directions.

According to different current directions, correct small vector selection three can be classed into cases, which are elaborated as follows.

Case 1: $i_a > 0, i_b < 0$ and $i_c < 0$
As shown in Fig. 6.47a with dotted green line, the medium vector [PON] discharge the upper capacitor, and the NPV is increased. While, the lower capacitor is discharged when the small vector [OON] is employed and i_c is smaller than 0. In this case, the neutral current flows out the neutral-point as depicted in Fig. 6.47d with dotted green line. Similarly, the small vector [ONO] discharges the lower capacitor with i_b smaller than 0, as dotted green line shows in Fig. 6.47c. When vector [OON] or [ONO] is adopted, the NPV can be decreased. Thus, they are selected as the correct small vector, which has opposite effect on NPV with medium vector. The vectors that synthesizing the voltage V_{ref} include large vector [PPN], medium vector [PON], zero vector [OOO] and small vector [OON] or [ONO].

Case 2: $i_a > 0, i_b > 0$ and $i_c < 0$ (The Range of θ Is from 0 to $\pi/6$)
Note that for this case, the lower capacitor is discharged by the medium vector [PON], which decreases the NPV as shown in Fig. 6.47a with solid red line. In contrast, owing to i_a and i_b are larger than 0, the upper and lower capacitor are discharged and charged with small vector [POO] and [ONO], respectively. Thus, the NPV is increased as

Table 6.5 The current directions and regions of the $3LT^2I$

Region of I_r	1	2	3	4	5	6
Three phase current direction	$(+--)$	$(++-)$	$(-+-)$	$(-++)$	$(--+)$	$(+-+)$

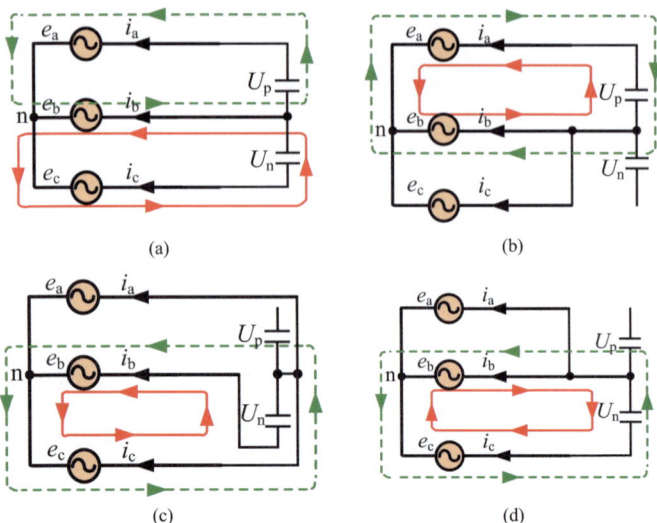

Fig. 6.47 Current paths of the medium and small vector. **a** Medium vector [PON]. **b** Small vector [POO]. **c** Small vector [ONO]. **d** Small vector [OON]

shown in Fig. 6.47b, c with solid red line. So, the correct small vectors are [POO] and [ONO]. In this case, the voltage V_{ref} can be synthesized with vectors [PPN], [PON], [OOO] and or [OOP] [ONO].

Case 3: $i_a > 0$, $i_b < 0$ and $i_c > 0$

Being similar with Case 1, the medium vector [PON] increases the NPV. Considering i_a and i_c are larger than 0 at the same time, the small vectors [POO], [OON] cannot be used to suppress the NPV oscillation, since they have same effect on NPV as [PON]. Nevertheless, as shown in Fig. 6.47c with green dotted line, the small vector [ONO] has the ability to decrease NPV. Consequently, only the vector [ONO] can be selected as the correct small vector to mitigate the NPV oscillation induced by [PON], since it discharges the lower capacitor with $i_b < 0$. Thus, the vectors [PPN], [PON], [OOO] and [OON] are adopted to synthesize the reference voltage vector.

The afore-mentioned analysis show that the NPV oscillation can be effectively suppressed under different operations. The key to achieve this objective is the correct small vectors selection. As analyzed above, two correct small vectors can be selected in some area, which is different from that in normal operation. However, in PV inverters, only one optimal small vector can be used in each switching cycle. The optimal small vector having the most effective impact on NPV among the correct small vectors should be used in LVRT operation. Hence, how to determine the optimal small vector and arrange the switching sequence is developed as follows.

Step 3: Optimal Small Vector Determination and Switching Sequence Arrangement

The reference voltage vector can be expressed as

6.3 Advanced Control Technology for Neutral-Point Voltage Balance Under ...

$$V_{\text{ref}} = V_R(t)e^{j\omega t} \tag{6.136}$$

where $V_R(t)$ is the magnitude of the reference voltage vector. In LVRT operation, the $V_R(t)$ is not constant due to the feedforward component of the negative-sequence grid voltage. Meanwhile, the grid currents are controlled to be balanced and the power factor varies with injected reactive current value. The balanced three-phase grid currents are presented as

$$\begin{cases} i_a = I_m \cos(\omega t - \theta) \\ i_b = I_m \cos(\omega t - \theta - \frac{2\pi}{3}) \\ i_c = I_m \cos(\omega t - \theta + \frac{2\pi}{3}) \end{cases} \tag{6.137}$$

Then the reference grid current I_r can be obtained as

$$I_r = I_m e^{j(\omega t - \theta)} = \frac{2}{3}\left(i_a e^{j0} + i_b e^{j\frac{2\pi}{3}} + i_c e^{j\frac{4\pi}{3}}\right) \tag{6.138}$$

It is known that the NPV is caused by NP current flowing out or into the neutral-point. In one switching cycle, ΔU_{np} presents the NPV variation and is expressed as

$$\Delta U_{np} = \frac{1}{C}\int_0^{T_s} i_{np}(t)dt = \frac{1}{C}(D_M i_M + D_S i_S)T_s \tag{6.139}$$

where D_M, D_S, i_M, and i_S are duty cycles and corresponding currents of medium and small vector in each switching cycle, respectively. It is evident that the NPV is influenced by the duty cycles and grid currents of medium and small vectors. Thus, the NPV oscillation caused by medium vector should be effectively compensated by the optimal small vector in LVRT operation.

In order to determine the optimal small vector among the correct small vectors, the force of small vector SVF on NPV that composed of the grid current and duty cycle is proposed and defined as

$$SVF = D_S|i_S| \tag{6.140}$$

Still taking Sector 2 as an example, the vectors of V_{L2}[PPN], V_{M1}[PON], V_{z0}[OOO] and V_{S1p}[POO], V_{S2n}[OON] or V_{S6n}[ONO] are adopted to synthesize the V_{ref}. As revealed by the above analysis, the small vectors are used to suppress the NPV oscillation generated by medium vector. The $SVFs$ of V_{S1p}[POO], V_{S2n}[OON] and V_{S6n}[ONO] are represented as SVF_{S1p}, SVF_{S2n}, and SVF_{S6n}, respectively. Corresponding to the previous part, this part is also divided into three cases.

Case 1: $i_a > 0$, $i_b < 0$ and $i_c < 0$

As analyzed above, in this case, the correct small vectors include V_{S2n}[OON] and V_{S6n}[ONO], which has opposite effects on NPV with medium vector. When [OON] is selected, the V_{ref} is synthesized with V_{L2}[PPN], V_{M1}[PON], V_{z0}[OOO] and V_{S2n}[OON]. Based on the volt-second balance principle, the relationship between the duty cycles and voltage vectors is expressed as

$$\begin{cases} V_{L2}D_{L2} + V_{M1}D_{M1} + V_{S2n}D_{S2n} + V_{Z0}D_{Z0} = V_{ref} \\ D_{L2} + D_{M1} + D_{S2n} + D_{Z0} = 1 \end{cases} \quad (6.141)$$

where D_{L2}, D_{M1}, D_{S2n}, and D_{Z0} are the duty cycle of V_{L2}, V_{M1}, V_{S2n} and V_{z0}, respectively. Moreover, in order to suppress the NPV oscillation, the NPV increment caused by medium vector V_{M1}[PON] should be equal to that of decrement with small vector V_{S2n}[OON].

$$\Delta U_{np\text{-}M}1 = \Delta U_{np\text{-}s}2n \quad (6.142)$$

where $\Delta U_{np\text{-}M1}$ and $\Delta U_{np\text{-}S2n}$ present NPV variation induced by V_{M1}, V_{S2n}, respectively. They are expressed as follows

$$\begin{cases} \Delta U_{np\text{-}M}1 = \frac{1}{C}\int_0^{T_s} i_b(t)dt \\ \Delta U_{np\text{-}s}2n = \frac{1}{C}\int_0^{T_s} i_c(t)dt \end{cases} \quad (6.143)$$

In each switching cycle, assuming that there is no change in the current, above expression is re-written as

$$\begin{cases} \Delta U_{np\text{-}M}1 = \frac{1}{C}i_b D_M 1 \\ \Delta U_{np\text{-}s}2n = \frac{1}{C}i_c D_S 2n \end{cases} \quad (6.144)$$

Thus, the relationship between D_{M1} and D_{S2n} can be derived as

$$i_c D_{S2n} = i_b D_{M1} \quad (6.145)$$

According to the above analysis, the duty cycles of D_{L2}, D_{M1}, D_{S2n}, and D_{Z0} are obtained as

$$\begin{cases} D_{L2} = \sqrt{3}\frac{V_R(t)}{V_{dc}}\sin(\omega t - \frac{\pi}{6}) - \frac{D_{S2n}}{2} \\ D_{M1} = 2\sqrt{3}\frac{V_R(t)}{V_{dc}}\sin(\frac{\pi}{3} - \omega t) \\ D_{S2n} = 2\sqrt{3}\frac{i_b}{i_c}\frac{V_R(t)}{V_{dc}}\sin(\frac{\pi}{3} - \omega t) \\ D_{Z0} = 1 - D_{L2} - D_{M1} - D_{S2n} \end{cases} \quad (6.146)$$

It should be noted that the duty cycles are affected by the grid dips for $V_R(t)$. To generate the correct switching sequence, the duty cycles in above equation should be larger than 0. The corresponding currents of V_{S2n}[OON] is equal to i_c. Substituting D_{S2n} and i_c into the expression of SVF, and considering the limitation of duty cycles, the small vector force of SVF_{S2n} is derived as

6.3 Advanced Control Technology for Neutral-Point Voltage Balance Under …

$$SVF_{S2n} = \min \begin{cases} 2\sqrt{3}\frac{i_b}{i_c}\frac{V_R(t)}{U_{dc}}\sin\left(\frac{\pi}{3} - \omega t\right) \\ 2\left[1 - \frac{\sqrt{3}V_R(t)}{U_{dc}}\sin\left(\frac{2\pi}{3} - \omega t\right)\right] \\ 6\frac{V_R(t)}{U_{dc}}\sin\left(\omega t - \frac{\pi}{6}\right) \end{cases} \left| I_m \cos\left(\omega t - \theta + \frac{2\pi}{3}\right) \right| \quad (6.147)$$

It can be observed that the *SVF* includes the information of grid dips and reactive power by introducing $V_R(t)$ and θ. The electrical degree ωt and power factor angle θ satisfy

$$\begin{cases} \frac{\pi}{6} + 2k\pi \leq \omega t \leq \frac{\pi}{3} + 2k\pi \\ \frac{\pi}{6} \leq \theta \leq \frac{\pi}{3} \quad (k = 0, 1, 2 \ldots) \end{cases} \quad (6.148)$$

Similarly, when [ONO] is adopted, the V_{ref} is synthesized as

$$\begin{cases} V_{L2}D_{L2} + V_{M1}D_{M1} + V_{S6n}D_{S6n} + V_{Z0}D_{Z0} = V_{\text{ref}} \\ D_{L2} + D_{M1} + D_{S6n} + D_{Z0} = 1 \end{cases} \quad (6.149)$$

where D_{S6n} is the duty cycle of V_{S6n}. The NPV variation caused by [ONO] is expressed as

$$\Delta U_{\text{np-S6n}} = \frac{1}{C} i_b D_{S6n} \quad (6.150)$$

Therefore, to reduce the NPV oscillation, the duty cycles of D_{M1} and D_{S6n} should be satisfied as

$$D_{S6n} = D_{M1} \quad (6.151)$$

According to above equations, the duty cycles of D_{L2}, D_{M1}, D_{S6n}, and D_{Z0} are derived as

$$\begin{cases} D_{L2} = \sqrt{3}\frac{V_R(t)}{U_{dc}}\sin\left(\omega t - \frac{\pi}{6}\right) + D_{S6n} \\ D_{M1} = \sqrt{3}\frac{V_R(t)}{U_{dc}}\sin\left(\frac{\pi}{3} - \omega t\right) \\ D_{S6n} = \sqrt{3}\frac{V_R(t)}{U_{dc}}\sin\left(\frac{\pi}{3} - \omega t\right) \\ D_{Z0} = 1 - D_{L2} - D_{M1} - D_{S6n} \end{cases} \quad (6.152)$$

The corresponding current of V_{S6n}[ONO] is i_b. Consequently, with the limitation of positive duty cycles, the small vector force of SVF_{S6n} is written as

$$SVF_{S6n} = \min \begin{cases} \sqrt{3}\frac{V_R(t)}{U_{dc}}\sin\left(\frac{\pi}{3} - \omega t\right) \\ 1 - \frac{\sqrt{3}V_R(t)}{U_{dc}}\sin\left(\omega t + \frac{\pi}{3}\right) \end{cases} \left| I_m \cos\left(\omega t - \theta - \frac{2\pi}{3}\right) \right| \quad (6.153)$$

The optimal small vector is determined by comparing the value of SVF_{S2n} and SVF_{S6n}. If SVF_{S2n} is larger than SVF_{S6n}, the impact of small vector V_{S2n}[OON] on NPV

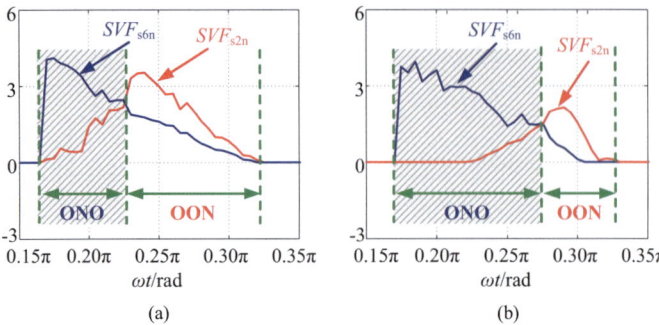

Fig. 6.48 SVF_{S2n} and SVF_{S6n} of Type A when V_{ref} locates in Sector II and current directions is (+--) (**a**) Voltage dip depth is 0.625 (**b**) Voltage dip depth is 0.517

is greater than that of $V_{S6n}[ONO]$. In other words, among correct small vectors, $V_{S2n}[OON]$ has the most effective capability to mitigate NPV oscillation. Thus, it is defined as the optimal small vector. Conversely, if SVF_{S2n} is less than SVF_{S6n}, the optimal small vector is $V_{S6n}[ONO]$. Thus, considering grid voltage dips and reactive power, the optimal small vector is determined, which can effectively mitigate NPV oscillation in LVRT operation.

To provide a more intuitive understanding for the reader, the *SVFs* waveform are given with the parameters in Step 4. The waveforms of SVF_{S2n} and SVF_{S6n} are shown in Fig. 6.48 under different grid voltage dip depth of Type A. In the shaded area, since the SVF_{S6n} is larger than SVF_{S2n}, the small vector [ONO] is adopted to suppress NPV oscillation caused by medium vector. Inversely, the small vector [OON] is used in the other area. Compared Fig. 6.48a with Fig. 6.48b, the shaded area is increased with power factor increasing. In the shade area, the NPV is uncontrollable since the small vector [ONO] cannot be selected with LMSZVM method. Fortunately, this issue can be effectively solved with the proposed method.

On the other hand, in order to reduce the total switching loss, the number of switching events should be minimized in one switching cycle and between switching cycles. Therefore, the switching sequences are arranged as follows.

1. [OON] is the optimal small vector:
   OOO-OON-PON-PPN-PON-OON-OOO
2. [ONO] is the optimal small vector:
   ONO-OOO-PON-PPN-PON-OOO-ONO

Case 2: $i_a > 0$, $i_b > 0$ and $i_c < 0$ (The Range of θ Is from 0 to $\pi/6$)
According to the above analysis, $V_{S1p}[POO]$ and $V_{S6n}[ONO]$ are selected as the correct small vectors. When [POO] is selected, the V_{ref} is synthesized as

$$\begin{cases} V_{L2}D_{L2} + V_{M1}D_{M1} + V_{S2n}D_{S1p} + V_{Z0}D_{Z0} = V_{ref} \\ D_{L2} + D_{M1} + D_{S1p} + D_{Z0} = 1 \end{cases} \quad (6.154)$$

6.3 Advanced Control Technology for Neutral-Point Voltage Balance Under ...

where D_{S1p} is the duty cycle of V_{S1p}. Meanwhile, to suppress NPV oscillation, the variation of NPV generated by [PON] and [POO] should be same, which is expressed as

$$i_a D_{S1p} = i_b D_{M1} \tag{6.155}$$

Solving above equations, and considering positive restriction of duty cycles, the small vector force SVF_{S1p} is calculated as

$$SVF_{S1p} = \min \left\{ \begin{array}{l} 2\sqrt{3}\frac{i_b}{i_a+i_b}\frac{V_R(t)}{U_{dc}} \sin\left(\frac{\pi}{3} - \omega t\right) \\ 2\left[1 - \frac{\sqrt{3}V_R(t)}{U_{dc}} \sin\left(\omega t + \frac{\pi}{3}\right)\right] \end{array} \right\} |I_m \cos(\omega t - \theta)| \tag{6.156}$$

where θ is lower than $\pi/6$. When V_{S6n}[ONO] is adopted, the expression of SVF_{S6n} is the same as that in Case 1 except for the range of θ.

Similarly, if SVF_{S1p} is larger than SVF_{S6n}, V_{S1p}[POO] is determined as the optimal small vector. Compared with [ONO], the NPV oscillation is more effectively suppressed with [POO]. Conversely, V_{S6n}[ONO] is selected as the optimal small vector when SVF_{S1p} is smaller than SVF_{S6n}.

Figure 6.49a shows the results of SVF_{S1p} and SVF_{S6n} when θ is equal to $\pi/15$ of Type B. Since the SVF_{S6n} is larger than SVF_{S1p} in the shaded area, the V_{S6n}[ONO] is used to compensate for the decreased NPV caused by V_{M1}[PON]. Otherwise, the small vector of V_{S1p}[POO] is adopted.

When V_{S1p}[POO] is adopted, to minimize the number of switches events, the switching sequence are arranged as: OOO-POO-PON-PPN-PON-POO-OOO.

Case 3: $i_a > 0$, $i_b < 0$ and $i_c > 0$

In this case, the range of θ is $5\pi/12 \sim \pi/2$. Only the small vector V_{S6n}[ONO] can be used to suppress the NPV oscillation generated by [PON]. The SVF_{S6n} is larger than

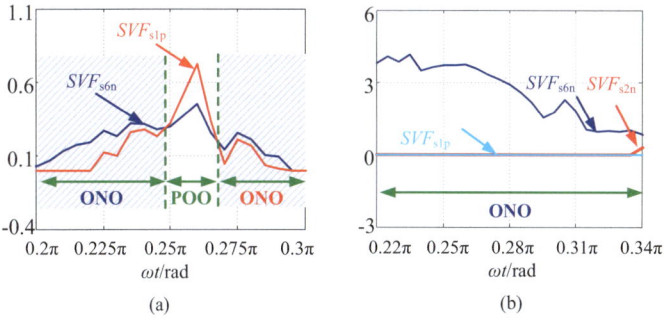

Fig. 6.49 SVF_{S1p}, SVF_{S2n}, SVF_{S6n} when V_{ref} locates in Sector 2 (**a**) current directions is (++−) and grid voltage dip depth is 0.85 of Type B (**b**) current directions is (+−+) and grid voltage dip depth is 0.5 of Type A

SVF_{S1p} and SVF_{S2n} as depicted in Fig. 6.49 when grid voltage dip depth is 0.5 of Type A. Meanwhile, the switching sequence is the same as that in case I.

The pictorial representations of switching sequences for synthesizing reference voltage and CMV waveforms with different optimal small vectors are depicted in Fig. 6.50. It can be seen that phase A and phase B are generated by output states [O] and [P], while phase C is generated by output states [O] and [N] when the optimal small vector is [POO] or [OON]. However, when the optimal small vector is [ONO], the output states in phase B include [N], [O] and [P]. In addition, the CMV has three changes from 0 to $-U_{dc}/6$ or from 0 to $U_{dc}/6$. Thus, the CMV magnitude is restricted within one-sixth of the dc-link voltage, which is half of the conventional SVM method.

A similar procedure is used to calculate the SVF and select the optimal small vector in other sectors. The correct small vector, optimal small vector and the switching sequence in different sectors are summarized in Table 6.6.

To suppress the dc NPV unbalance, a PI controller and optimal small vector selection process are used. Figure 6.51 shows the detailed block diagram of the proposed SSVM method in LVRT operation. First, based on V_{ref} and the direction

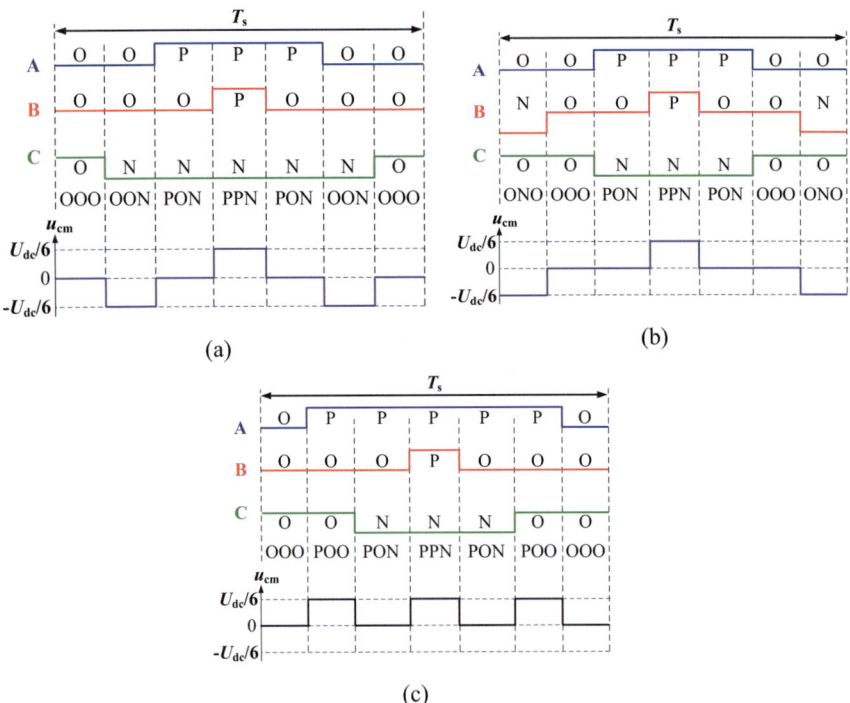

Fig. 6.50 The switching sequence and CMV waveforms with different optimal small vectors (**a**) Optimal small vector is [OON] (**b**) Optimal small vector is [ONO] (**c**) Optimal small vector is [POO]

6.3 Advanced Control Technology for Neutral-Point Voltage Balance Under ...   381

Table 6.6 Summarization of correct small vector, optimal small vector and the switching sequence according to the voltage sectors and current region

Sector	Current region	Correct small vector	SVF criteria	Optimal small vector	Switching sequence
1	6 (+−+)	[OPO]	SVF_{S3p}	[OPO]	OPO-OOO-PON-PNN-...
	1 (+−−)	[OPO] [OON]	$SVF_{S3p} > SVF_{S2n}$	[OPO]	OPO-OOO-PON-PNN-...
			$SVF_{S3p} < SVF_{S2n}$	[OON]	OOO-OON-PON-PNN-...
2	6 (+−+)	[ONO]	SVF_{S6n}	[ONO]	ONO-OOO-PON-PPN-...
	1 (+−−)	[ONO] [OON]	$SVF_{S6n} > SVF_{S2n}$	[ONO]	ONO-OOO-PON-PPN-...
			$SVF_{S6n} < SVF_{S2n}$	[OON]	OOO-OON-PON-PPN-...
	2 (++−)	[ONO] [POO]	$SVF_{S6n} > SVF_{S1p}$	[ONO]	ONO-OOO-PON-PPN-...
			$SVF_{S6n} < SVF_{S1p}$	[POO]	OOO-POO-PON-PPN-...
3	1 (+−−)	[NOO]	SVF_{S4n}	[NOO]	NOO-OOO-OPN-PPN-...
	2 (++−)	[NOO] [OPO]	$SVF_{S4n} > SVF_{S3p}$	[NOO]	NOO-OOO-OPN-PPN-...
			$SVF_{S4n} < SVF_{S3p}$	[OPO]	OOO-OPO-OPN-PPN-...
4	1 (+−−)	[POO]	SVF_{S1p}	[POO]	POO-OOO-OPN-NPN-...
	2 (++−)	[POO] [OPO]	$SVF_{S1p} > SVF_{S3p}$	[POO]	POO-OOO-OPN-NPN-...
			$SVF_{S1p} < SVF_{S3p}$	[OPO]	OOO-OPO-OPN-NPN-...
	3 (−+−)	[POO] [OON]	$SVF_{S1p} > SVF_{S2n}$	[POO]	POO-OOO-OPN-NPN-...
			$SVF_{S1p} < SVF_{S2n}$	[OON]	OOO-OON-OPN-NPN-...
5	2 (++−)	[OOP]	SVF_{S5p}	[OOP]	OOP-OOO-NPO-NPN-...
	3 (−+−)	[OOP] [NOO]	$SVF_{S5p} > SVF_{S4n}$	[OOP]	OOP-OOO-NPO-NPN-...
			$SVF_{S5p} < SVF_{S4n}$	[NOO]	OOO-NOO-NPO-NPN-...
6	2 (++−)	[OON]	SVF_{S2n}	[OON]	OON-OOO-NPO-NPP-...
	3 (−+−)	[OON] [NOO]	$SVF_{S2n} > SVF_{S4n}$	[OON]	OON-OOO-NPO-NPP-...
			$SVF_{S2n} < SVF_{S4n}$	[NOO]	OOO-NOO-NPO-NPP-...
	4 (−++)	[OON] [OPO]	$SVF_{S2n} > SVF_{S3p}$	[OON]	OON-OOO-NPO-NPP-...
			$SVF_{S2n} < SVF_{S3p}$	[OPO]	OOO-OPO-NPO-NPP-...
7	3 (−+−)	[ONO]	SVF_{S6n}	[ONO]	ONO-OOO-NOP-NPP-...
	4 (−++)	[ONO] [OOP]	$SVF_{S6n} > SVF_{S5p}$	[ONO]	ONO-OOO-NOP-NPP-...
			$SVF_{S6n} < SVF_{S5p}$	[OOP]	OOO-OOP-NOP-NPP-...
8	3 (−+−)	[OPO]	SVF_{S3p}	[OPO]	OPO-OOO-NOP-NNP-...
	4 (−++)	[OPO] [OOP]	$SVF_{S3p} > SVF_{S5p}$	[OPO]	OPO-OOO-NOP-NNP-...
			$SVF_{S3p} < SVF_{S5p}$	[OOP]	OOO-OOP-NOP-NNP-...
	5 (−−+)	[OPO] [NOO]	$SVF_{S3p} > SVF_{S4n}$	[OPO]	OPO-OOO-NOP-NNP-...
			$SVF_{S3p} < SVF_{S4n}$	[NOO]	OOO-NOO-NOP-NNP-...
9	4 (−++)	[POO]	SVF_{S1p}	[POO]	POO-OOO-ONP-NNP-...
	5 (−−+)	[POO] [ONO]	$SVF_{S1p} > SVF_{S6n}$	[POO]	POO-OOO-ONP-NNP-...
			$SVF_{S1p} < SVF_{S6n}$	[ONO]	OOO-ONO-ONP-NNP-...

(continued)

Table 6.6 (continued)

Sector	Current region	Correct small vector	SVF criteria	Optimal small vector	Switching sequence
10	4 (−++)	[NOO]	SVF_{S4n}	[NOO]	NOO-OOO-ONP-PNP- ...
	5 (−−+)	[NOO] [ONO]	$SVF_{S4n} > SVF_{S6n}$	[NOO]	NOO-OOO-ONP-PNP-...
			$SVF_{S4n} < SVF_{S6n}$	[ONO]	OOO-ONO-ONP-PNP-...
	6 (+−+)	[NOO] [OOP]	$SVF_{S4n} > SVF_{S5p}$	[NOO]	NOO-OOO-ONP-PNP-...
			$SVF_{S4n} < SVF_{S5p}$	[OOP]	OOO-OOP-ONP-PNP-...
11	5 (−−+)	[OON]	SVF_{S2n}	[OON]	OON-OOO-PNO-PNP-...
	6 (+−+)	[OON] [POO]	$SVF_{S2n} > SVF_{S1p}$	[OON]	OON-OOO-PNO-PNP-...
			$SVF_{S2n} < SVF_{S1p}$	[POO]	OOO-POO-PNO-PNP-...
12	5 (−−+)	[OOP]	SVF_{S5p}	[OOP]	OOP-OOO-PNO-PNN-...
	6 (+−+)	[OOP] [POO]	$SVF_{S5p} > SVF_{S1p}$	[OOP]	OOP-OOO-PNO-PNN-...
			$SVF_{S5p} < SVF_{S1p}$	[POO]	OOO-POO-PNO-PNN-...
	1 (+−−)	[OOP] [ONO]	$SVF_{S5p} > SVF_{S6n}$	[OOP]	OOP-OOO-PNO-PNN-...
			$SVF_{S5p} < SVF_{S6n}$	[ONO]	OOO-ONO-PNO-PNN-...

of I_r, the correct small vectors are selected under different operating conditions. Then, the *SVF* including the information of grid dips and reactive power on NPV is calculated. According to the *SVF* value, the optimal small vector that has the most effective impact on NPV is determined among the correct small vectors, and the switching sequence is arranged to reduce the switching loss. With the proposed method, the NPV oscillation and CMV can be effectively reduced in LVRT operation without increasing the dc-link capacitor values.

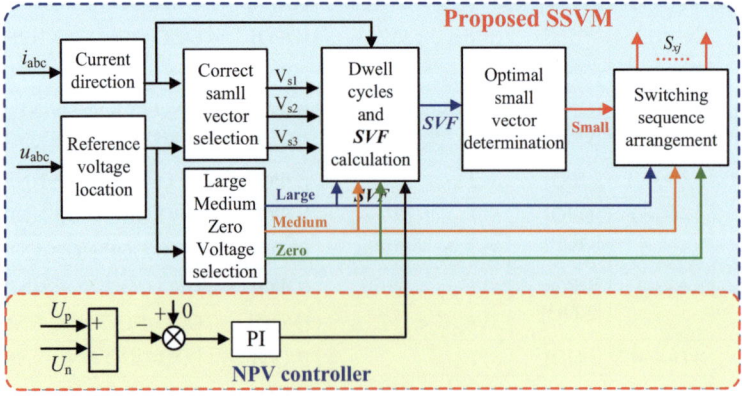

Fig. 6.51 The block diagram of the proposed SSVM method

Table 6.7 Experiment parameters for NPV balance control methods for normal power grid conditions

Parameters	Values
DC bus voltage	300 V
dc-link capacitor C_1, C_2	470 μF
AC line-to-line voltage	156 V
Grid frequency	50 Hz
Output power	10 kW
Peak grid current	20 A
L-filter	1.6 mH, 0.1 Ω
Switching frequency	5 kHz

Step 4: Experimental Verification for SSVM

In this section, only the experimental results based on common-mode voltage reduction modulation are provided as an example to demonstrate the effectiveness of NPV balance control methods.

To prove the correctness of the proposed scheme, Method-1 (LMZVM), Method-2 (LMSZVM), Method-3 (proposed method) and Method-4 (SVM) are applied to this inverter system for a better comparison. The experimental results are categorized into steady-state response and dynamic response. The experimental parameters are listed in Table 6.7.

1. **Steady-state response**

   Three cases of grid voltage dips are considered in steady state tests. The corresponding results and analysis is given as follows.

Case 1: Type A Operation (Three-Phase Grid Voltages Dip to 50% of the Nominal Value)

Figure 6.52 shows the experimental results for Type A ($i_d = 0$ A, $i_q = -20$ A) operation. Three-phase grid voltages and currents are shown in Fig. 6.52a, b, respectively. In this case, three-phase grid voltages dip to 50% of the nominal value. It can be observed that the phase difference between the grid voltage and the grid current is approximately $\pi/2$. As depicted in Fig. 6.52c, the peak-to-peak value of the NPV is 32 V when using the conventional Method-1. When Method-2 is applied, it can be seen from Fig. 6.52d that the peak-to-peak value of the NPV is 24 V. The NPV oscillation suppression performance is limited since the uncontrolled area is existed as the theoretical analysis. Similar experimental results are conducted by the proposed method with the optimal small vector and switching sequence. The NPV oscillation of Method-3 is approximately 2 V, and better suppression performance of NPV oscillation can be achieved in comparison with Method-2.

Case 2: Type B Operation (Grid Voltage of Phase A Dips to 50%)

Figure 6.53 shows the experimental results for Type B ($i_d = 18.8$ A, $i_q = -6.8$ A) operation. Grid voltages and grid-side currents are shown in Fig. 6.53a, b, where the

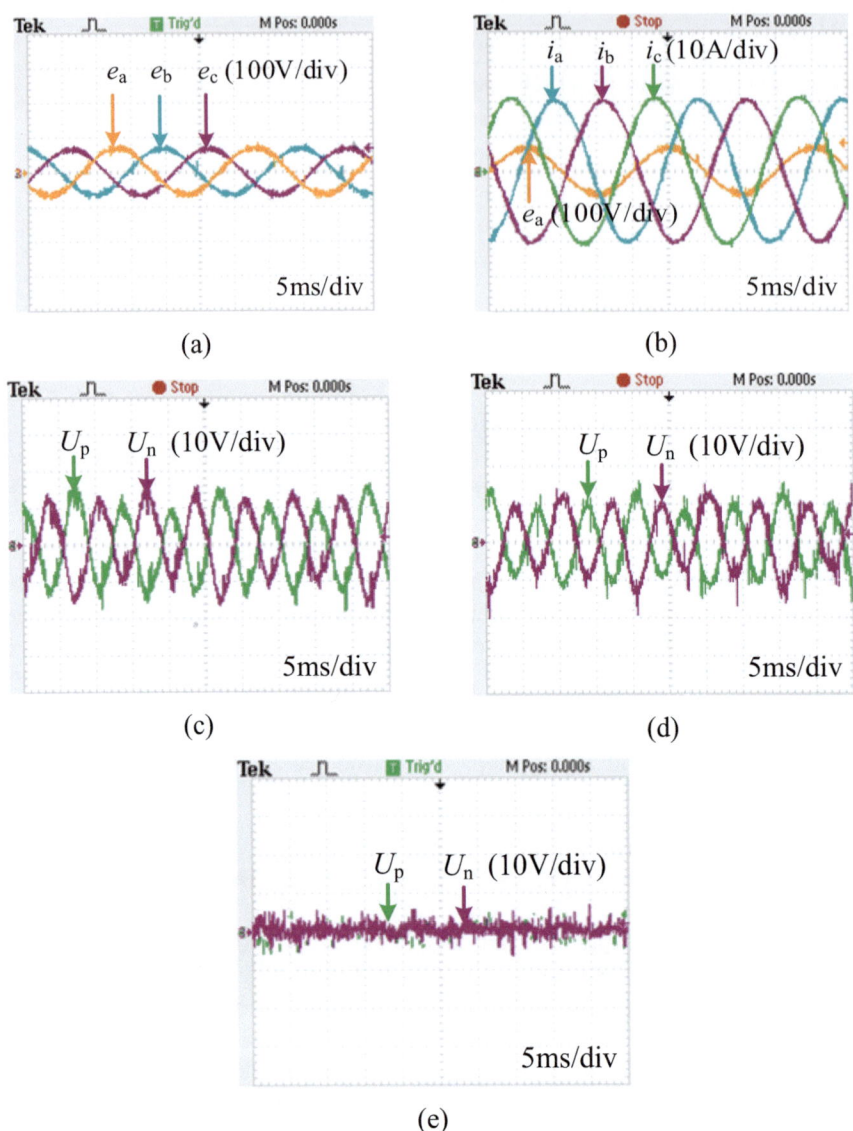

Fig. 6.52 Experimental results in Type A operations. **a** Grid voltages. **b** Grid-side currents. **c** Method-1. **d**. Method-2. **e**. Method-3

grid voltage of phase A dips to 50%. As shown in Fig. 6.53c that the peak-to-peak value of the NPV using the conventional Method-1 has relatively higher value at 24 V, due to a fact that the grid voltage is unbalanced and the power factor is low. On the other hand, the peak-to-peak value of the NPV is mitigated to 6 V when

6.3 Advanced Control Technology for Neutral-Point Voltage Balance Under … 385

using Method-2 in Fig. 6.53d. Finally, the peak-to-peak value of the NPV is further reduced to 3 V by using the proposed method as shown in Fig. 6.53e.

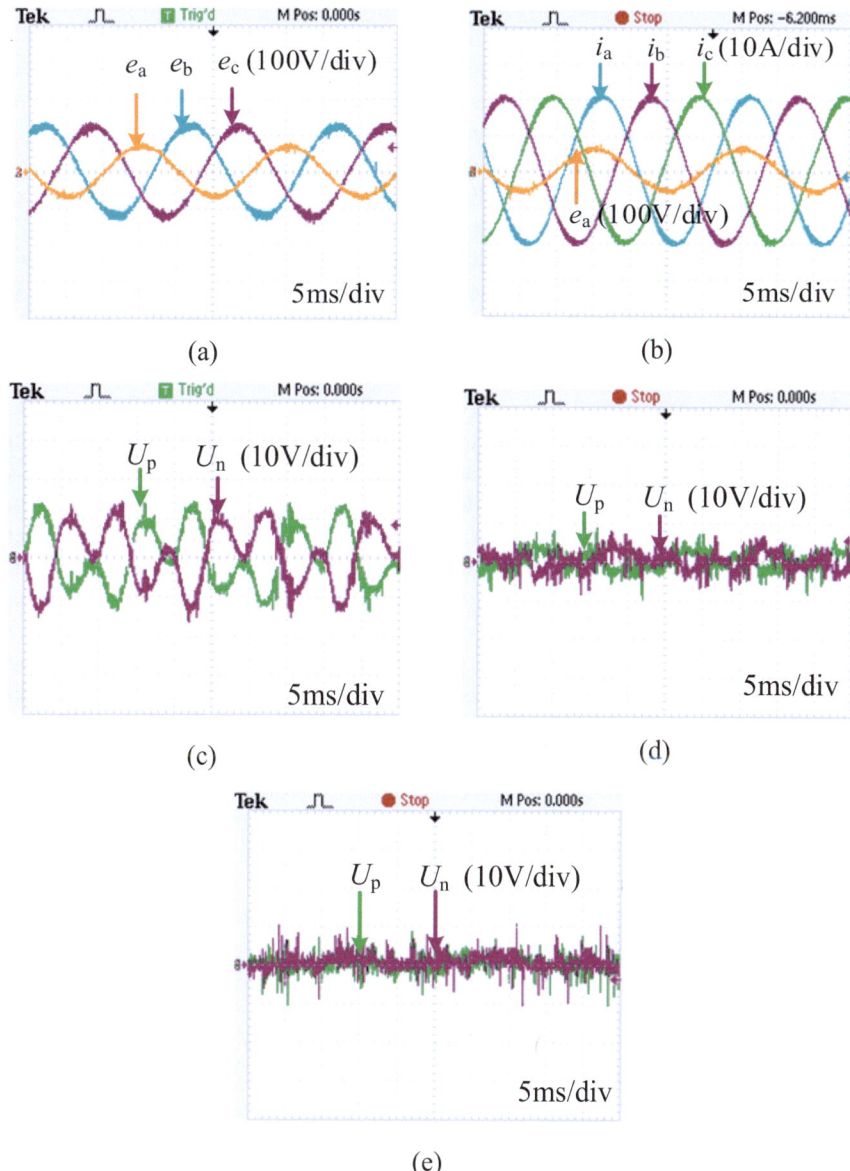

Fig. 6.53 Experimental results in Type B operations. **a** Grid voltages. **b** Grid-side currents. **c** Method-1. **d** Method-2. **e** Method-3

Case 3: Type E Operation (Two-Phase Grid Voltages of B and C Dip to 50%)
Figure 6.54 shows the experimental results for Type E ($i_d = 18.3$ A, $i_q = -8$ A) operation. Grid voltages and currents are shown in Fig. 6.54a, b, where the two-phase grid voltages of B and C dip to 50%. Figure 6.54c shows NPV with the conventional Mehtod-1. The peak-to-peak value of NPV is 31 V. It is important to note that Method-2 can reduce the peak NPV to 10 V in Fig. 6.54d. However, the reduction of NPV oscillation is limited. This is mainly because the optimal small vector is not adopted. Fortunately, this limitation is solved by the proposed method. The peak-to-peak value of NPV is significantly reduced to approximately 2 V as shown in Fig. 6.54e.

Figure 6.55 shows the comparative results of CMV with two modulation methods. When conventional Method-4 is applied, it is obviously seen that the magnitude of CMV is 101 V as all voltage vectors are utilized. On the other hand, the CMV magnitude is attenuated to approximately 56 V, which is half of the value using the conventional Method-4. In addition, line-to-line voltages are also provided. As observed, when the conventional Method-4 and proposed method are applied, the line-to-line voltages consist of five voltages level.

2. **Dynamic response**

The transient process of NPV is shown in Fig. 6.56 during the transition between different modulation methods. It can be observed that the NPV fluctuation is 24 V and 18 V using the conventional Method-2 when θ is equal to $\pi/4$ and $5\pi/12$. The reason is that the NPV cannot be controlled in the uncontrollable area. When the proposed method is applied, the NPV oscillation is effectively controlled and reduced to 2 V. The dynamic control performance of NPV is consistent with that in the steady state.

In order to clearly evaluate the performance of the proposed scheme, total harmonic distortion (THD) characteristics of output currents for different methods are investigated. The obtained results are listed in Table 6.8. It is clear that the THDi of the conventional SVM method is the lowest, as it always adopts nearest three vectors to synthesize the reference voltage vector. However, the CMV magnitude is one-third of dc-link voltage, which is extremely high. Since Method-1, Method-2, and Method-3 avoid using the basic vectors with high CMV magnitudes, the current THDis are slightly higher than that of the conventional SVM method. Method-1 adopts large, medium, and zero vectors to generate the output voltage, but it cannot control the neutral-point voltage balance actively. Method-3 can effectively control the neutral-point voltage balance while reducing CMV, and the THDi is increased to some extent for the non-nearest small vectors are adopted. Fortunately, the THDi remains at an acceptable level.

When the PV 3LT^{2}I operates in LVRT mode, the power factor decreases and the CMV reduction and NPV balance should be achieved simultaneously using conventional modulation method. The SSVM strategy is developed to solve above issues. In this method, the correct small vectors which have opposite effect on NPV with medium vector are selected under different power factor angles based on the location of output voltage and the region of grid currents. Then, the *SVF* including the information of grid dips and reactive power on NPV is proposed. According to the value of *SVF*, the optimal small vector having the most significant impact on NPV among the correct vectors is determined. The

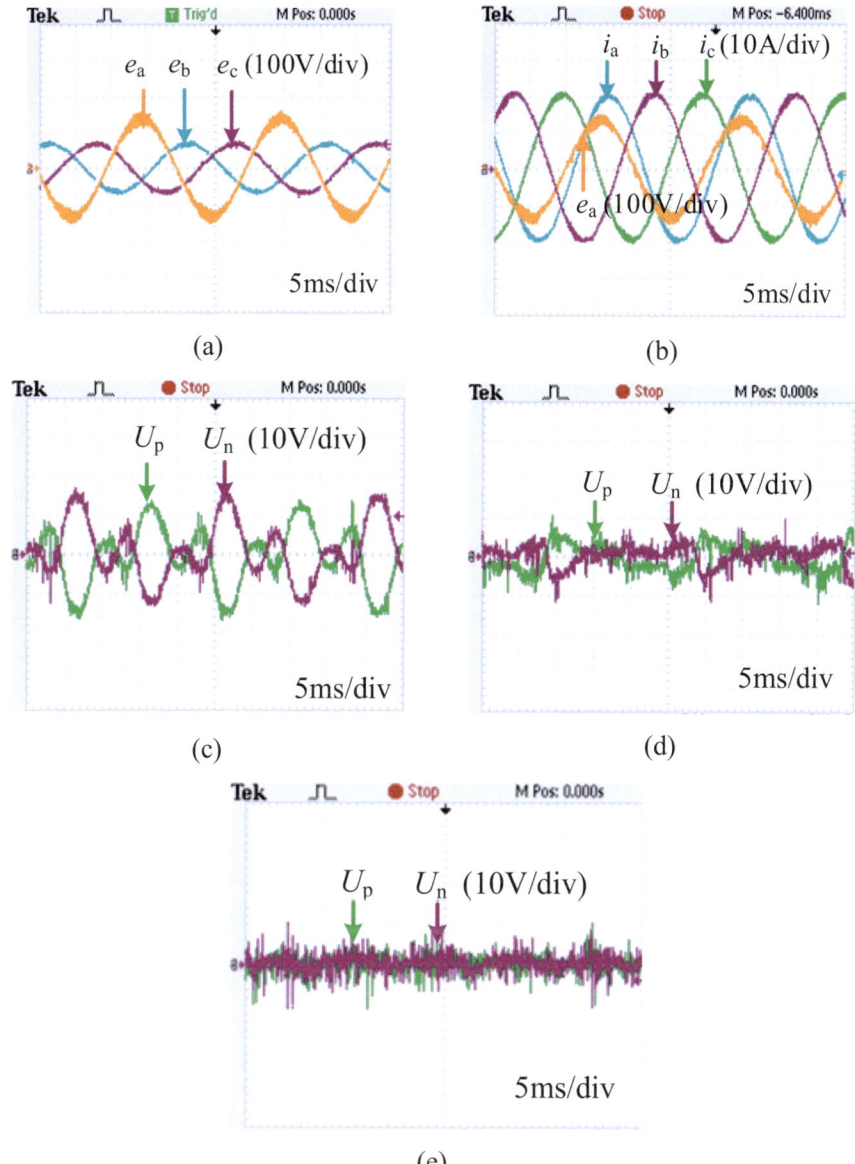

Fig. 6.54 Experimental results in Type E operations. **a** Grid voltages. **b** Grid-side currents. **c** Method-1. **d** Method-2. **e** Method-3

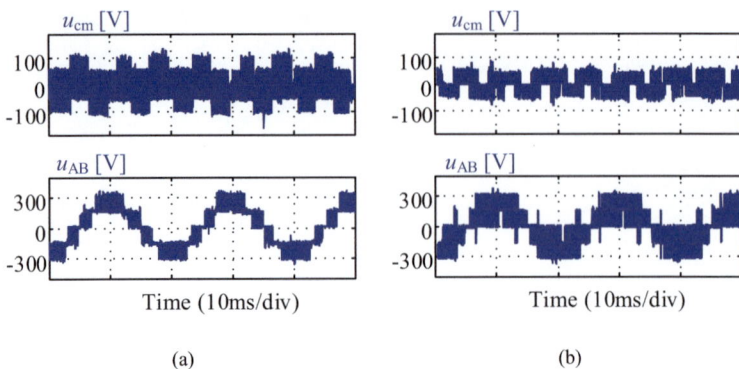

Fig. 6.55 Experimental results of CMV and line-to-line voltage. **a** Method-4. **b** Method-3

validity of the proposed method has been demonstrated through the experimental results.

In this section, the NPV balance control technology is introduced, which is integrated into the PWM modulator and auxiliary loop. Three methods are presented based on different modulation method. Unlike the NPV balance control methods that only use NPV information under normal grid conditions, the above three methods consider both current direction and NPV information under the grid fault conditions.

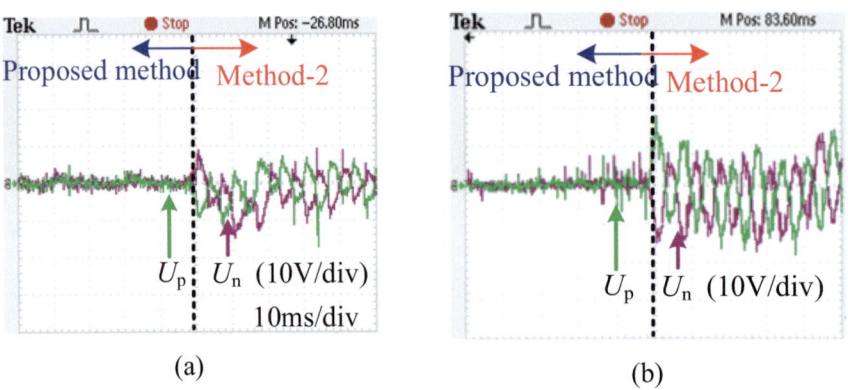

Fig. 6.56 Dynamic response of NPV during modulation algorithm switching in Type A operations. **a** θ is equal to $\pi/4$. **b** θ is equal to $5\pi/12$

Table 6.8 THDs of grid currents with different methods

Method	Type A (%)	Type B (%)	Type E (%)
Method-1	3.31	3.14	3.18
Method-2	2.62	3.26	3.21
Method-3	3.56	3.46	3.49
Method-4	2.31	2.13	2.03

6.4 Summary

With the integration of large-scale renewable power generation and renewable power consumption system into the power grid, the power system exhibits significant "weak inertia" characteristics, leading to frequent grid faults. Under grid fault conditions, the traditional control technology is no longer applicable, which results in the dc voltage unstable and grid current distortion of PV inverter. Additionally, the grid standard requirements for PV inverter are imposed for grid safe operation. To ensure the normal operation of PV inverters and to satisfy the grid standard requirements under grid fault conditions, the advanced control technology is proposed. First, the advanced control technology for dc voltage and grid current control is elaborated, which are integrated into middle loop and inner loop. This control technology includes two control schemes: dual-loop voltage and current control scheme and dual-loop voltage and power control scheme. Each control scheme has its own advantages and meets different control requirements in practical applications. Next, the technology for NPV balance under grid fault conditions is presented, which is integrated into PWM modulator and auxiliary loop. Furthermore, the NPV balance control technology including three methods based on different modulation methods. A notable feature of these three methods is that they consider both current direction and NPV information.

To facilitate reader's understanding of this technology, its theoretical analysis, implementation process, and output waveforms are provided for each control method.

References

1. H. Geng, C. Liu, X. Zhang, G. Yang, *Low Voltage Ride-Through of New Energy Grid-Connected Power Generation Systems* (China Machine Press, Beijing, 2014)
2. X. Zhang, R. Cao, *Grid-Connected Photovoltaic Power Generation and Inverter Control Technologies* (China Machine Press, Beijing, 2013)
3. Z. Shao, X. Zhang, F. Wang, R. Cao, H. Ni, Analysis and control of neutral-point voltage for transformerless three-level PV inverter in LVRT operation. IEEE Trans. Power Electron. **32**(3), 2347–2359 (2017)
4. China Electricity Council, *GB/T 19964-2012: technical regulations for connecting photovoltaic power stations to power systems* (China Electricity Council, 2012)
5. E.ON Netz GmbH, Grid code: high and extra high voltage, Tech. Rep. Status:1, 2006

6. C. Qin, C. Zhang, A. Chen, X. Xing, G. Zhang, Circulating current suppression for parallel three-level inverters under unbalanced operating conditions. IEEE J. Emerg. Sel. Top Power Electron. **7**(1), 480–492 (2019)
7. Y. Jiang, C. Qin, X. Xing, X. Li, C. Zhang, A hybrid passivity-based control strategy for three-level t-type inverter in LVRT operation. IEEE J. Emerg. Sel. Top Power Electron. **8**(4), 4009–4024 (2020)
8. X. Li, X. Xing, C. Qin, B. Duan, C. Zhang, G. Zhang, Generalised model predictive control scheme for three-level converter under unbalanced grid voltage considering full power factor operation. IET Renew. Power Gener. **14**(16), 3115–3125 (2020)
9. J. Wang, X. Mu, Q. Li, Study of passivity-based decoupling control of T-NPC PV grid-connected inverter. IEEE Trans. Ind. Electron. **64**(9), 7542–7551 (2017)
10. P. Rodríguez, R. Teodorescu, I. Candela, A.V. Timbus, M. Liserre, F. Blaabjerg, New positive-sequence voltage detector for grid synchronization of power converters under faulty grid conditions, in *2006 37th IEEE Power Electronics Specialists Conference*, (2006), pp. 1–7
11. Z. Zhang, H. Fang, F. Gao, J. Rodríguez, R. Kennel, Multiple-vector model predictive power control for grid-tied wind turbine system with enhanced steady-state control performance. IEEE Trans. Ind. Electron. **64**(8), 6287–6298 (2017)
12. J. Rodriguez et al., State of the art of finite control set model predictive control in power electronics. IEEE Trans. Ind. Informat. **9**(2), 1003–1016 (2013)
13. Z. Zhang, C.M. Hackl, R. Kennel, Computationally efficient DMPC for three-level NPC back-to-back converters in wind turbine systems with PMSG. IEEE Trans. Power Electron. **32**(10), 8018–8034 (2017)
14. X. Xing, A. Chen, Z. Zhang, J. Chen, C. Zhang, Model predictive control method to reduce common-mode voltage and balance the neutral-point voltage in three-level T-type inverter, in *2016 IEEE Applied Power Electronics Conference and Exposition*, (2016), pp. 3453–3458
15. C. Fu, C. Zhang, G. Zhang, C. Zhang, Q. Su, Finite-time command filtered control of three-phase AC/DC converter under unbalanced grid conditions. IEEE Trans. Ind. Electron. **70**(7), 6876–6886 (2023)
16. Q. Ren et al., An improved passivity-based direct power control strategy for AC/DC converter under unbalanced and distorted grid conditions. IEEE Trans. Power Electron. **38**(10), 12153–12165 (2023)
17. E. Cruz-Zavala, J.A. Moreno, W. Luo, L.M. Fridman, Uniform robust exact differentiator. IEEE Trans. Automat. Control **56**(11), 2727–2733 (2011)
18. B. Chen, C. Lin, Finite-time stabilization-based adaptive fuzzy control design. IEEE Trans. Fuzzy Syst. **29**(8), 2438–2443 (2021)
19. Y. Zhang, J. Jiao, J. Liu, J. Gao, Direct power control of PWM rectifier with feedforward compensation of DC-bus voltage ripple under unbalanced grid conditions. IEEE Trans. Ind. Appl. **55**(3), 2890–2901 (2019)
20. Y. Jiang, X. Li, C. Qin, X. Xing, Z. Chen, Improved particle swarm optimization based selective harmonic elimination and neutral point balance control for three-level inverter in low-voltage ride-through operation. IEEE Trans. Ind. Informat. **18**(1), 642–652 (2022)
21. Y. Li, X. Yang, W. Chen, T. Liu, F. Zhang, Neutral-point voltage analysis and suppression for NPC three-level photovoltaic converter in LVRT operation under imbalanced grid faults with selective hybrid SVPWM strategy. IEEE Trans. Power Electron. **34**(2), 1334–1355 (2019)
22. X. Li et al., Neutral-point voltage oscillation mitigation scheme for transformerless three-level PV inverter in LVRT operation with selective space vector modulation. IEEE J. Emerg. Sel. Top Power Electron. **10**(3), 2776–2789 (2022)

Open Access This chapter is licensed under the terms of the Creative Commons Attribution 4.0 International License (http://creativecommons.org/licenses/by/4.0/), which permits use, sharing, adaptation, distribution and reproduction in any medium or format, as long as you give appropriate credit to the original author(s) and the source, provide a link to the Creative Commons license and indicate if changes were made.

The images or other third party material in this chapter are included in the chapter's Creative Commons license, unless indicated otherwise in a credit line to the material. If material is not included in the chapter's Creative Commons license and your intended use is not permitted by statutory regulation or exceeds the permitted use, you will need to obtain permission directly from the copyright holder.

Chapter 7
Control Technology of Photovoltaic Inverters for Multi-functional Operation

This chapter presents the control technology of photovoltaic (PV) inverter for multi-functional operation. Multi-functional modes of PV inverter mainly refer to the power quality control mode and the islanded mode.

To make full use of PV inverter capacity and save the costs of PV generation system, the PV inverter equipped with the power quality control mode are required. When PV inverter operates in the power quality control mode at night, it has the ability of current harmonic elimination and reactive power compensation. The control scheme for the power quality control mode including harmonics and reactive current detection, reference current synthesis, and reference current tracking control.

The PV inverters operating in islanded mode plays a vital role in maintaining continuity of power supply and ensuring grid stability during grid disturbances or fault. The control technology for the islanded mode is different from that for grid-connected mode, because the output characteristic and control objectives in these two modes are strikingly different. This chapter presents the nonlinear control methods embedded into single-loop voltage control scheme to improve the dynamic performance and stability of PV inverter system operating in islanded mode when there is a large operating range, load disturbance and parameter uncertainty.

7.1 Overview of Multi-functional Modes

PV generation is one of the most promising new energy power generation forms. The technology presented in Chaps. 1–6 realizes the targets maximizing power generation, safety operation of PV inverters and users, high robustness to multi-source disturbances, and high-quality power transmission, and other performance. However, PV generation system still face two important issues that need to be addressed. (1) Low utilization rate of inverter capacity and high content of current harmonic and reactive power in renewable power generation system. (2) PV inverter should be equipped with standalone power supply function, so as to maintain the continuity of power supply, especially for critical loads and ensure power system stability during grid disturbances or fault.

1. **Low utilization rate of inverter capacity and high content of harmonic current and reactive power**

    On one hand, the output power of PV generation system is highly dependent on irradiance. Due to the alternation of day and night and the change of weather conditions, PV generation system has the problems of intermittent output power and low utilization rate of inverter capacity, usually less than 30%. Especially, the PV inverters are shutdown at night. On the other hand, the content of current harmonics and reactive power are increasing with more and more power electronic equipment and nonlinear loads connected to the power grid. In this case, a large amount of current harmonics and reactive power are injected into the grid, which will cause grid voltage distortion, fluctuations and other electrical equipment failures. Thus, the power quality control equipment is required to install into PV generation system to copy with the above power quality issues. However, the additional power quality control equipment inevitably increases the costs of construction, operation and maintenance of PV generation system.

    As we all know that in traditional industrial applications, shunt active power filters are usually used to eliminate harmonic current and compensate reactive power. Since the main circuit structure of shunt active power filter and PV inverter is consistent, power quality problems such as current harmonics and reactive power can be controlled by PV inverters using the appropriate control scheme.

    Therefore, by adding the appropriate control scheme in the original control structure, the PV inverter not only has the function of converting solar energy into electricity power, but also has the function of power quality control, which make full use of PV inverter capacity and save the costs of PV generation system. The PV inverter operates in power quality control mode at night and on cloudy days, while operates in power generation mode when the irradiance is sufficient, which realizes the conversion and transmission of green energy as much as possible.

    The control scheme for the power quality control mode including harmonics and reactive current detection, reference current synthesis, and reference current tracking control, which are shown in Fig. 7.1.

2. **PV inverter equipped with standalone power supply function**

    Recent years, the PV power generation systems have been developed rapidly since it has low restrictions on the construction site, and are widely installed in mountains, islands, and the places without power grid coverage. Besides, the PV power generation systems are also usually used as the backup power in case of disaster or power outages. As above-mentioned application occasions, the PV inverter must work independently and not be connected to the power grid. This operation mode is called islanded mode, which provides an effective solution for the areas without power grid coverage or the occasion that require independent power supply. Additionally, with the development of microgrids, the PV generation system operated as the power supply unit play an important

7.1 Overview of Multi-functional Modes

role. In this case, the PV inverter operating in islanded mode provides voltage and frequency support. To summarize, the PV inverters in islanded mode plays a vital role in maintaining the continuity of power supply and ensuring grid stability during grid disturbances or fault. It is also essential for enhancing grid resilience, ensuring the reliability of the power supply, and mitigating safety risks.

The main circuits of PV inverter operating in grid-connected generation mode and islanded mode are similarly. However, the output characteristics are strikingly different in these two operation modes. When PV inverter operates in grid-connected generation mode, the output point characteristics of the PV inverter is a current source. The output current is decided by the input power of PV array. While, when PV inverter operates in islanded mode, the output characteristics is a voltage source. The output voltage has constant magnitude and frequency, which usually decided by the load characteristics. Besides, the transmitted power decided by the load demands.

Since the output characteristic and control objectives are different in the above two modes, the control technology for grid-connected generation mode is not capable for islanded mode. Thus, the control technology suitable for islanded mode are researched by the experts and scholars. The conventional control scheme is cascaded dual-loop structure, in which the outer loop governs AC voltage regulation and the inner loop achieves AC current regulation of the inductor. However, the current sensors are required to detect inductor currents, which increases costs and the probability of failure, thereby decreasing the reliability of control systems. Besides, the PV inverter operates in wide range and load disturbances and parameter uncertainties are still presence in islanded mode, the traditional linear control methods deteriorates system's dynamic performance and stability. To solve above issues, the author's research team has done a lot of researches and proposed the nonlinear control methods based on single-loop voltage control scheme. The details are elaborated in Sect. 7.3.2.

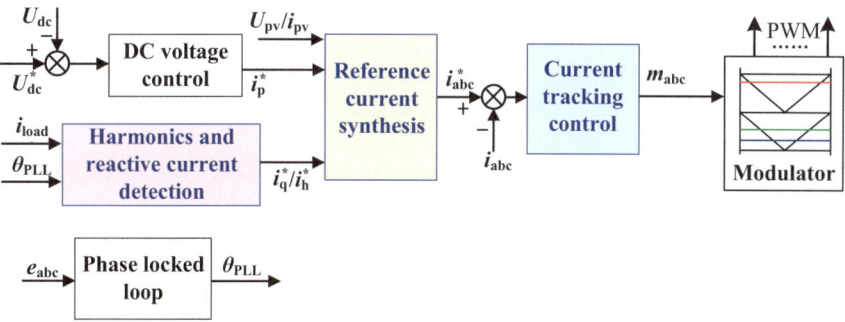

Fig. 7.1 The overall control diagram for power quality control mode

7.2 Photovoltaic Inverters in the Power Quality Control Mode

In this section, the power quality control mode of PV inverter is explored. The power quality issues include current harmonic and reactive power, which easily cause grid voltage distortion, fluctuations and other electrical equipment failures.

Since the main circuit of PV inverter consistent with that in shunt active power filter, the PV inverter has the ability to eliminate harmonic and compensation reactive power. By adding the appropriate control scheme, PV inverter can be used as power quality control equipment at night and on cloudy days, and as power generation equipment when the irradiance is sufficient, which make full use of the PV inverter capacity and save the investment costs of shunt active power filter.

The rest of this section is organized as follows. First, the advantages and the principles of PV inverters operating in power quality control mode are explained. Considering different actual working conditions, three working modes are elaborated. Then the control scheme for power quality control mode of PV inverter is illustrated. The control scheme includes harmonics and reactive current detection, reference current synthesis, and reference current tracking control.

7.2.1 Principle of Power Quality Control

As mentioned above, by adding the appropriate control scheme of the PV inverter, harmonic current and reactive power can be effectively suppressed. Compared with traditional passive power filters, PV inverters with power quality control capabilities have the following advantages:

1. The PV inverters can compensate current harmonics with varying frequencies and magnitudes, as well as fluctuating reactive power, and they respond extremely quickly to changes in the compensation targets.
2. The PV inverters can simultaneously compensate for both current harmonics and reactive power, and they can continuously compensate for reactive power.
3. The PV inverters are not significantly affected by grid impedance and are less prone to resonance with the grid.
4. Due to the ability to track changes in grid frequency, the compensation characteristics of PV inverters are not affected by variations in grid frequency.
5. The PV inverters can either compensate for a single current harmonic and reactive power source individually or collectively compensate for multiple current harmonics and reactive power sources.
6. No energy storage element is needed when compensating reactive power, and the value of energy storage element is also small when compensating harmonics.
7. When the PV inverter operates in power quality control mode, overload generally does not occur in the high-power application.

7.2 Photovoltaic Inverters in the Power Quality Control Mode

The basic structure and working principle of PV inverter with power quality control mode are shown in Figs. 7.2, 7.3, and 7.4. In which, e_x and i_{gx} (x = a, b, c) are the grid voltage and grid current. i_{Lx} is the load current. i_x is the output current of PV inverter. According to the actual working conditions, the PV inverters can be divided into three working modes. Mode 1: only output active power. Mode 2: only output harmonic and reactive power. Mode 3: output active power, reactive power and harmonic at the same time.

Mode 1: Only Output Active Power

Figure 7.2 shows the power flowing directions when the PV inverter only outputs active power. When sunlight is sufficient, the full capacity of the PV inverter is used

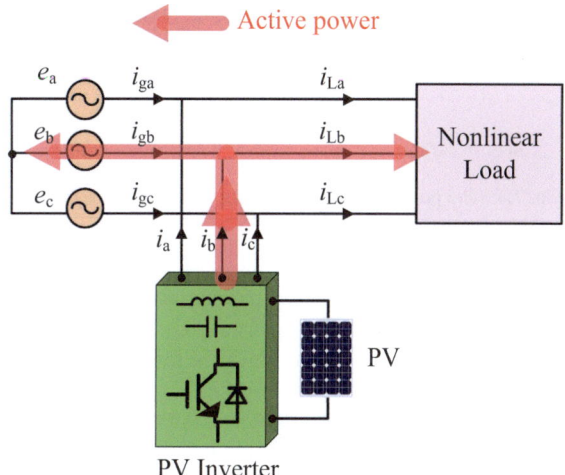

Fig. 7.2 The power flowing directions when the PV inverter only outputs active power

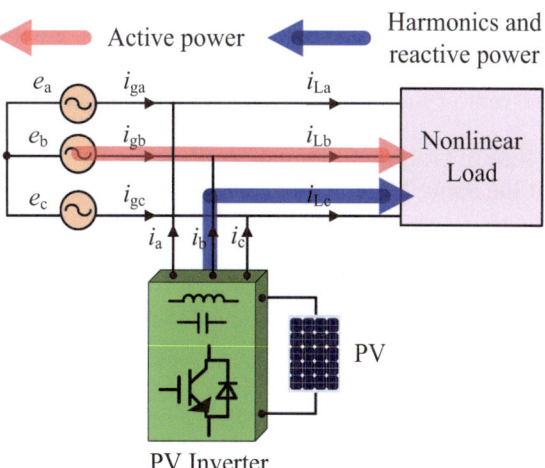

Fig. 7.3 The power flowing directions when the PV inverter only outputs harmonic current and reactive power

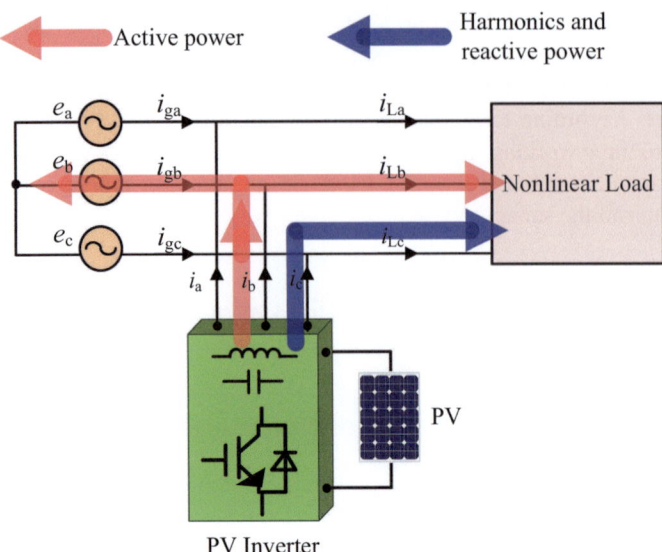

Fig. 7.4 The power flowing directions when the PV inverter outputs active power, reactive power, and harmonic at the same time

to converter the solar energy into electricity power and transmit active power to the grid. At this time, the PV inverter does not have enough capacity to compensate for the harmonics and reactive power in the load current.

Mode 2: Only Output Harmonic Current and Reactive Power

Figure 7.3 shows the power flowing directions when the PV inverter only outputs harmonic current and reactive power mode. At night and on cloudy days, the PV inverters cannot transmit active power to the grid. At this time, since the solar irradiance is insufficient, PV inverter cannot convert and transmit solar energy. Thus, the active power required by the load is provided only by the power grid. Furthermore, the PV inverter output harmonic current and reactive power to the load by adjusting the control strategy to achieve power quality control.

Mode 3: Output Active Power, Reactive Power, and Harmonic Current at the Same Time

Figure 7.4 shows the directions of power flow when the PV inverter outputs active power, reactive power, and harmonic current at the same time. When the sunlight is not very sufficient, the PV inverter may not use the full capacity after achieving maximum power point tracking. At this time, the remaining capacity of the PV inverter can be used to output reactive power and harmonics to achieve power quality control of the load current.

In order to achieve power quality control function, the appropriate control scheme with various control methods is required for PV inverters. A detailed control scheme and various control methods are described in the following section.

7.2.2 Control Schemes for Power Quality Control

Based on the introduction of the above power quality control principle, as shown in Fig. 7.5, the technology for power quality control mode mainly includes three parts: harmonics and reactive current detection, reference current synthesis, and reference current tracking control. This section mainly elaborates the implementation process or design rules of each parts.

Harmonics and Reactive Current Detection

Harmonic and reactive current detection is a crucial step of power quality control mode. This section primarily introduces two commonly used harmonics and reactive current detection methods.

Method 1: Based on Instantaneous Reactive Power Theory

The instantaneous reactive power theory in three-phase circuit was first proposed by Hirofumi Akagi in 1983 [1], and has been continuously studied and gradually improved. This section mainly introduces the principle of the instantaneous reactive power theory and the harmonic current and reactive power detection method based on the instantaneous reactive power theory.

The instantaneous voltage and current in each phase of three-phase circuit are denoted as e_a, e_b, e_c and i_a, i_b, i_c respectively. For convenience of analysis, the instantaneous voltage and current are transformed into $\alpha\beta$ two-phase stationary coordinate frame, which are expressed as (7.1) and (7.2). e_α, e_β and i_α, i_β are the instantaneous voltage and current in $\alpha\beta$ two-phase stationary coordinate frame.

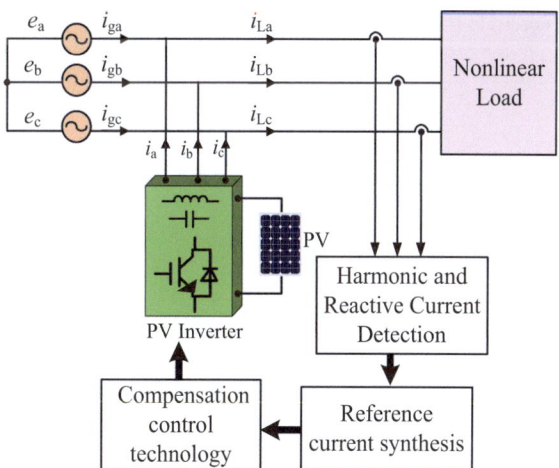

Fig. 7.5 Control block diagram of PV inverter for power quality control mode

$$\begin{bmatrix} e_\alpha \\ e_\beta \end{bmatrix} = \sqrt{\frac{2}{3}} \begin{bmatrix} 1 & -\frac{1}{2} & -\frac{1}{2} \\ 0 & \frac{\sqrt{3}}{2} & -\frac{\sqrt{3}}{2} \end{bmatrix} \begin{bmatrix} e_a \\ e_b \\ e_c \end{bmatrix} \quad (7.1)$$

$$\begin{bmatrix} i_\alpha \\ i_\beta \end{bmatrix} = \sqrt{\frac{2}{3}} \begin{bmatrix} 1 & -\frac{1}{2} & -\frac{1}{2} \\ 0 & \frac{\sqrt{3}}{2} & -\frac{\sqrt{3}}{2} \end{bmatrix} \begin{bmatrix} i_a \\ i_b \\ i_c \end{bmatrix} \quad (7.2)$$

Figure 7.6 shows the voltage and current vectors in the $\alpha\beta$ stationary coordinate frame. It can be seen that the vectors e_α, e_β and i_α, i_β can synthesize rotated voltage vector $\boldsymbol{e}$ and rotated current vector $\boldsymbol{i}$ in the $\alpha\beta$ stationary coordinate frame respectively, which are expressed as

$$\begin{cases} \boldsymbol{e} = e_\alpha + e_\beta = E_m \angle \varphi_e \\ \boldsymbol{i} = i_\alpha + i_\beta = I_m \angle \varphi_i \end{cases} \quad (7.3)$$

where E_m and I_m are magnitudes of vectors $\boldsymbol{e}, \boldsymbol{i}$; φ_e, φ_i are phase angles of vectors $\boldsymbol{e}$, $\boldsymbol{i}$.

Definition 1 As shown in Fig. 7.6, the instantaneous active current i_p and instantaneous reactive current i_q are the projections of vector $\boldsymbol{i}$ onto vector $\boldsymbol{e}$ and its normal, respectively, which are expressed as

$$\begin{cases} i_p = I_m \cos \varphi \\ i_q = I_m \sin \varphi \end{cases} \quad (7.4)$$

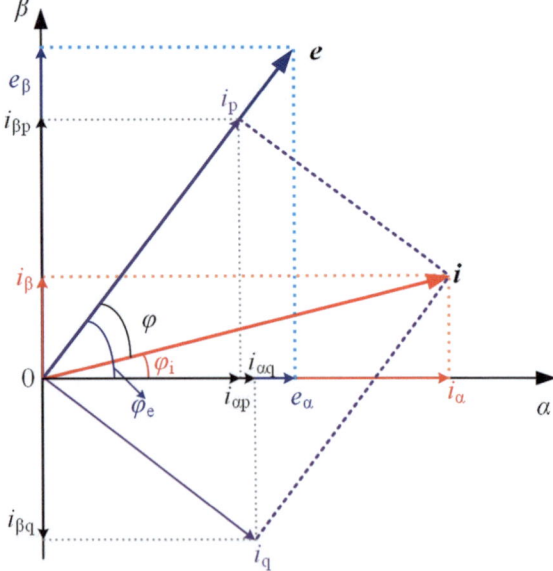

Fig. 7.6 Voltage and current vectors in the $\alpha\beta$ stationary coordinate frame

7.2 Photovoltaic Inverters in the Power Quality Control Mode

where $\varphi = \varphi_e - \varphi_i$.

Definition 2 The instantaneous reactive power q (instantaneous active power p) is the products of the magnitude of the voltage vector e and the instantaneous reactive current i_q (instantaneous active current i_p), which can be expressed as

$$\begin{cases} p = E_m i_p \\ q = E_m i_q \end{cases} \quad (7.5)$$

According to above definition and relationships among vectors e_α, e_β, i_α, i_β and rotated voltage vector e and rotated current vector i, the instantaneous active and reactive power can also be expressed

$$\begin{bmatrix} p \\ q \end{bmatrix} = \begin{bmatrix} e_\alpha & e_\beta \\ e_\beta & -e_\alpha \end{bmatrix} \begin{bmatrix} i_\alpha \\ i_\beta \end{bmatrix} = \mathbf{C}_{pq} \begin{bmatrix} i_\alpha \\ i_\beta \end{bmatrix} \quad (7.6)$$

Besides, according to the relationship of e_a, e_b, e_c and e_α, e_β, i_a, i_b, i_c and i_α, i_β, the expression of p and q for three-phase voltage and current are expressed as

$$\begin{cases} p = e_a i_a + e_b i_b + e_c i_c \\ q = \frac{1}{\sqrt{3}}[(e_b - e_c)i_a + (e_c - e_a)i_b + (e_a - e_b)i_c] \end{cases} \quad (7.7)$$

From (7.7), it can be seen that the instantaneous active power of the three-phase circuit is the instantaneous power of the three-phase circuit.

Definition 3 As shown in Fig. 7.6, the instantaneous reactive current $i_{\alpha q}$, $i_{\beta q}$ (instantaneous active current $i_{\alpha p}$, $i_{\beta p}$) are the projections of instantaneous reactive current i_q (Instantaneous active current i_p) onto the α-axis, β-axis respectively, which can be expressed as

$$\begin{cases} i_{\alpha p} = i_p \cos \varphi_e = \frac{e_\alpha}{e} i_p = \frac{e_\alpha}{e_\alpha^2 + e_\beta^2} p \\ i_{\beta p} = i_p \sin \varphi_e = \frac{e_\beta}{e} i_p = \frac{e_\beta}{e_\alpha^2 + e_\beta^2} p \end{cases} \quad (7.8)$$

$$\begin{cases} i_{\alpha q} = i_q \sin \varphi_e = \frac{e_\beta}{e} i_q = \frac{e_\beta}{e_\alpha^2 + e_\beta^2} q \\ i_{\beta q} = -i_q \cos \varphi_e = \frac{-e_\alpha}{e} i_q = \frac{-e_\alpha}{e_\alpha^2 + e_\beta^2} q \end{cases} \quad (7.9)$$

According to above expression, the following mathematical relationships are obtained as

$$\begin{cases} i_{\alpha p}^2 + i_{\beta p}^2 = i_p^2 \\ i_{\alpha q}^2 + i_{\beta q}^2 = i_q^2 \end{cases} \quad (7.10)$$

$$\begin{cases} i_{\alpha p} + i_{\alpha q} = i_\alpha \\ i_{\beta p} + i_{\beta q} = i_\beta \end{cases} \quad (7.11)$$

The above expression displays the current orthogonality in the α-phase and β-phase. Furthermore, the instantaneous active currents and instantaneous reactive currents can be referred to as the active and reactive components of the instantaneous current, respectively.

Definition 4 Instantaneous reactive powers of q_α, q_β (p_α, p_β) are the products of the instantaneous voltage and instantaneous reactive current (instantaneous active current), respectively, which can be expressed as

$$\begin{cases} p_\alpha = e_\alpha i_{\alpha p} = \dfrac{e_\alpha^2}{e_\alpha^2 + e_\beta^2} p \\ p_\beta = e_\beta i_{\beta p} = \dfrac{e_\beta^2}{e_\alpha^2 + e_\beta^2} p \end{cases} \tag{7.12}$$

$$\begin{cases} q_\alpha = e_\alpha i_{\alpha q} = \dfrac{e_\alpha e_\beta}{e_\alpha^2 + e_\beta^2} q \\ q_\beta = e_\beta i_{\beta q} = \dfrac{-e_\alpha e_\beta}{e_\alpha^2 + e_\beta^2} q \end{cases} \tag{7.13}$$

Thus, the following mathematical relationship are obtained as

$$\begin{cases} p_\alpha + p_\beta = p \\ q_\alpha + q_\beta = 0 \end{cases} \tag{7.14}$$

Definition 5 For the three-phase system, the instantaneous reactive currents i_{aq}, i_{bq}, i_{cq} and active currents i_{ap}, i_{bp}, i_{cp} of each phase are obtained by applying inverse Clarke transformation to the instantaneous reactive current $i_{\alpha q}$, $i_{\beta q}$ and active current $i_{\alpha p}$, $i_{\beta p}$, which can be expressed as

$$\begin{bmatrix} i_{ap} \\ i_{bp} \\ i_{cp} \end{bmatrix} = \sqrt{\dfrac{2}{3}} \begin{bmatrix} 1 & 0 \\ -\dfrac{1}{2} & \dfrac{\sqrt{3}}{2} \\ -\dfrac{1}{2} & -\dfrac{\sqrt{3}}{2} \end{bmatrix} \begin{bmatrix} i_{\alpha p} \\ i_{\beta p} \end{bmatrix} \tag{7.15}$$

$$\begin{bmatrix} i_{aq} \\ i_{bq} \\ i_{cq} \end{bmatrix} = \sqrt{\dfrac{2}{3}} \begin{bmatrix} 1 & 0 \\ -\dfrac{1}{2} & \dfrac{\sqrt{3}}{2} \\ -\dfrac{1}{2} & -\dfrac{\sqrt{3}}{2} \end{bmatrix} \begin{bmatrix} i_{\alpha q} \\ i_{\beta q} \end{bmatrix} \tag{7.16}$$

Substituting the expression of instantaneous reactive current $i_{\alpha q}$, $i_{\beta q}$, instantaneous active current $i_{\alpha p}$, $i_{\beta p}$, instantaneous reactive current and active current of each phase are obtained that

$$\begin{cases} i_{ap} = 3e_a \dfrac{P}{A} \\ i_{bp} = 3e_b \dfrac{P}{A} \\ i_{cp} = 3e_c \dfrac{P}{A} \end{cases} \tag{7.17}$$

7.2 Photovoltaic Inverters in the Power Quality Control Mode

$$\begin{cases} i_{aq} = (e_b - e_c)\frac{q}{A} \\ i_{bq} = (e_c - e_a)\frac{q}{A} \\ i_{cq} = (e_a - e_b)\frac{q}{A} \end{cases} \quad (7.18)$$

where $A = (e_a - e_b)^2 + (e_b - e_c)^2 + (e_c - e_a)^2 = 2(e_a^2 + e_b^2 + e_c^2 - e_a e_b - e_b e_c - e_c e_a)$.

As the above expression, the following mathematical relationship are obtained as

$$\begin{cases} i_{ap} + i_{bp} + i_{cp} = 0 \\ i_{aq} + i_{bq} + i_{cq} = 0 \end{cases} \quad (7.19)$$

$$\begin{cases} i_{ap} + i_{aq} = i_a \\ i_{bp} + i_{bq} = i_b \\ i_{cp} + i_{cq} = i_c \end{cases} \quad (7.20)$$

The above expressions demonstrate the current symmetry in the *a*-phase, *b*-phase, and *c*-phase. It can be observed that the current orthogonality in the *α*-phase and *β*-phase corresponds to the current symmetry in the *a*-phase, *b*-phase, and *c*-phase.

Definition 6 The instantaneous reactive power q_a, q_b, q_c (instantaneous active power p_a, p_b, p_c) of each phase are the products of the instantaneous voltage and the instantaneous reactive current (instantaneous active current) in each phase, respectively, which can be expressed as

$$\begin{cases} P_a = e_a i_{ap} = 3e_a^2 \frac{P}{A} \\ P_b = e_b i_{bp} = 3e_b^2 \frac{P}{A} \\ P_c = e_c i_{cp} = 3e_c^2 \frac{P}{A} \end{cases} \quad (7.21)$$

$$\begin{cases} q_a = e_a i_{aq} = e_a(e_b - e_c)\frac{q}{A} \\ q_b = e_b i_{bq} = e_b(e_c - e_a)\frac{q}{A} \\ q_c = e_c i_{cq} = e_c(e_a - e_b)\frac{q}{A} \end{cases} \quad (7.22)$$

Based on the expressions of instantaneous reactive power and active power of each phase, the relationship of them are obtained as

$$\begin{cases} p_a + p_b + p_c = p \\ q_a + q_b + q_c = 0 \end{cases} \quad (7.23)$$

Traditional power theory is only applicable when both voltage and current are sinusoidal waves. Conversely, the instantaneous reactive power theory is applicable not only to sinusoidal waves but also to non-sinusoidal waves. It follows from the above definitions that the concepts in instantaneous reactive power theory are very similar to those in traditional power theory, which can be regarded as an extension and generalization of traditional power theory.

When the three-phase voltage and current are sinusoidal, the three-phase voltage and current can be expressed as

$$\begin{cases} e_a = E_m \sin \omega t \\ e_b = E_m \sin\left(\omega t - \frac{2\pi}{3}\right) \\ e_c = E_m \sin\left(\omega t + \frac{2\pi}{3}\right) \end{cases} \quad (7.24)$$

$$\begin{cases} i_a = I_m \sin(\omega t - \varphi) \\ i_b = I_m \sin\left(\omega t - \varphi - \frac{2\pi}{3}\right) \\ i_c = I_m \sin\left(\omega t - \varphi + \frac{2\pi}{3}\right) \end{cases} \quad (7.25)$$

The three-phase voltage and current are transformed into $\alpha\beta$ two-phase stationary coordinate frame, which are expressed as

$$\begin{bmatrix} e_\alpha \\ e_\beta \end{bmatrix} = E_{m2} \begin{bmatrix} \sin \omega t \\ -\cos \omega t \end{bmatrix} \quad (7.26)$$

$$\begin{bmatrix} i_\alpha \\ i_\beta \end{bmatrix} = I_{m2} \begin{bmatrix} \sin(\omega t - \varphi) \\ -\cos(\omega t - \varphi) \end{bmatrix} \quad (7.27)$$

where $E_{m2} = \sqrt{\frac{3}{2}} E_m$, $I_{m2} = \sqrt{\frac{3}{2}} I_m$.

By substituting the voltage and current in $\alpha\beta$ two-phase stationary coordinate frame into instantaneous active and reactive power expressions, it can be obtained that

$$\begin{cases} p = \frac{3}{2} E_m I_m \cos \varphi \\ q = \frac{3}{2} E_m I_m \sin \varphi \end{cases} \quad (7.28)$$

Then, the root mean square (RMS) values of phase voltage and current are defined as $E = \frac{E_m}{\sqrt{2}}$, $I = \frac{I_m}{\sqrt{2}}$, the instantaneous active and reactive power can be rewritten as

$$\begin{cases} p = 3EI \cos \varphi \\ q = 3EI \sin \varphi \end{cases} \quad (7.29)$$

From the above equation, it can be seen that when three-phase voltage and current are sinusoidal, p, q are constants, which are consistent with the traditional theory.

Moreover, the instantaneous active current and instantaneous reactive current can also be obtained as

$$\begin{cases} i_{\alpha p} = I_{m2} \cos \varphi \sin \omega t \\ i_{\alpha q} = I_{m2} \sin \varphi \sin\left(\omega t - \frac{\pi}{2}\right) \end{cases} \quad (7.30)$$

7.2 Photovoltaic Inverters in the Power Quality Control Mode

It can be seen that the expressions for the instantaneous active current and instantaneous reactive current in phase α are exactly the same as the traditional power theory. The same conclusion can be drawn for β-phase. Besides, the instantaneous reactive power theory has a broader scope of application compared to the traditional power theory.

The harmonics and reactive current detection methods based on instantaneous reactive power theory exhibit superior real-time performance. When only reactive currents are detected, the results can be obtained without any delay. For harmonic current detection, the delay varies depending on the harmonic composition of the current and the type of filters used, but it does not exceed one fundamental cycle. For example, in the case of a three-phase bridge rectifier, which is a typical harmonic source in power grids, the detection delay is approximately one-sixth of fundamental cycle. Thus, the harmonics and reactive current detection methods based on instantaneous reactive power theory demonstrate excellent real-time characteristics, making them highly suitable for practical applications.

Depending on the calculation process, the harmonics and reactive current detection methods based on instantaneous reactive power theory can be divided into p, q method [2] and i_p, i_q method [3].

1. **p, q Method**

    The block diagram of the p, q method is shown in Fig. 7.7. In this context, C_{32} represents the transformation matrix from abc stationary coordinate frame to $\alpha\beta$ stationary coordinate frame, while C_{23} is the transformation matrix from $\alpha\beta$ stationary coordinate frame to abc stationary coordinate frame. C_{pq} is the power calculation matrix, and C^{-1}_{pq} is the inverse matrix of C_{pq}. The expressions for these transformations are expressed as

    $$C_{32} = \sqrt{\frac{2}{3}} \begin{bmatrix} 1 & -\frac{1}{2} & -\frac{1}{2} \\ 0 & \frac{\sqrt{3}}{2} & -\frac{\sqrt{3}}{2} \end{bmatrix} \quad (7.31)$$

    $$C_{23} = C^{-1}_{32} = \sqrt{\frac{2}{3}} \begin{bmatrix} 1 & 0 \\ -\frac{1}{2} & \frac{\sqrt{3}}{2} \\ -\frac{1}{2} & -\frac{\sqrt{3}}{2} \end{bmatrix} \quad (7.32)$$

    $$C_{pq} = \begin{bmatrix} e_\alpha & e_\beta \\ e_\beta & -e_\alpha \end{bmatrix} \quad (7.33)$$

    When the active power p and reactive power q are calculated, the DC components $\bar{p}, \bar{q}$ are obtained by low-pass filter (LPF). When there is no distortion in the grid voltage waveform, $\bar{p}$ is generated by the fundamental active current and voltage, and $\bar{q}$ is generated by the fundamental reactive power and voltage. Thus, the fundamental components i_{af}, i_{bf}, i_{cf} of the detected currents i_a, i_b, i_c can be calculated as

$$\begin{bmatrix} i_{af} \\ i_{bf} \\ i_{cf} \end{bmatrix} = \mathbf{C}_{23}\mathbf{C}_{pq}^{-1}\begin{bmatrix} \bar{p} \\ \bar{q} \end{bmatrix} = \frac{1}{e^2}\mathbf{C}_{23}\mathbf{C}_{pq}\begin{bmatrix} \bar{p} \\ \bar{q} \end{bmatrix} \qquad (7.34)$$

Then, by subtracting i_{af}, i_{bf}, i_{cf} from i_a, i_b, i_c, the harmonic components i_{ah}, i_{bh}, i_{ch} of i_a, i_b, i_c can be obtained.

When the PV inverters are used to compensate for both current harmonics and reactive power, it is necessary to detect both harmonics and reactive current in the compensation object. In this case, it is sufficient to disconnect the channel for calculating q in the block diagram of the p, q method. At this point, the fundamental active components i_{apf}, i_{bpf}, i_{cpf} of the detected currents i_a, i_b, i_c can be calculated from $\bar{p}$ as

$$\begin{bmatrix} i_{apf} \\ i_{bpf} \\ i_{cpf} \end{bmatrix} = \mathbf{C}_{23}\mathbf{C}_{pq}^{-1}\begin{bmatrix} \bar{p} \\ 0 \end{bmatrix} \qquad (7.35)$$

By subtracting i_{apf}, i_{bpf}, i_{cpf} from i_a, i_b, i_c, respectively, the harmonic and the fundamental reactive components of i_a, i_b, i_c are obtained, denoted as i_{ad}, i_{bd}, i_{cd}. The subscript d denotes the detection result derived from the detection circuit.

Due to the use of a LPF to obtain $\bar{p}$, $\bar{q}$, when the detected current changes, it takes a certain delay time to get the accurate $\bar{p}$, $\bar{q}$, so that the detection results have a certain delay. However, when only detecting the reactive current, there is no need for LPF, and only the inverse transformation of q can be directly derived from the reactive current, so that there is no delay, the obtained reactive current is expressed as

$$\begin{bmatrix} i_{aq} \\ i_{bq} \\ i_{cq} \end{bmatrix} = \frac{1}{e^2}\mathbf{C}_{23}\mathbf{C}_{pq}\begin{bmatrix} 0 \\ q \end{bmatrix} \qquad (7.36)$$

2. i_p, i_q **Method**

The block diagram of the i_p, i_q harmonics and reactive current detection method is shown in Fig. 7.8, where

$$\mathbf{C} = \begin{bmatrix} \sin \omega t & -\cos \omega t \\ -\cos \omega t & -\sin \omega t \end{bmatrix}$$

In this method, the sinusoidal signal $\sin\omega t$ and the corresponding cosine signal $-\cos\omega t$, which are in phase with the phase grid voltage e_a, are required. These signals are obtained by a phase locked loop (PLL) and a sinusoidal and cosinusoidal signal generating circuit. The DC components $\bar{i}_p$, $\bar{i}_q$ are derived through LPF filtering, which are generated by i_{af}, i_{bf}, i_{cf}. Thus i_{af}, i_{bf}, i_{cf} and i_{ah}, i_{bh}, i_{ch} can be calculated from $\bar{i}_p$, $\bar{i}_q$.

7.2 Photovoltaic Inverters in the Power Quality Control Mode

Similar to the p, q method, when the sum of harmonics and reactive current is to be detected, it is only necessary to disconnect the channel for calculating i_q in the block diagram of the i_p, i_q method. If only the reactive current is to be detected, the inverse transformation of i_q is sufficient.

The above two harmonics and reactive current detection methods can be realized with both analog and digital circuits. When realized with analog circuits, the p, q method requires 10 multipliers and 2 dividers. i_p, i_q method requires only 8 multipliers. In order to ensure the accuracy of detection, it is best to use high performance four quadrant analog multiplier chip.

3. **The influence of grid voltage waveform distortion**
The ideal grid voltage waveform should be sinusoidal, but the actual grid voltage waveform will have a certain distortion due to various factors, and this distortion is tolerated within certain limits. Generally, the total harmonic distortion of the grid voltage has been reached 2–3% on average, and its value is higher when the waveform distortion is severe. Therefore, it is necessary to study the effect of grid voltage waveform distortion on the detection method.

Both p, q method and i_p, i_q method are applicable to three-phase, three-wire circuits. For simplicity, it is assumed here that the three phases are symmetrical and the currents with harmonics are expressed as

$$\begin{cases} i_a = \sum_{n=1}^{\infty} \sqrt{2} I_n \sin(n\omega t + \varphi_n) \\ i_b = \sum_{n=1}^{\infty} \sqrt{2} I_n \sin\left[n\left(\omega t - \frac{2\pi}{3}\right) + \varphi_n\right] \\ i_c = \sum_{n=1}^{\infty} \sqrt{2} I_n \sin\left[n\left(\omega t + \frac{2\pi}{3}\right) + \varphi_n\right] \end{cases} \quad (7.37)$$

where $n = 3k \pm 1$ and k is an integer (when $k = 0$, only the '+' sign applies, i.e. $n = 1$). ω is the grid angular frequency. I_n and φ_n represent the RMS value and initial phase angle of each current, respectively.

First, the grid voltage without distortion is analyzed. Based on this, the impact of grid voltage distortion is then analyzed.

(a) **Analysis of the grid voltage without distortion**
When the three-phase grid voltages are symmetrical, the grid voltages can be expressed as

$$\begin{cases} e_a = \sqrt{2} E_1 \sin \omega t \\ e_b = \sqrt{2} E_1 \sin\left(\omega t - \frac{2\pi}{3}\right) \\ e_c = \sqrt{2} E_1 \sin\left(\omega t + \frac{2\pi}{3}\right) \end{cases} \quad (7.38)$$

where E_1 represents the RMS value of the grid voltage when $n = 1$. The three-phase voltages are transformed into $\alpha\beta$ two-phase stationary coordinate frame, and are expressed as

$$\begin{bmatrix} e_\alpha \\ e_\beta \end{bmatrix} = \sqrt{3}E_1 \begin{bmatrix} \sin\omega t \\ -\cos\omega t \end{bmatrix} \tag{7.39}$$

Similarly, the symmetrical three-phase currents with harmonics are transformed into $\alpha\beta$ two-phase stationary coordinate frame, and are expressed as

$$\begin{bmatrix} i_\alpha \\ i_\beta \end{bmatrix} = \sqrt{3} \begin{bmatrix} \sum_{n=1}^{\infty} I_n \sin(n\omega t + \varphi_n) \\ \sum_{n=1}^{\infty} \mp I_n \cos(n\omega t + \varphi_n) \end{bmatrix} \tag{7.40}$$

where the '−' sign is taken for $n = 3k + 1$ and the '+' sign for $n = 3k - 1$.

According to the p, q method, the active power and reactive power can be derived as

$$\begin{aligned} \begin{bmatrix} p \\ q \end{bmatrix} &= \sqrt{3}E_1 \begin{bmatrix} \sin\omega t & -\cos\omega t \\ -\cos\omega t & -\sin\omega t \end{bmatrix} \begin{bmatrix} i_\alpha \\ i_\beta \end{bmatrix} \\ &= 3E_1 \begin{bmatrix} \sum_{n=1}^{\infty} I_n \cos[(1 \mp n)\omega t \mp \varphi_n] \\ \sum_{n=1}^{\infty} \pm I_n \sin[(1 - n)\omega t - \varphi_n] \end{bmatrix} \end{aligned} \tag{7.41}$$

The DC components of active power and reactive power are obtained by LPF as

$$\begin{bmatrix} \bar{p} \\ \bar{q} \end{bmatrix} = 3 \begin{bmatrix} E_1 I_1 \cos(-\varphi_1) \\ E_1 I_1 \sin(-\varphi_1) \end{bmatrix} \tag{7.42}$$

Thus, when the grid voltage without distortion, the fundamental components of the three-phase currents are obtained as

$$\begin{aligned} \begin{bmatrix} i_{af} \\ i_{bf} \\ i_{cf} \end{bmatrix} &= \frac{\mathbf{C_{23}C_{pq}}}{3E_1^2} \begin{bmatrix} \bar{p} \\ \bar{q} \end{bmatrix} = \mathbf{C_{23}} \begin{bmatrix} \sqrt{3}I_1 \sin(\omega t + \varphi_1) \\ -\sqrt{3}I_1 \cos(\omega t + \varphi_1) \end{bmatrix} \\ &= \begin{bmatrix} \sqrt{2}I_1 \sin(\omega t + \varphi_1) \\ \sqrt{2}I_1 \sin\left(\omega t - \frac{2\pi}{3} + \varphi_1\right) \\ \sqrt{2}I_1 \sin\left(\omega t + \frac{2\pi}{3} + \varphi_1\right) \end{bmatrix} \end{aligned} \tag{7.43}$$

At this point, $e^2 = 3E_1^2$. It can be seen that i_{af}, i_{bf}, i_{cf} are accurately derived, and the resulting harmonic components i_{ah}, i_{bh}, i_{ch} are also accurate.

Based on the i_p, i_q method, i_p and i_q can be derived from the block diagram of the i_p, i_q method, and are expressed as

7.2 Photovoltaic Inverters in the Power Quality Control Mode

$$\begin{bmatrix} i_p \\ i_q \end{bmatrix} = CC_{32} \begin{bmatrix} i_a \\ i_b \\ i_c \end{bmatrix} = \begin{bmatrix} \sin\omega t & -\cos\omega t \\ -\cos\omega t & -\sin\omega t \end{bmatrix} C_{32} \begin{bmatrix} i_a \\ i_b \\ i_c \end{bmatrix}$$

$$= \begin{bmatrix} \sin\omega t & -\cos\omega t \\ -\cos\omega t & -\sin\omega t \end{bmatrix} \begin{bmatrix} i_\alpha \\ i_\beta \end{bmatrix} \quad (7.44)$$

It can be seen that i_p and i_q differ from p and q only by a factor $\sqrt{3}E_1$, which is consistent with the definitions of p and q. Thus, i_p and i_q can be further expressed as

$$\begin{bmatrix} i_p \\ i_q \end{bmatrix} = \sqrt{3} \begin{bmatrix} \sum_{n=1}^{\infty} I_n \cos[(1 \mp n)\omega t \mp \varphi_n] \\ \sum_{n=1}^{\infty} \pm I_n \sin[(1 - n)\omega t - \varphi_n] \end{bmatrix} \quad (7.45)$$

i_p, i_q are obtained by LPF as

$$\begin{bmatrix} \overline{i_p} \\ \overline{i_q} \end{bmatrix} = \sqrt{3} \begin{bmatrix} I_1 \cos(-\varphi_1) \\ I_1 \sin(-\varphi_1) \end{bmatrix} \quad (7.46)$$

Based on the block diagram of the i_p, i_q method, the fundamental component of the current is derived as

$$\begin{bmatrix} i_{af} \\ i_{bf} \\ i_{cf} \end{bmatrix} = C_{23} C \begin{bmatrix} \overline{i_p} \\ \overline{i_q} \end{bmatrix} = C_{23} \begin{bmatrix} \sin\omega t & -\cos\omega t \\ -\cos\omega t & -\sin\omega t \end{bmatrix} \begin{bmatrix} \overline{i_p} \\ \overline{i_q} \end{bmatrix}$$

$$= \begin{bmatrix} \sqrt{2}I_1 \sin(\omega t + \varphi_1) \\ \sqrt{2}I_1 \sin(\omega t - \frac{2\pi}{3} + \varphi_1) \\ \sqrt{2}I_1 \sin(\omega t + \frac{2\pi}{3} + \varphi_1) \end{bmatrix} \quad (7.47)$$

It is evident that the i_p, i_q method can accurately determine the values of i_{af}, i_{bf}, i_{cf}, and i_{ah}, i_{bh}, i_{ch}.

(b) **Analysis of the grid voltage with distortion**

When the grid voltage is distorted, the grid voltage may be symmetrical or asymmetrical. It is assumed that the distorted grid voltage are symmetrical, and can be expressed as follow

$$\begin{cases} e_a = \sum_{n=1}^{\infty} \sqrt{2}E_n \sin(n\omega t + \theta_n) \\ e_b = \sum_{n=1}^{\infty} \sqrt{2}E_n \sin\left[n\left(\omega t - \frac{2\pi}{3}\right) + \theta_n\right] \\ e_c = \sum_{n=1}^{\infty} \sqrt{2}E_n \sin\left[n\left(\omega t + \frac{2\pi}{3}\right) + \theta_n\right] \end{cases} \quad (7.48)$$

where E_n and θ_n are the RMS value and initial phase angle of grid voltages, and $\theta_1 = 0$. The three-phase voltages with distortion are transformed into $\alpha\beta$ two-phase stationary coordinate frame, and expressed as

$$\begin{bmatrix} e_\alpha \\ e_\beta \end{bmatrix} = \sqrt{3} \begin{bmatrix} \sum_{n=1}^{\infty} E_n \sin(n\omega t + \theta_n) \\ \sum_{n=1}^{\infty} \mp E_n \cos(n\omega t + \theta_n) \end{bmatrix} \quad (7.49)$$

According to the p, q method, the active power and reactive power are obtained, which are expressed as

$$\begin{bmatrix} p \\ q \end{bmatrix} = 3 \begin{bmatrix} \sum_{n=1}^{\infty} E_n I_n \cos(\theta_n - \varphi_n) \\ + \sum_{n=1}^{\infty} \sum_{m(m \neq n)=1}^{\infty} E_n I_m \cos[(n \mp m)\omega t + (\theta_n \mp \varphi_n)] \\ \sum_{n=1}^{\infty} \pm E_n I_n \sin(\theta_n - \varphi_n) \\ + \sum_{n=1}^{\infty} \sum_{m(m \neq n)=1}^{\infty} \mp E_n I_m \sin[(n - m)\omega t + (\theta_n - \varphi_n)] \end{bmatrix} \quad (7.50)$$

where m is introduced to distinguish the voltage and current of different orders and is taken in the same way as n.

The DC components of p and q are expressed as

$$\begin{bmatrix} \bar{p} \\ \bar{q} \end{bmatrix} = 3 \begin{bmatrix} \sum_{n=1}^{\infty} E_n I_n \cos(\theta_n - \varphi_n) \\ \sum_{n=1}^{\infty} \pm E_n I_n \sin(\theta_n - \varphi_n) \end{bmatrix} \quad (7.51)$$

At this point, $e^2 = \sum_{n=1}^{\infty} E_n^2$. Thus, when the grid voltage is distorted, the fundamental components of the three-phase currents are derived as

$$\begin{bmatrix} i_{af} \\ i_{bf} \\ i_{cf} \end{bmatrix} = \frac{1}{3 \sum_{n=1}^{\infty} E_n^2} C_{23} \begin{bmatrix} e_\alpha & e_\beta \\ e_\beta & -e_\alpha \end{bmatrix} \begin{bmatrix} \bar{p} \\ \bar{q} \end{bmatrix} \quad (7.52)$$

Comparing the fundamental components of the three-phase currents derived without grid voltage distortion and with grid voltage distortion, the errors can be obtained as

$$\begin{bmatrix} \Delta i_{af} \\ \Delta i_{bf} \\ \Delta i_{cf} \end{bmatrix} = C_{23} \begin{bmatrix} e_{\alpha h} \frac{\bar{P}}{e^2} + e_{\beta h} \frac{\bar{Q}}{e^2} \\ e_{\beta h} \frac{\bar{P}}{e^2} - e_{\alpha h} \frac{\bar{Q}}{e^2} \end{bmatrix}$$

7.2 Photovoltaic Inverters in the Power Quality Control Mode

$$+ C_{23} \begin{bmatrix} e_{\alpha f}\left(\frac{\bar{p}}{e^2} - \frac{I_1 \cos\varphi_1}{E_1}\right) + e_{\beta f}\left[\frac{\bar{q}}{e^2} - \frac{I_1 \sin(-\varphi_1)}{E_1}\right] \\ e_{\beta f}\left(\frac{\bar{p}}{e^2} - \frac{I_1 \cos\varphi_1}{E_1}\right) + e_{\alpha f}\left[\frac{\bar{q}}{e^2} - \frac{I_1 \sin(-\varphi_1)}{E_1}\right] \end{bmatrix} \quad (7.53)$$

where Δ is the error. Subscripts f, h are the fundamental and harmonic components.

Comparing fundamental components of the three-phase currents derived without grid voltage distortion and with grid voltage distortion, it can be seen that the reasons for the error can be summarized as

① When grid voltage has distortion, e_α, e_β contain harmonics, so that the i_{af}, i_{bf}, i_{cf} also contain harmonics.
② When grid voltage has no distortion, the $\bar{P}$ and $\bar{Q}$ in are only related to the fundamental voltage and current. However, when grid voltage has distortion $\bar{P}$ and $\bar{Q}$ are related to all harmonic voltages and currents.
③ $e^2 = \sum_{n=1}^{\infty} E_n^2$ is larger than $e^2 = 3E_1^2$.

Based on the analyses above, it can be generalized that for a three-phase three-wire system, as long as there is a distortion in the grid voltages, regardless of whether the three-phase voltages and currents are symmetrical, the detection results using the p, q method will have errors. However, when using the i_p, i_q method for detection, only $\sin\omega t$ and $-\cos\omega t$ are involved in the computation. The harmonic components of the distorted voltage do not appear during the calculation process. Thus, the detection results are not affected by the distortion of the grid voltages, ensuring the accuracy of the detection results.

Method 2: Based on Time-Domain Transformation Algorithm
Based on the orthogonality property of trigonometric functions, a harmonic current and reactive power detection method utilizing a Time-domain based Transform Algorithm (TTA) is introduced in this section, which has the advantages of clear physical significance, flexible detection objectives, and ease of implementation on digital processors. Additionally, this method is applicable to both three-phase three-wire and three-phase four-wire systems.

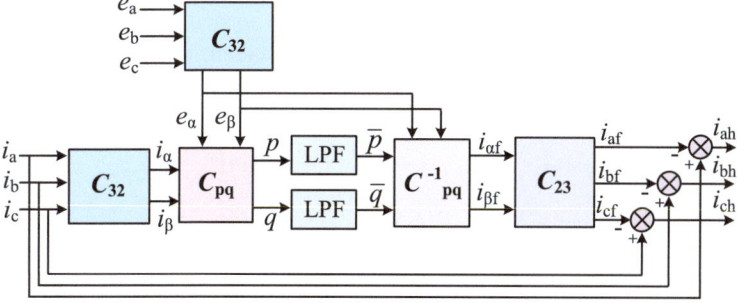

Fig. 7.7 The block diagram of the p, q method

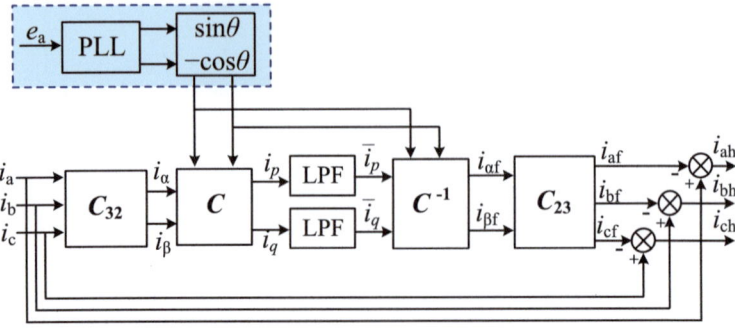

Fig. 7.8 The block diagram of the i_p, i_q method

1. **The principle of harmonic and reactive power detection method using TTA**
   According to the symmetric component method, the instantaneous currents in the three-phase three-wire system can be decomposed into a positive-sequence component and a negative-sequence component, which can be expressed as:

$$i_x(n) = \sum_{k=1}^{\infty}\left[I_{1k}\sin\left(\frac{2\pi n}{N}k + \varphi_{1k} - \frac{2\pi l}{3}\right) + I_{2k}\sin\left(\frac{2\pi n}{N}k + \varphi_{2k} + \frac{2\pi l}{3}\right)\right]$$
(7.54)

where the subscript 1 denotes the positive-sequence component, and the subscript 2 denotes the negative-sequence component. k represents the harmonic order; φ is the initial phase angle; N is the number of sampling points per fundamental frequency cycle; n ($n = 0, 1, \ldots, N-1$) is the count value of the sampling points; x ($x = $ a, b, c) is the three-phase notation; l ($x = $ a, $l = 0$; $x = $ b, $l = 1$; $x = $ c, $l = 2$) represents the corresponding number of phase x.

Generally, the specific harmonic is compensated or not compensated, rather to compensate the positive- and negative-sequence of the specific harmonic. However, for the fundamental component, it is often necessary to separately compensate the positive-sequence, negative-sequence, active, and reactive components. Therefore, further decomposition is required for the fundamental component, whereas no further decomposition is needed for the other harmonics.

The fundamental components of the instantaneous currents i_a, i_b, i_c can be further decomposed as

$$\begin{aligned}i_{x1}(n) &= I_{11}\sin\left(\frac{2\pi}{N}n + \varphi_{11} - \frac{2l\pi}{3}\right) + I_{21}\sin\left(\frac{2\pi}{N}n + \varphi_{21} + \frac{2l\pi}{3}\right)\\ &= I_{11}\sin\left(\frac{2\pi}{N}n - \frac{2l\pi}{3}\right)\cos\varphi_{11} + I_{11}\cos\left(\frac{2\pi}{N}n - \frac{2l\pi}{3}\right)\sin\varphi_{11}\\ &\quad + I_{21}\sin\left(\frac{2\pi}{N}n - \frac{2l\pi}{3}\right)\cos\left(\varphi_{21} + \frac{4l\pi}{3}\right)\end{aligned}$$

7.2 Photovoltaic Inverters in the Power Quality Control Mode

$$+ I_{21} \cos\left(\frac{2\pi}{N}n - \frac{2l\pi}{3}\right) \sin\left(\varphi_{21} + \frac{4l\pi}{3}\right) \tag{7.55}$$

where the first term $I_{11}\sin(2n\pi/N - 2l\pi/3) \cos\varphi_{11}$ represents the fundamental positive-sequence active current i_{px11}, with equal three-phase amplitudes and phase sequentially differing by $2\pi/3$. The second term $I_{11}\cos(2n\pi/N - 2l\pi/3) \sin\varphi_{11}$ represents the fundamental positive-sequence reactive current i_{qx11}, with equal three-phase amplitudes and phase sequentially differing by $2\pi/3$. The third term $I_{21}\sin(2n\pi/N - 2l\pi/3) \cos(\varphi_{21} + 4l\pi/3)$ represents the fundamental negative-sequence active current i_{px21} (which is in the same direction as the fundamental voltage). Although the phase differences between the three phases current still follow the sequence of $2\pi/3$, the amplitudes are no longer equal. The fourth term $I_{21}\cos(2n\pi/N - 2l\pi/3) \sin(\varphi_{21} + 4l\pi/3)$ represents the fundamental negative-sequence reactive current i_{qx21} (which is orthogonal to the fundamental voltage), and similar to the third term, although the phase shifts between the three phases are still $2\pi/3$ apart, the amplitudes are no longer equal. The above mathematical relationship is shown in the Fig. 7.9.

As shown in (7.54) and (7.55), the current is decomposed into the fundamental positive-sequence active component, the fundamental positive-sequence reactive component, the fundamental negative-sequence active component, the fundamental negative-sequence reactive component, the harmonic negative-sequence component, and the harmonic positive-sequence component. In order to separately extract each of these components to satisfy the different compensation purposes, the structure of the harmonic and reactive power detection method based on ATT is introduced in this section as shown in Fig. 7.10.

In Fig. 7.10, when $k = 1$, the term of $\sin(2n\pi/N - 2l\pi/3)$ is the synchronous signal synchronized with the fundamental positive-sequence component of the phase voltage, which is obtained by soft phase locking.

The three-phase voltage can be written as $U\sin(2n\pi/N - 2l\pi/3)$, so the instantaneous fundamental power S_1 can be expressed as

$$\begin{aligned}
S_1 &= i_{x1}(n) * U \sin\left(\frac{2\pi}{N}n - \frac{2l\pi}{3}\right) \\
&= \frac{1}{2}U\left\{I_{11}\cos\varphi_{11}\left[1 - \cos\left(\frac{4\pi}{N}n - \frac{4l\pi}{3}\right)\right] + I_{11}\sin\varphi_{11}\sin\left(\frac{4\pi}{N}n - \frac{4l\pi}{3}\right)\right. \\
&\quad + I_{21}\cos\left(\varphi_{21} + \frac{4l\pi}{3}\right)\left[1 - \cos\left(\frac{4\pi}{N}n - \frac{4l\pi}{3}\right)\right] \\
&\quad \left. + I_{21}\sin\left(\varphi_{21} + \frac{4l\pi}{3}\right)\sin\left(\frac{4\pi}{N}n - \frac{4l\pi}{3}\right)\right\}
\end{aligned} \tag{7.56}$$

It can be derived that:

(a) The first term $UI_{11}\cos\varphi_{11}[1 - \cos(4\pi n/N - 4\pi l/3)]/2$ is caused by the positive sequence active component of the fundamental current, which consists of a DC component and a second-harmonic component. The DC component is the same across all three phases, representing the active power transmitted from the power source to the load. The second-harmonic component of the three phases has the same amplitude, the phase difference is $4\pi/3$ in sequence, and the sum of the three phases is zero, indicating that this part of the energy flows between the three phases.

(b) The second term $UI_{11}\sin\varphi_{11}\sin(4\pi n/N - 4\pi l/3)/2$ is caused by the positive-sequence reactive component of the fundamental current, where the amplitude of each phase is equal, and the sum of the three phases is zero, indicating that the energy flows between the three phases, and this part of the energy will not flow through the DC-side energy storage element when the active power filter compensates for this part.

(c) The third term $UI_{21}\cos(\varphi_{21} + 4l\pi/3)[1 - \cos(4\pi n/N - 4\pi l/3)]/2$ is caused by the negative-sequence "active" component of the fundamental current, which also consists of a DC component and an AC component. The DC components of the three phases are not the same, but the sum of the three phases is zero, indicating that this part of the energy flows out of one phase and into the other two, or out of two phases and into one phase. The AC component integrates to zero over one period, indicating that there is energy exchange between the source and load without reflecting the energy consumed by the load. The amplitude of the AC component differs, and the sum of the three phases is not zero. When compensating for this part, the active power filter requires energy storage elements.

(d) The fourth term $UI_{21}\sin(\varphi_{21} + 4l\pi/3)\sin(4\pi n/N - 4\pi l/3)/2$ is caused by the negative-sequence "reactive" component of the fundamental current, which integrates to zero over one period, indicating that this part does not contribute to the active energy consumed by the load.

(e) The sum of the third and fourth terms represents the instantaneous power of the fundamental negative-sequence, and the sum of the two parts is equal to $UI_{21}[\cos(\varphi_{21} + 4l\pi/3) - \cos(4\pi n/N + \varphi_{21})]/2$, and the sum of the instantaneous power of the three-phase fundamental negative-sequence is equal to $-3UI_{21}\cos(4\pi n/N + \varphi_{21})/2$. When compensating for this part, the active power filter requires energy storage elements.

Similarly, the instantaneous harmonics power can be obtained as

$$S_n = i_{xk}(n) * U \sin\left(\frac{2\pi}{N}n - \frac{2l\pi}{3}\right)$$

$$= \frac{1}{2} + U\left\{I_{1k}\left[\cos\left(\frac{2n\pi}{N}(k-1) + \varphi_{1k}\right)\right.\right.$$

$$\left.- \cos\left(\frac{2n\pi}{N}(k+1) + \varphi_{1k} - \frac{4l\pi}{3}\right)\right]$$

$$+ I_{2k}\left[\cos\left(\frac{2n\pi}{N}(k-1) + \varphi_{2k} + \frac{4l\pi}{3}\right)\right.$$

7.2 Photovoltaic Inverters in the Power Quality Control Mode

$$-\cos]\left(\frac{2n\pi}{N}(k+1) + \varphi_{2k}\right)\right\} \tag{7.57}$$

It can be concluded that whether it is the instantaneous power of the negative sequence component of the harmonics or the instantaneous power of the positive sequence component, both of them are AC, and the sum of the three phases is not zero. When compensating for this part, the active power filter requires energy storage elements.

When $k = 1$, multiplying (7.54) by $\sin(2\pi n/N - 2\pi l/3)$ yields

$$i_x(n)\sin\left(\frac{2\pi n}{N} - \frac{2\pi l}{3}\right)$$
$$= \frac{1}{2}\left\{I_{11}\cos\varphi_{11}\left[1 - \cos\left(\frac{4\pi n}{N} - \frac{4\pi l}{3}\right)\right] + I_{11}\sin\varphi_{11}\sin\left(\frac{4\pi n}{N} - \frac{4\pi l}{3}\right)\right.$$
$$+ I_{21}\cos\left(\varphi_{21} + \frac{4\pi l}{3}\right)\left[1 - \cos\left(\frac{4\pi n}{N} - \frac{4\pi l}{3}\right)\right]$$
$$\left. + I_{21}\sin\left(\varphi_{21} + \frac{4\pi l}{3}\right)\sin\left(\frac{4\pi n}{N} - \frac{4\pi l}{3}\right)\right\} + i_h \tag{7.58}$$

By adding an appropriate LPF with a cut-off frequency less than twice the fundamental frequency, the DC component can be extracted. By doubling this DC component, we have

$$B_{x1} = I_{11}\cos\varphi_{11} + I_{21}\cos\left(\varphi_{21} + \frac{4\pi l}{3}\right) \tag{7.59}$$

Multiplying both sides of (7.54) by $\cos(2\pi n/N - 2\pi l/3)$ yields

$$i_x(n)\cos\left(\frac{2\pi n}{N} - \frac{2\pi l}{3}\right)$$
$$= \frac{1}{2}\left\{I_{11}\left[\sin\varphi_{11} + \sin\left(\frac{4\pi n}{N} + \varphi_{11} - \frac{4\pi l}{3}\right)\right]\right.$$
$$\left. + I_{21}\left[\sin\left(\frac{4\pi n}{N} + \varphi_{21}\right) + \sin\left(\varphi_{21} + \frac{4\pi l}{3}\right)\right]\right\} + i'_h \tag{7.60}$$

By applying the same processing strategy, as in (7.58)–(7.60), it yields

$$A_{x1} = I_{11}\sin\varphi_{11} + I_{21}\sin\left(\varphi_{21} - \frac{2\pi l}{3}\right) \tag{7.61}$$

Then, the following relationship can be obtained:

$$\begin{pmatrix} A_{11} & B_{11} \\ A_{21} & B_{21} \end{pmatrix} = \frac{1}{3}\begin{pmatrix} A_{al} + A_{bl} + A_{cl} & B_{al} + B_{bl} + B_{cl} \\ 2A_{al} - A_{bl} - A_{cl} & 2B_{al} - B_{bl} - B_{cl} \end{pmatrix}$$

$$= \begin{pmatrix} I_{11}\sin\varphi_{11} & I_{11}\cos\varphi_{11} \\ I_{21}\sin\varphi_{21} & I_{21}\cos\varphi_{21} \end{pmatrix} \quad (7.62)$$

The A_{11} and B_{11} are the fundamental positive-sequence reactive and active components, respectively; A_{21} and B_{21} are the fundamental negative-sequence reactive and active components of a-phase, respectively.

Similarly, by multiplying both sides of (7.54) by $2\sin[k(2\pi n/N - 2\pi l/3)]$ and $2\cos[k(2\pi n/N - 2\pi l/3)]$ as well as applying the LPF, A_{xk} and B_{xk} are derived as

$$\begin{cases} A_{xk} = I_{1k}\sin\left[\frac{2(k-1)\pi l}{3} + \varphi_{1k}\right] + I_{2k}\sin\left[\frac{2(k+1)\pi l}{3} + \varphi_{2k}\right] \\ B_{xk} = I_{1k}\cos\left[\frac{2(k-1)\pi l}{3} + \varphi_{1k}\right] + I_{2k}\cos\left[\frac{2(k+1)\pi l}{3} + \varphi_{2k}\right] \end{cases} \quad (7.63)$$

Define:

$$i_{x11} = A_{11}\cos\left(\frac{2\pi}{N}n - \frac{2l\pi}{3}\right) + B_{11}\sin\left(\frac{2\pi}{N}n - \frac{2l\pi}{3}\right) \quad (7.64)$$

$$i_{x21} = A_{21}\cos\left(\frac{2\pi}{N}n + \frac{2l\pi}{3}\right) + B_{21}\sin\left(\frac{2\pi}{N}n + \frac{2l\pi}{3}\right) \quad (7.65)$$

$$i_{x1} = A_{x1}\cos\left(\frac{2\pi}{N}n - \frac{2l\pi}{3}\right) + B_{x1}\sin\left(\frac{2\pi}{N}n - \frac{2l\pi}{3}\right) \quad (7.66)$$

$$i_{xk} = A_{xk}\cos\left[k\left(\frac{2\pi n}{3} - \frac{2\pi l}{3}\right)\right] + B_{xk}\sin\left[k\left(\frac{2\pi n}{3} - \frac{2\pi l}{3}\right)\right] \quad (7.67)$$

Combining (7.64)–(7.67) in different ways, different current components can be detected.

1. When only compensating for harmonics and fundamental negative-sequence components, the fundamental positive-sequence current i_{x11} can first be calculated from (7.64), and then i_{x11} can be subtracted from the load current to obtain the reference current.
2. When only compensating for the fundamental negative-sequence component, the fundamental negative-sequence current i_{x21} can be calculated directly from (7.65).
3. When only compensating for harmonics, the fundamental current i_{xf} can firstly be calculated from (7.66), and then i_{xf} can be subtracted from the load current to obtain the reference current.
4. When compensating for a specific order harmonic, the harmonic current i_{xk} can be calculated from (7.67) directly.

   Based on the above detection results, the purpose of detection and separation of different current components can be achieved through different combinations.

7.2 Photovoltaic Inverters in the Power Quality Control Mode

2. **The harmonic and reactive power detection method in three-phase four-wire systems**

   The above analyses are based on three-phase three-wire system, for three-phase four-wire system, the presence of the zero-sequence component introduces some differences.

   In a three-phase four-wire system, the current $i_x(n)$ can be decomposed as

$$i_x(n) = \sum_{k=1}^{\infty} \left[I_{1k} \sin\left(\frac{2\pi}{N} nk + \varphi_{1k} - \frac{2l\pi}{3}\right) \right.$$
$$\left. + I_{2k} \sin\left(\frac{2\pi}{N} nk + \varphi_{2k} + \frac{2l\pi}{3}\right) + I_{0k} \sin\left(\frac{2\pi}{N} nk + \varphi_{0k}\right) \right] \quad (7.68)$$

Compared to (7.54), there is an additional zero-sequence component with subscript 0.

Similarly, the instantaneous power of the zero-sequence component is expressed as

$$p_{0x} = I_{0k} \sin\left(\frac{2\pi}{N} nk + \varphi_{0k}\right) \sin\left(\frac{2\pi}{N} nk - \frac{2l\pi}{3}\right) \quad (7.69)$$

It can be seen that the instantaneous power of the zero-sequence component, when summed across the three phases, equals zero. Like the fundamental positive-sequence reactive power, this power flows between the three phases.

As with (7.59)–(7.63), the DC component can be obtained as

$$B_{x1} = I_{11} \cos \varphi_{11} + I_{21} \cos\left(\varphi_{21} + \frac{4l\pi}{3}\right) + I_{01} \cos\left(\varphi_{01} + \frac{2l\pi}{3}\right) \quad (7.70)$$

$$A_{x1} = I_{11} \sin \varphi_{11} + I_{21} \sin\left(\varphi_{21} + \frac{4l\pi}{3}\right) + I_{01} \sin\left(\varphi_{01} + \frac{2l\pi}{3}\right) \quad (7.71)$$

$$A_{xk} = I_{1k} \sin\left(\varphi_{1k} - \frac{2l\pi}{3}\right) + I_{2k} \sin\left(\varphi_{2k} + \frac{2l\pi}{3}\right) + I_{0k} \sin(\varphi_{0k}) \quad (7.72)$$

$$B_{xk} = I_{1k} \cos\left(\varphi_{1k} - \frac{2l\pi}{3}\right) + I_{2k} \cos\left(\varphi_{2k} + \frac{2l\pi}{3}\right) + I_{0k} \cos(\varphi_{0k}) \quad (7.73)$$

Let $k = 1$, (7.72) and (7.73) yields:

$$A'_{x1} = I_{11} \sin\left(\varphi_{11} - \frac{2l\pi}{3}\right) + I_{21} \sin\left(\varphi_{21} + \frac{2l\pi}{3}\right) + I_{01} \sin(\varphi_{01}) \quad (7.74)$$

$$B'_{x1} = I_{11} \cos\left(\varphi_{11} - \frac{2l\pi}{3}\right) + I_{21} \cos\left(\varphi_{21} + \frac{2l\pi}{3}\right) + I_{01} \cos(\varphi_{01}) \quad (7.75)$$

Define the following relationship:

$$\begin{pmatrix} A_{11} & B_{11} \\ A_{01} & B_{01} \\ A_{21} & B_{21} \end{pmatrix} = \frac{1}{3} \begin{pmatrix} A_{al} + A_{bl} + A_{cl} & B_{al} + B_{bl} + B_{cl} \\ A'_{al} + A'_{bl} + A'_{cl} & B'_{al} + B'_{bl} + B'_{cl} \\ 2A_{al} - A_{bl} - A_{cl} - 3A_{01} & 2B_{al} - B_{bl} - B_{cl} - 3B_{01} \end{pmatrix}$$

$$= \begin{pmatrix} I_{11} \sin \varphi_{11} & I_{11} \cos \varphi_{11} \\ I_{01} \sin \varphi_{01} & I_{01} \cos \varphi_{01} \\ I_{21} \sin \varphi_{21} & I_{21} \cos \varphi_{21} \end{pmatrix} \quad (7.76)$$

Define:

$$i_{x01} = A_{01} \cos\left(\frac{2\pi}{N}n\right) + B_{01} \sin\left(\frac{2\pi}{N}n\right) \quad (7.77)$$

By combining (7.64)–(7.67) and (7.77), the different current reference signals can be obtained.

From the above analysis, it can be seen that, whether it is a three-phase three-wire system or a three-phase four-wire system, there is a uniform expression when calculating the fundamental positive sequence active current, fundamental positive sequence reactive current, fundamental positive sequence current, fundamental current and harmonic current. The only difference arises when distinguishing between the fundamental negative-sequence and zero-sequence components. In the three-phase three-wire system, there is no zero-sequence component, and the fundamental negative-sequence components are identical. Therefore, the three-phase three-wire system can be regarded as a special case of the three-phase four-wire system, and both systems can be represented using the same mathematical equations.

Reference Current Synthesis

Due to the limited rated current of PV inverters, it is often not possible to fully compensate for the harmonics and reactive current of the load. Therefore, limiting the harmonics and reactive current is necessary [4]. The impact of harmonic limiting algorithms on harmonic suppression with different load types has been discussed in existing literature. However, the current methods for load type identification are limited to single-phase PV inverters and cannot be directly applied to three-phase PV inverters [5]. To address this, based on the analyses of PV power generation and the load conditions at the point of common coupling, this section primarily introduces a reference current synthesis method for three-phase PV inverters with power quality control. This method consists of three steps: load weight judgment, load type identification, and reference current synthesis.

7.2 Photovoltaic Inverters in the Power Quality Control Mode

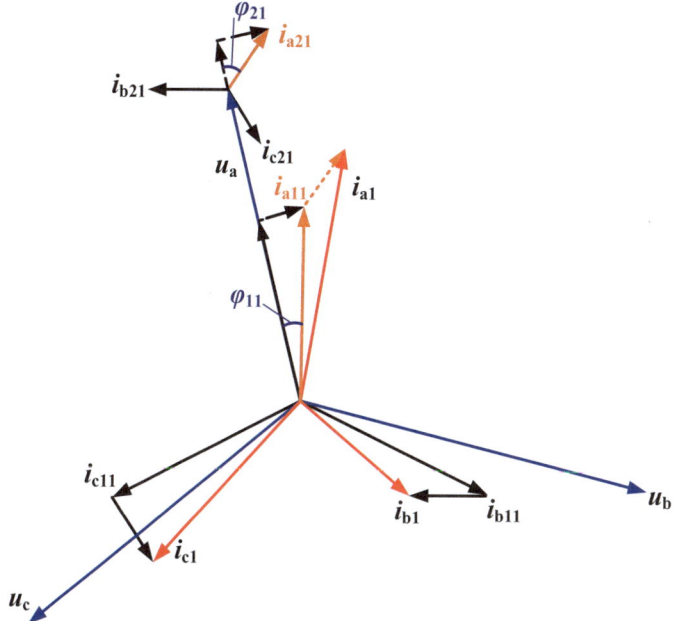

Fig. 7.9 Schematic diagram of the fundamental current after decomposition by the symmetric component method

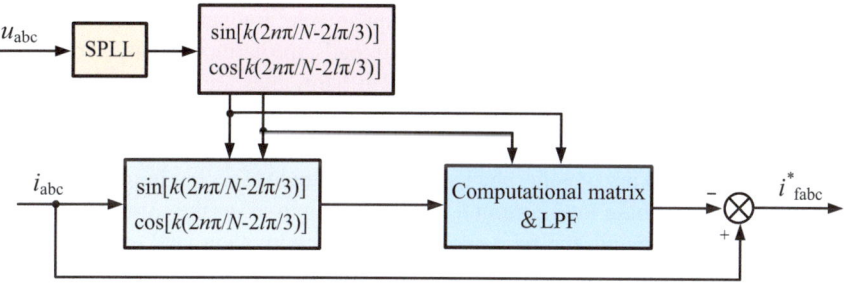

Fig. 7.10 Structure of the time domain based current detection method

Step 1: Load Weight Judgment

The load weight judgment is to compare the load fundamental active power P_1 and PV power P_{PV}. Based on the comparison results, it is determined whether the load harmonic current needs to be detected.

The load fundamental power P_1 is calculated as follow

$$P_1 = P_{1a} + P_{1b} + P_{1c} = (E_a I_{La1} + E_b I_{Lb1} + E_c I_{Lc1})\cos\varphi \tag{7.78}$$

where $\cos\varphi$ is the power factor, E_x and I_{Lx1} ($x = a, b, c$) are respectively the RMS value of grid voltage and the RMS value of load fundamental current.

The output power of the PV array is obtained through the maximum power point tracking control. The output power of the PV array P_{PV} is expressed as

$$P_{PV} = V_{PV} I_{PV} \tag{7.79}$$

where V_{PV} and I_{PV} is the voltage and current of the PV array.

Due to the instability of PV output power and the uncertainty of load variations, in order to prevent the converter from frequently switching between light-load and heavy-load modes, an appropriate threshold $\Delta P \neq 0$ is selected within the allowable range of power fluctuation. After a delayed comparison, the load is judged to be light or heavy, based on this threshold. If $P_1 < (P_{PV} - \Delta P)$ holds true, it is considered as a light load; otherwise, it is regarded as a heavy.

Step 2: Load Type Identification

In general, the linear load current contains only the fundamental wave component of the power supply, while the nonlinear load current contains the fundamental wave component and odd harmonics. For most three-phase nonlinear loads, the harmonic contents of 5th, 7th, and 11th harmonic currents are relatively high. For example, the typical three-phase bridge uncontrolled rectifier load has the fifth harmonic content is the largest. Therefore, the linearity or nonlinearity of the load can be determined by judging whether the load current contains a specified number of harmonics.

Generally, linear load currents only contain fundamental frequency components, while nonlinear load currents contain both fundamental frequency components and odd harmonics. For most three-phase nonlinear loads, the harmonic content of lower-order harmonics such as the 5th, 7th, and 11th harmonics are relatively high. For example, in a typical three-phase bridge uncontrolled rectifier load, the 5th harmonic content is the largest. Therefore, the linearity or nonlinearity of the load can be determined by checking whether the load current contains specific harmonic orders.

Since the three-phase load current $i_{Lx}(t)$ can be expressed as the sum of the fundamental active component, the fundamental reactive component, and each harmonic component. The load current of phase A is expressed as

$$i_{La}(t) = \sqrt{2} I_{p1} \sin(\omega t + \theta_{p1}) + \sqrt{2} I_{q1} \cos(\omega t + \theta_{q1}) + \sqrt{2} \sum_{n=2m+1}^{\infty} I_n \sin(n\omega t + \theta_n) \tag{7.80}$$

where $m = 1, 2, 3, \ldots$.

It can be seen that after processing the nonlinear load current $i_{La}(t)$ using instantaneous reactive power theory, the harmonic components $i_{Lah}(t)$ can be obtained as

7.2 Photovoltaic Inverters in the Power Quality Control Mode

$$i_{Lah}(t) = \sqrt{2} \sum_{n=2m+1}^{\infty} I_n \sin(n\omega t + \theta_n) \tag{7.81}$$

Taking the detection of the fifth harmonic as an example, the identification method of load characteristics is illustrated by multiplying the five times frequency unit sinusoidal component with $i_{lah}(t)$, which is expressed as:

$$\Delta i_5(t) = \sqrt{2} \sin(5\omega t) \times i_{Lah}(t) = I_{p5} \cos(\theta_{q5}) - I_{q5} \sin(\theta_{q5})$$

$$+ \sqrt{2} \sin(5\omega t) \sum_{n=2i+1}^{\infty} I_n \sin(n\omega t + \theta_n) = \Delta I_5 + \Delta \tilde{i}_5(t) \tag{7.82}$$

where $i = 2, 3, \ldots$; ΔI_5 is the constant in $\Delta i_5(t)$, which is expressed as:

$$\Delta I_5 = I_{p5} \cos(\theta_{q5}) - I_{q5} \sin(\theta_{q5}) \tag{7.83}$$

From (7.83), it can be inferred that if the fifth harmonic is present in $i_{La}(t)$, then ΔI_5 is a non-zero constant. Taking a very small positive constant ε (which is related to current sampling accuracy and digital low-pass filtering), if $\Delta I_5 > \varepsilon$, it indicates that the 5th harmonic is present in $i_{La}(t)$, and the load can be judged as nonlinear. Similarly, the 7th, 11th, 13th, and other odd-order harmonics or their linear combinations can be used for load type identification, but the computational load will increase correspondingly.

Step 3: Current Synthesis

The magnitude of the load power affects the reference current of the PV inverter. When the load is light, the fundamental and harmonic currents required by the load are both supplied by the PV inverter. At this time, harmonic and reactive power detection are not required, which eliminates the delay of harmonic and reactive power detection and enhances the system response speed. When the load is heavy, the PV inverter cannot supply all the power required by the load. At this point, the PV inverter compensates for the harmonics and reactive current of the load according to the inverter margin while ensuring the maximum PV power is connected to the grid.

Since the grid-connected current margin of the inverter may not necessarily be greater than the peak value of the reactive and harmonic currents of the load, amplitude limiting control is necessary. There are two main types of amplitude limiting control: amplitude clamping algorithm (ACA) and amplitude scaling algorithm (ASA).

1. **ACA method**:
   This method clamps the reference current amplitude to a set value allowed by the switching device.
2. **ASA method**:

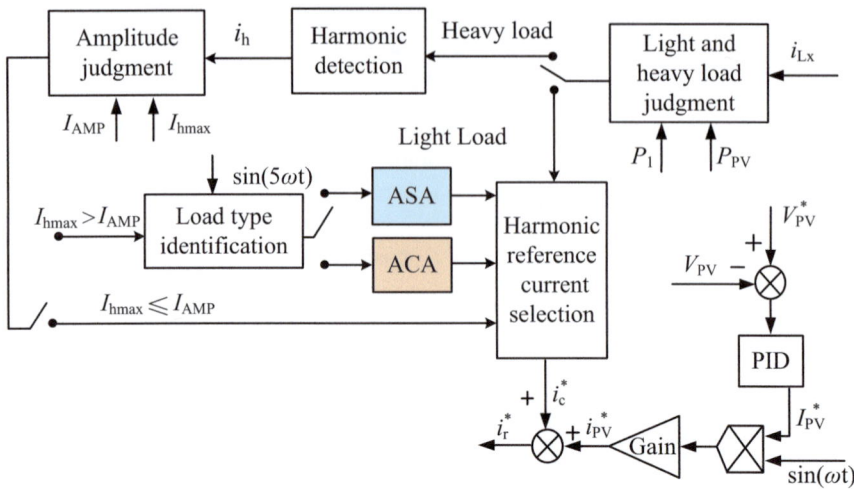

Fig. 7.11 The flow diagram of reference current synthesis method

This method reduces the reference current in proportion to the harmonic current linearly, so that the current peak equals the set value allowed by the switching device.

However, the amplitude limiting algorithm has a significant impact on the load harmonic suppression and reactive power compensation effects. In order to achieve optimal compensation for load harmonics and reactive currents, the load type must be considered when synthesizing the reference current. To this end, this section proposes a reference current synthesis method based on load analysis, and its working principle is shown in Fig. 7.11.

First, by detecting the voltage and current of the PV panel, the output power P_{PV} is calculated in real-time. The incremental conductance method combined with a DC-DC converter [6] is used to obtain the maximum power point voltage V_{PV}. The error between V_{PV} and the actual DC-link voltage is fed into the PID controller. The output of the PID controller is set as the amplitude of the inverter grid-connected current i^*_{PV}.

Subsequently, by detecting and phase-locking the load current and grid voltage, the phase of the inverter grid-connected current (i_{PV}^*) and the load active power (P_1) are obtained. If $P_1 \leq P_{PV}$, the load is considered light, and the harmonic reference current (i_c^*) is set as the load current (i_{Lx}). Conversely, if $P_1 > P_{PV}$, the load is considered heavy, and the load reactive and harmonic current (i_h) are calculated based on instantaneous reactive power theory. Furthermore, the maximum amplitude of i_h (i_{hmax}) is compared with the device rated current margin (I_{AMP}). If $I_{hmax} \leq I_{AMP}$, the load type does not need to be distinguished and $i_c^* = i_h$. However, if $I_{hmax} > I_{AMP}$, the ASA method should be applied for linear loads and the ACA method should be applied for nonlinear loads. The corresponding reference currents for different algorithms are expressed as

7.2 Photovoltaic Inverters in the Power Quality Control Mode

Table 7.1 Coefficients for instruction current

Control coefficient	Light load	Heavy load		
		$I_{hmax} \leq I_{AMP}$	$I_{hmax} > I_{AMP}$	
			Nonlinear	Linear
K_L	1	0	0	0
K_H	0	1	0	0
K_{ACA}	0	0	1	0
K_{ASA}	0	0	0	1

$$I_{AMP} = I_n - I_{PV}^* \tag{7.84}$$

$$i_{ASA}(t) = \frac{I_{AMP}}{I_{hmax}} i_h(t), \quad i_h(t) > I_{AMP} \tag{7.85}$$

$$i_{ACA}(t) = I_{AMP}, \quad i_h(t) > I_{AMP} \tag{7.86}$$

where I_n is the rated current of the switching device, I_{PV}^* is the amplitude of the inverter grid-connected current, I_{AMP} is the current margin of the switching device, I_{hmax} is the amplitude of the three-phase harmonic current i_h, $i_{ASA}(t)$ is the harmonic current value with the ASA method, and $i_{ACA}(t)$ is the harmonic current value with the ACA method.

Considering the different types of loads, the final synthesized reference current for the PV inverter is expressed as

$$\begin{aligned} i_r^*(t) &= i_{pv}^*(t) + i_c^*(t) = i_{pv}^*(t) + K_L i_l(t) \\ &+ K_H i_h(t) + K_{ASA} i_{ASA}(t) + K_{ACA} i_{ACA}(t) \end{aligned} \tag{7.87}$$

where i_r^* is the reference current of the PV inverter, i_{pv}^* is the MPPT reference current, i_c^* is the load control current, K_L is the light load coefficient, K_H is the harmonic current coefficient, K_{ACA} is the clamping amplitude limiting coefficient, and K_{ASA} is the scaling amplitude limiting coefficient. The values of K_{ACA} and K_{ASA} are shown in Table 7.1.

Reference Current Tracking Control

After synthesizing the reference current, it is necessary to select an appropriate harmonic and reactive current control method to obtain the PWM signals for the PV inverter. The commonly adopted control methods include direct transient current control, adaptive hysteresis band control, and repetitive control, which will be illustrated in this section.

Method 1: Direct Transient Current Control

By selecting the switching state and operation time of the inverter, direct transient current control can directly control the output current of the PV inverter. This method has the advantages of a simple control circuit, reduced output reactor and improved current tracking control performance [7].

In order to explain the principle of direct transient current control, the A-phase bridge of the two-level PV inverter in Fig. 7.12 is taken as an example to illustrate the influence of the inverter switching state on the output current. The working states of the inverter can be summarized as follow

$$S_{m(m=1,2,3,4)} = \begin{cases} S_{a1} \text{ on}, S_{a2} \text{ off}, i_a \geq 0 & m = 1 \\ S_{a1} \text{ off}, S_{a2} \text{ on } or \text{ off}, i_a \geq 0 & m = 2 \\ S_{a1} \text{ off}, S_{a2} \text{ on}, i_a < 0 & m = 3 \\ S_{a1} \text{ on } or \text{ off}, S_{a2} \text{ off}, i_a < 0 & m = 4 \end{cases} \quad (7.88)$$

When the inverter operates in S_1 state at time t, assume that the terminal voltages of the capacitor C_1 and C_2 are $v_{e1}(t)$ and $v_{e2}(t)$, the grid voltage is $e_a(t)$, the output current is $i_a(t)$, and the continuous on-time of the S_{a1} is τ, the following relationship can be obtained as

$$i_a(t+\tau) - i_a(t) = \frac{v_{e1}(t) - e_a(t)}{L} \tau \quad (7.89)$$

This expression can be abbreviated as

$$\Delta i_{a\_rp}(t, \tau) = \frac{v_{e1}(t) - e_a(t)}{L} \tau \quad (7.90)$$

where $\Delta i_{a\_rp}(t, \tau) = i_a(t+\tau) - i_a(t)$.

Similarly, the expression of the direct control effect of S_2, S_3, and S_4 working states on the output current can be obtained as

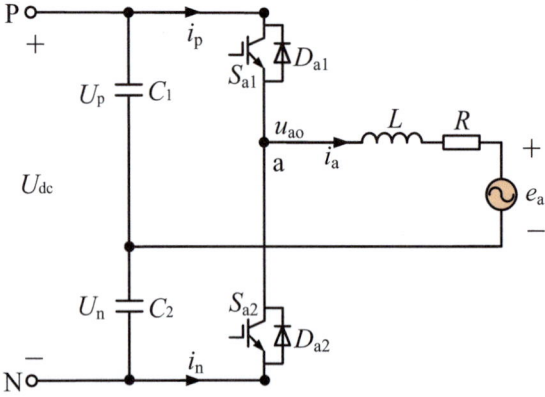

Fig. 7.12 Topology of A-phase bridge in the two-level PV inverter

7.2 Photovoltaic Inverters in the Power Quality Control Mode

$$\Delta i_{a\_dp}(t, \tau) = -\frac{v_{e2}(t) + e_a(t)}{L}\tau \qquad (7.91)$$

$$\Delta i_{a\_rn}(t, \tau) = -\frac{v_{e2}(t) + e_a(t)}{L}\tau \qquad (7.92)$$

$$\Delta i_{a\_dn}(t, \tau) = \frac{v_{e1}(t) - e_a(t)}{L}\tau \qquad (7.93)$$

where $\Delta i_{a\_dp}(t, \tau)$, $\Delta i_{a\_rn}(t, \tau)$, and $\Delta i_{a\_dn}(t, \tau)$ represent the current changes in the work states of S_2, S_3, and S_4 respectively.

Define the change in output current per unit time as the instantaneous current displacement factor δi_a. Assuming that the DC-side voltage of the inverter is sufficiently high, the detailed explanation of the control effect of each instantaneous current displacement factor on the current is as follows.

1. When the inverter works in S_1 state, the instantaneous current displacement factor is expressed as

$$\delta i_{a\_rp}(t) = \frac{v_{e1}(t) - e_a(t)}{L} \qquad (7.94)$$

It can be seen that the output current is positive and still increasing in the positive direction at this moment. The amount of increase in the current amplitude depends on the duration of action of this instantaneous current displacement factor.

2. When the inverter works in S_2 state, the instantaneous current displacement factor is expressed as

$$\delta i_{a\_dp}(t) = -\frac{v_{e2}(t) + e_a(t)}{L} \qquad (7.95)$$

The output current is positive but is decreasing in the positive direction at this moment. The current amplitude reduction depends on the duration of action of this instantaneous current displacement factor.

3. When the inverter works in S_3 state, the instantaneous current displacement factor is expressed as

$$\delta i_{a\_rn}(t) = -\frac{v_{e2}(t) + e_a(t)}{L} \qquad (7.96)$$

The output current is negative and still increasing in the negative direction at this moment. The amount of increasement in the current amplitude depends on the duration of action of this instantaneous current displacement factor.

4. When the inverter works in S_4 state, the instantaneous current displacement factor is expressed as

$$\delta i_{a\_dn}(t) = \frac{v_{e1}(t) - e_a(t)}{L} \qquad (7.97)$$

The output current is negative and still decreasing in the negative direction at this moment. The current amplitude reduction depends on the duration of action of this instantaneous current displacement factor. From (7.94) to (7.97), the following relationships can be obtained as

$$\delta i_{\text{a\_rp}}(t) = \delta i_{\text{a\_dn}}(t) \tag{7.98}$$

$$\delta i_{\text{a\_dp}}(t) = \delta i_{\text{a\_rn}}(t) \tag{7.99}$$

Therefore, (7.94)–(7.97) can be simplified as

$$\delta i_{\text{a\_r}}(t) = \frac{v_{\text{e1}}(t) - e_{\text{a}}(t)}{L} \tag{7.100}$$

$$\delta i_{\text{a\_d}}(t) = -\frac{v_{\text{e2}}(t) + e_{\text{a}}(t)}{L} \tag{7.101}$$

Under the premise that the control period T_s is sufficiently small, the energy storage capacity of the DC-side capacitor is sufficiently large, and the output inductance L is not saturated, it can be assumed that within one period, the values of v_{e1}, v_{e2}, e_{a}, and L remain constant. In this case, the instantaneous current displacement factors in (7.100) and (7.101) are constant within one cycle T_s. At this point, the change in current Δi_{a} is entirely dependent on the action time of the current displacement factor, which can be expressed as

$$\Delta i_{\text{a}} = \delta i_{\text{a}} \tau \quad \tau \in [0, T_s] \tag{7.102}$$

where τ is the action time of the current displacement factor.

In practical applications, when Δi_{a} is determined, the action time τ can be easily calculated according to (7.102) and controlled by PWM modulation, which is the basic idea of realizing direct transient current control in this section.

Based on the above basic theory of direct transient current control, the practical engineering application method is explained below. In a control period of T_s, the complementary conduction of the upper and lower switches on each bridge of the inverter is equivalent to the alternating action of the two current displacement factors. Figure 7.13 shows the principle of the output current control tracking. It can be seen that the actual current $i(k)$ at time k can be obtained through real-time sampling. Let $i^*(k+1)$ denote the reference current value at time $(k+1)$. Current tracking control aims to manipulate the action time of the current displacement factors, so that the output current shifts to $i^*(k+1)$ at time $(k+1)$.

Assuming that the current $i(k)$ of phase A is positive, the two displacement factors $\delta i_{\text{a\_rp}}$ and $\delta i_{\text{a\_dp}}$ alternate in their action within one period T_s. Let t_1 be the action time of $\delta i_{\text{a\_rp}}$, and t_2 be the action time of $\delta i_{\text{a\_rp}}$. In order to the actual output current value to reach the reference value $i^*(k+1)$ at time $(k+1)$, the following equation

7.2 Photovoltaic Inverters in the Power Quality Control Mode

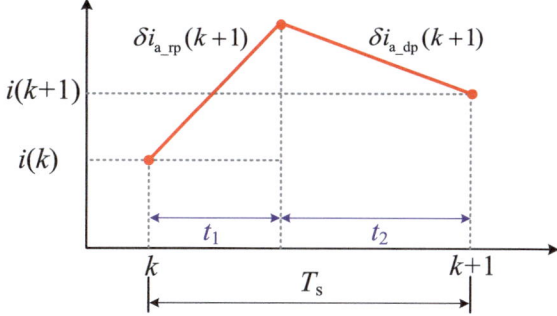

Fig. 7.13 Principle of the output current control tracking

should be satisfied

$$\delta i_{a\_rp} t_1 + \delta i_{a\_dp} t_2 = i^*(k+1) - i(k) \tag{7.103}$$

where $t_1 + t_2 = T_s$.

Based on this, t_1 is obtained as

$$t_1 = \frac{i^*(k+1) - i(k) - \delta i_{a\_dp} T_s}{\delta i_{a\_rp} - \delta i_{a\_dp}} \tag{7.104}$$

Combining the dead time t_d, turn-on delay time t_{on}, and turn-off delay time t_{off} of the switching device, the comprehensive compensation time is obtained as

$$t_0 = t_d + t_{on} - t_{off} \tag{7.105}$$

The actual turn-on drive time T_{1on} of S_{a1} is adjusted to

$$T_{1on} = t_1 + t_0 = \frac{i^*(k+1) - i(k) - \delta i_{a\_dp} T_s}{\delta i_{a\_rp} - \delta i_{a\_dp}} + t_0 \tag{7.106}$$

Similarly, when the output current $i(k)$ is negative, assuming the action time of δi_{a_rn} is t_1 and the action time of δi_{a_dn} is t_2, the following mathematical relationships can be derived

$$\delta i_{a\_rn} t_1 + \delta i_{a\_dn} t_2 = i^*(k+1) - i(k) \tag{7.107}$$

t_2 is obtained as

$$t_2 = \frac{i^*(k+1) - i(k) - \delta i_{a\_rn} T_s}{\delta i_{a\_dn} - \delta i_{a\_rn}} \tag{7.108}$$

To facilitate the implementation of PWM control, it is still converted into the actual turn-on time of S_{a1}. Combining the comprehensive delay time compensation, the actual turn-on drive time T_{1on} of S_{a1} is obtained as

$$T_{1on} = t_2 - t_0 = \frac{i^*(k+1) - i(k) - \delta i_{a\_rn} T_s}{\delta i_{a\_dn} - \delta i_{a\_rn}} - t_0 \qquad (7.109)$$

Combining Eqs. (7.106) and (7.109), T_{1on} is obtain as

$$T_{1on} = \frac{i^*(k+1) - i(k) - \delta i_{a\_d} T_s}{\delta i_{a\_r} - \delta i_{a\_d}} + \lambda t_0 \qquad (7.110)$$

where $\lambda = 1 \ i_c \geq 0; \lambda = -1 \ i_c < 0$.

Thus, a complete set of control algorithms for implementing pulse width modulated current tracking control can be obtained as

$$\begin{cases} \delta i_{a\_r}(t) = \frac{v_{c1}(t) - e_a(t)}{L} \\ \delta i_{a\_d}(t) = -\frac{v_{c2}(t) + e_a(t)}{L} \end{cases} \qquad (7.111)$$

In practical applications, it is only necessary to set the PWM period in the DSP according to the predetermined current control period T_s. Then, the pulse width is calculated according to (7.111). In the next control period, the control of the PV inverter switching device will not take up additional DSP running time. Therefore, compared to traditional methods, direct transient current control does not need to increase the size, weight, and cost of the equipment. Meanwhile, the complexity of the control circuit and the output inductance can be further reduced.

Method 2: Adaptive Hysteresis Band Control

Amongst different current control techniques, the hysteresis band (HB) current control is preferred for power quality control due to its unassailable advantages such as appropriate stability, fast transient response, simple implementation and operation, high accuracy, inherent current peak limitation, and overload rejection [8]. Meanwhile, the 3LT²I has the advantages of high output current quality, high efficiency, and low loss, which is widely used in PV inverters [9–11]. Thus, this section will present an adaptive HB control strategy for 3LT²I based on the single HB control.

Firstly, the switching functions and model of the 3LT²I in Fig. 7.14 are discussed. Define S_x ($x = a, b, c$) as the switching function, which is also the output of the HB comparator. In phase A, S_a may be written as

$$S_a = \begin{cases} 1, & \text{when } S_{a3} \text{ and } S_{4a} \text{ are closed} \\ 0, & \text{when } S_{a1} \text{ and } S_{a4} \text{ are closed} \\ -1, & \text{when } S_{a1} \text{ and } S_{a2} \text{ are closed} \end{cases} \qquad (7.112)$$

According to Fig. 7.14, based on Kirchhoff's law, the relationship between output currents i_{cx} ($x = a, b, c$) and terminal voltages V_{cx} is obtained as

7.2 Photovoltaic Inverters in the Power Quality Control Mode

$$V_{cx} = L\frac{di_{cx}}{dt} + Ri_{cx} + e_x \quad (7.113)$$

The operation principle of the HB control in three-level PV inverters is depicted in Fig. 7.15. Furthermore, if H indicates the width of the hysteresis band, Δi represents the error between the command current i_{cx}^* and the inductor current i_{cx}, the logic relation of the HB control method is

$$S_x = \begin{cases} 1, & \Delta i > H \left(\Delta i = i_{cx}^* - i_{cx} \right) \\ 0, & |\Delta i| < H \\ -1, & \Delta i < -H. \end{cases} \quad (7.114)$$

To keep the switching frequency fixed, the width of HB will be designed to adjust automatically as a function of modulation frequency f_c, supply voltage e_x, dc-link voltage U_{dc}, slope of the reference compensator current di_{cx}^*/dt and the coupling inductance L. The calculation formula of H can be deduced based on the current waveform in Fig. 7.15. Figure 7.16 shows the geometric relationship between the reference current and the compensation current. It is seen that since system parameters such as switching frequency, grid voltage, DC bus voltage, and output filter inductance remain constant, the width of the hysteresis band H will automatically adjust according to the rate of change of the reference current.

Moreover, distinguished from the two-level inverter, the output voltage of the utilized inverter has three levels ($+U_{dc}/2$, 0, $-U_{dc}/2$). Indeed, the HB band is not associated with a specific voltage level but the voltage level pair is changed during the process of tracing the reference current [12]. There are two level pairs: ($+U_{dc}/2$, 0) and (0, $-U_{dc}/2$) for a three-level HB. Consequently, these two-level pairs must be individually considered when deriving the formula of H. The ascendant and descendant slopes of the inductor currents generated by two different voltage groups are defined as di_{cx}^+/dt and di_{cx}^-/dt.

According to (7.113), the following equations can be derived for the voltage levels ($+U_{dc}/2$, 0)

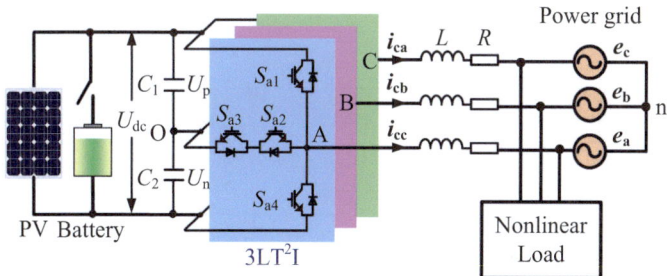

Fig. 7.14 Structure of 3LT²I with Nonlinear load

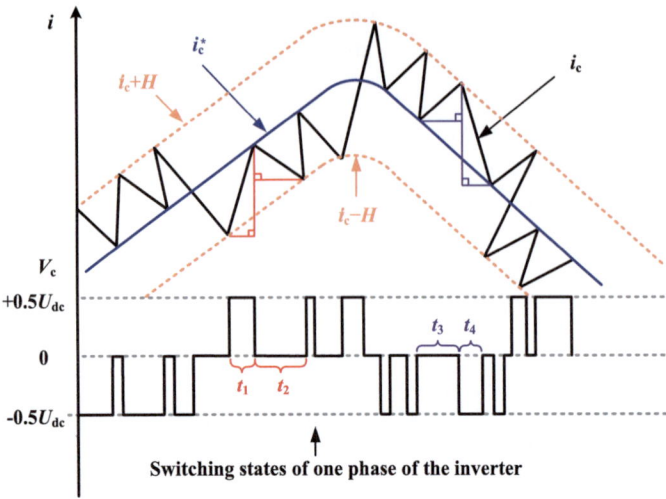

Fig. 7.15 Operation principle of the HB control in three-level PV inverters

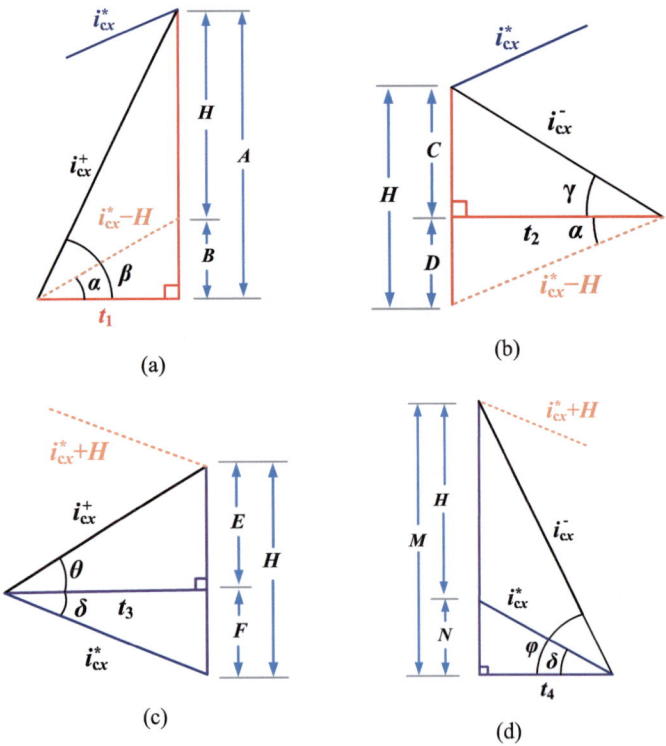

Fig. 7.16 Geometric relationship between the reference current and the compensation current

7.2 Photovoltaic Inverters in the Power Quality Control Mode

$$\frac{di_{cx}^+}{dt} = \frac{1}{L}\left(\frac{U_{dc}}{2} - e_x\right) \tag{7.115}$$

$$\frac{di_{cx}^-}{dt} = -\frac{1}{L}e_x \tag{7.116}$$

Through geometric analyses of the current waveforms (a) and (b) in Fig. 7.16, we can obtain

$$H = A - B = t_1 \cdot \tan\beta - t_1 \cdot \tan\alpha = t_1 \cdot \frac{di_{cx}^+}{dt} - t_1 \cdot \frac{di_{cx}^*}{dt} \tag{7.117}$$

$$H = C + D = t_2 \cdot \tan\gamma + t_2 \cdot \tan\alpha = -t_2 \cdot \frac{di_{cx}^-}{dt} + t_2 \cdot \frac{di_{cx}^*}{dt} \tag{7.118}$$

Equation (7.119) can be deduced by adding and subtracting (7.117) and (7.118)

$$(t_2 - t_1)\frac{di_{cx}^*}{dt} + t_1\frac{di_{cx}^+}{dt} - t_2\frac{di_{cx}^-}{dt} = 2H \tag{7.119}$$

$$(t_1 + t_2)\frac{di_{cx}^*}{dt} - t_1\frac{di_{cx}^+}{dt} - t_2\frac{di_{cx}^-}{dt} = 0 \tag{7.120}$$

The relationship between switching frequency and $t_1 + t_2$ is expressed as

$$(t_1 + t_2) = T_s = \frac{1}{f_s} \tag{7.121}$$

Thus, the calculation formula of H for the voltage group $(+U_{dc}/2, 0)$ can be deduced as

$$H = \frac{0.5U_{dc}\left(L\frac{di_c^*}{dt} + e_x\right) - \left(L\frac{di_c^*}{dt} + e_x\right)^2}{0.5f_s U_{dc} L} \tag{7.122}$$

Similarly, through geometric analyses (c) and (d) in Fig. 7.16, the formula of H for voltage group $(0, -U_{dc}/2)$ can be derived as

$$H = \frac{-0.5U_{dc}\left(L\frac{di_c^*}{dt} + e_x\right) - \left(L\frac{di_c^*}{dt} + e_x\right)^2}{0.5f_s U_{dc} L} \tag{7.123}$$

Compared with two-level HB control, the three-level HB control can output three voltage levels. As a result, more than one switching combination can be chosen to increase or reduce the output current. For example, to increase i_{ca}, S_a can be set to 1 from 0, or be set to 0 from -1. These two cases should use different formulas to obtain H. Which formula to use is determined by the switching states at the moment and the variation direction of the output current. What's more, the switching states are controlled to change between the two adjacent voltage levels to reduce the switching losses. Therefore, the logic selection of the two formulas can be considered separately for the following three situations.

1. Switching state is 1: The output voltage is $+U_{dc}/2$, Eq. (7.122) will be selected to calculate H, because the switching state will be changed between 0 and 1.
2. Switching state is -1: The output voltage is $-U_{dc}/2$, Eq. (7.123) will be selected, because the switching state will be changed between -1 and 0.
3. Switching state is 0: The output voltage is 0, the right equation will be chosen after detecting the direction of the output current.

    (a) The current needs to be increased: The output current needs to be increased, Eq. (7.122) will be selected to calculate H as the switching state will be changed from 0 to 1.
    (b) The current needs to be reduced: The output current needs to be reduced, Eq. (7.123) will be selected to calculate H as the switching state will be changed from 0 to -1.

According to the calculation of H described above. The value of H will vary automatically along with the system parameters to achieve a constant switching frequency. Take H and Δi as the input of the HB comparator and the HB comparator can generate PWM signals for the inverter.

Method 3: Repetitive Control

Repetitive control (RC) is very suitable for tracking or attenuating periodic signals [13] and is successfully utilized in active power filter [14], power factor correction converters [15] and some PV inverters to achieve harmonic compensation and zero-error output [16]. The mechanism of repetitive control is derived from the internal model principle, which can achieve perfect tracking performance by including the model of the reference signal. To employ the RC method for harmonic compensation in PV inverters, the internal model of the harmonics need to be extracted from the periodic signal. Furthermore, due to the delay unit of RC, the dynamic response will be postponed.

This section presents an enhanced plug-in repetitive controller (PIRC) designed for the harmonic compensation of three-phase PV inverters. Firstly, the PIRC is constructed in synchronous rotating frame, where the $6k \pm 1$ orders harmonics in the line-side are transformed to $6k$ orders harmonics and the delay unit of RC is reduced to 1/6. So, the dynamic response of the system can be significantly improved. Then, noticing that the PIRC is usually implemented in a DSP, FPGA or dSPACE with fixed sampling frequency, a fractional delay (FD) generally exists in the controller, leading to elevated steady-state errors. An FD filter is employed to approximate the delay and eliminate the error. The PIRC and the conventional repetitive controller (CRC) are shown in Fig. 7.17.

Compared with CRC, PIRC has a forward path and a conventional feedback controller which can improve the dynamic response performance of the control

7.2 Photovoltaic Inverters in the Power Quality Control Mode

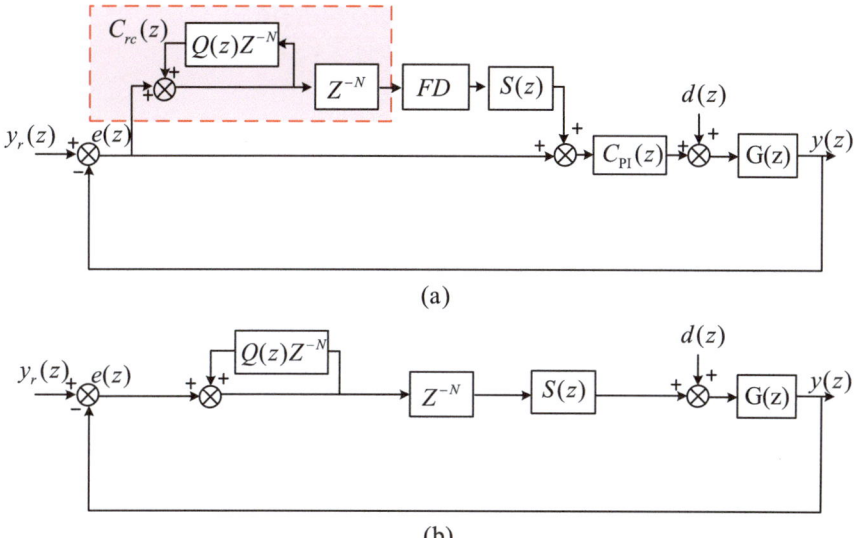

Fig. 7.17 Control diagram of RC controllers **a** PIRC, **b** CRC

system. In the repetitive controller $C_{rc}(z)$, the delay unit z^{-N} can extract the reference signal from the periodic signal of last fundamental period and result in zero tracking error, where N is the number of samples for each cycle with $N = f_s/f_f$, f_s is the sampling frequency and f_f is the fundamental frequency of reference signal. $Q(z)$ is a constant less than 1 or a low-pass filter, which can improve the system stability, fractional delay (FD) is a fractional delay filter used to improve the steady-state performance of the control system when N is not integer. $S(z)$ is phase lead compensator to stabilize the system. $C_{PI}(z)$ is a conventional feedback proportional and integral (PI) controller that used to improve the dynamic response performance. $y_r(z)$ and $y(z)$ are the reference signal and the output of the system. $e(z)$ is the tracking error between $y_r(z)$ and $y(z)$. $d(z)$ is the disturbance in the system, including the harmonics generated by inverter. $G(z)$ is the transfer function of inverter.

Based on the internal model theory, a model of the reference signal is needed in the feedback control loop to achieve zero error tracking of the reference signal. Therefore, the core of the repetitive controller is a period signal generating system called modified internal model which is incorporated in the positive feedback path. The transfer function of repetitive controller is expressed as

$$C_{rc}(z) = \frac{z^{-N}}{1 - Q(z)z^{-N}} \quad (7.124)$$

From Fig. 7.17, the equivalent plant of the control system can be expressed as $P(z) = C_{PI}(z)G(z)$ and the tracking error $e(z)$ can be derived by

$$e(z) = y_r(z) - y(z) = \frac{y_r(z) - d(z)G(z)}{1 + [1 + C_{rc}(z)P(z)]} \quad (7.125)$$

Substituting (7.124) into (7.125) yields

$$e(z) = \frac{\left(1 - Q(z)z^{-N}\right)(y_r(z) - d(z)G(z))}{[1 + P(z)]\left[1 - z^{-N}(Q(z) - S(z)P_c(z))\right]} \quad (7.126)$$

where $P_c(z) = \frac{P(z)}{1+P(z)}$.

From (7.126), we can get the stability conditions of the overall control system as following:

1. $P(z)$ is stable, that is, the roots of the polynomial $1 + P(z) = 0$ are inside the unit circle;
2. The transfer function $1 - z^{-N}(Q(z) - S(z)P_c(z)) = 0$ is stable.

Condition (1) can be regarded as the stable condition of a closed-loop control system with PI controller alone. By choosing appropriate parameters of PI controller, condition (1) can be satisfied.

In order to meet condition (2), the following inequality must be satisfied.

$$|Q(z) - S(z)P_c(z)| \leqslant 1 \quad (7.127)$$

If $Q(z) = 1$, the PIRC is critical stable. To improve the stability of the system, $Q(z)$ can be designed as a constant between 0 and 1. However, with the decreasing of $Q(z)$, the system is more stable, while the compensation effect is getting worse. In practical applications, $Q(z)$ can be chosen from 0.85 to 0.95.

If $P_c(z)$ is the minimum phase system, $S(z)$ is equal to $P^{-1}{}_c(z)$, so that the condition (2) holds in the full frequency domain and allows for perfect compensation of harmonics. But $S(z)$ will be fully dependent on the parameters of the plant. Due to the parameter and model uncertainty of $P(z)$, $S(z)$ is unrealizable. Due to the fact that the amplitude of the high frequency harmonics in PV inverters is very low, only the low frequency harmonics need to be considered. Therefore, $S(z)$ can be designed as a low-pass filter which can filter the high frequency harmonics of the system and has the ability to compensate for phase angle. The compensator $S(z)$ can be constructed as

$$S(z) = z^m K_{rc} G_f(z) \quad (7.128)$$

where z^m is the phase lead compensation part which is used to compensate for the system lag phase, K_{rc} is the gain of the controller and $G_f(z)$ is a second order low-pass filter. At the low and middle frequencies, $|S(z)P(z)| \approx 1$, while at the high frequencies, $|S(z)P(z)| < 1$. Therefore, the control system is stable in the full frequency domain.

When there are $6k \pm 1$ orders harmonic in the grid, to simplify the analysis, the harmonics can be transformed to $6k$ orders harmonics. Therefore, the delay unit z^{-N} can be designed as $z^{-N/6}$, which can fully compensate for the harmonic signals. In this

7.2 Photovoltaic Inverters in the Power Quality Control Mode

way, the repetitive period of repetitive control can be reduced to 1/6 of the usual value, significantly improving the dynamic response speed and reducing the utilization of system resources. However, in physical applications, $z^{-N/6}$ is not promised to be an integer due to fixed sampling frequency, potentially causing the controller to fail to accurately track the reference signal and increasing the steady-state error of the PIRC. For instance, when the grid frequency is 50 Hz and the fixed sampling frequency of the control system is 10 kHz, we have $N/6 = 100/3$.

To solve this problem, a FD filter is used to approximate the fractional delay [17]. Let $N = N_i + D$, where N_i is the integer part of N and D is the fractional part. Compared with the finite-impulse-response (FIR) filter, the infinite-impulse-response (IIR) filter does not attenuate the amplitude at low frequencies. Therefore, it is not necessary to consider the limiting factor of amplitude in the filter design when choosing IIR filters. The corresponding transfer function can be designed as

$$H(z) = \frac{z^{-M}A(z^{-1})}{A(z)} \tag{7.129}$$

where $A(z) = \sum_{n=0}^{M} a_n z^{-n}$.

where M is the order of the filter. a_n are the filter coefficients, which can be chosen as

$$a_n = (-1)^n \frac{M!}{n!(M-n)!} \prod_{i=0}^{M} \frac{D+i}{D+i+n} \tag{7.130}$$

where $n = 0, 1, 2, \ldots, M$ and $a_0 = 1$.

The overall control diagram is shown in Fig. 7.18. The system consists two feedback control loops. The control objective of the voltage control loop is to stabilize the output voltage, and the output of voltage loop is used as reference signal for the current control loop. The current control loop is aiming at tracking the reference current signals. The PLL is used to obtain the phase angle θ_{pll} of the power grid voltage, which is employed in Park transformation. The harmonics are compensated by the introduced PIRC controller.

In conclusion, the main circuit of PV inverter consistent with that in shunt active power filter. Thus, the PV inverter has the ability to eliminate harmonic and compensate reactive power. By adding the appropriate control scheme, PV inverter can be used as the power quality control equipment, which make full use of the PV inverter capacity and save the investment costs.

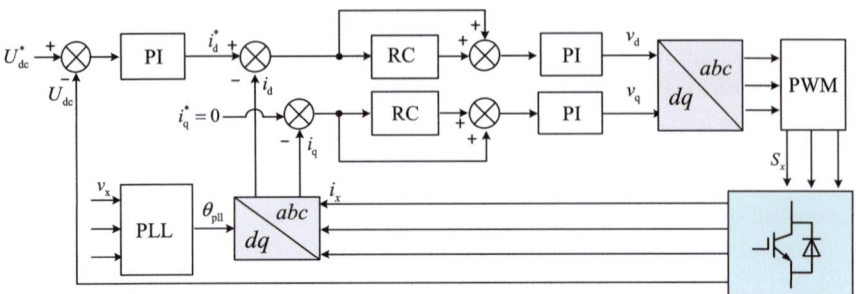

Fig. 7.18 Overall control diagram of the improved PIRC

7.3 Photovoltaic Inverters in the Islanded Mode

In this section, the application of inverters in the islanded mode is explored. The islanded mode becomes crucial when the inverter must operate independently from the power grid. The inverters in the islanded mode plays a vital role in maintaining the continuity of power supply and ensuring grid stability during grid disturbances or fault. It is essential for enhancing grid resilience, ensuring the reliability of the power supply, and mitigating safety risks.

7.3.1 Principle of the Islanded Mode

In the islanded mode, the primary function of the PV inverter is to maintain a stable power supply, monitoring and adjusting the output power or voltage to match the fluctuating load demands. The structure of $3LT^2I$ operating in the islanded mode is shown in Fig. 7.19.

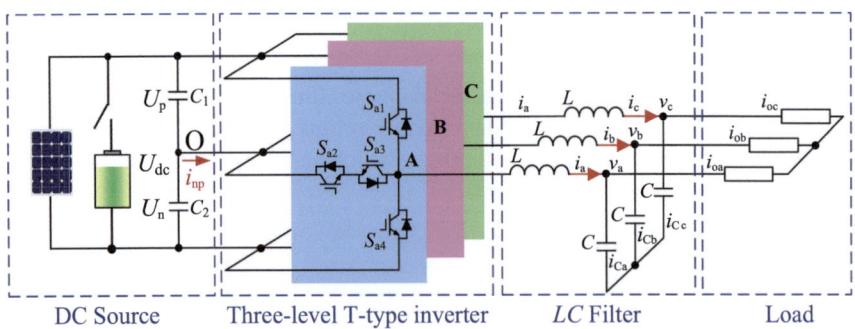

Fig. 7.19 The structure of $3LT^2I$ operating in the islanded mode

7.3 Photovoltaic Inverters in the Islanded Mode

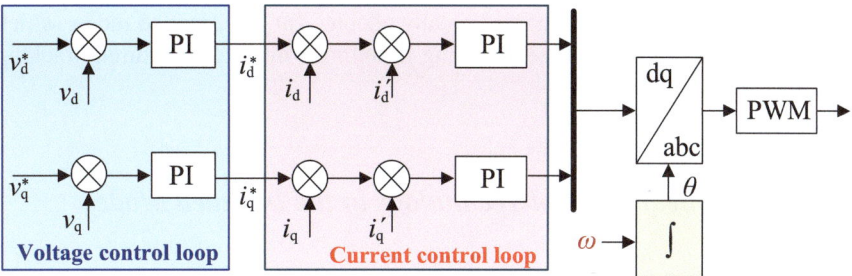

Fig. 7.20 Dual-loop voltage and current control scheme

Being different from the grid-connected mode, 3LT²I operating in the islanded mode is equipped with the *LC* filter at the AC side. By utilizing the appropriate control scheme, the inverter maintains the voltage across the filter capacitor with constant magnitude and frequency and supplies to the load.

As typical control scheme for the islanded mode, the PV inverter employs a cascaded dual-loop structure, comprising the outer loop for AC voltage regulation and the inner loop for AC current regulation. PI control is commonly utilized due to its ease of implementation and the satisfactory steady-state performance.

The cascaded dual-loop structure with PI controller is illustrated in Fig. 7.20, wherein two cascades of PI controllers are employed. The outer loop governs the direct and quadrature components of voltage, while also providing reference current points for the inner loop to regulate both amplitude and frequency at the common coupling point. The current control loop regulates the output currents, incorporating feedforward terms for current compensation. Moreover, this controller exclusively regulates voltage amplitude with the designated frequency of 50 Hz, which is fed into Park transformations through an integrator.

The reference voltages are provided as

$$\begin{cases} v_d^* = \frac{U_{dc}}{\sqrt{3}} \\ v_q^* = 0 \\ \theta = \int \omega dt \end{cases} \quad (7.131)$$

where v_d^* and v_q^* are the direct and quadrature components of the reference voltage of the inverter output voltage. The design of the PI controllers is conducted utilizing the analytical approach based on the canonical form of the transfer function for a second-order closed-loop system. The resultant controller gains are presented as

$$\begin{cases} K_P = \frac{2\zeta \omega L - R}{K} \\ K_I = \frac{L\omega^2}{K} \end{cases} \quad (7.132)$$

In conclusion, in islanded operation mode, the PV inverter can provide stable power supply to the local power system according to the practical load demand.

The cascaded dual-loop structure is usually adopted for the islanded mode, which maintain the AC output voltage operating in setting point and AC current tracking performance.

7.3.2 Nonlinear Control Technology in the Islanded Mode

As the introduction in previous section, in the typically cascaded dual-loop structure, the outer loop to regulate AC output voltages and the inner loop to regulate inductor currents. Thus, the current sensors are needed to detect the inductor currents in inner loop. In practice, the use of multiple sensors increases the costs and the probability of failure, thereby decreasing the reliability of control systems. Therefore, this section aims to develop a current sensorless control scheme that can offer good steady-state and dynamic performance for PV inverter operating in the islanded mode under load disturbances and parameter uncertainties.

Additionally, the PI control is commonly utilized due to its ease of implementation and satisfactory steady-state performance. However, its dynamic performance deteriorates in the presence of load disturbances and parameter uncertainties. Besides, the PI controller and other linear controller are designed with the small-signal linearized inverter model around a predefined operating point. In another word, the global large-signal stability of the inverter cannot be guaranteed, and the control performance is affected by different operating points since the controller gains are tuned at a certain operating point.

While the nonlinear control method is based on large-signal nonlinear model, which is suitable for controlling the inverter to achieve a wide operation range. Several nonlinear control approaches have been proposed for power converters in recent years. The typically nonlinear control method includes model predictive control (MPC), sliding-mode control (SMC), and so on.

Thus, the control technology that imbedding the nonlinear control methods into the single-loop voltage control scheme is proposed by the author's team. The nonlinear control methods consist improved MPC (IMPC) [18] and observer-based continuous sliding-mode control (OCSMC) [19], which will be illustrated in this section.

First, the inverter dynamic model can be established as

$$\frac{d}{dt}\begin{bmatrix} i \\ v \end{bmatrix} = \begin{bmatrix} 0 & -\frac{1}{L} \\ \frac{1}{C} & 0 \end{bmatrix}\begin{bmatrix} i \\ v \end{bmatrix} + \begin{bmatrix} \frac{1}{L} & 0 \\ 0 & -\frac{1}{C} \end{bmatrix}\begin{bmatrix} v_i \\ i_o \end{bmatrix} \quad (7.133)$$

where i, v, v_i, and i_o as the inductor current, output voltage, inverter voltage, and output current of one-phase, respectively.

In the stand-alone mode, the inverter needs to maintain stable output voltages and its control objectives are as follows:

1. Construct $v_{i\alpha}$ such that v_α track its reference v_α^*.
2. Construct $v_{i\beta}$ such that v_β track its reference v_β^*.

7.3 Photovoltaic Inverters in the Islanded Mode

where the references v_α^* and v_β^* are usually constants.

Improved Model Predictive Control (OCSMC) Method

As shown in Fig. 7.21, the IMPC with voltage feedback only is proposed to enhance output voltage performance, which consists three steps. The desired voltage v is calculated by the simplified discrete model considering voltage fluctuation. The capacitor current is estimated ($\hat{i}_C$) by the strong tracking Kalman file (STKF)-based observer. Finally, the inverter is controlled by the selected vectors. The detailed description of each part is given in the following text.

Step 1: Simplified Discrete-Model Considering Voltage Increment

To simplify the digital control implementation during the per-sampling period, the discrete-model using Backward-Euler approximation is established as

$$\begin{cases} v(k+1) = v(k) + \frac{T_s}{C}[i(k+1) - i_o(k+1)] \\ i(k+1) = i(k) + \frac{T_s}{L}[v_i(k+1) - v(k+1)] \end{cases} \quad (7.134)$$

Through algebraic calculations, the predicate output voltage in $\alpha\beta$-reference frame can be obtained as

$$X_v(k+1) = A_d X_v(k) + C_d[X_1(k) - X_o(k+1)] + B_d U(k+1) \quad (7.135)$$

where the $X_v = [v_\alpha, v_\beta]^T$, $U = [v_{i\alpha}, v_{i\beta}]^T$, $X_1 = [i_\alpha, i_\beta]^T$, $X_o = [i_{o\alpha}, i_{o\beta}]^T$, $A_d = \text{diag}(CL/\Phi; CL/\Phi)$, $B_d = \text{diag}(T_s^2/\Phi; T_s^2/\Phi)$, $\Phi = T_s^2 + CL$.

Since the sampling time T_s is very small, the corresponding load current i_o changes very slowly [20]. Thus, assuming that load current does not change during one sampling period, the difference of load current $\Delta i_o(k+1)$ is almost zero.

$$i_o(k+1) - i(k) = [i_o(k+1) - i_o(k)] - [i(k) - i_o(k)]$$
$$= \Delta i_o(k+1) - i_C(k)$$

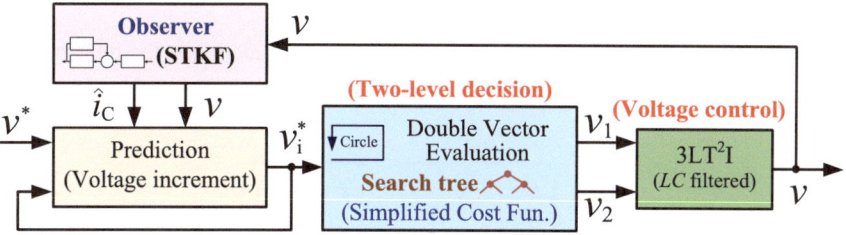

Fig. 7.21 The diagram of the proposed MPC scheme for voltage control

$$= \Delta i_{o\alpha\beta}(k+1) + \frac{C[v(k+1) - v(k+1)]}{T_s} \quad (7.136)$$

where i_C is the capacitor current.

According to above expressions, the prediction $X_v(k+1)$ yields

$$X_v(k+1) = A_d X_v(k) + B_d U(k+1) + C_d X_i(k) \quad (7.137)$$

where $X_i = [i_{C\alpha}, i_{C\beta}]^T$.

To achieve the objective of reference voltage $v^*(k+1)$ tracking, the cost function J_{C1} with prediction $X_v(k+1)$ is defined as

$$\begin{aligned} J_{C1} &= \left\| X_v^*(k+1) - X_v(k+1) \right\|_2^2 \\ &= [X_v^*(k+1) - X_v(k+1)]^T [X_v^*(k+1) - X_v(k+1)] \end{aligned} \quad (7.138)$$

Correspondingly, the desired voltage $U^*(k+1)$ (optimal control state) can be updated as

$$U^*(k+1) = B_d^{-1}[X_v^*(k+1) - A_d X_v(k) - C_d X_i(k)] \quad (7.139)$$

Then, the cost function with desired voltage $U^*(k+1)$ can be defined as

$$J_{CV} = \left\| U^*(k+1) - [v_{c\alpha}(k+1), v_{c\beta}(k+1)]^T \right\|_2^2 \quad (7.140)$$

where $[v_{c\alpha}(k+1), v_{c\beta}(k+1)]^T$ is the candidate vector.

However, since the output voltage v is mainly decided by the inverter voltage v_i, fluctuations of v_i can impact the output voltage performance. Thus, weighting factor Γ of the increment ΔU is introduced to reduce the output voltage distortion. Then the multi-objectives cost function, which considers voltage increment, can be expressed as

$$J_d = \|X(k+1) - X_v(k+1)\|_2^2 + \|\Gamma \Delta U(k+1)\|_2^2 \quad (7.141)$$

where $\Delta U(k+1) = U(k+1) - U(k)$, $\Gamma = \text{diag}(\gamma, \gamma)$.

As the min J_d is the Quadratic-Programming (QP) issue, define the intermediate variable ρ as

$$\rho = \overbrace{\begin{bmatrix} B_d \\ \Gamma \end{bmatrix}}^{A_{bz}} U(k+1) - \overbrace{\begin{bmatrix} E(k+1) \\ \Gamma U(k) \end{bmatrix}}^{b} \quad (7.142)$$

where $E(k+1) = X(k+1) - A_d X_v(k) - C_d X_i(k)$.

The cost function evaluation min J_d can be written as

7.3 Photovoltaic Inverters in the Islanded Mode

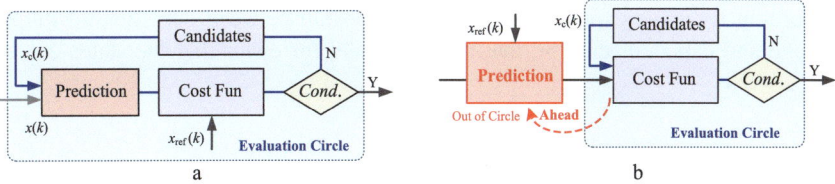

Fig. 7.22 Evaluation structure of MPC for PV inverter in the islanded mode **a** conventional MPC structure, **b** proposed structure (prepositive prediction)

$$\min\ J_d = \min(\rho^T \rho) = \min\left[(A_b z - b)^T (A_b z - b)\right] \tag{7.143}$$

The extremal solution z^* can be obtained as

$$z^* = \left(A_b{}^T A_b\right)^{-1} A_b{}^T b \tag{7.144}$$

Therefore, considering the fluctuation of inverter voltage, z^* becomes the new desired voltage ($U_d^*(k+1) = z^*$), which can be updated as

$$U_d^*(k+1) = \left(B_d^T B_d + \Gamma^T \Gamma\right)^{-1}\left[B_d^T E(k+1) + \Gamma^T \Gamma U(k)\right] \tag{7.145}$$

Then, the cost function with desired voltage $U(k+1)$ can be defined as

$$J_{dV} = \left\|U_d(k+1) - \left[v_{c\alpha}(k+1), v_{c\beta}(k+1)\right]\right\|^2 = f\left(v_i^*, v_c\right) \tag{7.146}$$

The advantage of proposed method is that it optimizes the solution of MPC considering the voltage fluctuation reduction. Moreover, compared to conventional method, the proposed method has a linear rather than exponential or trigonometric expression style and requires fewer calculations, as shown in Fig. 7.22.

The feature of the proposed method is that the reference voltage v^* evaluation of the vector is replaced with the desired voltage v_i^* evaluation of the vector and prepositive prediction. The cost function with desired voltage is calculated before the evaluation cycle for the proposed structure, which only needs once calculation during the per-sampling period. Despite adding an objective related to voltage fluctuation, the proposed structure maintains simplified cost function with improved computational efficiency.

Step 2: State Estimation Based on Strong Tracking Kalman Filter
With the capacitor current obtained from the proposed observer instead of the measurement, the derivative of capacitor current can be written as

$$i_C(k+1) - i_C(k) = [i(k+1) - i(k)] - [i_o(k+1) - i_o(k)] = \frac{T_s}{L}[v_i(k) - v(k)] \tag{7.147}$$

According to above expression and the relationship between derivative of output voltage and capacitor current, the discrete-model for estimating the states can be expressed as

$$\begin{cases} X(k) = FX(k-1) + Bu(k-1) + w(k-1) \\ Z(k) = HX(k) + v(k) \end{cases} \quad (7.148)$$

where the $w(k-1)$ and the $v(k)$ represent the process noise matrix and the measurement noise matrix respectively, and $X = [v_\alpha \ v_\beta \ i_{C\alpha} \ i_{C\beta}]^T$, $u = [v_{i\alpha} \ v_{i\beta}]^T$, $Z = [v_\alpha \ v_\beta]^T$,

$$B = \begin{bmatrix} 0 & 0 & \frac{T_s}{L} & 0 \\ 0 & 0 & 0 & \frac{T_s}{L} \end{bmatrix}^T, H = \begin{bmatrix} 1 & 0 & 0 & 0 \\ 0 & 1 & 0 & 0 \end{bmatrix}, F = \begin{bmatrix} 1 & 0 & \frac{T_s}{C} & 0 \\ 0 & 1 & 0 & \frac{T_s}{C} \\ -\frac{T_s}{L} & 0 & 1 & 0 \\ 0 & \frac{T_s}{L} & 0 & 1 \end{bmatrix}.$$

As the observer, the STKF algorithm can handle system parameter mismatches and model errors caused by the mutations, which improves system robustness. The STKF algorithm's recursive process contains the prediction process and the correction process, which are described below.

Prediction Process Calculation of state prediction value can be expressed as

$$\hat{X}^-(k) = F\hat{X}^-(k-1) + Bu(k-1) \quad (7.149)$$

where $\hat{X}^-(k)$ is the predicted value of the estimated states, $\hat{X}^-(k-1)$ is the estimated state values.

To overcome the deficiency that Kalman file (KF) heavily relies on past data, the fading factor λ is introduced into the state prediction covariance $\hat{X}^-(k)$ [21], which can be expressed as

$$\hat{P}^-(k) = \lambda(k)F(k)\hat{P}(k-1)F(k)^T + Q(k-1) \quad (7.150)$$

where Q represents the process noise covariance matrix, $\lambda(k) = \text{diag}(\lambda(1, k), \lambda(2, k), \lambda(3, k), \lambda(4, k))$, which can be expressed as

$$\lambda(i, k) = \begin{cases} \lambda_0 & \lambda_0 \geq 1 \\ 1 & \lambda_0 < 1 \end{cases} \quad (7.151)$$

where $\lambda_0 = tr[N(k)]/tr[M(k)]$, and $tr[\cdot]$ denotes the trace of a matrix, the matrices $N(k)$ and $M(k)$ are defined as

$$\begin{cases} N(k) = S_0(k) - HQ(k-1)H^T - \xi R(k) \\ M(k) = HF\hat{P}(k-1)F^T H^T \end{cases} \quad (7.152)$$

$$S_0(k) = \begin{cases} d(1)d^T(1) & k = 1 \\ \frac{[\rho S_0(k) + d(k)d^T(k)]}{1+\rho} & k > 1 \end{cases} \quad (7.153)$$

7.3 Photovoltaic Inverters in the Islanded Mode

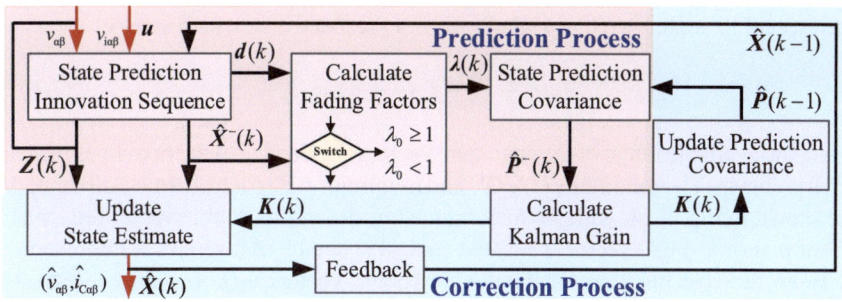

Fig. 7.23 Flowchart of STKF algorithm

where the forgetting factor $\rho \in (0, 1)$ and is typically 0.95, the weakening factor $\xi \in [0, \infty)$ makes the state estimate smoother, $R(k)$ represents the measurement noise covariance matrix. And the $d(k) = Z(k) - H\hat{X}^-(k)$.

Correction Process Due to the state prediction covariance $P^-(k)$ varies, the Kalman gain $K(k)$ can be adjusted in real-time, which can be expressed as

$$K(k) = \hat{P}^-(k)H^T\left[H\hat{P}^-(k)H^T + R(k)\right]^{-1} \quad (7.154)$$

The key of the STKF algorithm is to update the estimate state, which can be expressed as

$$\hat{X}(k) = \hat{X}^-(k) + K(k)\left[Z(k) - H\hat{X}^-(k)\right] \quad (7.155)$$

The implementation of the STKF algorithm is realized through iterative operation. Eventually, the capacitor current and output voltage can be estimated as shown in Fig. 7.23.

Since the capacitor current can be estimated by the STFK-based observer, the hardware cost for current sensors is saved, which reduces system maintenance. Meanwhile, the reliability of the inverter is improved because the possibility of sensor damage is reduced. In addition, the STFK algorithm has the ability for restraining effects of parameter errors, measurement errors, disturbances and noises.

Step 3: Low Complexity Evaluation for the Double Vector

In order to track the desired voltage v_i^* with improving control accuracy, the cost function for the double vector (v_1, v_2) based on the desired voltage v_i^* is defined as

$$J = [v_i^*(k+1)T_s - v_1(k+1)T_1(k+1) - v_2(k+1)T_2(k+1)]^2 \quad (7.156)$$

where main vector $v_1(k+1)$ and subvector $v_2(k+1)$ synthesize the virtual vector to follow the desired voltage $v(k+1)$, switching times $T_1(k+1)$ and $T_2(k+1)$

correspond to vectors $v_1(k+1)$ and $v_2(k+1)$ respectively, which satisfy

$$T_1(k+1) + T_2(k+1) = T_s \tag{7.157}$$

Then, the available range of the candidate vectors is extended to the bold lines instead of 19 points in. The minimum $J1/2$ CV and minimum $J^{1/2}$ (control errors) distribution are shown in Fig. 7.24, with the maximum control error of double vector reduced to 1/2 of that of a single vector. Thus, the control accuracy of inverters is improved.

From the cost function for the double vector, vectors $v_1(k+1)$ and $v_2(k+1)$, as adjacent vectors, can be obtained by calculating the nearest distance between the desired point $\mathbf{v}^*(k+1)$ and the line of equilateral triangle, as shown in Fig. 7.25a ($d_3 < d_2 < d_1$). It is difficult to find the optimal solution ($\mathbf{V}_{M6}, \mathbf{V}_{PN1}$) by simultaneously changing four variables (v_1, T_1, v_2, T_2).

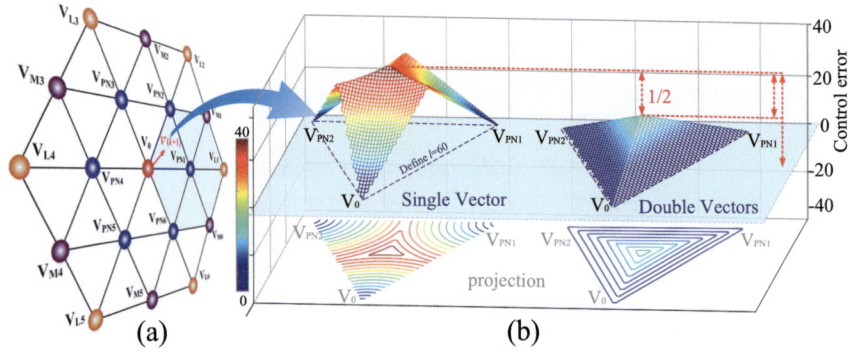

Fig. 7.24 The vector range and control error distribution

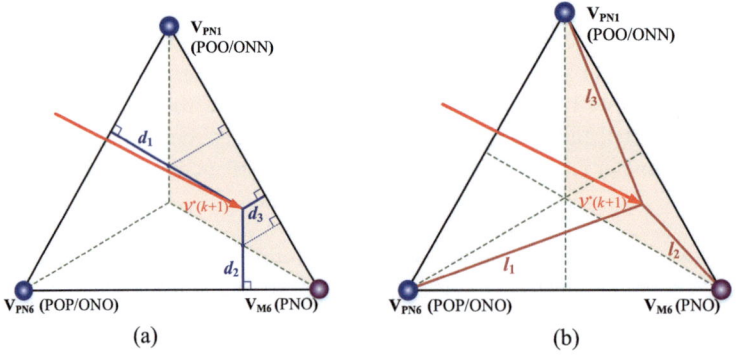

Fig. 7.25 The vector range and selection principles **a** conventional principle of vectors selection, **b** simplify principle of vectors selection

7.3 Photovoltaic Inverters in the Islanded Mode

The straight-line distance between the desired voltage and the candidate vector is defined as l_d, which can be written as

$$l_d^2 = \left[v_i^*(k+1) - v_T(k+1)\right]^2 \quad (d = 1, 2, 3) \tag{7.158}$$

where $v_T(k+1)$ is the candidate vector of the triangle.

In Fig. 7.25b, the distance relationship can be expressed as

$$\begin{cases} l_2|_{\mathbf{V}_{M6}} < l_1|_{\mathbf{V}_{PN6}} \\ l_3|_{\mathbf{V}_{PN1}} < l_1|_{\mathbf{V}_{PN6}} \end{cases} \tag{7.159}$$

Thus, instead of the complex calculation of the cost function with the double vector, double vector selection is equivalent to selecting two smaller line distances ($l_2|\mathbf{V}_{M6}$, $l_3|\mathbf{V}_{PN1}$) from three candidates based on the geometrical theory. It is worth noting that the selection rule in the triangle could be easily extended to hexagon (sector). Then, the simplified cost function for selecting vector is defined as

$$\begin{aligned} J_A &= \left\| v_i^2(k+1) - v_c(k+1) \right\|_2^2 \\ &= \left[v_\alpha(k+1) - v_{c\alpha}(k+1)\right]^2 + \left[v_\beta(k+1) - v_{c\beta}(k+1)\right]^2 \end{aligned} \tag{7.160}$$

The J_A is convex, and the minimum value exists in the finite set. Even though the cost function is simplified, sphere decoding algorithm (SDA) as a maximum-likelihood detection method, is introduced to significantly reduce the number of searched control inputs. To simplify the double vector evaluation, a two-layer decision (TLD) algorithm is proposed based on the sphere decoding principle, which can find the optimal vectors by traversing the search tree sequentially in a top-to-bottom manner, where each node has six branches in the two-layer search tree.

The overall control block diagram is depicted in Fig. 7.26, where the inverter with LC filter operates in voltage control mode, which includes output voltage control and NP voltage balance control. For the AC side control, the STKF-based observer estimates the capacitor current with voltage feedback only. Then, the desired voltage is calculated using the simplified discrete-model considering voltage increment. The TLD determines the double vector to accurately make the output voltage track the reference value.

Observer-Based Continuous Sliding-Mode Voltage Control (OCSMC) Method

In this section, the OCSMC is proposed without requiring any current sensors. Subsequently, the stability analysis and tuning control parameters are given.

In the dq-reference frame, the inverter dynamic model can be established as the d-axis model and the q-axis model, which are given as follows

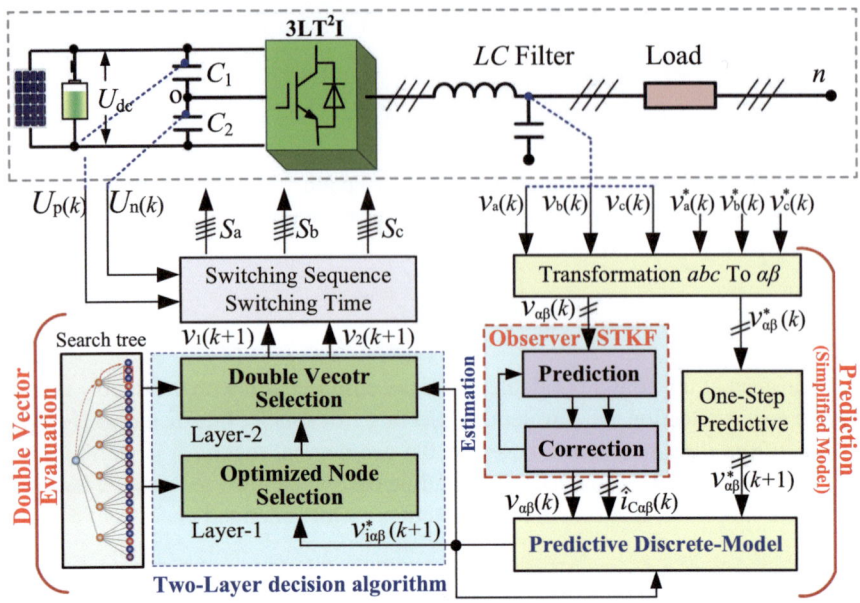

Fig. 7.26 The overall control block diagram of single-loop voltage control scheme using IMPC

$$\begin{cases} \frac{dv_d}{dt} = \omega v_q + \frac{1}{C} i_d + d_1(t) \\ \frac{di_d}{dt} = -\frac{1}{L} v_d + \omega i_q + \frac{1}{L} v_{id} + d_2(t) \end{cases} \quad (7.161)$$

$$\begin{cases} \frac{dv_q}{dt} = -\omega v_d + \frac{1}{C} i_q + d_3(t) \\ \frac{di_q}{dt} = -\frac{1}{L} v_q - \omega i_d + \frac{1}{L} v_{iq} + d_4(t) \end{cases} \quad (7.162)$$

where $d_1(t) = -1/C\ i_{od}(t)$ and $d_3(t) = -1/C\ i_{oq}(t)$ are taken as external load disturbances. $d_2(t)$ and $d_4(t)$ are unmodeled dynamics caused by parameter uncertainties of L, and ω is the angular frequency. Based on the model in d-axis and q-axis, the control scheme includes two parts: d-axis output voltage control design and q-axis output voltage control design.

Part 1: d-Axis Output Voltage Control Design

The d-axis control design includes the following steps. Firstly, a finite-time observer (FTO) is designed to estimate the lumped disturbances. Based on the estimations from the FTO, a continuous sliding-mode controller is constructed to achieve the d-axis output voltage control. Finally, stability analysis is conducted to demonstrate that the proposed strategy can achieve fast and accurate tracking control.

New state variables are introduced as

$$x_1 = v_d - v_d^*, x_2 = \frac{dv_d}{dt} \quad (7.163)$$

7.3 Photovoltaic Inverters in the Islanded Mode

And then, the d-axis model can be written as

$$\begin{cases} \frac{dx_1}{dt} = x_2 \\ \frac{dx_2}{dt} = -\frac{1}{CL}x_1 - \frac{1}{CL}v_d^* + \frac{1}{CL}v_{id} + x_3(t) \\ x_3(t) = \omega\frac{dv_q}{dt} + \frac{\omega}{C}i_q + \frac{1}{C}d_2(t) + \frac{dd_1(t)}{dt} \end{cases} \quad (7.164)$$

where x_3 is the matched lumped disturbance induced by load disturbances, parameter uncertainties, unmeasured inductor currents, etc. Here, x_3 and its first derivative are assumed to be bounded. This assumption is reasonable due to physical limitations on the voltages, currents, and their rates of change.

The defined states x_1 and x_2 are used to transform the model in d-axis into a new one with matched lumped disturbances, making it possible to reduce the difficulty of control design.

Step 1: The Design of FTO for the Lumped Disturbances Estimation

Disturbance observer technique provides a feasible way to realize feedforward disturbance compensation so as to improve control performance and robustness [22]. Inspired by this idea, an FTO is designed to estimate the states and lumped disturbances, which is given as

$$\begin{cases} \frac{d\hat{x}_1}{dt} = \hat{x}_2 + g_1|e_1|^{\frac{2}{3}}\operatorname{sgn}(e_1) \\ \frac{d\hat{x}_2}{dt} = \hat{x}_3 - \frac{1}{CL}x_1 - \frac{1}{CL}v_d^* + \frac{1}{CL}v_{id} + g_2|e_1|^{\frac{1}{3}}\operatorname{sgn}(e_1) \\ \frac{d\hat{x}_3}{dt} = g_3 \operatorname{sgn}(e_1) \end{cases} \quad (7.165)$$

where $\hat{x}_i$ is the estimation value of x_i, e_i is the estimation error, $g_i > 0$ is the parameter of the FTO, and $i = 1, 2, 3$. In particular, the estimation errors are defined as

$$\begin{cases} e_1 = x_1 - \hat{x}_1 \\ e_2 = x_2 - \hat{x}_2 \\ e_3 = x_3 - \hat{x}_3 \end{cases} \quad (7.166)$$

Then, the error dynamics can be written as

$$\begin{cases} \frac{de_1}{dt} = -g_1|e_1|^{\frac{2}{3}}\operatorname{sgn}(e_1) + e_2 \\ \frac{de_2}{dt} = -g_2|e_1|^{\frac{1}{3}}\operatorname{sgn}(e_1) + e_3 \\ \frac{de_3}{dt} = -g_3 \operatorname{sgn}(e_1) + \frac{dx_3}{dt} \end{cases} \quad (7.167)$$

The above equation is finite-time stable [22]. Therefore, it is obtained that the estimation errors e_1, e_2 and e_3 will converge to zero in a finite time $t > T_o$, achieving a precise estimation of states and lumped disturbances.

Step 2: Design of Continuous Sliding-Mode Controller

Using the estimations obtained from the FTO, a continuous sliding-mode control will be designed to achieve the tracking control of the d-axis output voltage. Considering a sliding surface as

$$\hat{s}_d = c_1 x_1 + \hat{x}_2 \quad c_1 > 0 \tag{7.168}$$

Taking its time derivative, it follows that

$$\begin{aligned} \frac{d\hat{s}_d}{dt} &= c_1 \frac{d\hat{x}_1}{dt} + \frac{d\hat{x}_2}{dt} \\ &= c_1 \hat{x}_2 + c_1 e_2 + \hat{x}_3 - \frac{1}{CL} x_1 - \frac{1}{CL} v_{od}^* + \frac{1}{CL} u_d + g_2 |e_1|^{\frac{1}{3}} \operatorname{sgn}(e_1) \end{aligned} \tag{7.169}$$

In order to regulate the d-axis output voltage, a continuous sliding-mode controller is designed as

$$v_{id} = CL\left(-k_1|\hat{s}_d|^{\frac{1}{2}} \operatorname{sgn}(\hat{s}_d) - k_2 \hat{s}_d - c_1 \hat{x}_2 - \hat{x}_3 - g_2|e_1|^{\frac{1}{3}} \operatorname{sgn}(e_1)\right) + x_1 + v_d^* \tag{7.170}$$

where k_1 and k_2 are positive constants.

According to above equations, time derivative of sliding surface is simplified as

$$\frac{d\hat{s}_d}{dt} = -k_1|\hat{s}_d|^{\frac{1}{2}} \operatorname{sgn}(|\hat{s}_d|) + c_1 e_2 - k_2 \hat{s}_d. \tag{7.171}$$

It is discussed earlier that the estimation errors e_1 and e_2 converge to zero in finite-time. Since the trajectories of time derivative of sliding surface cannot escape to infinity in finite time, thus time derivative of sliding surface is rewritten as

$$\frac{d\hat{s}_d}{dt} = -k_1|\hat{s}_d|^{\frac{1}{2}} \operatorname{sgn}(\hat{s}_d) - k_2 \hat{s}_d \tag{7.172}$$

Selecting $k_1 > 0$ and $k_2 > 0$, we have $\hat{s}_d = 0$ in finite time. Further, it is obtained that x_1 tends to zero exponentially, i.e., a fast and accurate tracking control of the d-axis output voltage is achieved.

Step 3: Stability Analysis for the Proposed Strategy

The proof is given as follows.

Proof Choose the Lyapunov function as

$$V = \frac{1}{2}\hat{s}_d^2. \tag{7.173}$$

Then, taking the derivative of V yields that

$$\frac{dV}{dt} = -k_1|\hat{s}_d|^{\frac{1}{2}} \operatorname{sgn}(\hat{s}_d)\hat{s}_d - k_2 \hat{s}_d^2 \leq -2^{\frac{3}{4}} k_1 V^{\frac{3}{4}} - 2k_2 V. \tag{7.174}$$

7.3 Photovoltaic Inverters in the Islanded Mode

According to the Lyapunov function, V is negative definite. Using the fast finite-time Lyapunov stability theory in [23], V tends to zero in a finite time, i.e., $\hat{s}_d = 0$. It further implies that $dx_1/dt = -c_1 x_1$, then x_1 tends to zero exponentially. This completes the proof.

Remark 1 As discussed earlier, the discontinuous nature of the traditional sliding-mode control (SMC) leads to chattering problems [24]. In the expression of the time derivative of sliding surface, a continuous sliding-mode voltage controller is constructed by using the fractional-power terms instead of the "sgn" function in the traditional SMC, which avoids the chattering problem and guarantees good steady-state performance. Meanwhile, this controller achieves finite-time convergence of sliding surfaces to improve the dynamic response of output voltages.

Part 2: q-Axis Output Voltage Control Design
Similar with part of the d-axis output voltage control design, introduce new state variables as

$$x_4 = v_q - v_q^*, \quad x_5 = dv_q/dt. \tag{7.175}$$

Then transform the q-axis model into one with matched lumped disturbances, which is described as

$$\begin{cases} \frac{dx_4}{dt} = x_5 \\ \frac{dx_5}{dt} = -\frac{1}{CL}x_4 - \frac{1}{CL}v_q^* + \frac{1}{CL}v_{iq} + x_6(t) \\ x_6(t) = -\omega \frac{dv_d}{dt} - \frac{\omega}{C}i_d + \frac{1}{C}d_4(t) + \frac{dd_3(t)}{dt} \end{cases} \tag{7.176}$$

where $x_6(t)$ is the matched lumped disturbance induced by load disturbances and parameter uncertainties, along with unmeasured currents, etc. Here, x_6 and its first derivative are assumed to be bounded in practice.

Then, an FTO is employed to estimate the states and lumped disturbances, given as

$$\begin{cases} \frac{d\hat{x}_4}{dt} = \hat{x}_5 + g_4|e_4|^{\frac{2}{3}} \operatorname{sgn}(e_4) \\ \frac{d\hat{x}_5}{dt} = \hat{x}_6 - \frac{1}{CL}x_4 - \frac{1}{CL}v_q^* + \frac{1}{CL}v_{iq} + g_5|e_4|^{\frac{1}{3}} \operatorname{sgn}(e_4) \\ \frac{d\hat{x}_6}{dt} = g_6 \operatorname{sgn}(e_4) \end{cases} \tag{7.177}$$

where $e_4 = x_4 - \hat{x}_4$ is the estimation error, g_i is the FTO parameter with $i = 4, 5, 6$.
The sliding surface is designed as

$$\hat{s}_q = c_2 x_4 + \hat{x}_5, \quad c_2 > 0 \tag{7.178}$$

And construct a q-axis output voltage controller as

$$v_{iq} = CL\left(-k_3|\hat{s}_q|^{\frac{1}{2}}\text{sgn}(\hat{s}_q) - k_4\hat{s}_q - c_1\hat{x}_5 - \hat{x}_6 - g_5|e_4|^{\frac{1}{3}}\text{sgn}(e_4)\right) + x_4 + v_q^*$$
(7.179)

If selecting $k_3 > 0$ and $k_4 > 0$, we have $\hat{s}q = 0$ in finite time, which leads to x_4 tending to zero exponentially. This achieves a fast and accurate tracking control of the q-axis output voltage. The proof is omitted here as it is the same as that in Part 1.

Theorem For the LC filtered inverter system, using the continuous sliding mode controllers and the FTOs, the output voltage tracking errors x_1 and x_4 tend to zero exponentially, thereby fulfilling the control objectives.

Remark 2 Figure 7.27 shows the block diagram of the OCSMC strategy for the inverter. Benefiting from the newly established dynamic models and the designed FTOs, this strategy only requires measuring the output voltage, eliminating the need for load and inductor current measurements. Consequently, it saves current sensors, reduces costs and improves the reliability of the inverter control system.

Based on the theoretical analysis and practical experience, the suggestions for parameter tuning are given as follows:

1. Referring to [22], set the FTO parameters as $g_1 = 6\omega_0$, $g_2 = 11\omega_0^{3/2}$, $g_3 = 6\omega_0^3$, where ω_0 is a positive parameter related to the bandwidth of the FTO.
2. Select the control parameters $k_1 > 0$ and $k_2 > 0$ incrementally with given $c_1 > 0$.
3. For the q-axis control system, parameters are selected to be the same as those used in the d-axis control.

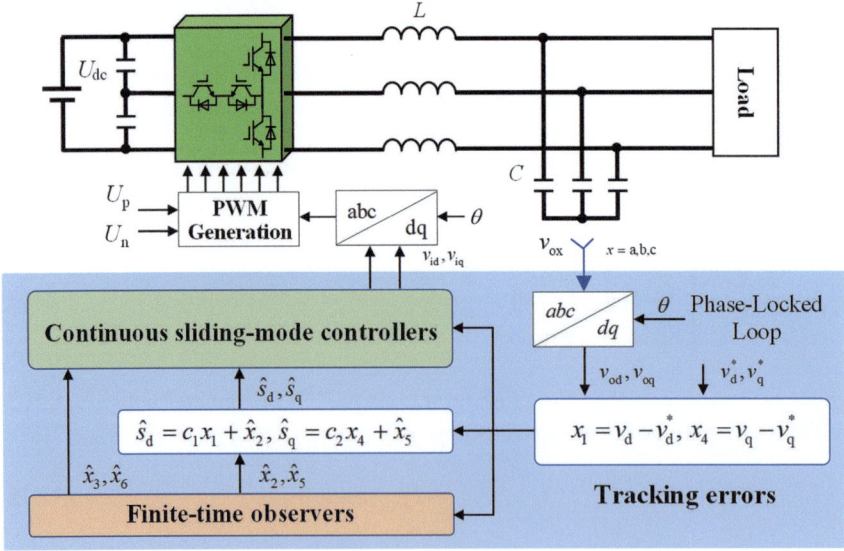

Fig. 7.27 Control diagram single-loop voltage control scheme using OCSMC

7.3.3 Experimental Validations for the Islanded Mode Operation

To illustrate the effectiveness of the control schemes for PV inverter operated in the islanded mode, comprehensive simulation and experimental results are given. It should be noted that the control methods based on dual-loop voltage and current control scheme have been extensively studied and relatively mature. While the control methods based on single-loop voltage control scheme are developed in recent years and relatively new. Thus, only the control methods based on single-loop voltage control scheme are validated in this section.

Experimental Results for the IMPC

To validate the proposed improved method, an experimental system has been built, which consists of a DC power source (Chroma 62150H-1000S), a digital controller (based on dSPACE DS1005 and FPGA DS5203), main power circuit of inverter, gate drivers, LC filter and three-phase resistive load. The THD is measured by the 'TPS2PWR1' application with average acquire model. The main experimental parameters are shown in Table 7.2.

For convenience analysis results, three methods are compared and defined to easily distinguish in Table 7.3.

As the output voltage v is 100 V, Fig. 7.28 depicts the experimental waveforms of the MPC-II and the IMPC when the resistive load is 12 Ω. The output voltage waveforms demonstrate that the IMPC performs lower voltage fluctuation compared to MPC-II, with the THDv_a of 2.57% for the IMPC and 3.14% for the MPC-II.

Table 7.2 The main parameters for the experimental system

Parameters/Components	Values
DC input voltage (U_{dc})	200 V
Inductors (L_a, L_b, L_c)	5 mH
Capacitors (C_a, C_b, C_c)	20 µF
DC-link capacitors (C_1, C_2)	3760 µF
Sampling time (T_s)	100 µs
Output frequency (f)	50 Hz
IGBT module	10-FZ12NMA080SH01-M260F

Table 7.3 The comparison of three methods definition

Methods	Current sensors	Vectors/T_s	Evaluation value
MPC-single	Need (i, i_o)	One	$X_v^*(k+1)$
MPC-II	Need (i, i_o)	Double	$U^*(k+1)$
IMPC	Not need	Double	$U_d^*(k+1)$

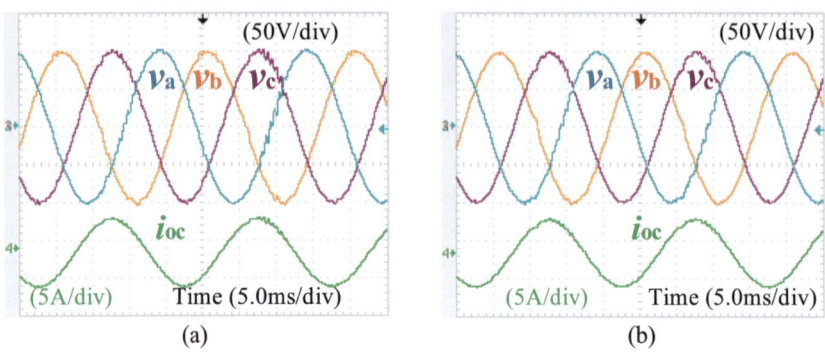

Fig. 7.28 Experimental waveform when the resistive load is 12 Ω **a** MPC-II, **b** IMPC

In addition, as shown in Fig. 7.29, the THDv_a of the IMPC is 2.23% when the output voltage v is 80 V. The output voltage quality of the IMPC significantly improves compared to that of the MPC-II, due to considering the voltage fluctuations reduction.

The voltage THD comparisons under various cases are shown in Fig. 7.30. When the resistive load is 16 Ω, it can be observed form Fig. 7.30a that the voltage THD of the IMPC remains lower than that of the MPC-II with different voltages, and the voltage THD of the IMPC is less than 2.5%. When the resistive load varies from 12 to 48 Ω (output voltage maintains 100 V), the voltage THD of the MPC-II exceeds 2.8%, while the THD of the IMPC does not exceed 2.6% as shown in Fig. 7.30b. This indicates that the IMPC has better voltage control performance than MPC-II.

The experimental dynamic waveforms of the MPC-II and the IMPC are shown in Fig. 7.31. As the resistive load step changed from 12 to 16 Ω, the voltage is not affected and maintains 100 V. It can be observed that the dynamic response time of the proposed MPC is the same as that of the MPC-II, which is less than 1.5 ms. Therefore,

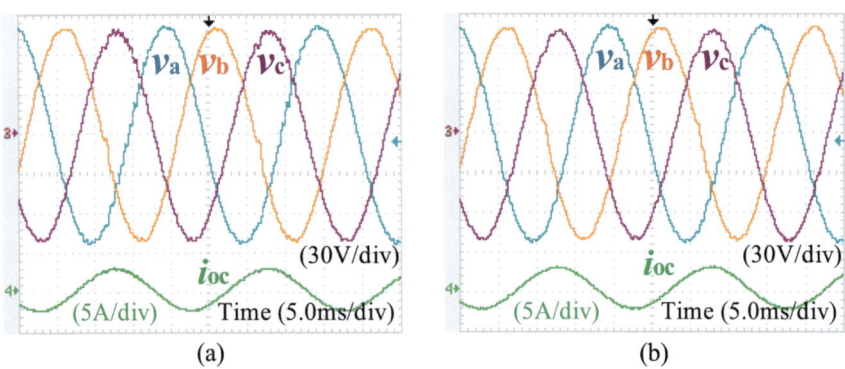

Fig. 7.29 Experimental waveform when the output voltage is 80 V **a** MPC-II, **b** IMPC

7.3 Photovoltaic Inverters in the Islanded Mode

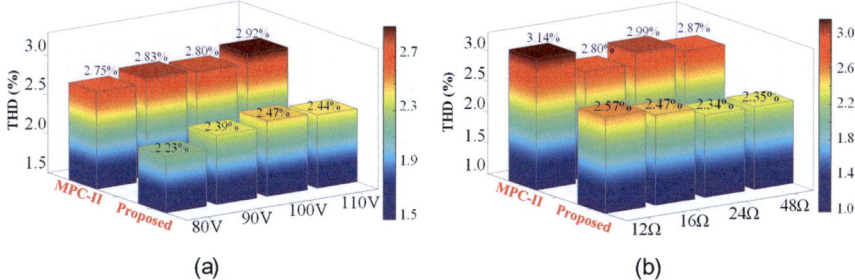

Fig. 7.30 Measurement voltage THD of IMPC with different cases **a** in the case of various output voltage, **b** in the case of various load

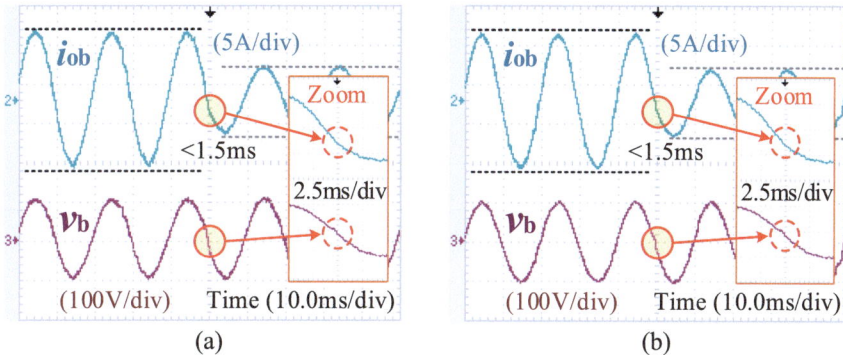

Fig. 7.31 Experimental dynamic waveforms when the load step changed from 12 Ω steps to 16 Ω **a** MPC-II, **b** IMPC

the constraints on ΔU have little impact on the dynamic response speed, which can be ignored. In conclusion, the IMPC has a satisfactory dynamic performance.

To demonstrate the correctness of the IMPC the measured capacitor current i_{Ca} ($i_a - i_{oa}$) and estimated capacitor current $\hat{i}_{Ca}$ are shown in the Fig. 7.32. It can be observed that the estimated capacitor current $\hat{i}_C$ is basically consistent with the measured capacitor current i_{Ca} under steady state in Fig. 7.32a. In addition, when the resistive load changed form 12 Ω steps to 16 Ω (as shown in Fig. 7.32b), the performance of the capacitor current estimation is satisfactory. The significance of the STKF-based observer algorithm for estimating the capacitor current is proved.

Figure 7.33 depicts experimental voltage waveforms of MPC-single and the proposed MPC with the same resistive load (16 Ω). The voltage fluctuation (> 15 V) of the MPC-single is large than that (< 5 V) of the IMPC obviously. Due to the adoption of a double vector, the proposed method has better performance with the increased switching frequency.

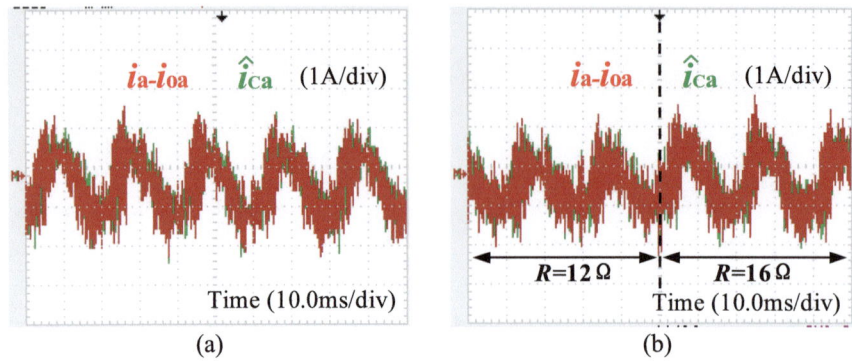

Fig. 7.32 Experimental waveform of capacitor current **a** load $= 12\,\Omega$ in the steady state, **b** load changed from 12 to 16 Ω

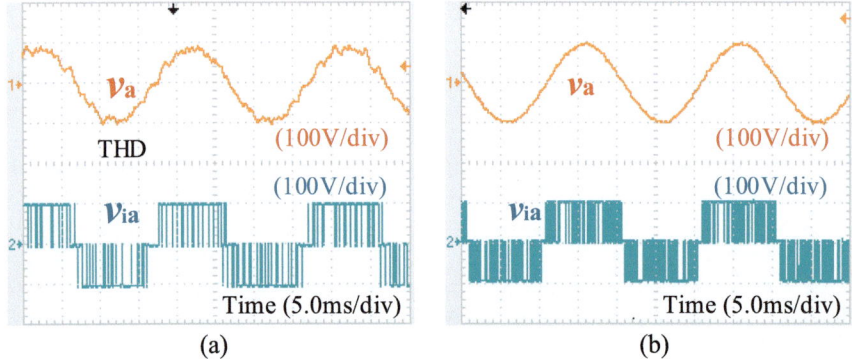

Fig. 7.33 Experimental voltage waveform of two MPC method in steady state **a** MPC-single, **b** IMPC

Experimental Results for OCSMC

In this section simulation and experimental results are given to verify its effectiveness of the single-loop voltage control scheme using OCSMC. Table 7.4 summarizes the system parameters of the tested inverter. Meanwhile, a dual-loop structure adaptive method in [25] is implemented for comparison, and its parameters are well-tuned using the corresponding selection criteria. The OCSMC parameters are selected as $k_1 = k_3 = 3 \times 10^4$, $k_2 = k_4 = 1.5 \times 10^4$, $c_1 = c_2 = 2000$, $\omega_0 = 1200$.

1. **Simulation Results**

    The simulation results consist of steady-state performance and dynamic performance.

    (a) **Steady-state performance**: Figure 7.34a, b show the steady-state performance with a liner *RL* load (i.e., Load 1) for both the adaptive and OCSMC

7.3 Photovoltaic Inverters in the Islanded Mode

Table 7.4 The system parameters of the tested inverter

Parameters/Components	Values
Rated power S_r	3 kVA
DC-link voltage (U_{dc})	300 V
DC-link capacitor (C_1, C_2)	2200 μF
Filter inductance (L_a, L_b, L_c)	3 mH
Filter capacitor C	47 μF
d-axis output voltage reference v_d^*	110 V
q-axis output voltage reference v_q^*	0 V
Load 1 (RL load)	15 Ω, 5 mH
Load 2 (RL load)	15 Ω, 1 mH
Switching frequency f	10 kHz

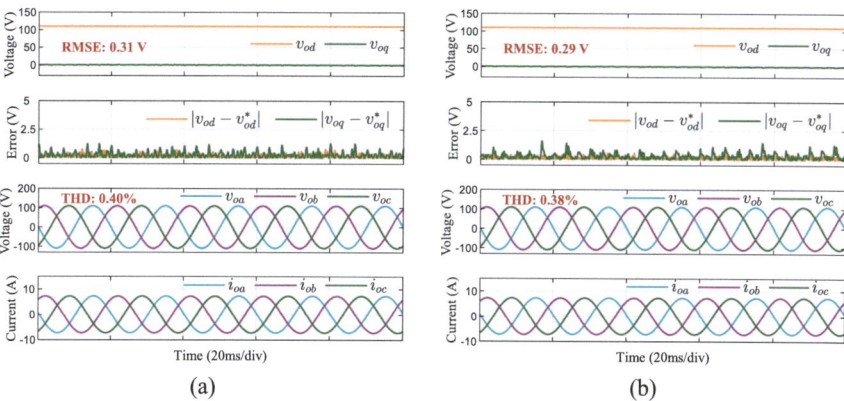

Fig. 7.34 Steady-state performance with an RL load **a** adaptive method, **b** OCSMC

methods. The corresponding results include dq-axis output voltages, voltage tracking errors, three-phase output voltages, and load currents. To evaluate the steady-state performance, the root mean square error (RMSE) and THD of output voltages are introduced. The RMSE is defined as RMSE = 1/2(RMSE (v_d) + RMSE (v_q)). The THD is calculated by the fast Fourier transform analysis.

Under the RL load, the adaptive method achieves good voltage tracking performance and sinusoidal three-phase output voltages, with an RMSE of 0.31 V and a THD of 0.40%. Meanwhile, the proposed OCSMC strategy obtains good steady-state control performance with slightly improved indices, as shown in Fig. 7.34b. Besides, the steady-state performance with the nonlinear load in Fig. 7.35b (i.e., a three-phase diode rectifier) is shown in Fig. 7.36. Compared to the results with the RL load, both methods suffer from larger voltage tracking errors under this critical load, and an increase

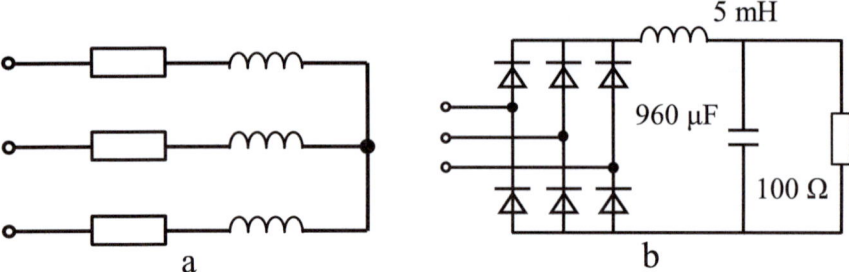

Fig. 7.35 Two types of load circuits **a** linear *RL* load, **b** nonlinear load with a three-phase diode rectifier

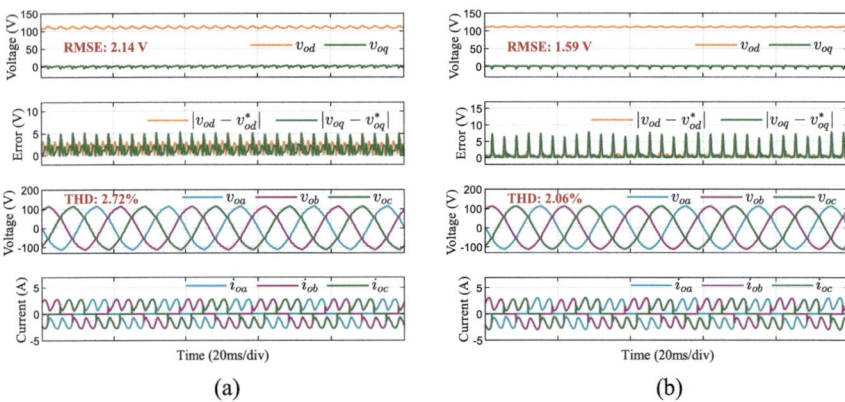

Fig. 7.36 Steady-state performance with the nonlinear load **a** adaptive method, **b** OCSMC

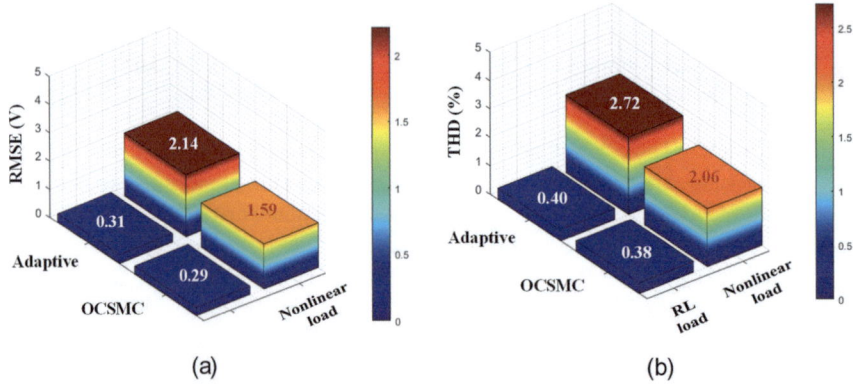

Fig. 7.37 Comparisons for steady-state performance **a** RMSE, **b** THD

7.3 Photovoltaic Inverters in the Islanded Mode

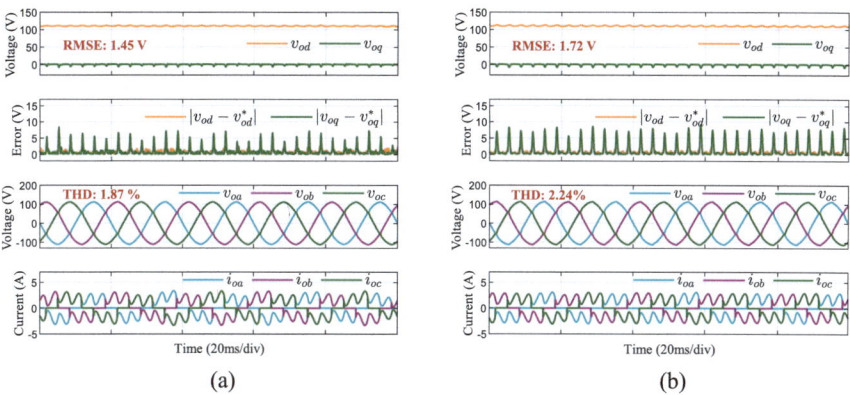

Fig. 7.38 Steady-state performance of the OCSMC with the nonlinear load and parameter uncertainties **a** − 15% variations of L, **b** + 15% variations of L

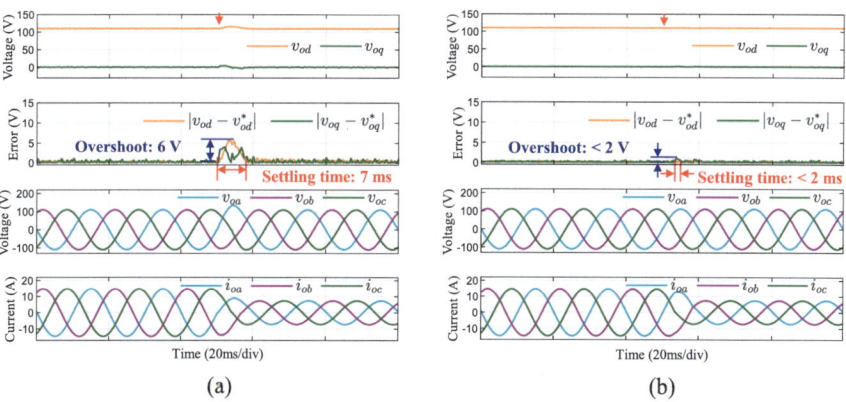

Fig. 7.39 Dynamic performance with the load disturbances **a** adaptive method, **b** OCSMC

in THD. However, compared to the adaptive method, the OCSMC strategy exhibits smaller RMSE and reduces THD from 2.72 to 2.06%. Figure 7.37 summarizes the comparison results.

In addition, the robustness of the OCSMC against parameter uncertainties is tested when the L has ±15% variations in the control implementation. Figure 7.38 shows the corresponding results with the nonlinear load. The output voltage RMSE and THD of the OCSMC only exhibit slight changes due to the filter parameter variations, which indicates the proposed strategy has good robustness against parameter uncertainties.

(b) **Dynamic performance evaluation**: Initially, the inverter operates in the steady state with Loads 1 and 2. Then, Load 2 is disconnected to verify the dynamic performance and robustness against load disturbances. Figure 7.39 compares the response waveforms of the adaptive and OCSMC methods.

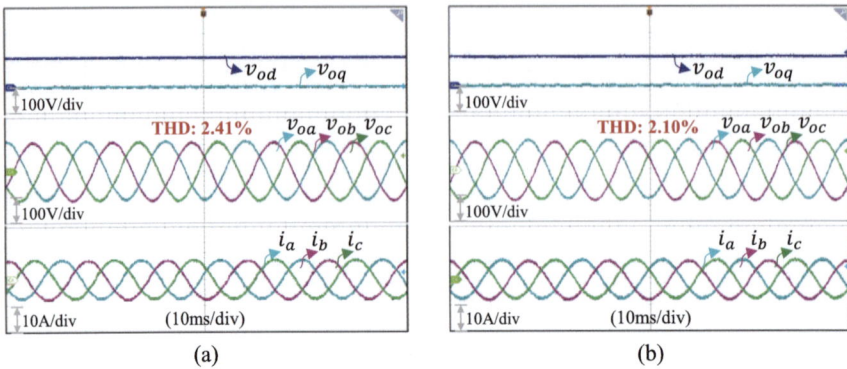

Fig. 7.40 Experimental results of steady-state performance with an *RL* load **a** adaptive method, **b** OCSMC

The load currents decrease after cutting off Load 2. Although both methods recover the output voltages to their references after load disturbances, the adaptive method suffers from a large voltage overshoot of 6 V with a settling time of 7 ms. In comparison, the proposed OCSMC performs enhanced dynamic response, with a shorter settling time of 2 ms and a smaller overshoot of 2 V.

2. **Experimental Results**

   The proposed OCSMC strategy of the inverter was further verified on a real-time testing platform, where system parameters are the same as the simulation. Figure 7.40 depicts the comparative experimental results of the adaptive and OCSMC methods when the inverter operates in a steady state with Load 1. In Fig. 7.40a, the adaptive method achieves good voltage tracking control and produces sinusoidal three-phase voltages with a THD of 2.41%. The OCSMC strategy performs better with a lower THD of 2.10%.

   Figure 7.41 compares the dynamic responses of the adaptive and OCSMC methods when the Load 2 is added to the inverter. As shown in Fig. 7.41a, the adaptive method has a larger overshoot of 40 V and a slower settling time of 11 ms. In contrast, as shown in Fig. 7.41b, the OCSMC achieves improved dynamic performance with a reduction of 72.7% in settling time compared to the adaptive method. Overall, both experimental and simulation results show similar trends. It should be noted that the simulation environment is idealized, thus resulting in better performance compared to the experimental results.

   In conclusion: this section mainly concretes on the advanced control technologies of PV inverter operating in the islanded mode. Firstly, a mathematical model of PV inverter operating in the islanded mode is constructed and analyzed, laying a solid foundation for controller design. Based on the dynamic model, advanced control methods are gradually evolved, encompassing the design of IMPC schemes and OCSMC controllers. Through the utilization of nonlinear control methods, the quantity of sensors is reduced, the robustness

7.4 Summary

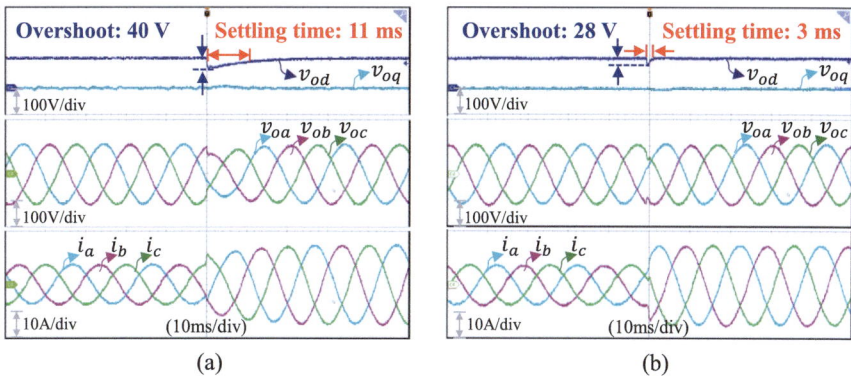

Fig. 7.41 Experimental results of dynamic performance with the load disturbances **a** adaptive method, **b** OCSMC

of control is enhanced, and the influence of uncertain factors under diverse operating conditions is effectively mitigated, maintaining system reliability and enhancing system performance. Finally, comprehensive simulation and experimental outcomes are presented. The results not only validate the correctness and effectiveness of the proposed models and control methods but also furnish readers with a distinct comprehension of the phenomena presented in this section.

7.4 Summary

In this chapter, the control technology is presented for PV inverter to realize multi functional operation. Multi-functional modes include the power quality control mode and the islanded mode.

To address the power quality issues such as harmonics and reactive power in renewable energy generation systems, the power quality control technology for PV inverters is introduced. First, based on actual working conditions, the principles of power quality control under different working modes are explained. Second, the power quality control scheme is elaborated in detail, which includes harmonics and reactive current detection, reference current synthesis, and reference current tracking control. Using the control scheme of PV inverters, the power quality of renewable energy generation systems is improved.

In islanded operation mode, the PV inverter can provide stable power supply to the local power system according to the practical demand. To realize islanded operation mode, and further improve steady-state and dynamic performance of PV inverter under load disturbances and parameter uncertainties. The control technology that imbedding the nonlinear control methods into the single-loop voltage control scheme is proposed by the author's team. With the proposed control technology, the quantity of sensors is reduced, the robustness and reliability are improved.

It needs to be emphasized that, this chapter not only introduces the control schemes of PV inverter for multi-functional operation but also delineates the design process of controllers, offering readers valuable guidance and insights into the subject matter.

References

1. H. Akagi, Y. Kanazawa, A. Nabae, Generalized theory of the instantaneous reactive power in three-phase circuits, in *IEEE & JIEE Proceedings IPEC* (1983), pp. 1375–1386
2. H. Akagi, Y. Kanazawa, A. Nabae, Instantaneous reactive power compensators comprising switching devices without energy storage components. IEEE Trans. Ind. Appl. **20**(3), 625–630 (1984)
3. A.E. Emanuel, J.A. Orr, D. Cyganski et al., A survey of harmonic voltages and currents at distribution substations. IEEE Trans. Power Deliv. **6**(4), 1883–1890 (1991)
4. C. Du, C. Zhang, X. Liu et al., Control strategy on the three-phase grid-connected PV generation system with shunt active power filter. Trans. China Electrotech. Soc. **25**(9), 163–169 (2010)
5. T.F. Wu, H.S. Nien, H.M. Hsieh et al., PV power injection and active power filtering with amplitude-clamping and amplitude-scaling algorithms. IEEE Trans. Ind. Appl. **43**(3), 731–741 (2007)
6. V. Salas, E. Olías, A. Lázaro et al., New algorithm using only one variable measurement applied to a maximum power point tracker. Sol. Energy Mater. Sol. Cells. **87**(1–4), 675–684 (2005)
7. Y. Chunjie, Z. Chenghui, C. Alian et al., A novel current tracking method for active power filter based on direct transient current control. Trans. China Electrotech. Soc. **27**(4), 117–122 (2012)
8. Y. Ping, A. Chen, C. Zhang, A novel adaptive hysteresis band current control method for three-level based active power filter, in *2020 7th International Power Electronics and Motion Control Conference* (2012), pp. 2716–2720
9. Y. Jiang, C. Qin, X. Xing et al., A hybrid passivity-based control strategy for three-level T-type inverter in LVRT operation. IEEE J. Emerg. Sel. Topics Power Electron. **8**(4), 4009–4024 (2020)
10. X. Xing, X. Li, F. Gao et al., Improved space vector modulation technique for neutral-point voltage oscillation and common-mode voltage reduction in three-level inverter. IEEE Trans. Power Electron. **34**(9), 8697–8714 (2019)
11. X. Li, X. Xing, C. Zhang et al., Simultaneous common-mode resonance circulating current and leakage current suppression for transformerless three-level T-type PV inverter system. IEEE Trans. Ind. Electron. **66**(6), 4457–4467 (2019)
12. F. Zare, S. Zabihi, G. Ledwich, An adaptive hysteresis current control for a multilevel inverter used in an active power filter, in *2007 European Conference on Power Electronics and Applications*, (2007), pp. 1–8
13. K. Zhou, W. Lu, Y. Yang et al., Harmonic control: A natural way to bridge resonant control and repetitive control, in *2013 American Control Conference* (2013), pp. 3189–3193
14. Z.X. Zou, K. Zhou, Z. Wang et al., Frequency-adaptive fractional-order repetitive control of shunt active power filters. IEEE Trans. Ind. Electron. **62**(3), 1659–1668 (2015)
15. Y. Cho, J.S. Lai, Digital plug-in repetitive controller for single-phase bridgeless PFC converters. IEEE Trans. Power Electron. **28**(1), 165–175 (2013)
16. Y. Li, J. Song, B. Duan et al., Repetitive control for harmonic compensation in three-phase isolated matrix rectifier, in *IECON 2020 The 46th Annual Conference of the IEEE Industrial Electronics Society* (2020), pp. 1443–1448

References

17. G. Escobar, G.A. Catzin-Contreras, M.J. Lopez-Sanchez, Compensation of variable fractional delays in the $6k \pm 1$ repetitive controller. IEEE Trans. Ind. Electron. **62**(10), 6448–6456 (2015)
18. T. Liu et al., An improved model predictive control to enhance voltage performance for LC filtered three-level inverters with voltage feedback only. IEEE Trans. Ind. Inform. **19**(9), 9809–9820 (2023)
19. C. Fu, C. Zhang, G. Zhang, Z. Zhang, Current sensorless sliding-mode voltage control for LC filtered three-level t-type inverters. IEEE Trans. Circuits Syst. II Exp. Brief. **71**(4), 2264–2268 (2024)
20. C. Fu, C. Zhang, G. Zhang, Q. Su, Finite-time adaptive fuzzy control for three-phase PWM rectifiers with improved output performance. IEEE Trans. Circuits Syst. II Exp. Brief. **70**(8), 3044–3048 (2023)
21. T. Jin, X. Shen, T. Su, R.C.C. Flesch, Model predictive voltage control based on finite control set with computation time delay compensation for PV systems. IEEE Trans. Energy Convers. **34**(1), 330–338 (2019)
22. Y. Cheng, Q. Chang, A carrier tracking loop using adaptive strong tracking Kalman filter in GNSS receivers. IEEE Commun. Lett. **24**(12), 2903–2907 (2020)
23. A. Chalanga, S. Kamal, L.M. Fridman, B. Bandyopadhyay, J.A. Moreno, Implementation of super-twisting control: super-twisting and higher order sliding-mode observer-based approaches. IEEE Trans. Ind. Electron. **63**(6), 3677–3685 (2016)
24. H. Xu, G. Cui, Q. Ma, Z. Li, W. Hao, Fixed-time disturbance observer-based distributed formation control for multiple QUAVs. IEEE Trans. Circuits Syst. II Exp. Brief. **70**(6), 2181–2185 (2023)
25. Y. Song, H. Ye, F.L. Lewis, Prescribed-time control and its latest developments. IEEE Trans. Syst. Man Cybern. Syst. **53**(7), 4102–4116 (2023)

Open Access This chapter is licensed under the terms of the Creative Commons Attribution 4.0 International License (http://creativecommons.org/licenses/by/4.0/), which permits use, sharing, adaptation, distribution and reproduction in any medium or format, as long as you give appropriate credit to the original author(s) and the source, provide a link to the Creative Commons license and indicate if changes were made.

The images or other third party material in this chapter are included in the chapter's Creative Commons license, unless indicated otherwise in a credit line to the material. If material is not included in the chapter's Creative Commons license and your intended use is not permitted by statutory regulation or exceeds the permitted use, you will need to obtain permission directly from the copyright holder.

Chapter 8
Fault-Tolerant Control Technology of Photovoltaic Inverter for Reliability Improvement

Nowadays, improving the reliability of the PV inverters is getting increased interest, since it is closely related to the cost and efficiency of the entire system. In the PV inverter, power semiconductor switches are the core components, which realize the transformation of electrical form and determines the flow direction of the system energy. However, the power semiconductor switch is one of the most fragile components due to the limitations of the material and the manufacturing process. Statistics indicate that about 34% of the power converter system's failures are caused by the power device and soldering failures [1]. Therefore, it is important to develop the fault-tolerant control technology to improve system reliability.

Generally, the fault of power semiconductor switches can be divided into two cases: a short-circuit fault (SCF) and an open-circuit fault (OCF). When a SCF occurs, an abnormal overcurrent will be produced within a very short time, which leads to serious damage of the entire system. Hence, the SCF tolerant control strategies are usually based on hardware circuits that is not cost effective. For instance, the SCF is turned into OCF in [2] by adding fast fuses. The OCF is usually caused by lifting and cracking of bonding wires in the power device modules due to the thermal cycling and a gate drive fault. The open-circuit fault will lead to current distortion and secondary damage to the other components through induced noise and vibration. Compared with short-circuit fault, the OCF does not cause serious damage and its fault-tolerant control strategies can be realized without additional devices. Hence, many researchers pay more attention to the fault-tolerant control strategies when OCF occurs. So, the fault-tolerant control strategies for power semiconductor switches occurring OCF in the PV inverter is the topic in this chapter.

At present, the fault-tolerant control methods can be classified into two categories: hardware-based methods and software-based methods. Commonly, software-based fault-tolerant control strategies are proposed from the perspective of changing pulse width modulation (PWM). The essence of fault-tolerant control is to avoid using the current paths affected by the faulty power semiconductor switches, so as to ensure the correct output current. Besides, other control objectives for performance

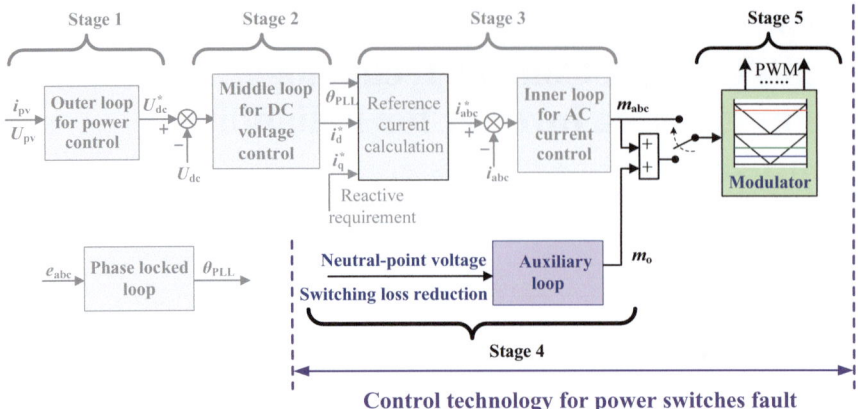

Fig. 8.1 The modules associated with fault-tolerant control and other control objectives in overall control diagram

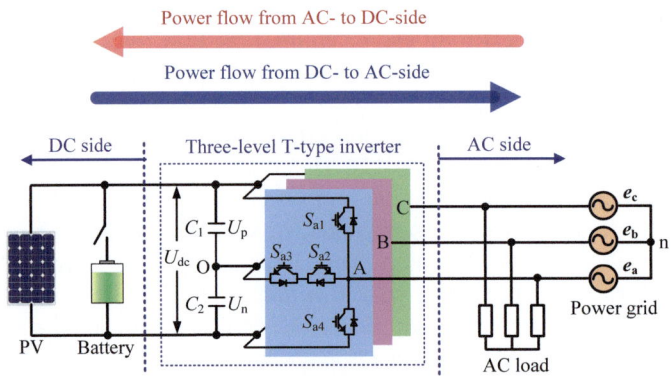

Fig. 8.2 PV generation system using 3LT^{2}I with both power flowing directions

improvement should also be satisfied for specific the PV inverter topology by adding appropriate control methods. The other control objectives including switching loss reduction, and neutral-point voltage (NPV) control for three-level T-type inverter (3LT^{2}I). The modules related to above objectives in overall control diagram are PWM modulator and auxiliary control loop as shown in Fig. 8.1.

It is well known that current paths of the PV inverters are decided by both the topology and the direction of power flowing. Thus, the fault-tolerant control technology is proposed based on the specific topology and the power flowing direction. The 3LT^{2}I is a typical topology in PV generation systems. Therefore, in this chapter, the topology of 3LT^{2}I with both power flowing directions shown in Fig. 8.2 is taken as a typical example to illustrate and analyze the fault-tolerant control strategies.

In this chapter, the author's research team propose three fault-tolerant control strategies according to different PWM methods, which are methods based on space vector pulse width modulation (SVPWM), based on continuous carrier-based pulse width modulation (CB-PWM) and based on discontinuous CB-PWM, respectively. Besides, control objectives including NPV oscillation suppression and switching loss reduction are also realized by adding appropriate control methods in the auxiliary control loop.

This chapter is organized as follows: The OCF diagnosis is given firstly since it is the primary condition to achieve the fault-tolerant control. According to specific power flowing direction, the detailed derivation, implementation process and control performance of three fault-tolerant control strategies and control methods for other objectives are elaborated, respectively.

8.1 Fault Diagnosis Methods

The fault detection and fault power semiconductor switches location are the great important issues, which is necessary to achieve the fault-tolerant control. In [2], a fault diagnosis method is proposed for the inverter system. However, it needs additional voltage sensors, which increase the system cost and complexity. In [3], an open-switch fault detection method for T-type converter is proposed. This method is realized through the grid voltage angle and the phase current, but it can only be utilized under unity power factor. In [4], for the rectifier based on neutral-point clamped (NPC) topology, an instant voltage error-based fault diagnosis method is proposed, where the boundary value is difficult to select.

In this section, the power flowing from AC side to DC side is as the example to illustrate a real-time OCF diagnosis method. Not only the fault detection but also the faulty switch identification are realized in the proposed fault diagnosis algorithm [5]. The presented fault diagnosis method is based on the instant voltage error of the line-to-line voltages and the dc-link voltage, which are already available in the control system. Hence, the proposed method can be utilized under different power factors and no more additional hardware is needed. Besides the proposed method has the advantages of fast response, and the faulty switch can be identified within two fundamental periods. The theory analysis and implementation process of the fault diagnosis method are first introduced in detail. Then the simulation waveforms are given to verify the effectiveness of the proposed fault diagnosis method. With the diagnosis results, readers can intuitively understand the effects of fault power semiconductor switches on output voltage and current, as well as the relationship between each logical factor in the presented fault diagnosis method and fault power semiconductor switches.

8.1.1 Theory Analysis and Implementation Process

To facilitate readers' understanding of the following elaborated method, the structure of $3LT^2I$ with power flowing from AC side to DC side is re-depicted in Fig. 8.3. In this structure, the current flowing from AC side to DC side is defined as positive. The operating state of each phase leg can be represented by the three switching states: [P], [O] and [N], which means that the phase is connected to the positive P, neutral O and negative N point of the dc-link, respectively.

Given the combination of operating state and current direction, there are six current paths, which can be seen in Fig. 8.4. However, if OCF occurs, the supposed pole voltage of the faulty phase will differ from its actual value. For example, when OCF occurs at S_{a1} and $i_a < 0$, the output voltage of phase A is equal to 0 instead of $U_{dc}/2$. On the other hand, if OCF occurs at S_{a2}, the output voltage will increase to $U_{dc}/2$ when $i_a > 0$. This feature can be utilized to detect the OCF and identify the faulty switch.

The instant voltage error between the estimated output voltage u_{JO_e} and measured output voltage u_{JO_m} in phase J is given by

$$\Delta u_{JO} = u_{JO\_e} - u_{JO\_m} \tag{8.1}$$

where $J = (A, B, C)$. The estimated pole voltage can be expressed as

$$u_{JO\_e} = \begin{cases} \frac{U_{dc}}{2} & (S_{x1}, S_{x2}, S_{x3}, S_{x4}) = (1, 1, 0, 0) \\ 0 & (S_{x1}, S_{x2}, S_{x3}, S_{x4}) = (0, 1, 1, 0) \\ -\frac{U_{dc}}{2} & (S_{x1}, S_{x2}, S_{x3}, S_{x4}) = (0, 0, 1, 1) \end{cases} \tag{8.2}$$

where $x = (a, b, c)$. Besides, to avoid additional voltage sensors, the measured pole voltage can be calculated as

$$u_{JO\_m} = e_x - Ri_x - L\frac{di_x}{dt} \tag{8.3}$$

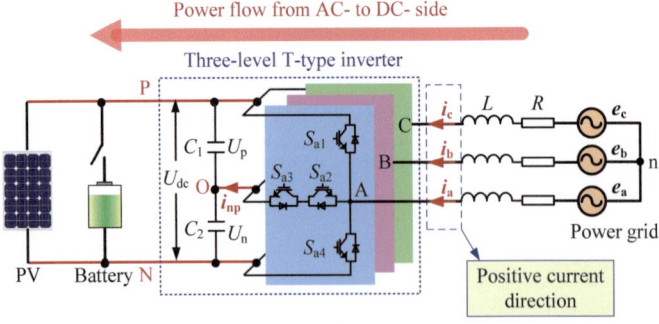

Fig. 8.3 The structure of $3LT^2I$ with the power flowing from AC side to DC side

8.1 Fault Diagnosis Methods

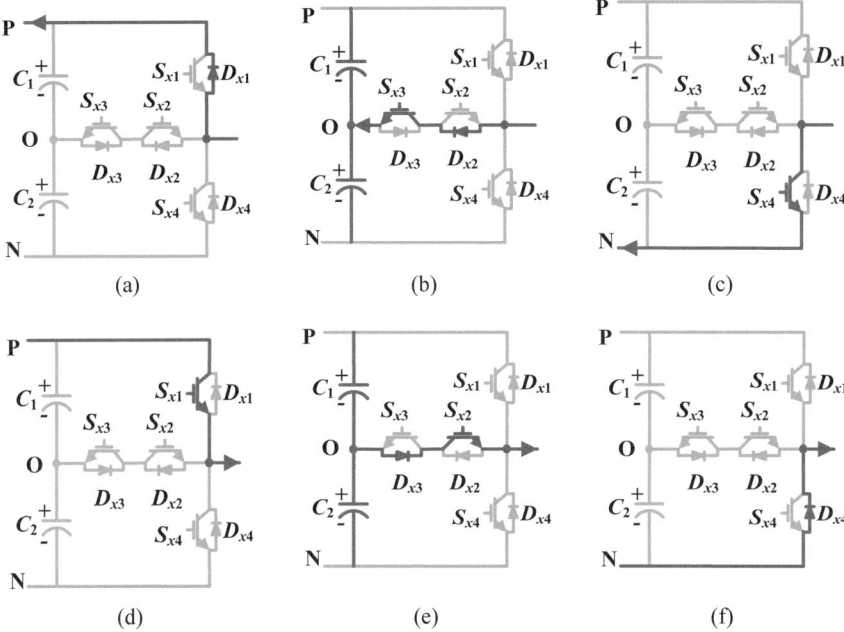

Fig. 8.4 Current paths according to the operating state and current direction **a** [P], $i_x > 0$, **b** [O], $i_x > 0$, **c** [N], $i_x > 0$, **d** [P], $i_x < 0$, **e** [O], $i_x < 0$, **f** [N], $i_x < 0$

where $I = (A, B, C)$, $I \neq J$. $y = (a, b, c)$, $y \neq x$. However, the output voltage of phase-to-neutral is influenced by the common mode voltage, therefore, the line-to-line voltage is considered. Then, the voltage error between the line-to-line voltages can be calculated as

$$\Delta u_{xy} = u_{JO\_e} - u_{IO\_e} - \left[e_{xy} - R(i_x - i_y) - L\frac{d(i_x - i_y)}{dt} \right] \quad (8.4)$$

The measured and estimated voltages should be filtered by low-pass filter to reduce the effect of high-frequency components. In order to identify the OCF, three logic factors are used and defined as

$$\xi_{xy} = \begin{cases} +1 & \Delta u_{xy} > U_{th} \\ 0 & \Delta u_{xy} = U_{th} \\ -1 & \Delta u_{xy} < U_{th} \end{cases} \quad (xy \in ab, bc, ca) \quad (8.5)$$

where U_{th} is a threshold value to improve the accuracy. The value of U_{th} is selected as two times of the largest magnitude of the voltage deviations of the measured and estimated line-to-line voltages under the corresponding normal conditions.

The sum of the three logic factors absolute value can be expressed as

$$\xi_{sum} = |\xi_{ab}| + |\xi_{bc}| + |\xi_{ca}| \tag{8.6}$$

Once $\xi_{sum} = 2$, a fault indicator (F) will be set to one, which indicates that an OCF occurs in the rectifier. Meanwhile, to identify the faulty switch, at the rising edge of F, another three diagnosis logic factors are assigned as

$$\varepsilon_{xy} = \xi_{xy} \text{ at the rising edge of fault indictor F} \tag{8.7}$$

Using the above three logic factor ε_{xy}, the faulty switch cannot be identified. In order to locate the faulty switch, another logic factor ξ_{dc} should be introduced and defined as

$$\xi_{dc} = \begin{cases} 1 & \Delta U_{dc} > U_{th\_dc} \text{ or } \Delta U_{dc} < -U_{th\_dc} \\ 0 & -U_{th\_dc} \leq \Delta U_{dc} \leq U_{th\_dc} \end{cases} \tag{8.8}$$

where $\Delta U_{dc} = U_{dc\_m} - U_{dc\_ref}$, U_{dc_m} is the measured dc-link voltage; U_{dc_ref} is the reference dc-link voltage, U_{th_dc} is threshold value of the dc-link voltage.

Once OCF occurs at the inner switches (S_{x2} or S_{x3}), the whole negative or positive phase current cannot be generated. Then, the dc-link voltage is largely changed. However, if OCF occurs at the outer switches (S_{x1} or S_{x4}), the current is affected within a short interval during a fundamental period, and therefore, the dc-link voltage has a much smaller change. Here, in order to distinguish the outer switch OCF and inner switch OCF, the threshold value U_{th_dc} is selected as 5 V, about 5% of the dc-link voltage. However, the value of ξ_{dc} varies between 1 and 0 during one fundamental period, making it difficult to detect the faulty switch. Hence, another logic factor ε_{dc} is used. If the OCF is detected by sum of the three logic factors absolute values, once ξ_{dc} equals to 1, the logic factor ε_{dc} will be assigned to 1. After a fundamental period when the OCF is detected, if ξ_{dc} is still 0, then ε_{dc} will be set to 0.

According to the above analysis, the OCF can be detected and the faulty switch can be located by using logic factors ε_{ab}, ε_{bc}, ε_{ca} and ε_{dc}, as shown in Table 8.1.

The flag indicator FLAG is assigned based on the status of logic factors. Taking S_{b1} OCF as an example, the logic factor ε_{ab}, ε_{bc}, ε_{ca} and ε_{dc} are $(-1, 1, 0, 0)$, then FLAG is defined as 5 which means the S_{b1} OCF is detected. The block diagram of the proposed diagnosis method is shown in Fig. 8.5.

8.1.2 Diagnosis Results for Different Faults of Power Semiconductor Switches

In this section, the simulation and experimental waveforms are given, so that readers can intuitively understand effects of fault power semiconductor switches on output voltage and current. Meanwhile, the corresponding relationships between each logical factor and fault power semiconductor switches are clearly demonstrated.

8.1 Fault Diagnosis Methods

Table 8.1 Diagnosis logic factor for different fault switches

Fault switches	u_{ab}	u_{bc}	u_{ca}	u_{dc}	FALG
S_{a1}	1	0	−1	0	1
S_{a2}	1	0	−1	1	2
S_{a3}	−1	0	1	1	3
S_{a4}	−1	0	1	0	4
S_{b1}	−1	1	0	0	5
S_{b2}	−1	1	0	1	6
S_{b3}	1	−1	0	1	7
S_{b4}	1	−1	0	0	8
S_{c1}	0	−1	1	0	9
S_{c2}	0	−1	1	1	10
S_{c3}	0	1	−1	1	11
S_{c4}	0	1	−1	0	12

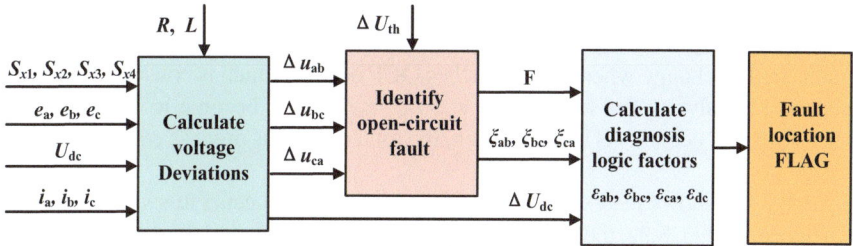

Fig. 8.5 Block diagram of the proposed diagnosis method

The parameters for simulation are as follows: the dc-link voltage is 100 V, the dc-link capacitance is 1175 μF, the switching frequency is 5 kHz, the L filter is 4 mH. To simulate the power flowing from AC side to DC side, a resistor with the value of 17 Ω is connected to the dc-link.

Figure 8.6 shows that when OCF occurs at 0.3 s, the phase current, pole voltage and line-to-line voltage are distorted. About 14 ms later, the value of the logic factors ε_{ab}, ε_{bc} and ε_{ca} are (1, 0, − 1), and the fault indicator F is set to 1 which means the OCF is detected. After a fundamental period when F equals 1, the logic factor is still 0, therefore ε_{dc} is set to 0. Then, the faulty switch indicator FLAG is set to 1, and the S_{a1} OCF is located. The total diagnosis process consumes about 34 ms.

On the other hand, Fig. 8.7 shows the diagnosis results when S_{a2} OCF occurs. About 16 ms later, the value of the logic factor ε_{ab}, ε_{bc} and ε_{ca} are (1, 0, − 1), and the fault indicator F is set to 1 which means the OCF is detected. Meanwhile, the logic factor ε_{dc} becomes to 1, therefore the faulty switch indicator FLAG is set to 2. Then, the S_{a2} OCF is located. The total diagnosis process consumes about 16 ms.

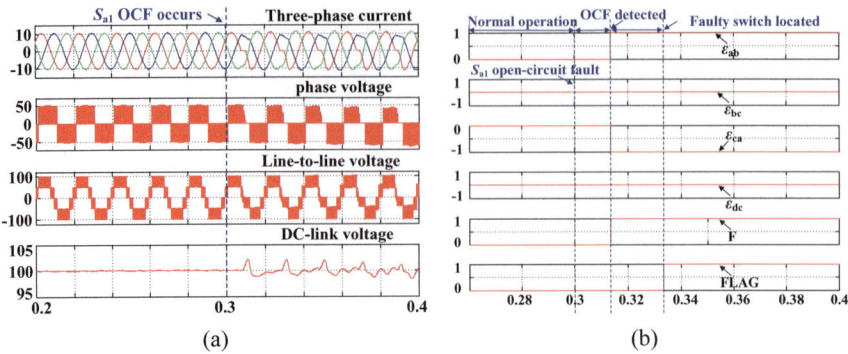

Fig. 8.6 Simulation waveforms when S_{a1} OCF and $pf = 0.8$, $m = 1$ **a** waveforms of voltage and current, **b** fault diagnosis logic factor

Compared with the S_{a2} OCF, the dc-link voltage has a much smaller ripple when S_{a1} OCF occurs.

The experimental waveforms are also given. Figure 8.8 shows the experimental results when S_{a1} OCF occurs. As shown in Fig. 8.8a, the dc-link voltage has a much smaller change when outer switches OCF occurs, which is consistent with the theoretical analysis. The logic factor ε_{ab}, ε_{bc}, ε_{ca} and ε_{dc} become to (1, 0, − 1, 0) when S_{a1} OCF occurs. As it can be seen, the total diagnosis process consumes about 32 ms.

Figure 8.9 shows the experimental results when OCF occurs at switch S_{a2}. As it can be seen in Fig. 8.9a, the dc-link voltage has a large ripple, and then the logic factor ε_{dc} becomes to 1, and thus, switch S_{a2} OCF is located. The total diagnosis process consumes about 13 ms which is similar as the simulation. As shown in Fig. 8.9b, the logic factor ε_{ab}, ε_{bc} and ε_{ca} become to (1, 0, − 1) when OCF occurs to S_{a2}.

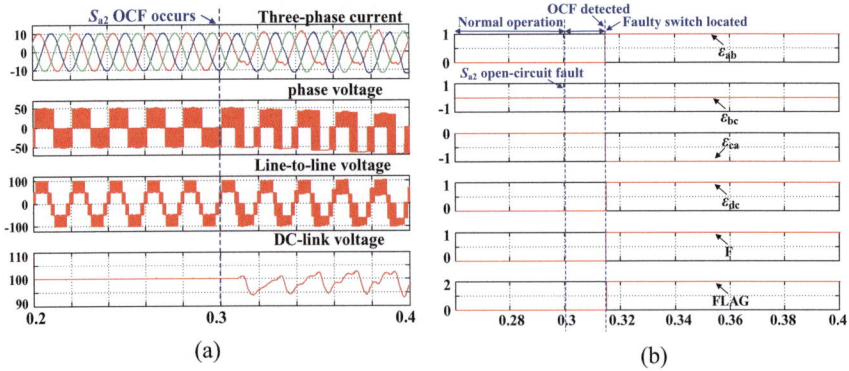

Fig. 8.7 Simulation waveforms when S_{a2} OCF and $pf = 0.8$, $m = 1$ **a** waveforms of voltage and current, **b** fault diagnosis logic factor

8.1 Fault Diagnosis Methods

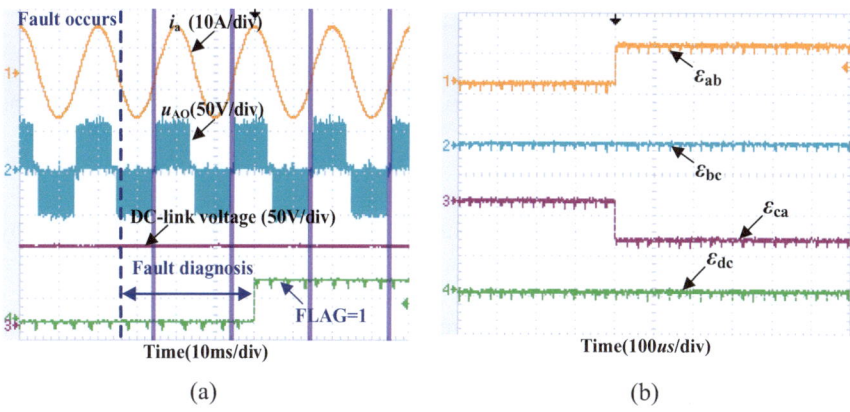

Fig. 8.8 Experimental waveforms when S_{a2} OCF. **a** Waveforms of voltage and current. **b** Fault diagnosis logic factor ($pf = 0.8, m = 1$)

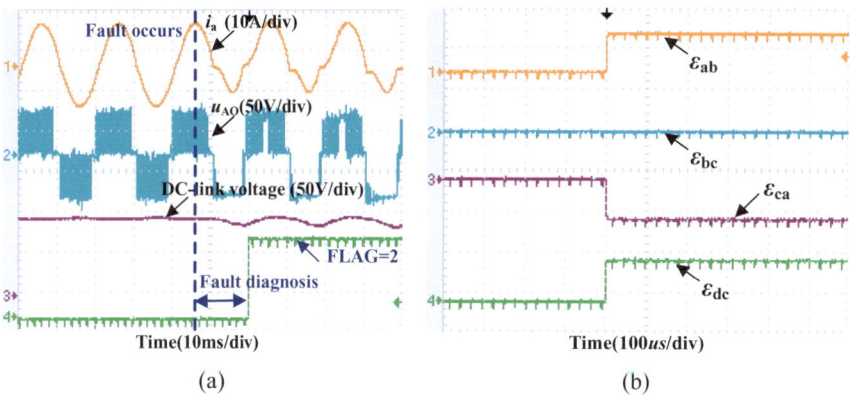

Fig. 8.9 Experimental waveforms when S_{a2} OCF. **a** Waveforms of voltage and current. **b** Fault diagnosis logic factor ($pf = 0.8, m = 1$)

To summarize, this section presents a fault diagnosis method, which is based on the instant voltage error of the line-to-line voltages and the dc-link voltage. The proposed method can be utilized under different power factors and no additional hardware is needed. Besides the proposed method has the advantages of fast response, and the faulty switch can be identified within two fundamental periods.

8.2 Fault-Tolerant Control Technology

As the above analyses reveal, the fault-tolerant control strategies for OCF are the main research area rather than those for SCF. The main crucial reasons are presented as follows. (1) Since the SCF usually leads to serious damage to the entire system, the hardware circuits are adopted to improve the response speed for the SCF, which will increase the system cost and the volume. (2) The SCF can be turned into OCF by adding fast fuses, thus the SCF problem can be solved by appropriate fault-tolerant control strategies based on OCF.

At present, the fault-tolerant control methods can be classified into two categories. One hardware-based methods are realized by adding additional components such as a redundant fourth-leg [6]. However, it is not cost-effective. To solve this issue, another category software-based method implemented by changing the modulation algorithm is proposed [7]. Thus, for the OCF occurring in $3LT^2I$, the fault-tolerant control methods based on modulation algorithm is the topic in this chapter.

Specifically, according to the characteristics of PWM methods, three types of fault-tolerant control methods are elaborated in this section. These three types of fault-tolerant control methods are the methods based on space vector pulse width modulation (SVPWM), based on carrier-based pulse width modulation (CB-PWM) and based on reduced switching losses pulse width modulation (RSL-PWM), respectively. For specific power flowing directions of the PV inverter, the detailed theory and implementation process of fault-tolerant control methods are elaborated as follows.

8.2.1 Based on Space Vector Pulse Width Modulation

In this section, the fault-tolerant control strategy based on SVPWM is presented. The $3LT^2I$ operating in the mode of power flowing from DC- to AC-side is taken as the application example to illustrate this fault-tolerant control strategy. As shown in Fig. 8.10, in this operation mode, the current flows from the DC side to AC side is defined as positive.

To easily understand the fault-tolerant control method for $3LT^2I$ from the perspective of SVPWM, the space vector diagram is divided into six sectors according to the position of six medium vectors shown in Fig. 8.11.

The OCF can be divided into two conditions: the faulty condition of half-bridge (HB) switches and neutral-point (NP) switches. In case of HB faulty condition, a fault-tolerant control strategy is proposed by redefining the three-phase turn-on times. However, in order to generate the output voltage without the impossible switching states, the modulation index should be reduced. Consequently, the magnitude of the output voltage is decreased, which is not suitable in grid-tied inverters. Besides, the NPV is difficult to adjust, since there are no substitute voltage vectors. While when OCF occurs in NP switches, the modulation index can still maintain at a high level since the output point can be connected to positive and negative dc-link.

8.2 Fault-Tolerant Control Technology

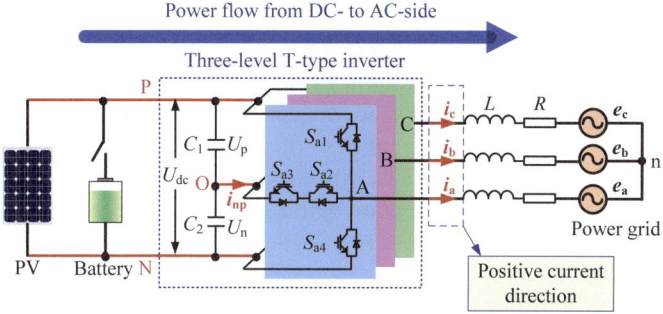

Fig. 8.10 The positive current definition 3LT^{2}I with the power flowing from DC- to AC- side

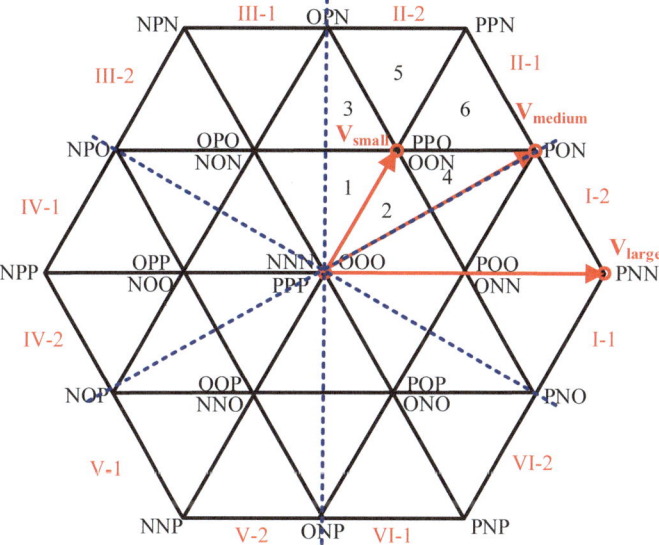

Fig. 8.11 The space vector diagram of 3LT^{2}I including six sectors according to the position of six medium vectors

Additionally, the NPV can be adjustable because the voltage vectors affecting the NPV can be selected. Thus, a fault-tolerant control strategy is illustrated for the NP faulty condition when the 3LT^{2}I operates in the mode of power flowing from DC- to AC-side.

Conventionally, two-level switching fault-tolerant control (2LS-TC) and hybrid discontinuous switching fault-tolerant control (HDS-TC) [8] have been proposed. In the 2LS-TC, the reference voltage of the faulty leg is generated without the switching state [O], which means that the phase is connected to the neutral point of the dc-link. The HDS-TC maintains three-level modulation method in all sectors as much as possible when the open-circuit fault occurs. Nevertheless, the NPV oscillations are never considered in the previously proposed fault-tolerant control strategies, which will increase the voltage stress on power devices and reduce the lifetime of the dc-link capacitors. Consequently, the volume of the converter will be increased since the oversized dc-link capacitors are typically required to suppress the NPV oscillations [9].

To suppress the NPV oscillations when an OCF occurs to NP switches, an improved two-level switching tolerant control (I2LS-TC) and a hybrid switching tolerant control (HS-TC) are proposed by the author's research team in [10], which include four steps.

Step 1: The NPV mathematic model is established when the $3LT^2I$ operates under normal conditions with the power flowing from DC- to AC-side. **Step 2**: the current paths are analyzed for [N] [P] faulty condition. **Step 3**: on the basis of these analyses, the deriving and implementation process of I2LS-TC and HS-TC are given, which restore the distorted output current and suppress the NPV oscillations. **Step 4**: The performance and comparison of the fault-tolerant controls with respect to the NPV oscillations, current total harmonic distortion (THD) and dynamic response are illustrated and analyzed for easier understanding by readers.

Step 1: Operating Principle and NPV Mathematic Modeling

As shown in the structure of $3LT^2I$ with the power flowing from DC- to AC- side, according to Kirchhoff's current law, current equations of nodes O can be derived as:

$$\begin{cases} i_{np} = -(i_{c1} + i_{c2}) = C_1 \frac{dU_p}{dt} - C_2 \frac{dU_n}{dt} \\ i_{np} = s_{ao}i_a + s_{bo}i_b + s_{co}i_c \end{cases} \quad (8.9)$$

where s_{xo} represents the switching functions and $s_{xo} = \{0, 1\}$ for $x = $ (a, b, c). Further, s_{xo} equals to "1" when the corresponding output phase is connected to NP, otherwise, they take the value of "0".

In order to simplify the analysis, let $C_1 = C_2 = C$ and ignore the initial voltage of the capacitor. NPV deviation is expressed as

$$U_p - U_n = \frac{1}{C} \int_0^t i_{np} dt = \frac{1}{C} \int_0^t (s_{ao}i_a + s_{bo}i_b + s_{co}i_c) dt \quad (8.10)$$

Further, the NPV deviation in a switching period can be expressed as

$$\Delta U_{np} = U_p - U_n = \frac{1}{C} \int_0^{T_s} \overline{i_{np}} dt = \frac{1}{C} \int_0^{T_s} (d_a i_a + d_b i_b + d_c i_c) dt \quad (8.11)$$

where d_a, d_b and d_c are the dwell time of [O] state of the corresponding phase in a switching period. The value of d_a, d_b and d_c are calculated as

8.2 Fault-Tolerant Control Technology

$$d_{x(x=a,b,c)} = \begin{cases} 2T_x & [O] \to [P] \\ T_s - 2T_x & [N] \to [O] \end{cases} \quad (8.12)$$

where T_a, T_b, T_c are the three phase turn-on times. Moreover, $[O] \to [P]$ means that the reference voltage is generated by [O] and [P] state, while $[N] \to [O]$ means that the reference voltage is generated by [N] and [O] state. From the NPV deviation expression in a switching period, the NPV fluctuation is caused by the nonzero NP current. Hence, the average value of the NP current must be zero in a switching period to eliminate the NPV oscillation.

Step 2: Current Paths Analyzing for NP Faulty Condition

The current paths of $3LT^2I$ are shown in Fig. 8.12, when NP switches occur OCF. The current paths with solid green lines and dotted red lines represent current paths under normal condition and NP switches faulty condition, respectively.

If open-circuit fault occurs at S_{a2} when the switching state is [O] with the positive phase current $i_a > 0$, the phase output will be connected to the negative dc-link instead of the O point, as shown in Fig. 8.12a. It results in a distortion of phase current. If the phase current is $i_a < 0$, during [O] state, the current path under the switch S_{a2} fault condition is the same with the current path under normal condition, as shown in Fig. 8.12b. Hence, there are no distortion in phase current and pole voltage. When open-circuit fault occurs at S_{a3} under the condition that the switching state is [O] and

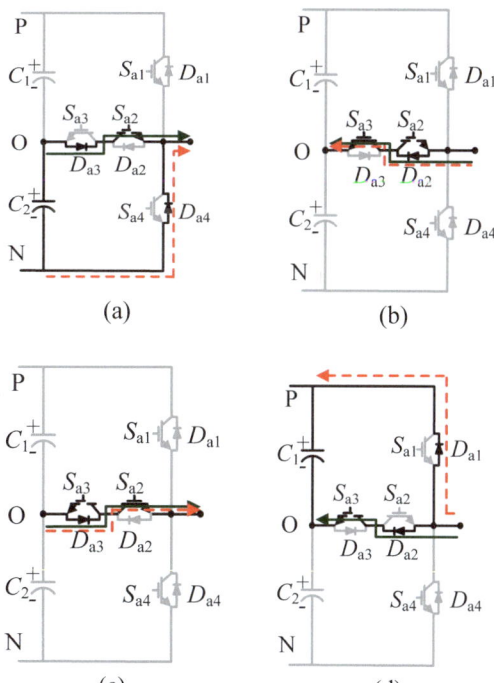

Fig. 8.12 Current paths **a** S_{a2} fault, $i_a > 0$. **b** S_{a2} fault, $i_a < 0$. **c** S_{a3} fault, $i_a > 0$. **d** S_{a3} fault, $i_a < 0$

the phase current is $i_a > 0$, there will be no distortion, since the current flows through the switch S_{a2} and the diode D_{a3}, instead of S_{a3} as shown in Fig. 8.12c. Figure 8.12d shows the current paths when phase current is $i_a < 0$. The [O] state is replaced by the [P] state. Therefore, the phase currents and voltage are distorted by the changed switching state.

Step 3: Deriving and Implementation Process of 2LS-TC and HS-TC

To restore the distorted output current and minimize the NPV oscillations when an open-circuit fault occurs to NP switches, two novel fault-tolerant control strategies named as I2LS-TC and HS-TC are proposed, respectively. In this section, the S_{a2} open-circuit fault is considered as an example for the explanation of fault-tolerant controls.

1. **Improved two-level switching tolerant control (2LS-TC)**

   If the open-circuit fault occurs in S_{a2}, as aforementioned, the switching state [O] of phase A becomes impossible. The reference voltage of the faulty leg should be generated without the impossible switching state [O]. Consequently, the value of d_a becomes zero. When $d_a = 0$, NPV deviation in a switching period is calculated as

$$\Delta U_{np} = U_p - U_n = \frac{1}{C} \int_0^{T_s} \overline{i_{np}} dt = \frac{1}{C} \int_0^{T_s} (d_b i_b + d_c i_c) dt \quad (8.13)$$

It means that the NPV is influenced by the other two phases, phase B and phase C. Thus, the suppression of NPV oscillations can be realized by modifying the dwell times of d_b and d_c, which is equivalent to adding a time-offset into the turn-on times of T_b and T_c. To assist the understanding of the time-offset calculation, the switching sequence of phases B and C in Sector I is considered. In this case, the normal switching sequence of phase B and C is shown in Fig. 8.13.

Both phases are generated by switching state [O] and [N]. According to the definition of d_x, the average value of NP current can be represented as

$$\overline{i_{np}} = 2(T_s - T_b)i_b + 2(T_s - T_c)i_c \quad (8.14)$$

In order to suppress the NPV oscillations, the average NP current should be zero. It can be achieved by adding a time-offset to the three-phase turn-on times. In this case, the average NP current in a switching period can be expressed as

$$\overline{i_{np}} = [T_s - 2(T_b + T_{off})]i_b + [T_s - 2(T_c + T_{off})]i_c \quad (8.15)$$

To make $\overline{i_{np}}$ equal to zero, T_{off} is calculated as

$$T_{off} = \frac{T_s i_b - 2T_b i_b + T_s i_c - 2T_c i_c}{2(i_b + i_c)} \quad (8.16)$$

8.2 Fault-Tolerant Control Technology

Using the similar manner, the optimal time-offset in the other sectors can be obtained as shown in Table 8.2.

It should be noted that in order to keep the principle of volt-second balance, the same adjustment should be applied to the original turn-on time of phase A. Therefore, the compensated value of the three-phase turn-on times are expressed as

$$T'_x = T_x + T_{\text{off}} \quad (x = \text{a, b, c}) \tag{8.17}$$

In order to avoid distortion, T_{off} should satisfy the following requirement.

$$-T_{\min} \leq T_{\text{off}} \leq \frac{T_s}{2} - T_{\max} \tag{8.18}$$

where $T_{\min}$ refers to the minimum turn-on time and $T_{\max}$ is the maximum turn-on time among the T_a, T_b, T_c.

As described earlier, in case of S_{a2} OCF, the reference voltage of the faulty leg should be generated using the switching states [P] and [N]. In order to keep the principle of volt-second balance, the turn-on time of the faulty leg should be revised. Combined with the expression of three-phase turn-on times, the redefined turn-on time of leg A is expressed as

$$T''_a = \begin{cases} \frac{T'_a}{2} = \frac{(T_a + T_{\text{off}})}{2} & \text{if } (V_a^* > 0) \\ \frac{T'_a + \frac{T_s}{2}}{2} = \frac{(T_a + T_{\text{off}} + \frac{T_s}{2})}{2} & \text{if}(V_a^* < 0) \end{cases} \tag{8.19}$$

The switch S_{a1} is turned ON using the redefined turn-on time. While the switch S_{a4} is operated complementary to S_{a1}. The switches S_{a2} and S_{a3} are always turned off.

The same fault-tolerant control for the S_{a2} open-circuit fault can also be applied when a fault occurs in S_{a3}. In summary, if an open-circuit fault occurs in NP switches of leg A, the redefined three-phase turn-on times are expressed as

$$\begin{cases} T''_a = \begin{cases} \frac{T'_a}{2} = \frac{(T_a + T_{\text{off}})}{2} & \text{if } (V_a^* > 0) \\ \frac{T'_a + \frac{T_s}{2}}{2} = \frac{(T_a + T_{\text{off}} + \frac{T_s}{2})}{2} & \text{if}(V_a^* < 0) \end{cases} \\ T''_b = T'_b = T_b + T_{\text{off}} \\ T''_c = T'_c = T_c + T_{\text{off}} \end{cases} \tag{8.20}$$

2. **Hybrid switching tolerant control (HS-TC)**

As discussed in the previous section, when an open-circuit fault occurs in S_{a2}, the switching state [O] is impossible if $i_a > 0$. On the contrary, if $i_a < 0$, the [O] state will not be influenced by the S_{a2} fault. It means that the inverter has both a healthy half cycle and an unhealthy half cycle. Consequently, in this section, a hybrid switching tolerant control method is proposed, in which a revised three-level switching method (R3LS) is used in the healthy half cycle and the I2LS-TC method is applied in the unhealthy half cycle. To further improve the performance, a dc deviation elimination algorithm and a dc bias are used in the HS-TC.

The R3LS algorithm is presented to minimize the NPV oscillations with time-offset injection when the inverter operates in the healthy half cycle. For the explanation of R3LS method, the reference voltage located in region 5 of Sector II is considered as an example. In this case, as shown in Fig. 8.14, phase A and phase B are both generated by switching states [O] and [P]. While phase C is made by switching states [N] and [O].

In this condition, the average NP current can be expressed as

$$\overline{i_{np}} = 2T_a i_a + 2T_b i_b + (T_s - 2T_c)i_c \quad (8.21)$$

Thus, the reduction of the NPV oscillations can be realized by adding a compensation value to the three-phase turn-on times. The average NP current can be rewritten as

$$\overline{i_{np}} = 2(T_a + T_{off})i_a + 2(T_b + T_{off})i_b + [T_s - 2(T_c + T_{off})]i_c \quad (8.22)$$

To make the NP current equal to zero, T_{off} can be obtained as

$$T_{off} = \frac{2T_a i_a + 2T_b i_b + T_s i_c - 2T_c i_c}{2(-i_a - i_b + i_c)} \quad (8.23)$$

The optimal time-offset T_{off} in other sectors when the inverter operates in healthy half cycle can be calculated using the same operating principle. Table 8.3 shows the calculation of T_{off} in the HS-TC when the open-circuit fault occurs in S_{a2}. It can be seen that in the unhealthy half cycle ($i_a > 0$), the value of T_{off} is same as I2LS-TC method. On the contrary, T_{off} is calculated using R3LS method in the healthy half cycle ($i_a < 0$).

However, for example, in case of S_{a2} fault, if the HS-TC is implemented based on Table 8.3, the NPV will be unbalanced. The simulation waveforms are shown in Fig. 8.15.

This is caused by the asymmetry of the modulation strategy and the limitation of T_{off}. When $i_a > 0$, the I2LS-TC is applied, and Fig. 8.15b shows the calculated value of T_{off} of I2LS-TC. We can see that in the red shaded area of Fig. 8.15b, in order to make the average NP current equal to zero, a T_{off} which is smaller than the lower limit is used. In this area, the T_{off} is set to equal the lower limit to avoid distortion. As a consequence, a dc deviation of NPV is introduced in this area. If the I2LS-TC is applied when $i_a < 0$, the dc deviation of NPV will be eliminated, because the T_{off} will be set to equal to the upper limit in the diametrically opposite position. However, in the HS-TC, the R3LS algorithm is used when $i_a < 0$. Due to the asymmetry of the T_{off} in a switching period, a dc deviation of NPV is

8.2 Fault-Tolerant Control Technology

introduced. Hence, it is necessary to adopt the NPV dc deviation elimination algorithm to balance the NPV.

To simplify the analysis, the NPV dc deviation elimination algorithm proposed in this paper is applied in the healthy half cycle. Once $U_p = U_n$, the dc deviation elimination algorithm will be removed immediately, then the R3LS method will be applied to suppress the NPV oscillations in the healthy half cycle.

Assuming the reference voltage is in region 5 of Sector II, if $U_p < U_n$, a positive value of average NP current is needed to eliminate the NPV dc deviation. Thus, the average NP current is obtained as

$$\overline{i_{np}} = 2(T_a + T_{off})i_a + 2(T_b + T_{off})i_b + [T_s - 2(T_c + T_{off})]i_c > 0 \quad (8.24)$$

Then, T_{off} can be calculated as

$$2(i_a + i_b - i_c)T_{off} > -(2T_a i_a + 2T_b i_b + T_s i_c - 2T_c i_c) \quad (8.25)$$

Furthermore, it is known that in a three-leg three-wire system, $i_a + i_b + i_c = 0$. Therefore, T_{off} can be rewritten as

$$-4i_c T_{off} > -(2T_a i_a + 2T_b i_b + T_s i_c - 2T_c i_c) \quad (8.26)$$

Thus, if $i_c > 0$, a negative T_{off} is needed to eliminate the NPV dc unbalance, while if $i_c < 0$, a positive T_{off} is needed. To balance the NPV as soon as possible, if a positive T_{off} is needed, the T_{off} is set to equal the upper limit, in the other conditions the T_{off} is set to the lower limit. Similar to the calculation of T_{off} in Sector II, the T_{off} in other sectors can be obtained as shown in Table 8.4.

When the dc deviation elimination algorithm is applied, the NPV oscillations are shown in Fig. 8.16.

It can be seen that the NPV oscillations can be further suppressed if the voltages across the two capacitors are interleaved which can be realized by adding a proper dc bias to the difference between U_p and U_n. However, under different operating conditions, the value of u_{off} is different in each case. In order to obtain a more general value of u_{off} for better balancing the NPV, an online estimation scheme is proposed. The scheme is realized by the following formula

$$u_{off} = \begin{cases} U_{p,min} - U_{p,max} & \text{if } U_{p,min} + U_{p,max} > U_{dc} \\ U_{p,max} - U_{p,min} & \text{if } U_{p,min} + U_{p,max} \leq U_{dc} \end{cases} \quad (8.27)$$

where $U_{p,max}$ and $U_{p,min}$ are the maximum and minimum value of the upper capacitor voltage in the last fundamental period of output, respectively. Then, the condition to choose which T_{off} should be used in the dc deviation elimination algorithm is changed as shown in Table 8.4.

Figure 8.17 shows the block diagram of the proposed HS-TC when S_{a2} open-circuit fault occurs. All the three kinds of T_{off} must satisfy the limitation of volt-second balance to avoid distortion.

Step 4: Performance and Comparison of Fault-Tolerant Controls

In this section, the performances and comparison of the four fault-tolerant control strategies (HDS-TC, 2LS-TC, I2LS-TC and HS-TC) for $3LT^2I$ with power flowing from DC- to AC-side are presented using the simulation results.

The main focus is on the following four cases. The four cases include NP voltage oscillations, output current THD, dynamic response and experimental waveforms when OCF occurs on S_{a2} and S_{a3}. The parameters for simulation and experiment are as follows. The switching frequency is 5 kHz, the control period is 200 μs, the upper

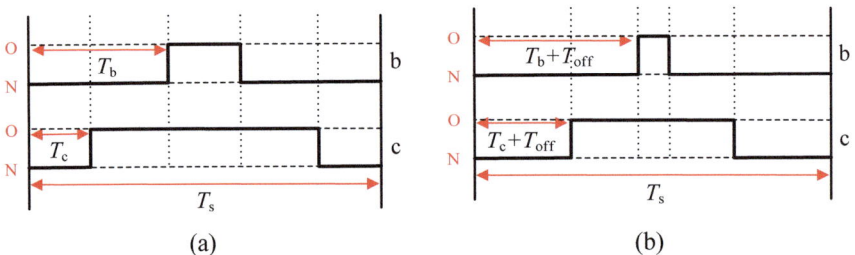

Fig. 8.13 Switching sequence of phase B and phase C **a** before T_{off} is added, **b** after T_{off} is added

Table 8.2 T_{off} of I2LS-TC when fault occurs in NP switches of leg A

Sector	T_{off}	Sector	T_{off}
I	$\frac{T_s i_b - 2T_b i_b + T_s i_c - 2T_c i_c}{2(i_b + i_c)}$	IV	$\frac{T_b i_b + T_c i_c}{i_b + i_c}$
II	$\frac{T_s i_c + 2T_b i_b - 2T_c i_c}{2(i_b + i_c)}$	V	$\frac{T_s i_b - 2T_b i_b + 2T_c i_c}{2(i_b + i_c)}$
III	$\frac{T_s i_c + 2T_b i_b - 2T_c i_c}{2(i_b + i_c)}$	VI	$\frac{T_s i_b - 2T_b i_b + 2T_c i_c}{2(i_b + i_c)}$

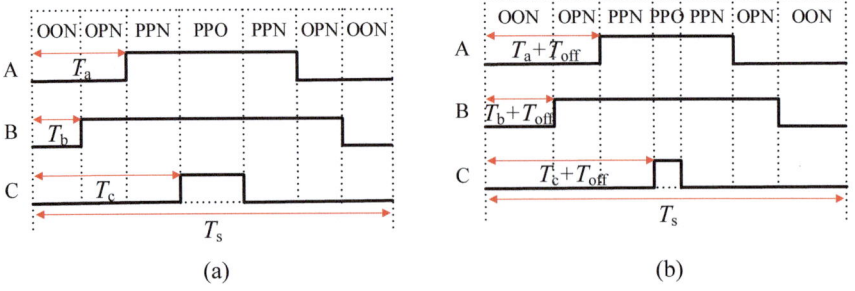

Fig. 8.14 Switching sequence in region 5 of Sector II (**a**) before T_{off} is added (**b**) after T_{off} is added

8.2 Fault-Tolerant Control Technology

Table 8.3 The calculation of T_{off} of HS-TC when S_{a2} fault occurs

Sector	$i_a > 0$ (I2LS-TC)	$i_a < 0$ (R3LS)
I	$\frac{T_s i_b - 2T_b i_b + T_s i_c - 2T_c i_c}{2(i_b + i_c)}$	$\frac{2T_a i_a + T_s i_b - 2T_b i_b + T_s i_c - 2T_c i_c}{2(-i_a + i_b + i_c)}$
II	$\frac{T_s i_c + 2T_b i_b - 2T_c i_c}{2(i_b + i_c)}$	$\frac{2T_a i_a + 2T_b i_b + T_s i_c - 2T_c i_c}{2(-i_a - i_b + i_c)}$
III	$\frac{T_s i_c + 2T_b i_b - 2T_c i_c}{2(i_b + i_c)}$	$\frac{T_s i_a - 2T_a i_a + 2T_b i_b + T_s i_c - 2T_c i_c}{2(i_a - i_b + i_c)}$
IV	$\frac{T_b i_b + T_c i_c}{i_b + i_c}$	$\frac{T_s i_a - 2T_a i_a + 2T_b i_b + 2T_c i_c}{2(i_a - i_b - i_c)}$
V	$\frac{T_s i_b - 2T_b i_b + 2T_c i_c}{2(i_b + i_c)}$	$\frac{T_s i_a - 2T_a i_a + T_s i_b - 2T_b i_b + 2T_c i_c}{2(i_a + i_b - i_c)}$
VI	$\frac{T_s i_b - 2T_b i_b + 2T_c i_c}{2(i_b + i_c)}$	$\frac{2T_a i_a + T_s i_b - 2T_b i_b + 2T_c i_c}{2(-i_a + i_b - i_c)}$

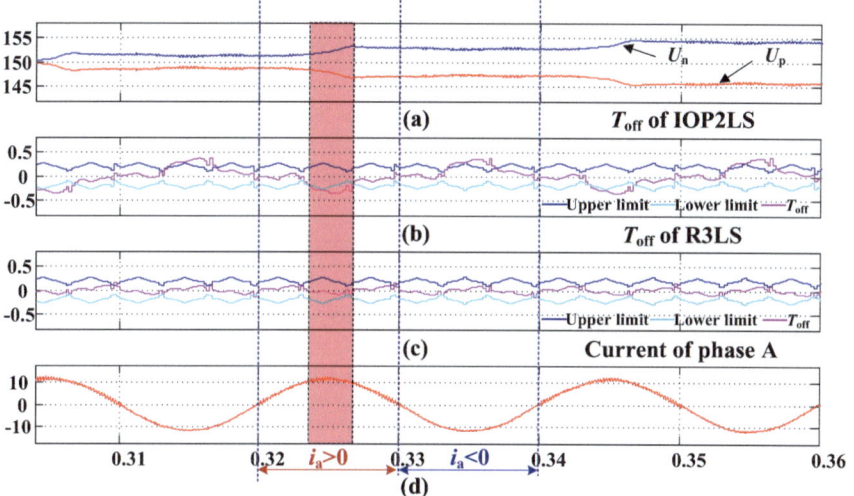

Fig. 8.15 Simulation results under unity power factor in case of S_{a2} fault **a** DC voltage when R3LS and I2LS method are applied, **b** T_{off} of I2LS method, **c** T_{off} of R3LS method, **d** current of phase A

Table 8.4 Selection of T_{off} of DC deviation elimination algorithm

	$U_p - U_n + u_{\text{off}} < 0$		$U_p - U_n + u_{\text{off}} > 0$	
I	$i_a > 0$ (+)	$i_a < 0$ (−)	$i_a > 0$ (−)	$i_a < 0$ (+)
II	$i_c > 0$ (−)	$i_c < 0$ (+)	$i_c > 0$ (+)	$i_c < 0$ (−)
III	$i_b > 0$ (+)	$i_b < 0$ (−)	$i_b > 0$ (−)	$i_b < 0$ (+)
IV	$i_a > 0$ (−)	$i_a < 0$ (+)	$i_a > 0$ (+)	$i_a < 0$ (−)
V	$i_c > 0$ (+)	$i_c < 0$ (−)	$i_c > 0$ (−)	$i_c < 0$ (+)
VI	$i_b > 0$ (−)	$i_b < 0$ (+)	$i_b > 0$ (+)	$i_b < 0$ (−)

+: T_{off} = upper limit; −: T_{off} = lower limit

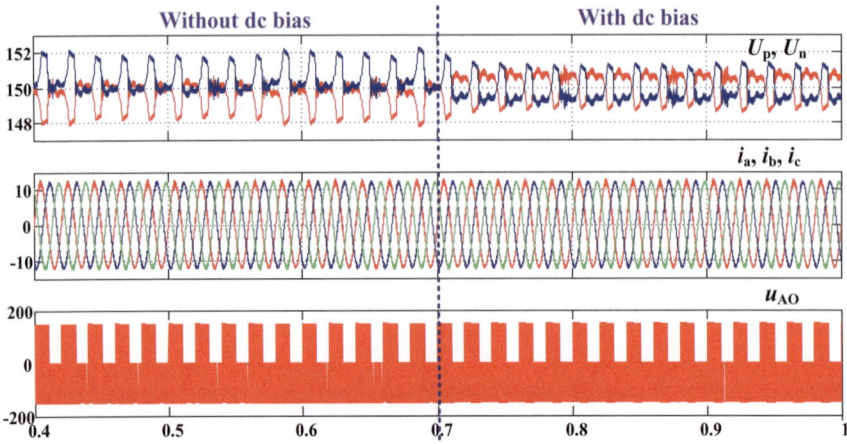

Fig. 8.16 NPV oscillations of HS-TC without and with dc bias

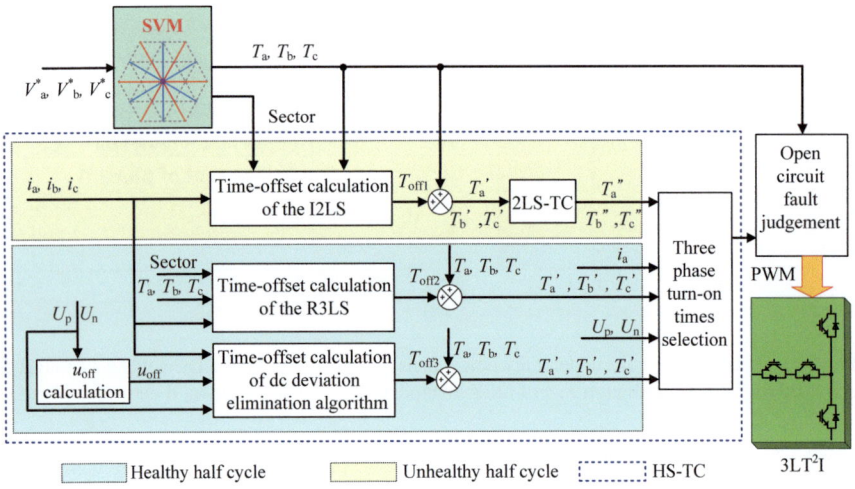

Fig. 8.17 Block diagram of the proposed HS-TC in case of S_{a2} open-circuit fault

and lower dc-link capacitors are both 940 µF, the dc-link voltage is 300 V, the L-filter value is 4 mH and the peak value of current is 12 A.

Case 1: NPV Oscillations

Figure 8.18 shows the simulation results of the phase current, phase voltage and NPV waveforms of the proposed two fault-tolerant control strategies under the condition of S_{a2} fault. As discussed earlier, when a fault occurs to S_{a2}, the output phase current is distorted due to the impossible switching state [O] when $i_a > 0$. Besides, the NPV will be unbalanced. When the proposed method is applied after 0.5 s, the distortion of the phase current is eliminated and the NPV is balanced again. I2LS-TC makes the output phase voltage of the faulty phase change to two levels as shown in Fig. 8.18a. While in the HS-TC method, the [O] state is maintained in the faulty phase when $i_a < 0$ as shown in Fig. 8.18b.

Figure 8.19 shows the simulation results of the NPV waveform of the four fault-tolerant controls. Under the condition of $m_a = 0.96$ and power factor equals to 1, the I2LS-TC and HS-TC are advantageous in reducing NPV oscillations. The magnitude

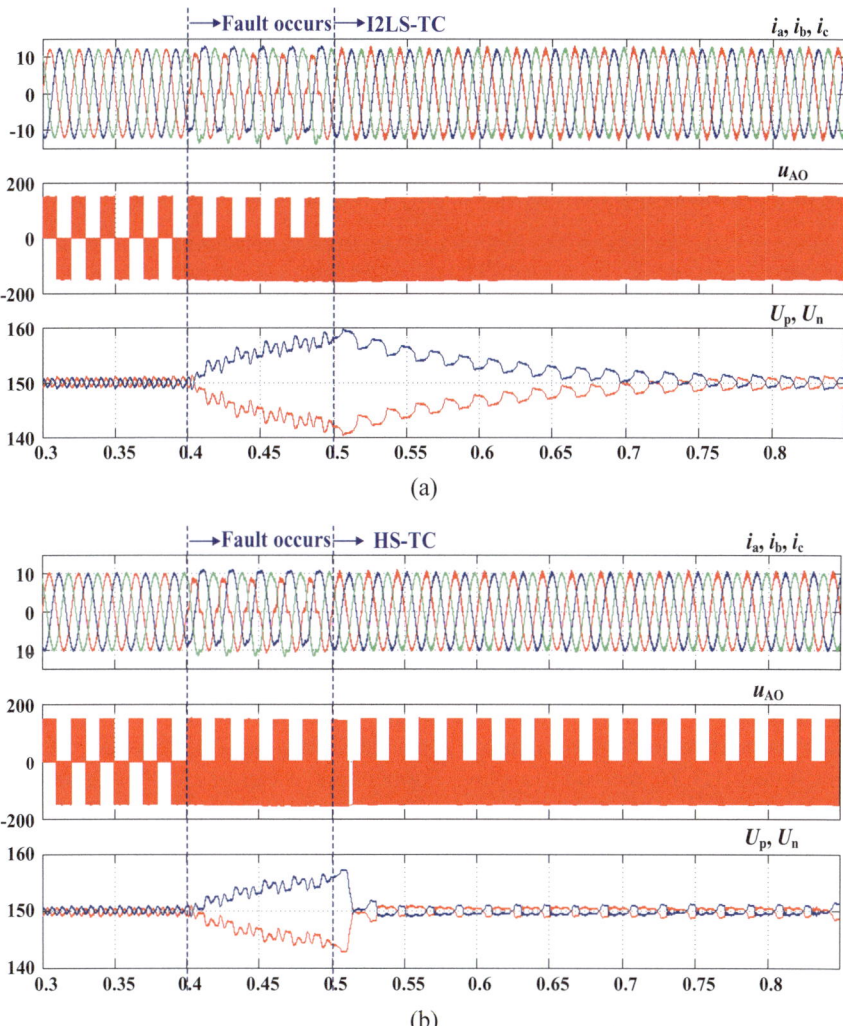

Fig. 8.18 Waveforms with the proposed fault-tolerant control strategies when S_{a2} OCF **a** I2LS-TC, **b** HS-TC

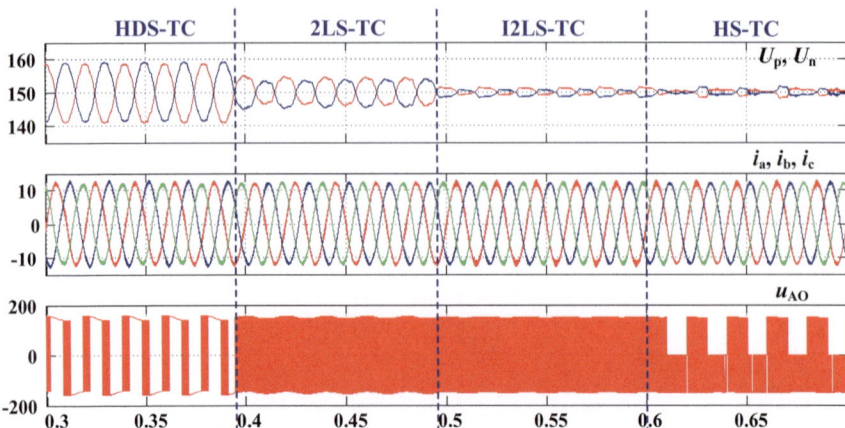

Fig. 8.19 Waveforms with four fault-tolerant control strategies when S_{a2} OCF, $pf = 1$ and $m_a = 0.96$

of ΔU_{np} in the proposed two methods is reduced to 1/6 of that in the HDS-TC and 1/3 of that in the 2LS-TC.

From the previous researches, the NPV oscillation is affected by the modulation index and the power factor. Therefore, to further help readers understand NPV oscillation, the analysis and investigation on the NPV oscillations at various operating conditions are carried out. The normalized amplitude of the oscillations is defined as $\Delta U_{NPn}/2 = \Delta U_{np} f_1 C / 2 I_{rms}$, in which $\Delta U_{np}/2$ is the peak-to-peak NPV oscillation, I_{rms} is the rms value of the AC side current, f_1 is the fundamental output frequency.

Figure 8.20 illustrates the maximum NPV oscillations that occur at different operating conditions of the four fault-tolerant control strategies. As shown in Fig. 8.20a, when the inverter operates with a higher modulation index ($m_a > 0.5$), the largest NPV oscillations will be generated in the HDS-TC compared with the rest three methods. Figure 8.20b, c show the maximum magnitude of the NPV oscillation of the 2LS-TC and I2LS-TC, respectively. Figure 8.20d is the same with Fig. 8.20c, since the maximum NPV oscillations appeared during the unhealthy half cycle in which the I2LS-TC method is applied. Furthermore, it can be clearly seen that the proposed two methods, I2LS-TC and HS-TC, have the smallest NPV oscillations.

Case 2: Current THD

In the 2LS-TC, the reference voltage of the faulty phase is generated by [P] and [N] states. In the HDS-TC, the same operating principle of the 2LS-TC is implemented in Sector II-2, III-1, V-2 and VI-1. However, the reference voltage of the faulty phase is clamped to [P] state in Sector I, II-1 and VI-2, and [N] state in sectors III-2, IV and V-1. Hence, the 2LS-TC is advantageous in current THD compared with HDS-TC.

The I2LS-TC is realized by adding an optimal time-offset value to the original three-phase turn-on times, and then the operating principle of the 2LS-TC is applied in the faulty phase. When the time-offset is added to the three-phase reference voltage,

8.2 Fault-Tolerant Control Technology

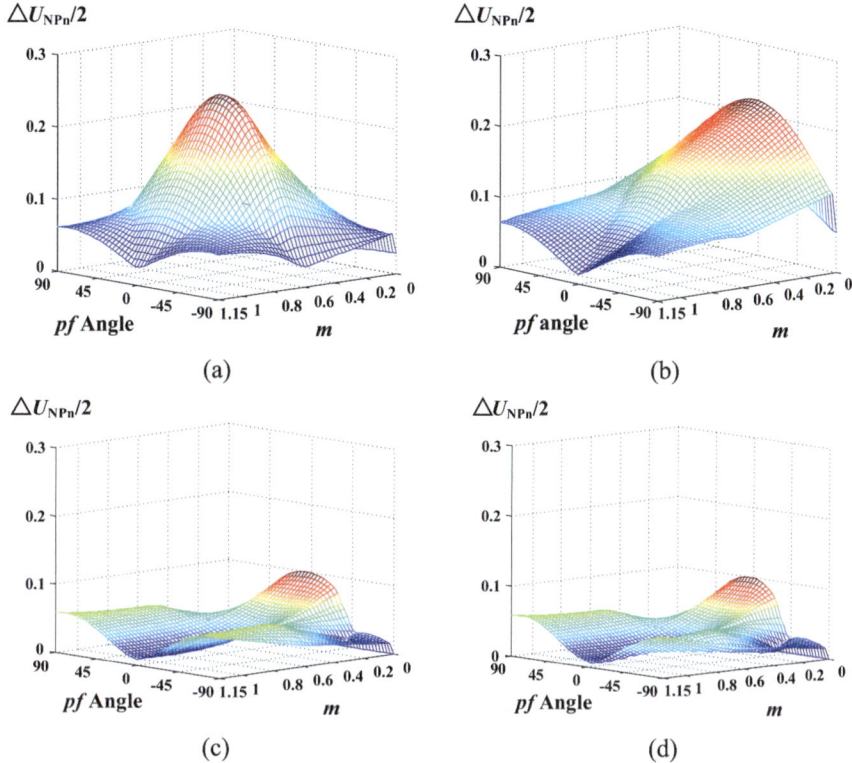

Fig. 8.20 Normalized amplitude of NPV under different *pf* and *m* with four fault-tolerant control strategies **a** HDS-TC, **b** 2LS-TC, **c** I2LS-TC, **d** HS-TC

in order to avoid over-modulation, the normal phase will be clamped to [P] or [N] state in some areas. The NPV oscillations are reduced when the I2LS-TC is applied, however, the magnitude of the NPV oscillations reduction is too small compared to the dc-link voltage. Hence, the NPV oscillation suppression does a little contribute to the reduction of current THD. Therefore, the output current THD will be slightly increased compared with the 2LS-TC.

In the I2LS-TC, the faulty phase is generated by two switching states, which is similar with the modulation of a two-level inverter. While in the HS-TC, the reference voltage of the faulty phase can be generated with three switching states in the healthy half cycle, which is similar with the modulation of a three-level inverter. Since the advantages in terms of the THD reduction of a three-level inverter are noticeable in comparison with a two-level inverter, the proposed HS-TC has a better performance than the I2LS-TC.

The above analysis is verified through Table 8.5 in which the average current THD of the four methods according to the different operating conditions is shown. It can be also seen that, the HS-TC maintains the best performance in current THD among the four methods in some circumstances.

Case 3: Dynamic Response

When OCF occurs on NP switches, the NP voltage will be unbalanced. Therefore, each method should have the ability to obviate the NP voltage dc unbalance. At the beginning of the simulation, the NP voltage deviation is set to 30 V. Then at 0.3 s, the four methods, HDS-TC, 2LS-TC, I2LS-TC and HS-TC, are implemented respectively. From Fig. 8.21, it can be seen that a relatively short dynamic process is achieved by the HS-TC method owing to the use of a dc deviation elimination algorithm. However, compared with the rest three methods, the I2LS-TC has the slowest dynamic response. Nevertheless, each method has the ability to eliminate the NP voltage dc unbalance.

Case 4: Experimental Waveforms When OCF Occurs on S_{a2} and S_{a3}

Figure 8.22 shows the experimental results when OCF occurs on S_{a2}. As shown in Fig. 8.22a, when OCF occurs in S_{a2}, the phase output will be connected to the negative instead of the NP dc bus. Hence, the positive phase current is distorted. Figure 8.22b shows that the distortion of phase current disappears and the output phase voltage of the faulty leg is generated by two switching states. The reason is that the impossible [O] state in the faulty leg is not used when I2LS-TC is adopted. When the HS-TC is implemented, the experimental waveforms are shown in Fig. 8.22c. Compared with the I2LS-TC, the phase voltage of the faulty leg has the [O] state when $i_a < 0$, since the [O] state is not influenced by the S_{a2} fault in the healthy half cycle.

The experimental results when OCF occurs on S_{a3} are depicted in Fig. 8.23. When S_{a3} OCF, the negative current is distorted since the switching state [O] is replaced with [P] as shown in Fig. 8.23a. Meanwhile, the NP voltage will be unbalanced in this condition. Figure 8.23b, c shows the experimental results of the proposed I2LS-TC and HS-TC when OCF occurs in S_{a3}. It can be seen that in the HS-TC, the pole voltage is generated by the states [O] and [P] when $i_a > 0$, which is different from that with OCF in S_{a2}. Moreover, both methods can eliminate the dc unbalance of the NP voltage, which maintains the $3LT^2I$ continuously operating.

Table 8.5 Average current THD of the four fault-tolerant control strategies under different operating conditions

Operating condition	HDS (%)	2LS (%)	I2LS (%)	HS (%)
$m_a = 0.96, pf = 1$	3.97	3.13	3.47	2.98
$m_a = 0.8, pf = 1$	4.50	4.06	4.14	3.77
$m_a = 0.96, pf = 0.8$	4.62	3.45	4.88	3.96
$m_a = 0.8, pf = 0.8$	4.98	3.80	5.13	4.21

8.2 Fault-Tolerant Control Technology

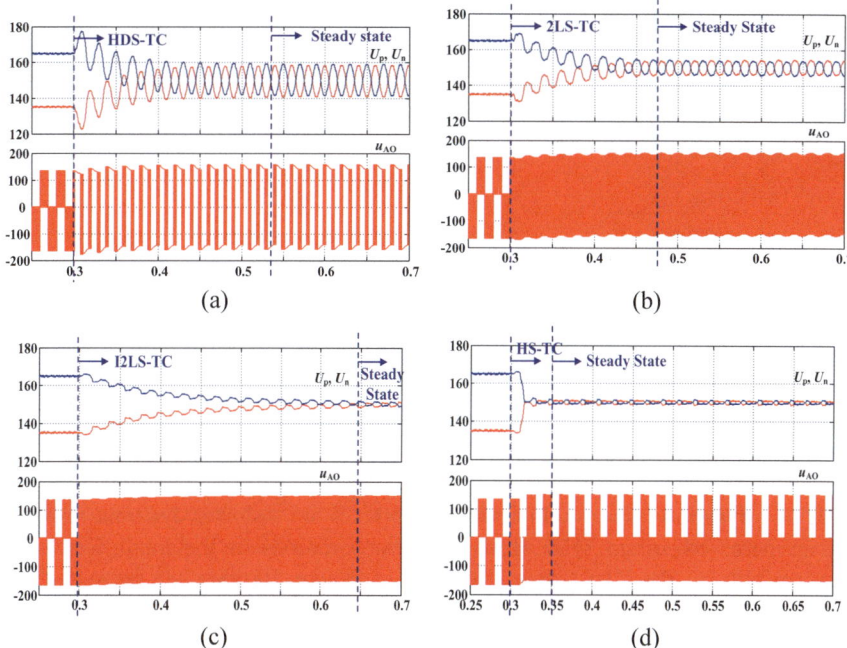

Fig. 8.21 Dynamic response for OCF on NP switches with different fault-tolerant control strategies **a** HDS-TC, **b** 2LS-TC, **c** I2LS-TC, **d** HS-TC

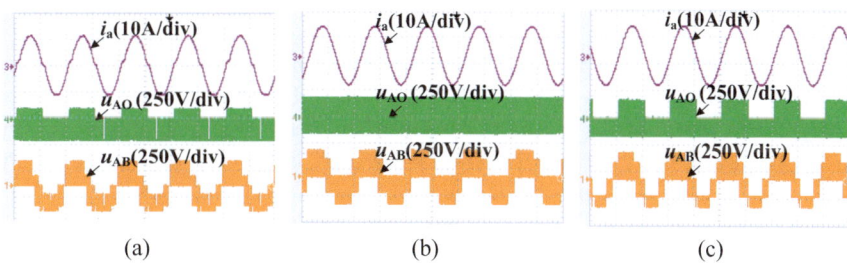

Fig. 8.22 Experimental results when OCF occurs on S_{a2} under different control conditions. **a** Without tolerant control strategies. **b** I2LS-TC. **c** HS-TC

Conclusion

All in all, to suppress the NPV oscillations when OCF occurs to NP switches, two novel two novel fault-tolerant control strategies named as I2LS-TC and HS-TC are proposed by the author team. These two methods are based on the space vector modulation. When OCF occurs to NP switches, after the proposed fault-tolerant control strategy is applied, the distortion of phase current is eliminated. Meanwhile, the NPV oscillations are considerably decreased by injecting a time-offset to the three-phase

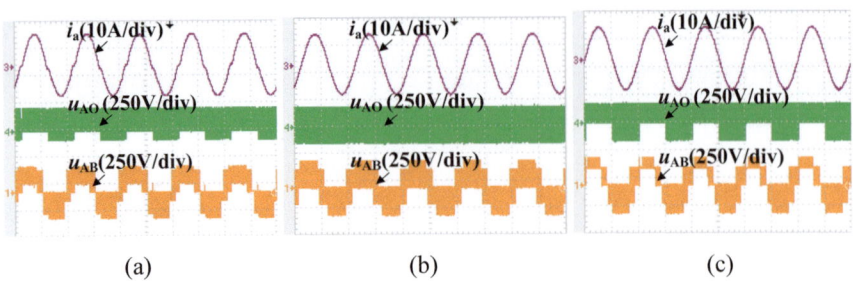

Fig. 8.23 Experimental results when OCF occurs on S_{a3} under different control conditions. **a** Without tolerant control strategies. **b** I2LS-TC. **c** HS-TC

turn-on times compared with the previously proposed fault-tolerant control strategies. The proposed two methods can suppress the NPV oscillations effectively under any operating conditions. Furthermore, the HS-TC has a good performance in current THD and dynamic response. In contrast, the I2LS has a slower dynamic response and an increased current THD. However, these variations are too small. Compared with the HS-TC, the I2LS-TC is simpler to calculate. Besides, the presented two methods do not require any additional hardware.

8.2.2 Based on Continuous Carrier-Based Pulse Width Modulation

The storage battery is usually installed in dc-link of the PV inverter to increase the absorption rate and suppress fluctuation of power in PV generation system. Especially, during the night, the energy of the power grid is stored into the storage battery to reduce the electricity cost and the power fluctuation. In this case, the storage battery is charged by the power grid, and the power flows from AC side to DC side.

Since the current paths and output states are different from that those when the power flow from DC side to AC side. Thus, fault-tolerant control strategies in the previous section cannot be directly used. Moreover, when the power flows from AC side to DC side, not only the OCF in HB switches but also the OCF in NP switches should be considered in the corresponding fault-tolerant control strategies.

When the OCF occurs to S_{a2}, the fault-tolerant controls of three-level switching-oriented fault-tolerant control (3LSO-TC) and one-phase two-level fault-tolerant control (OP2LS-TC) to realize the fault-tolerant controls under unity *pf* are proposed in [11]. However, this method is based on SVPWM, which is difficult to implement because of the requirement of complex calculations compared to the CB-PWM methods. Moreover, it is not always true due to the filter impedance and the non-unity *pf*. In [12] The reactive current is injected to eliminate the current distortion caused by the OCF of the HB switches. However, the *pf* of this method is changed which is not expected. In [13], a compensation value is added to the reference voltages to avoid

8.2 Fault-Tolerant Control Technology

the use of the switch with an OCF. For convenient comparison, this fault-tolerant control method is called CVA-TC. However, the trigonometric calculation is needed to obtain the compensation range of the compensation value in CVA-TC, which will increase the complexity of the system. In addition, in all the above fault-tolerant control for HB or NP switches OCF, the NPV oscillations are never considered.

Therefore, when 3LT2I operates in charging mode, the fault-tolerant control strategy with NPV oscillation suppression is proposed [14] by the author and elaborated in this section. This method is derived from the perspective of continuous CB-PWM, which includes four steps.

Step 1: The operating principle to adjust the NPV based on CB-PWM is illustrated. **Step 2**: The current distortion caused by the OCF is analyzed when the power of $3LT^2I$ flows from AC- to DC-side. **Step 3**: The deriving process of fault-tolerant control strategy based on CB-PWM. This method is demonstrated by dividing fault into two conditions: the faulty condition of HB switches and NP switches. Under the condition of HB switches OCF, a novel carrier-based outer switch fault-tolerant control (CB-OSTC) is presented. Under the condition of NP switch OCF, a carrier-based hybrid switching fault-tolerant control (CB-HSTC) is proposed. **Step 4**: The performance and comparison of the fault-tolerant controls are illustrated and analyzed for easier understanding by readers. The detailed analyses and deriving process in each step are illustrated as follows.

Step 1: Operating Principle to Adjust the NPV Based on Continuous CB-PWM

In order to facilitate the reader's understanding, this section first re-elaborates the CB-PWM modulation principle of $3LT^2I$. The three phase sinusoidal reference voltages are used and normalized as

$$\begin{cases} u_{a,ref} = m\cos(2\pi f_1 t) \\ u_{b,ref} = m\cos\left(2\pi f_1 t - \frac{2\pi}{3}\right) \\ u_{c,ref} = m\cos\left(2\pi f_1 t + \frac{2\pi}{3}\right) \end{cases} \quad (8.28)$$

where m is the modulation index, f_1 is the fundamental frequency of the AC source.

In order to match the range of the modulation index in space vector PWM, a third harmonic characterized zero-sequence voltage (u_o) injection is added to the three-phase reference voltage. The zero-sequence voltage u_o and redefined reference voltage $u_{x,ref,off}$ are expressed as

$$u_o = -\frac{u_{ref,max} + u_{ref,min}}{2} \quad (8.29)$$

$$\begin{cases} u_{a,ref,off} = u_{a,ref} + u_o \\ u_{b,ref,off} = u_{b,ref} + u_o \\ u_{c,ref,off} = u_{c,ref} + u_o \end{cases} \quad (8.30)$$

Fig. 8.24 The waveforms of zero-sequence voltage and redefined three phase reference voltage

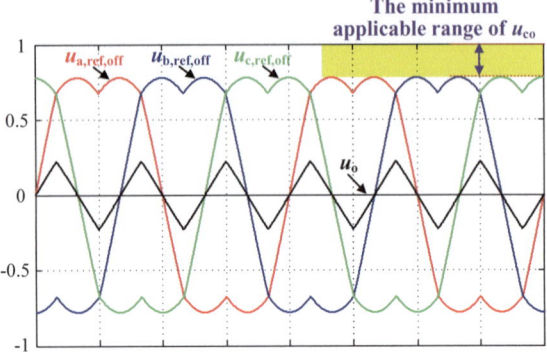

where the $u_{ref,max}$ and $u_{ref,min}$ are the maximum and minimum values of the three phase sinusoidal reference voltage ($u_{a,ref}$, $u_{b,ref}$, $u_{c,ref}$), respectively; $u_{x,ref,off}$ is the redefined three phase reference voltage and $x = (a, b, c)$.

The waveforms of zero-sequence voltage and redefined reference voltage are shown in Fig. 8.24.

Figure 8.25 shows the traditional phase disposition carrier-based approach, conventionally known to produce a lower harmonic content. In this approach, two vertically disposed in-phase carriers are used. There are three switching states ([P], [O], and [N]) in the 3LT²I. In a switching period, if the reference voltage is bigger than the carrier 1, the phase will be connected to the positive point of the dc-link ("P"), while if reference voltage is smaller than the carrier 2, the phase will be connected to the negative point of the dc-link ("N"). On the other hand, if the reference voltage is smaller than the carrier 1, but bigger than the carrier 2, the phase will be connected to the neutral point of the dc-link ("O").

It can be seen that the three output states are obtained from the comparison between carriers and three-phase modulation waveforms. Moreover, the three-phase modulation waveforms are continuous. Thus, this CB-PWM method is called continuous CB-PWM method. The following fault-tolerant method is derived on the basis of continuous CB-PWM.

Fig. 8.25 The principle of traditional phase disposition CB-PWM of three-level inverter

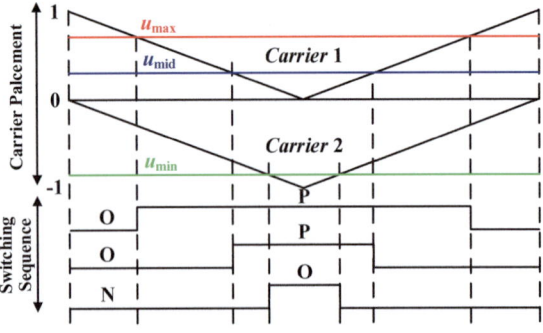

8.2 Fault-Tolerant Control Technology

One phase current contributes to the neutral current only when this phase is connected to the NP. Therefore, according to the polarity of the redefined reference voltage, the NP current in a switching period (T_s) can be expressed as

$$i_{np} = (1 - u_{max})i_{max} + [1 - \text{sgn}(u_{mid})u_{mid}]i_{mid} + (1 + u_{min})i_{min} \quad (8.31)$$

where u_{max}, u_{mid}, u_{min} are defined as the maximum, medium, minimum value of the three-phase redefined reference voltage, and i_{max}, i_{mid}, i_{min} as the current of the corresponding phase.

The NPV variation can be calculated as

$$\Delta U_{np} = U_p - U_n = -\frac{1}{C} \int_0^{T_s} i_{np} dt \quad (8.32)$$

It can be seen that if i_{np} is positive, the upper capacitor voltage U_p is decreased and the lower capacitor voltage U_n is increased. Otherwise, when i_{np} is negative, U_p is increased and U_n is decreased. Hence, the NPV can be adjusted by modifying the NP current.

Without current distortion, the NP current can be modified by adding a compensation value u_{co} to the three-phase reference voltage. When the maximum reference voltage u_{max} is larger than 0.5, the modified NP current can be rewritten as

$$\begin{aligned} i_{np} = &[1 - (u_{max} + u_{co})]i_{max} + [1 + (u_{min} + u_{co})]i_{min} \\ &+ [1 - \text{sgn}(u_{mid} + u_{co})(u_{mid} + u_{co})]i_{mid} \end{aligned} \quad (8.33)$$

In order to avoid over modulation under this condition, the u_{co} should satisfy the following requirement.

$$-1 - u_{min} \leq u_{co} \leq 1 - u_{max} \quad (8.34)$$

On the other hand, if the maximum reference voltage u_{max} is smaller than 0.5, the modified NP current is represented as

$$\begin{cases} i_{\mathrm{np}} = [1 - (u_{\max} + u_{\mathrm{co}})]i_{\max} \\ \qquad + \left[1 - \mathrm{sgn}(u_{\mathrm{mid}} + u_{\mathrm{co}})(u_{\mathrm{mid}} + u_{\mathrm{co}})\right]i_{\mathrm{mid}} \\ \qquad + [1 + (u_{\min} + u_{\mathrm{co}})]i_{\min} \quad \text{if } -u_{\max} \leq u_{\mathrm{co}} \leq -u_{\min} \\[4pt] i_{\mathrm{np}} = [1 - (u_{\max} + u_{\mathrm{co}})]i_{\max} \\ \qquad + [1 - (u_{\mathrm{mid}} + u_{\mathrm{co}})]i_{\mathrm{mid}} \\ \qquad + [1 - (u_{\min} + u_{\mathrm{co}})]i_{\min} \quad \text{if } u_{\mathrm{co}} > -u_{\min} \\[4pt] i_{\mathrm{np}} = [1 + (u_{\max} + u_{\mathrm{co}})]i_{\max} \\ \qquad + [1 + (u_{\mathrm{mid}} + u_{\mathrm{co}})]i_{\mathrm{mid}} \\ \qquad + [1 + (u_{\min} + u_{\mathrm{co}})]i_{\min} \quad \text{if } u_{\mathrm{co}} < -u_{\max} \end{cases} \qquad (8.35)$$

It should be noted that the three-phase reference voltage can be all non-negative or non-positive when u_{co} is added if $u_{\max} \leq 0.5$. However, in these two cases, the NPV will not be influenced after an additional compensation value is added if $u_{\min} + u_{\mathrm{co}} > 0$ or reduced if $u_{\max} + u_{\mathrm{co}} < 0$. Hence, from the standpoint of adjusting the NPV, the modified NP current and the limitation of u_{co} are expressed as

$$\begin{aligned} i_{\mathrm{np}} &= [1 - (u_{\max} + u_{\mathrm{co}})]i_{\max} \\ &\quad + \left[1 - \mathrm{sgn}(u_{\mathrm{mid}} + u_{\mathrm{co}})(u_{\mathrm{mid}} + u_{\mathrm{co}})\right]i_{\mathrm{mid}} \\ &\quad + [1 + (u_{\min} + u_{\mathrm{co}})]i_{\min} \end{aligned} \qquad (8.36)$$

$$\begin{aligned} &-1 - u_{\min} \leq u_{\mathrm{co}} \leq 1 - u_{\max} \quad \text{if } u_{\max} > 0.5 \\ &-u_{\max} \leq u_{\mathrm{co}} \leq -u_{\min} \quad \text{if } u_{\max} \leq 0.5 \end{aligned} \qquad (8.37)$$

Step 2: Current Paths Analyzing When Power Flow from AD- to DC-Side

Depending on the polarity of the reference voltage ($u_{x,\mathrm{ref,off}}$) and input current i_x, one fundamental period of the input current is divided into four time intervals which are represented as time interval A, B, C and D, which are shown in Fig. 8.26. The range of time interval B and D are caused by filter impedance and the power factor of the inverter.

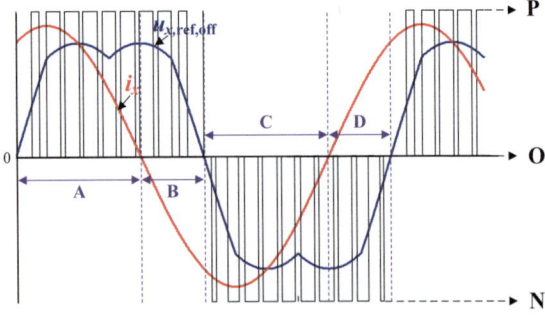

Fig. 8.26 Four time intervals depending on the polarity of reference voltage and input current in one fundamental period

8.2 Fault-Tolerant Control Technology

In each time interval, the current paths are different under normal condition. According to the current paths illustrated in Sect. 8.1, the relationships between and current paths and the polarity of reference voltage and input current are summarized in Table 8.6.

When an OCF occurs to the switch, the normal current path related to the faulty switch is influenced, which leads to the input and output current distortion. The reasons of current distortion are analyzed in detail, which includes the conditions of fault occurring in HB and fault occurring in NP switches.

Condition 1: Fault Occurring in HB Switches

When open-circuit fault occurs in S_{x1}, and $i_x > 0$ the current path of (d) becomes infeasible, which belongs to time interval B. Thus, the S_{x1} open-circuit fault causes input current distortion in time interval B. On the other hand, the input current i_x is distorted in interval D when open-circuit fault occurs in S_{x4}, and $i_x > 0$, since the current path (c), which is related to the S_{x4} open-circuit fault belongs to time interval D. In addition, we know that if the current paths (c) and (d) are used, the dc-link capacitor will be discharged, and thus, once an open-circuit fault occurs to outer switches, the output dc-link voltage will be increased in the area where the current paths (c) or (d) are used. Hence, the output current will also be distorted.

Condition 2: Fault Occurring in NP Switches

If the current path is (e), when $i_x < 0$, during the switching state [O], the input current will flow through the switch S_{x2}. If OCF occurs in S_{x2}, the input current will be distorted in time intervals B and C. Using the similar manner, we know that the current in time intervals A and D is distorted when OCF occurs in S_{x3}. Since the input current is distorted in some time intervals, the dc-link voltage will be affected.

In summary, the affected time intervals with the corresponding OCF switches are shown in Fig. 8.27.

Step 3: The Deriving Process of Fault-Tolerant Control Strategy Based on CB-PWM

In this section, the proposed CB-PWM fault-tolerant control strategy aims to restore the distorted input current and suppress the NPV oscillations when OCF occurs in HB switches and NP switches. In the condition of HB switches fault, the CB-OSTC is proposed. While the CB-HSTC is presented if NP switches fault. For the explanation, the analyses are implemented only considering the OCF occurring in phase leg A.

Table 8.6 The relationships between and current paths and the polarity of reference voltage and input current

Time interval	$u_{x,\text{ref,off}}$	I_x	Current path
A	Positive	Positive	(a), (b)
B	Positive	Negative	(d), (e)
C	Negative	Negative	(e), (f)
D	Negative	Positive	(b), (c)

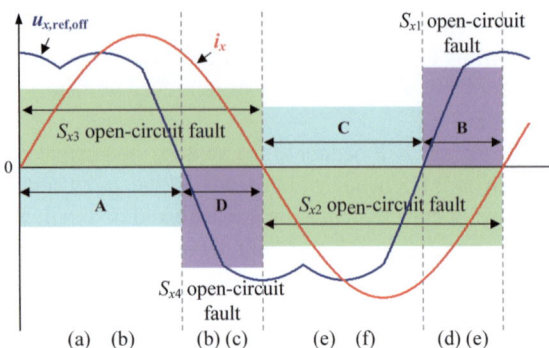

Fig. 8.27 The affected range of each open-circuit fault according to the time interval

Condition 1: CB-OSTC for Fault Occurs in HB Switches

The fault occurring in S_{a1} is taken as an example, and the input current is distorted in time interval B, since the current path becomes impossible. In order to restore the distorted current, the current path should be prohibited, which means that the reference voltage of the faulty phase should be generated without the switching state [P] in time interval B. It can be simply realized by adding an offset value to the reference voltage in time interval B. Based on the analysis, the requirement of the offset value in time interval B of the faulty phase A to maintain the normal operation with sinusoidal current can be expressed as

$$u_{\text{tole,off}} \leq -u_{\text{a,ref,off}} \quad \text{if } u_{\text{a,ref,off}} > 0 \text{ and } i_a < 0 \tag{8.38}$$

The $u_{\text{tole,off}}$ is a one-side voltage, and thus, it leads to the NPV unbalance. In order to keep the NP potential stable and suppress the NPV oscillations, a hybrid NPV control method is proposed. The proposed hybrid NPV control method is easily implemented by adding a compensation value ($u_{\text{o,com}}$) to the three phase reference voltage and the compensation value comprises two parts: a proportional NPV dc deviation ($k\Delta U_{\text{np}}$) and a dynamic adjustment factor (u_d) which can be expressed as

$$u_{\text{o,com}} = k\Delta U_{\text{np}} + u_d \tag{8.39}$$

The proportional NPV dc deviation is mainly used to eliminate the NPV dc deviation. While the dynamic adjustment factor is used to suppress the NPV oscillations. The choice of the proportional allocation factor k and the calculation of u_d are deduced as follows.

(a) **The choice of k**

When the reference voltage (u_{mid}) is positive, the NP current under normal condition can be expressed as

$$i_{\text{np1}} = (1 - u_{\max})i_{\max} + (1 - u_{\text{mid}})i_{\text{mid}} + (1 - u_{\min})i_{\min} \tag{8.40}$$

8.2 Fault-Tolerant Control Technology

A compensation value should be added to the reference voltage to eliminate the NPV dc deviation. Without loss of generality, we assume that the polarity of u_{mid} is not changed after the compensation value is added. Substituting the NPV offset with the proportional dc voltage deviation ($k\Delta U_{np}$) as a compensation component into NP current expression, the modified NP current can be revised as

$$i_{np2} = [1 - (u_{max} + k\Delta U_{dc})]i_{max}$$
$$+ [1 - (u_{mid} + k\Delta U_{dc})]i_{mid} \qquad (8.41)$$
$$+ [1 + (u_{min} + k\Delta U_{dc})]i_{min}$$

Thus, the NP current difference can be obtained as

$$i_{np2} - i_{np1} = -(i_{max} + i_{mid} - i_{min})k\Delta U_{dc} \qquad (8.42)$$

where i_{np2} is the modified NP current when the compensation value is added to the reference voltage, while i_{np1} is the original NP current under the condition of $u_{mid} > 0$.

Considering that the sum of the three-phase current is zero in a three-leg three-wire system, NP current difference can be re-written as

$$i_{np2} - i_{np1} = 2i_{min}k\Delta U_{np} \qquad (8.43)$$

Using the similar operating principle, when $u_{mid} \leq 0$, the NP current difference can be derived as

$$i_{np4} - i_{np3} = -2i_{max}k\Delta U_{np} \qquad (8.44)$$

where i_{np4} is the modified NP current when the compensation value is added to the reference voltage, while i_{np3} is the original NP current when $u_{mid} \leq 0$.

In order to eliminate the NPV dc deviation, for example if $U_p < U_n$, the NP current i_{np2} (or i_{np4}) should be larger than the NP current i_{np1} (or i_{np3}). In contrast, the NP current i_{np2} (or i_{np4}) should be smaller than i_{np1} (or i_{np3}) when the NPV $U_p > U_n$. Then, combined with the polarity of the current (i_{max} and i_{min}) and the medium reference voltage (u_{mid}), to eliminate the NPV dc deviation, the sign of k is summarized in Table 8.7.

In this table, four modes are included according to operating conditions. It is always true that $i_{max} > 0$ and $i_{min} < 0$ if the 3LT^{2}I operates with a higher pf which is larger than 0.866. However, if the pf becomes smaller than 0.866, the other modes will arise. Consequently, the sign of k should be set according to operating conditions. It is noted that, in mode 2 and 4, depending on the polarity of u_{mid}, the sign of k can be positive or negative. Hence, to keep the ability of mitigating the NPV dc deviation, the polarity of the medium reference voltage cannot be changed when the compensation value $k\Delta U_{dc}$ is added in case 2 and 4, which can be expressed as

$$\text{sgn}(u_{\text{mid}} + k\Delta U_{\text{np}}) = \text{sgn}(u_{\text{mid}}) \tag{8.45}$$

Selecting the absolute value of k should consider the relationship between the $k\Delta U_{\text{np}}$ and $u_{\text{o,com}}$. When $|\Delta U_{\text{np}}|$ is large, the prime target is to eliminate the NPV dc deviation, which means that the $k\Delta U_{\text{np}}$ should occupy a large proportion of $u_{\text{o,com}}$. On the other hand, if ΔU_{np} is small, the main target is to suppress the NPV oscillations, which means that the $k\Delta U_{\text{np}}$ should occupy a small proportion of $u_{\text{o,com}}$. Without loss of generality, assuming that the $k\Delta U_{\text{np}}$ plays a more important role when $|\Delta U_{\text{np}}| > 2$. On the contrary, if $|\Delta U_{\text{np}}| \leq 2$, the $k\Delta U_{\text{np}}$ plays a less important role, which means that the $k\Delta U_{\text{np}}$ occupies a small part of $u_{\text{o,com}}$ (at most a half of the $u_{\text{o,com}}$). Based on the above analyses, the boundary condition can be simplified as

$$|2k| = \frac{1}{2}\text{Range}(u_{\text{o,com}}) \tag{8.46}$$

where Range $(u_{\text{o,com}})$ is the applicable range of $u_{\text{o,com}}$. According to the relationship of three phase reference voltage and zero-sequence voltage, the minimum applicable range of the $u_{\text{o,com}}$ can be expressed as

$$\text{Range}(u_{\text{o,com}})_{\min} = 1 - \frac{\sqrt{3}}{2}m \tag{8.47}$$

The absolute value of k can be calculated as

$$|k| = \frac{1}{4}\left(1 - \frac{\sqrt{3}}{2}m\right) \tag{8.48}$$

(b) **The calculation of u_{d}**

In order to suppress the NPV oscillations, i_{np} should be zero in a switching period which can be realized by adding a dynamic adjustment factor (u_{d}). Substituting the u_{d} as a compensation value (u_{co}) into NP current expression, the NP current can be expressed as

$$\begin{aligned} i_{\text{np}} = &-(u_{\max} + u_{\text{d}})i_{\max} \\ &- \text{sgn}(u_{\text{mid}} + u_{\text{d}})(u_{\text{mid}} + u_{\text{d}})i_{\text{mid}} \\ &+ (u_{\min} + u_{\text{d}})i_{\min} \end{aligned} \tag{8.49}$$

Considering that the polarity of $u_{\text{mid}} + u_{\text{d}}$ can be positive or negative, to make the NP current equal to zero, two types of u_{d} (u_{dl} and u_{ds}) can be obtained.

$$u_{\text{dl}} = \frac{-u_{\max}i_{\max} - u_{\text{mid}}i_{\text{mid}} + u_{\min}i_{\min}}{i_{\max} + i_{\text{mid}} - i_{\min}} \tag{8.50}$$

8.2 Fault-Tolerant Control Technology

$$u_{ds} = \frac{-u_{max}i_{max} + u_{mid}i_{mid} + u_{min}i_{min}}{i_{max} - i_{mid} - i_{min}} \quad (8.51)$$

To attenuate the NPV oscillations, the selection of the proper dynamic adjustment factor is explained in Fig. 8.28.

First of all, if the u_{dl} satisfies the following limitation

$$\begin{cases} -u_{mid} < u_{dl} < 1 - u_{max} & \text{if } u_{max} > 0.5 \\ -u_{mid} < u_{dl} < -u_{min} & \text{if } u_{max} \leq 0.5 \end{cases} \quad (8.52)$$

The u_{dl} will be chosen as the optimum value of u_d. If the u_{dl} does not accord with the limitation, while the u_{ds} satisfies the following requirement

$$\begin{cases} -1 - u_{min} < u_{ds} < -u_{mid} & \text{if } u_{max} > 0.5 \\ -u_{max} < u_{ds} < -u_{mid} & \text{if } u_{max} \leq 0.5 \end{cases} \quad (8.53)$$

The u_{ds} will be selected as u_d. However, these two cases may not satisfy the restriction simultaneously. Under this condition, the optimum value of u_d should be one of the boundary values. If $u_{max} > 0.5$, the optimum u_d is one of the three boundaries $-1 - u_{min}$, $-u_{mid}$ (if $-1 - u_{min} < -u_{mid} < 1 - u_{max}$) and $1 - u_{max}$. While if $u_{max} \leq 0.5$, the optimum u_d should be one of three values $-u_{max}$, $-u_{mid}$ and $-u_{min}$. When $u_{max} > 0.5$, for example, the NP current of each boundary value is calculated firstly, and the one that has the minimum absolute value can be considered as the optimum solution. Then, the corresponding boundary value is used as the optimal value of u_d.

In order to avoid over modulation, the calculated $u_{o,com}$ should satisfy the following condition

$$-1 - u_{min} \leq u_{o,com} = k\Delta U_{np} + u_d \leq 1 - u_{max} \quad (8.54)$$

In addition, to realize the fault-tolerant control for S_{a1} open-circuit fault, the $u_{o,com}$ should also be restricted to avoid infeasible current path using. Besides, in time interval A, the addition of $u_{o,com}$ should not reverse the polarity of the $u_{a,ref,off}$ since the current path cannot be affected. Thus, when S_{a1} OCF, the limitation of $u_{o,com}$ can be concluded as

$$\begin{cases} -1 - u_{min} \leq u_{o,com} \leq -u_{a,ref,off} & \text{if } i_a \leq 0 \\ -1 - u_{min} \leq u_{o,com} \leq 1 - u_{max} & \text{if } i_a > 0 \end{cases} \quad (8.55)$$

When the hybrid NPV control method and the limitation are used in the CB-OSTC, the waveform of the NPV is shown in Fig. 8.29.

The NPV has a large dc deviation in the abnormal time interval B, since the limitation is used. Under this condition, if a proper dc bias ($u_{dc,bias}$) is added to the difference of the two capacitor voltages to make the difference of the two capacitor voltages equal to the variation of the upper or lower capacitor voltage

Table 8.7 The sign of the proportional allocation factor k

Modes	Operating conditions		The sign of k
1	$i_{max} > 0, i_{min} \leq 0$	$u_{mid} > 0$	Positive
		$u_{mid} \leq 0$	Positive
2	$i_{max} > 0, i_{min} > 0$	$u_{mid} > 0$	Negative
		$u_{mid} \leq 0$	Positive
3	$i_{max} \leq 0, i_{min} > 0$	$u_{mid} > 0$	Negative
		$u_{mid} \leq 0$	Negative
4	$i_{max} \leq 0, i_{min} \leq 0$	$u_{mid} > 0$	Positive
		$u_{mid} \leq 0$	Negative

in the abnormal time interval B. The waveforms of the two capacitor voltages will be interleaved. And thus, the magnitude of the NPV oscillations will be reduced to a half of that when the $u_{dc,bias}$ is not adopted.

$$\Delta U_{np} = U_p - U_n + u_{dc,bias} \tag{8.56}$$

In order to obtain a more general value, an online estimation scheme is proposed. The scheme is realized by the following formula

$$u_{dc,bias} = \begin{cases} U_{p,max} - U_{p,min} & \text{if } U_{p,max} + U_{p,min} > U_{dc} \\ U_{p,min} - U_{p,max} & \text{if } U_{p,max} + U_{p,min} \leq U_{dc} \end{cases} \tag{8.57}$$

where the $U_{p,max}$ and $U_{p,min}$ are the maximum and minimum value of the upper capacitor voltage in the last fundamental period of the output, respectively.

Condition 2: CB-HSTC for Fault Occurs in NP Switches

For convenience of analysis, the S_{a2} open-circuit fault is considered as an example. As aforementioned, the switching state [O] properly cannot be produced when the $i_a < 0$, thus resulting in distorted input current. However, under the condition of $i_a > 0$, the normal operation of the rectifier will not be affected by the S_{a2} fault.

Figure 8.30 shows the proposed CB-HSTC when S_{a2} open-circuit fault occurs. Depending on the current polarity of faulty phase A, the fault-tolerant control strategy for S_{a2} OCF can be divided into two half cycles: normal time interval (NTI) ($i_a > 0$) and abnormal time interval (ATI) ($i_a < 0$). In the ATI, the two-level switching method is conducted in the faulty phase A to maintain the normal operation. The reference voltage of phase A is compared with one carrier signal, and thus, the reference voltage is generated without the switching state [O]. While in the NTI, the traditional phase disposition carrier-based modulation strategy is used.

In order to suppress the NPV oscillations when a fault occurs in S_{a2}, the proposed hybrid NPV control method and the $u_{dc,bias}$ are used in the NTI. In the ATI, the NPV will not be affected by phase A, since the switching state [O] is inhibited in the faulty phase A. Hence, to attenuate the NPV oscillations, a compensation value

8.2 Fault-Tolerant Control Technology

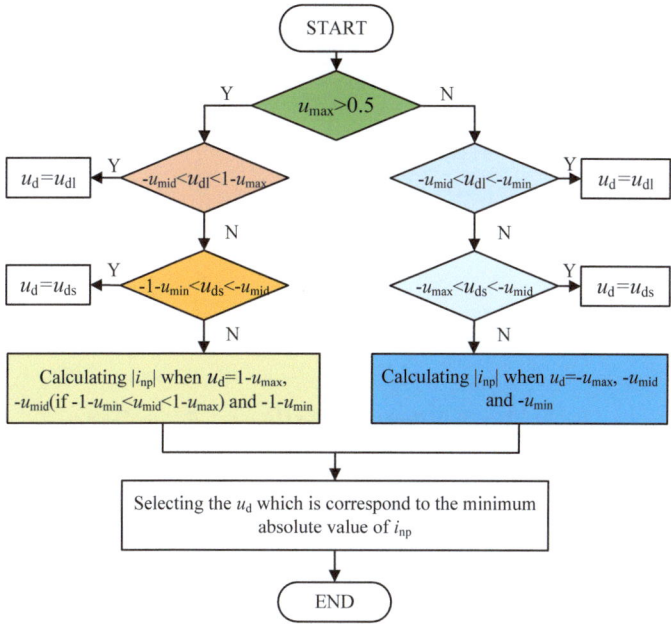

Fig. 8.28 Flowchart of the proposed algorithm to calculate an optimal u_d

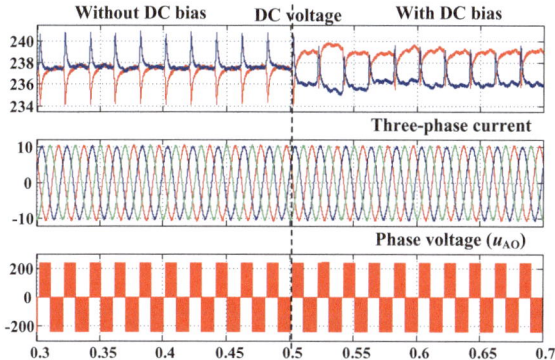

Fig. 8.29 The waveforms using CB-OSTC with and without dc bias

($u_{i,\text{com}}$), calculated based on phase B and C, is used. According to the magnitude of the faulty phase A, there are three modes.

Mode 1: the magnitude of phase A is the largest among the three phases.

By replacing u_{co} with $u_{i,\text{com}}$, the NP current can be represented as

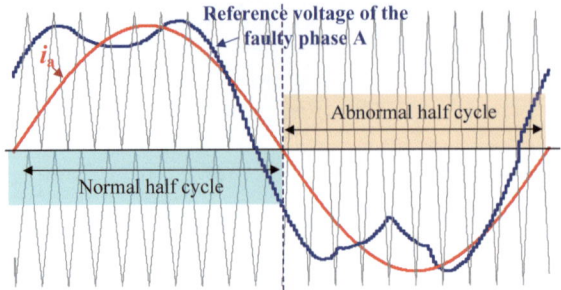

Fig. 8.30 Proposed CB-HSTC when S_{a2} open-circuit fault occurs

$$i_{np} = [1 - \text{sgn}(u_{mid} + u_{i,com})(u_{mid} + u_{i,com})]i_{mid} \\ + [1 + (u_{min} + u_{i,com})]i_{min} \quad (8.58)$$

In order to mitigate the NPV oscillations, the value of i_{np} should be zero. According to the polarity of $u_{mid} + u_{i,com}$, there are two kinds of $u_{i,com}$ ($u_{i,coml}$ and $u_{i,coms}$).

$$u_{i,coml} = \frac{i_{mid} + i_{min} - u_{mid}i_{mid} + u_{min}i_{min}}{i_{mid} - i_{min}} \quad (8.59)$$

$$u_{i,coms} = \frac{i_{mid} + i_{min} + V_{mid}i_{mid} + V_{min}i_{min}}{-i_{mid} - i_{min}} \quad (8.60)$$

Then, using the same operating principle of the calculation of u_d, the most optimal solution can be obtained as one of the above two kinds of $u_{i,com}$ and the boundary values. By substituting $u_{i,coml}$ and $u_{i,coms}$ as u_{dl} and u_{ds} into the flowchart of the optimal u_d calculation algorithm, the optimum value of $u_{i,com}$ can be obtained.

Mode 2: the magnitude of the faulty phase A is equal to u_{mid}.

Therefore, the NP current can be expressed as

$$i_{NP} = [1 - (u_{max} + u_{i,com})]i_{max} + [1 + (u_{min} + u_{i,com})]i_{min} \quad (8.61)$$

Similar to case 1, to make i_{np} equal to zero, $u_{i,com}$ is represented as

$$u_{i,com} = \frac{i_{max} + i_{min} - V_{max}i_{max} + V_{min}i_{min}}{i_{max} - i_{min}} \quad (8.62)$$

Mode 3: the magnitude of phase A is the smallest among the three phases.

The value of $u_{i,com}$ can be calculated as same as that demonstrated in case 1.

The block diagram of the proposed fault-tolerant control strategy for S_{a1} or S_{a2} OCF is shown in Fig. 8.31. The fault-tolerant control strategy for other switches OCF can be obtained using the control operating principle.

8.2 Fault-Tolerant Control Technology

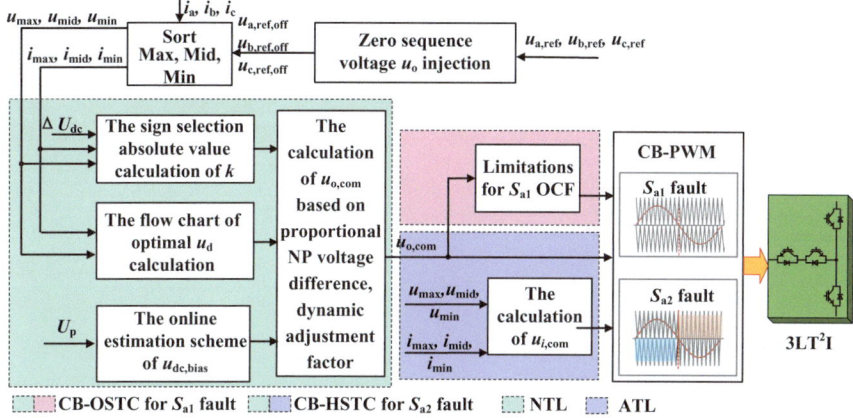

Fig. 8.31 Block diagram of the fault-tolerant control strategy based on continuous CB-PWM

Step 4: Performance and Comparison of Fault-Tolerant Controls Based on Continuous CB-PWM

Taking OCF occurring on S_{a1} and S_{a2} as an example, performance and comparison are conducted to illustrate the ability of fault-tolerant control and NPV oscillation suppression of the proposed method. The main performance includes the following four cases: the capability of fault-tolerant control, application range with different modulation index and power factor, the capability of NPV oscillation suppression, and average current THD.

The simulation parameters are as follows. The switching frequency is 10 kHz, the control period is 100 μs, the upper and lower dc-link capacitors are both 940 μF, the dc-link voltage is 300 V or 470 V, the L-filter value is 4 mH, the peak value of current is 10 A, and the phase peak voltage of power grid 150 V.

Case 1: Capability of Fault-Tolerant Control

Figure 8.32 shows the waveforms of the three-phase current, phase voltage u_{AO}, dc-link voltage and NPV waveforms in the case of S_{a1} open-circuit fault, where $m = 0.63$ and $pf = 0.8$. As shown in the blue shaded area of Fig. 8.32a, if OCF occurs in S_{a1}, the three-phase current is distorted seriously due to the impossible current path. Besides, the NPV is unbalanced. When the faulty switch is identified, and the proposed CB-OSTC is applied, the distortion of the three-phase current is eliminated and the NPV is balanced again, as shown in Fig. 8.32b.

Figure 8.33 shows the simulation results when OCF occurs to S_{a2}, under the condition of $m = 1$ and $pf = 0.8$. In Fig. 8.33a, the three-phase current is distorted, and the dc-link voltage has the ripple which is about 40 V. As shown in Fig. 8.33b, after the faulty switch S_{a2} is detected using the fault diagnosis method, the proposed CB-HSTC will be applied. The distortion of phase current disappears, and the reference

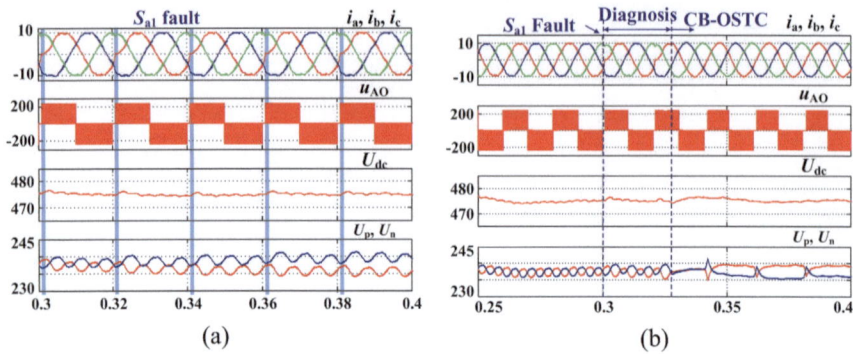

Fig. 8.32 Waveforms of S_{a1} OCF when $m = 0.63$, $pf = 0.8$ **a** before CB-OSTC is applied, **b** after CB-OSTC is applied

voltage of the faulty leg is generated by two switching states in the abnormal half cycle ($i_a < 0$). Meanwhile, the dc-link voltage as well as the NPV are well controlled.

Case 2: Application Range with Different Modulation Index and Power Factor
Figure 8.34 shows the compensation value and the changed reference voltage of the CVA-TC and the CB-OSTC method when S_{a1} OCF occurs. It can be seen that the hybrid NPV control method is used to calculate the compensation value ($u_{o,com}$) in the CB-OSTC. Furthermore, in order to avoid distortion, the $u_{o,com}$ should satisfy the requirement in time interval B. Hence, the reference voltage of the faulty phase in the CB-OSTC is clamped to zero in one-side. Besides, as shown in the red shaded area, the range of the reference voltage which is clamped to zero in CB-OSTC is smaller than that in CVA-TC.

When OCF occurs to outer switches, using the CVA-TC method, the time offset ($u_{o,com}$) should meet the following requirement in the abnormal time interval (time

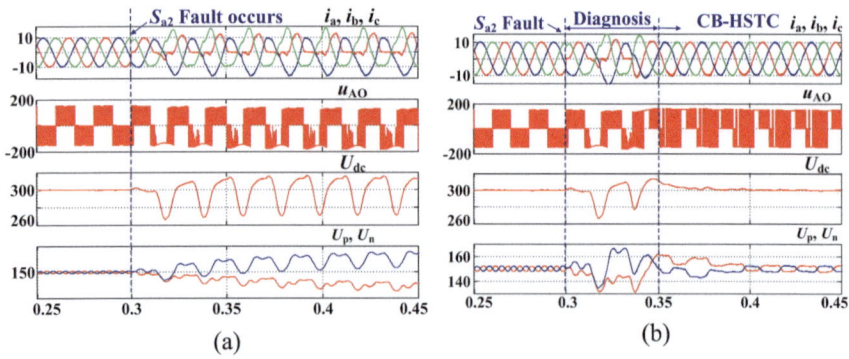

Fig. 8.33 Waveforms of S_{a2} OCF when $m = 1$, $pf = 0.8$ **a** before CB-HSTC is applied, **b** after CB-HSTC is applied

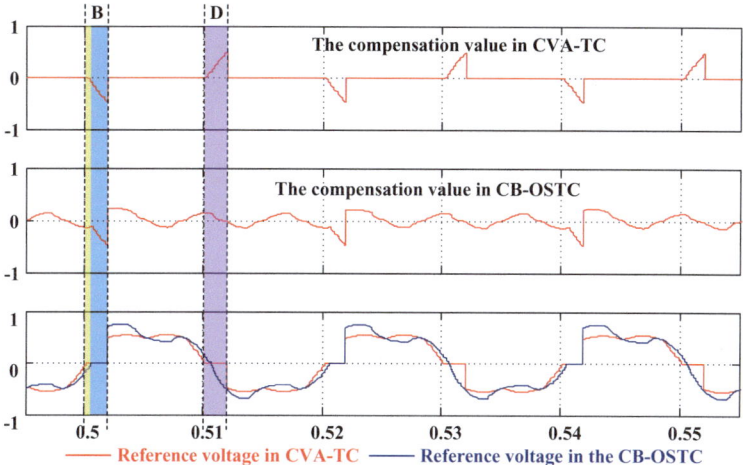

Fig. 8.34 Waveforms of compensation value and the changed reference voltage when S_{a1} OCF using the CVA-TC and CB-OSTC method

interval B or D)

$$u_{o,com} = -u_{x,ref,off} \quad (8.63)$$

where x is the phase containing OCF switches. In addition, the $u_{o,com}$ should be added to the three reference voltages to keep the principle of volt-second balance. To avoid over modulation, the redefined three phase reference voltage should satisfy the following requirement

$$-1 \leq u_{o,com} + u_{x,ref,off} \leq 1 \quad (x = a, b, c) \quad (8.64)$$

Based on the two limitations, the applicable range (AR) for various values of pf and m is shown in Fig. 8.35a, where the effect caused by the filter impedance is not considered. The blue shaded area represents the AR using CVA-TC. However, it is not comprehensive, and the red shaded area should be included in the AR. If m is smaller than 0.577, the proposed CB-OSTC is feasible in the whole pf range. On the contrary, if m is larger than 0.577, the AR decreases along with the increment of m.

As demonstrated before, the range of intervals B and D is influenced by the filter impedance. Hence, the AR considering the influence of filter impedance, is changed and shown in Fig. 8.35b. The filter impedance causes the AR to be reduced when the 3LT^{2}I operates with the leading pf. In contrast, if the 3LT^{2}I operates with the lagging pf, the filter impedance causes the increment of the applicable pf range. If OCF occurs to outer switches, when the combination of m and pf is not located in the AR, increasing the output dc-link voltage or decreasing the input three-phase voltage should be adopted. However, if OCF occurs to inner switches, the voltage, current or

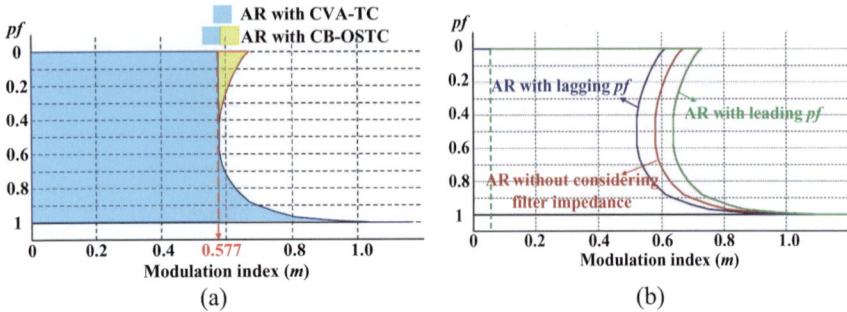

Fig. 8.35 Applicable range (AR) corresponding to modulation index and power factor when S_{a1} OCF **a** using the CVA-TC and CB-OSTC method without considering the effect of the filter impedance, **b** using the CB-OSTC method considering the effect of the filter impedance

kVA rating of $3LT^2I$ will not be influenced when the proposed fault-tolerant control strategy is applied.

Case 3: Capability of NPV Oscillation Suppression

Figure 8.36a, b shows the waveforms of the NPV, three phase current and phase voltage using different fault-tolerant control methods in the case of S_{a1} and S_{a2} fault, respectively. As shown in Fig. 8.36a, under the condition of $m = 0.63$ and $pf = 0.8$, the magnitude of the NPV oscillations is reduced to a half of that using CVA-TC when the CB-OSTC is applied. When the OCF occurs to S_{a2}, the 3LSO-TC and OP2LS-TC are used to make comparison with the proposed CB-HSTC in this section, which is shown in Fig. 8.36b. We can see that the proposed CB-HSTC is advantageous in terms of NPV oscillations suppression, in which the magnitude of the NPV oscillations equal to 1/6 of that in 3LSO-TC and 1/3 of that in OP2LS-TC, respectively.

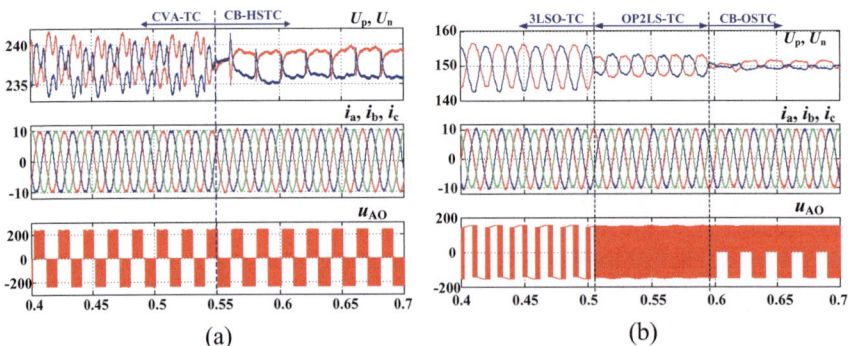

Fig. 8.36 The performance of NPV oscillations under different fault-tolerant control methods **a** when S_{a1} OCF occurs ($m = 0.63$, $pf = 0.8$), **b** when S_{a2} OCF occurs ($m = 1$, $pf = 1$)

8.2 Fault-Tolerant Control Technology

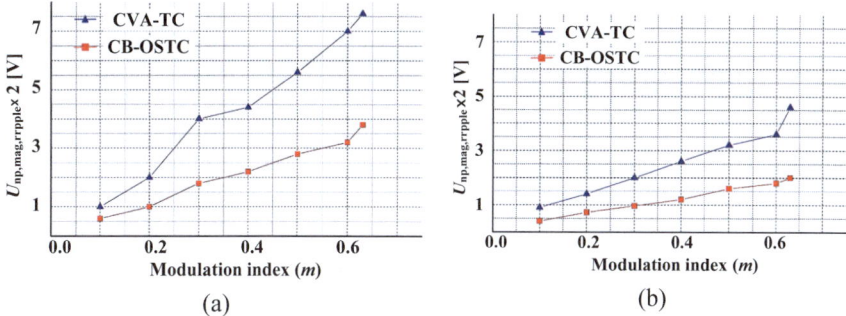

Fig. 8.37 The magnitude of the NPV oscillations using different fault-tolerant control methods under different modulation index m when S_{a1} OCF occurs **a** with leading pf ($pf = 0.8$), **b** with leading pf ($pf = 0.9$)

Since the NPV oscillation is affected by the modulation index and power factor, analyses and investigations on the NPV oscillations at various operating conditions are carried out for further understanding.

In the case of S_{a1} OCF, the performance of NPV oscillations using the proposed CB-OSTC and conventional CVA-TC methods are shown in Fig. 8.37a, b according to the modulation index when the pf is 0.8 and 0.9, respectively. The magnitude of the NPV oscillations in both methods increases along with the modulation index m. Compared with conventional CVA-TC, the proposed CB-OSTC has smaller NPV oscillations. The magnitude of the NPV oscillations in the CB-OSTC is reduced to about a half of that in CVA-TC under different operating conditions.

As shown in Fig. 8.38, the comparison of the three fault-tolerant controls (OP2LS-TC, 3LSO-TC and CB-HSTC) for S_{a2} open-circuit fault in terms of the NPV oscillations is conducted under different operating conditions. It can be seen that the proposed CB-HSTC can maintain the best performance in NPV oscillation suppression compared with the rest two fault-tolerant controls. As m increases, the range of $u_{o,com}$ becomes smaller, and hence, the magnitude of the NPV oscillation in the CB-HSTC are increased. Nevertheless, even in the worst case ($m = 1.15$), at least about 50% of the NPV oscillations in the OP2LS-TC or 3LSO-TC can be reduced when the CB-HSTC is applied.

Case 4: Average Current THD

The average input current THD of the three phases under different operating conditions is measured to verify the performance of the proposed fault-tolerant control strategy. When OCF occurs to S_{a1}, the averaged input current THD is shown in Table 8.8. Compared with the CVA-TC, the proposed CB-OSTC maintains a better performance in the current THD. This is because in the proposed CB-OSTC, the phase voltage of the faulty phase is clamped to [O] state in one side, while in CVA-TC, the phase voltage of the fault phase is also clamped to [O] state in the diametrically opposite position.

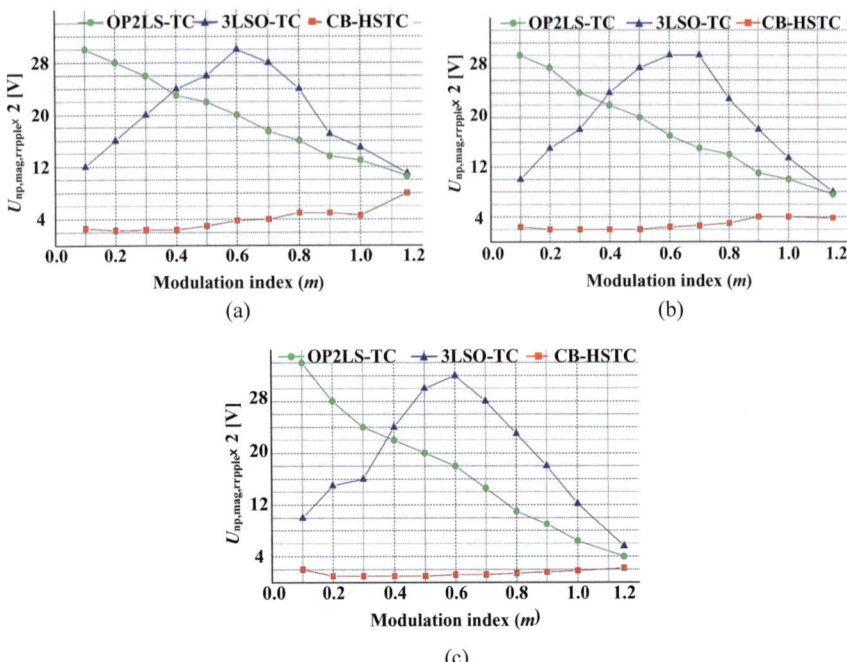

Fig. 8.38 The magnitude of the NPV oscillations using different fault-tolerant control methods under different modulation index m when S_{a2} OCF occurs **a** $pf = 0.8$, **b** $pf = 0.9$, **c** $pf = 1$

Table 8.8 The averaged current THD using different fault-tolerant control methods in case of S_{a1} fault

Condition	CVA-TC (%)	CB-OSTC (%)
$m = 0.63, pf = 0.8$	2.34	2.24
$m = 0.55, pf = 0.8$	2.85	2.71
$m = 0.63, pf = 0.6$	2.65	2.36

When OCF occurs to S_{a2}, the averaged input current THD of the three fault-tolerant controls under different operating conditions is shown in Table 8.9. In the OP2LS-TC, the reference voltage of the faulty phase is generated by switching state [P] and [N] during each switching period. However, in the 3LSO-TC, the reference voltage of the faulty phase will be clamped to [P] and [N] state in some switching period. Hence, the OP2LS-TC is advantageous in current THD compared with 3LSO-TC. In the proposed CB-HSTC, the reference voltage of the faulty phase is generated with three switching states in the normal half cycle, and thus, the proposed CB-HSTC maintains the best performance in current THD compared with the two methods.

To conclude, in this section, when $3LT^2I$ operated in charging mode, a continuous CB-PWM based fault-tolerant control strategy with NPV oscillation suppression is proposed for power semiconductor switches occurring OCF. The proposed fault-tolerant control method is illustrated by dividing fault into two conditions: the faulty

8.2 Fault-Tolerant Control Technology

Table 8.9 The averaged current THD using different fault-tolerant control methods in case of S_{a2} fault

Condition	3LSO-TC (%)	OP2LS-TC (%)	CB-HSTC (%)
$m = 1, pf = 1$	3.12	2.84	2.53
$m = 0.8, pf = 1$	4.42	4.01	3.3
$m = 0.8, pf = 0.8$	4.61	4.08	3.81

condition of HB switches and NP switches. In the case of the HB switches open-circuit fault, a CB-OSTC is proposed. Compared with CVA-TC, the magnitude of the NPV oscillations is dramatically reduced, because of the use of the hybrid NPV control strategy and the dc bias. Besides, a more comprehensive applicable *pf* range for various values of *m* is given in this paper. In the case of NP power semiconductor switches OCF, a CB-HSTC which is implemented by properly changing the reference voltage and the carrier is proposed. The CB-HSTC can suppress the NPV oscillations effectively compared to OP2LS-TC and 3LSO-TC. The proposed fault-tolerant control for outer or inner switches OCF is also advantageous in the dynamic response. Additionally, the proposed control maintains the best performance in terms of the averaged input current THD. The presented control can be simply implemented and do not require any additional hardware.

8.2.3 Based on Discontinues Carrier-Based Pulse Width Modulation

In the practical operation, besides the NPV oscillations, the loss should also be considered. The switching loss of the $3LT^2I$ is the major factor to meet the superior efficiency requirement and reduce the cost of the cooling system. In the previous literatures, some methods have been proposed to suppress the NPV oscillations and reduce the switching loss by adjusting the width of the positive and negative discontinuous sections [15, 16]. However, when an OCF occurs, the current paths are changed, resulting in these methods no longer applicable. Moreover, the extra switching events between the switching periods are never considered, which can further increase the switching losses.

To solve the above issues, the author conducted extensive research, and proposed a discontinuous CB-PWM fault-tolerant control strategy based on the theoretical basis presented in previous literature [17]. The $3LT^2I$ operating in charging mode and is still taken as the specific application to derive this fault-tolerant control strategy. Using this method, the fault-tolerant control, NPV oscillation suppression, and switching loss reduction are realized simultaneously when an OCF occurs in $3LT^2I$.

In the discontinuous CB-PWM fault-tolerant control strategy, one fundamental period is classified into two-time intervals: the NTI and ATI. In the NTI, a time-offset is utilized to achieve the NPV balance and loss reduction. Aside from this, the extra switching events between the switching periods are also reduced by employing an appropriate carrier. In the ATI, the proposed method is explained by dividing faults into two cases: the faulty condition of HB switches and NP switches. When OCF occurs at HB switches, a restriction of the time-offset is introduced. However, if an OCF occurs at NP switches, the two-level modulation strategy is applied in the faulty phase. Besides, the principle of carrier selection of the faulty phase is changed correspondingly to avoid the extra switching events.

This discontinuous CB-PWM fault-tolerant control strategy includes three steps. **Step 1**: The division of NTI and ATI in one fundamental period is given. **Step 2**: Deriving process of fault-tolerant control strategy based on discontinuous CB-PWM is illustrated according to NTI and ATI. **Step 3**: Performance and comparison of tolerant controls are given and analyzed for easier understanding by readers. The detailed analyses and deriving process in each step are illustrated as follows.

Step 1: Division of NTI and ATI for One Fundamental Period

To facilitate the division of NTI and ATI, the four-time intervals in one fundamental period are re-depicted again in Fig. 8.39.

When OCF occurs at S_{x1}, the current path which belongs to time interval B, becomes infeasible. The switching state [P] of the faulty phase will be changed into the switching state [O]. Hence, the current distortion appears in time interval B. While in other time intervals, the faulty switch is never used, and thus, the three switching states are not affected.

When 3LT^{2}I operates under normal condition, if $i_x \leq 0$ and flows through S_{x2}, the switching state is [O]. If OCF occurs at S_{x2}, the faulty phase will be connected to the negative instead of the NP dc-link when $i_x \leq 0$. Thus, the current will be distorted in time interval B and C. When OCF occurs at S_{x3}, the current distortion appears in time interval A and D. In addition, if OCF occurs at S_{x4}, the current will be distorted in time interval D, since the current path related to S_{x4} belongs to time interval D. In summary, according to the current direction and modulation wave of the faulty phase,

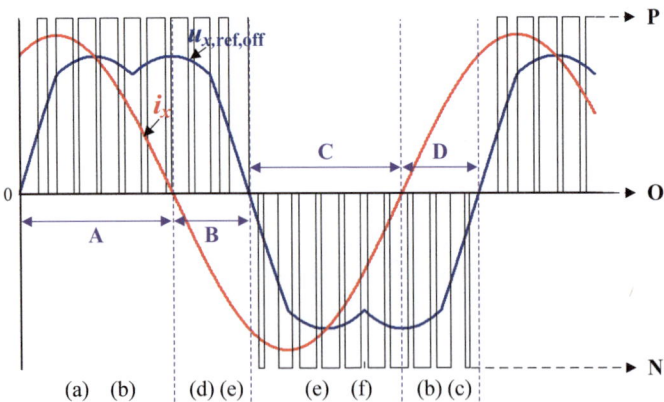

Fig. 8.39 Four-time intervals in one fundamental period

8.2 Fault-Tolerant Control Technology

Table 8.10 The division of NTI and ATI for one fundamental period

Faulty switch	NTI	ATI
S_{x1}	A, C, D	B
S_{x2}	A, D	B, C
S_{x3}	B, C	A, D
S_{x4}	A, B, C	D

the four time intervals can be classified into two types: the NTI in which the current is not influenced and the ATI in which the current will be distorted. Depending on the faulty switch, the division of NTI and ATI for one fundamental period is shown in Table 8.10.

Step 2: Deriving Process of Fault-Tolerant Control Strategy Based on Discontinuous CB-PWM

In this section, the deriving process of fault-tolerant control strategy is demonstrated by dividing the time intervals into two parts: fault-tolerant control in NTI and fault-tolerant control in ATI.

Part 1: Fault-Tolerant Control in NTI

The sort of the redefined reference voltage can be rearranged as

$$\begin{cases} u_{\max} = \max(u_{a,\text{ref,off}}, u_{b,\text{ref,off}}, u_{c,\text{ref,off}}) \\ u_{\text{mid}} = \text{mid}(u_{a,\text{ref,off}}, u_{b,\text{ref,off}}, u_{c,\text{ref,off}}) \\ u_{\min} = \min(u_{a,\text{ref,off}}, u_{b,\text{ref,off}}, u_{c,\text{ref,off}}) \end{cases} \quad (8.65)$$

In order to reduce the switching losses within one switching period, one phase can be clamped to switching state [P], [O] or [N], which can be realized by adding an appropriate time offset to the three phases without voltage distortion. According to the magnitude of $u_{\max}$, the clamping modes and the corresponding time offset are given in Table 8.11.

There are three possible clamping modes when $u_{\max} > 0.5$ and five possible clamping modes when $u_{\max} \leq 0.5$. The clamping modes are named as X_P, X_O, or X_N (X = MAX, MID or MIN), respectively. For instance, MAX_P denotes that the phase corresponding to $u_{\max}$ is clamped to switching state [P]. Note that, under the condition of $u_{\max} > 0.5$, to avoid over modulation, the clamping mode MID_O can only be used when $-1 - u_{\min} \leq -u_{\text{mid}} \leq 1 - u_{\max}$.

Depending on the NPV, the appropriate clamping mode can be obtained. Generally, the NPV is influenced by the NP current. The positive NP current will decrease

Table 8.11 Clamping modes and their corresponding time-offset

	MAX_P	MAX_O	MID_O	MIN_O	MIN_N
$u_{\max} > 0.5$	$1 - u_{\max}$	/	$-u_{\text{mid}}$	/	$-1 - u_{\min}$
$u_{\max} \leq 0.5$	$1 - u_{\max}$	$-u_{\max}$	$-u_{\text{mid}}$	$-u_{\min}$	$-1 - u_{\min}$

the upper dc-link capacitor voltage (U_p) and increase the lower dc-link capacitor voltage (U_n). Conversely, the negative NP current will increase the U_p and decrease the U_n. One phase makes contribution to the NP current only when the switching state [O] is used in this phase. When one of the clamping modes is used, the dwell time of switching state [O] ($d_{o,mo}$) in one switching period can be obtained by

$$d_{o,mo} = \begin{cases} 1 - u_{mo} & u_{mo} > 0 \\ 1 + u_{mo} & u_{mo} \leq 0 \end{cases} \quad mo \in (max, mid, min) \quad (8.66)$$

where u_{mo} denotes the modulation wave of the phase after the proper time-offset for clamping mode is injected to the three-phase reference voltages. Then, the average NP current (i_{np}) of the clamping mode in a switching period can be expressed as

$$i_{np} = \sum_{mo=max,mid,min} [1 - \text{sgn}(u_{mo}) \cdot u_{mo}] \cdot i_{mo} \quad (8.67)$$

where i_{mo} is the current corresponding to the phase of u_{mo}.

In order to balance the NPV as soon as possible, $\Delta U_{np} > 0$ for example, the clamping mode which has the maximum i_{np} should be applied. However, if $\Delta U_{np} \leq 0$, the clamping mode which has the minimum i_{np} should be used. It should be noted that the same NP current is generated by the clamping modes MAX_P and MIN_O and the clamping modes MIN_N and MAX_O when $u_{max} \leq 0.5$. Hence, in NTI, the clamping modes MAX_P and MIN_N are not used to avoid large dv/dt.

With the method illustrated above, the loss reduction within a switching period is achieved. In the traditional phase disposition carrier-based approach for 3LT^{2}I, the reference voltages are compared with two vertically disposed in-phase carriers (VDIPC). As shown in Fig. 8.40, there are two kinds of VDIPCs: the concave VDIPC and convex VDIPC.

If the same VDIPC is used to the three phases, as shown in Fig. 8.41, the extra switching event appears between the switching periods, since the different clamping modes are used.

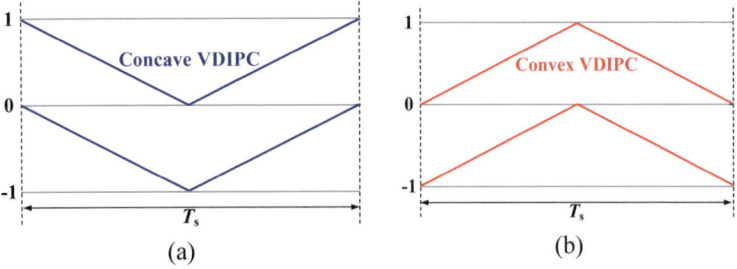

Fig. 8.40 Two types of VDIPCs for 3LT^{2}I **a** concave VDIPC, **b** convex VDIPC

8.2 Fault-Tolerant Control Technology

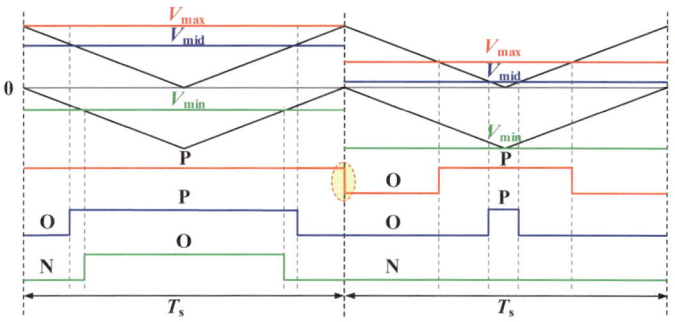

Fig. 8.41 The extra switching event appears between the switching periods

Hence, to eliminate the extra switching events in NTI, a simple algorithm by employing appropriate VDIPC for each phase is illustrated. The selection of appropriate VDIPC is divided into two modes depending on the magnitude of u_{max}.

Mode 1: $u_{max} > 0.5$

When $u_{max} > 0.5$, there are three possible clamping modes in NTI, which are MAX_P, MID_O and MIN_N. In the CB-PWM, depending on the magnitude of u_{mo}, the phase voltage will be generated by different combinations of the switching states which are summarized in Table 8.12.

If the clamping mode MAX_P is used, u_{max} will equal to one which means the voltage of the phase corresponding to u_{max} is generated by switching state [P]. However, if the other two clamping modes are applied, the voltage of the phase corresponding to u_{max} will be generated by switching state [P] and [O]. Thus, if the switching sequence of the phase corresponding to u_{max} is started and ended with switching state [P], which can be realized by employing convex VDIPC, as shown in Fig. 8.42a, the extra switching event in this phase will be eliminated.

As for the phase corresponding to u_{mid}, when different clamping modes are used, this phase voltage will be generated by switching state [O] and [N], switching state [O], or switching state [O] and [P]. As shown in Fig. 8.42b, if the switching sequence of the phase corresponding to u_{mid} is started and ended with switching state [O], the extra switching event of this phase will disappear. To achieve this, the concave VDIPC should be used when $u_{mid} > 0$, and the convex VDIPC should be used when $u_{mid} \leq 0$.

Similar with the analysis above, the voltage of the phase corresponding to u_{min} will be generated by the switching states [O] and [N] or the switching state [N]. And hence, the extra switching event of this phase will be eliminated if the concave VDIPC which makes the switching sequence be started and ended with the switching

Table 8.12 Relationship between the switching state and magnitude of modulation waves

	$u_{mo} = 1$	$0 < u_{mo} < 1$	$u_{mo} = 0$	$-1 < u_{mo} < 0$	$u_{mo} = -1$
Switching state	[P]	[P], [O]	[O]	[O], [N]	[N]

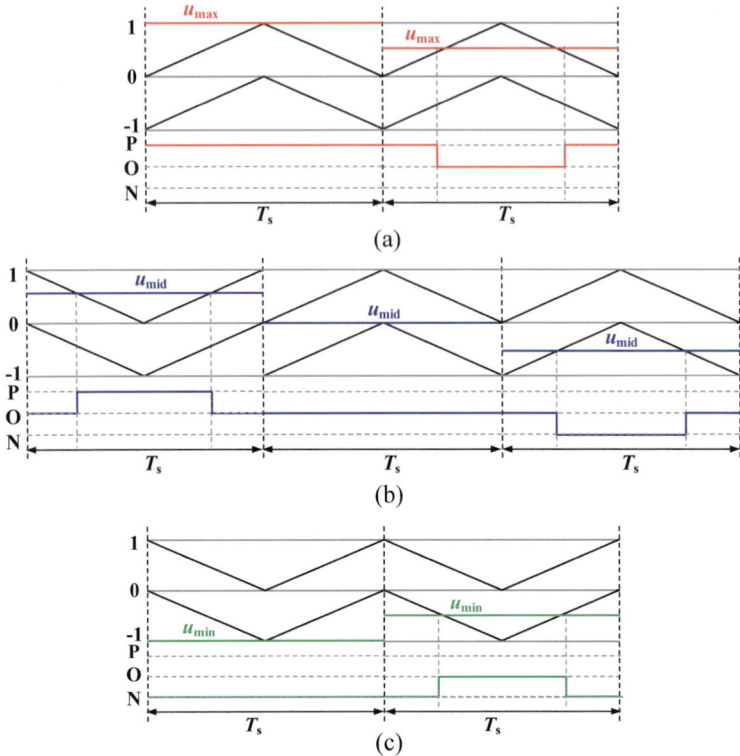

Fig. 8.42 When $u_{max} > 0.5$ the improved switching sequence of the phase corresponding to **a** u_{max}, **b** u_{mid} and **c** u_{mid}

state [N] is utilized. The changed switching sequence for the phase corresponding to u_{min} is shown in Fig. 8.42c.

It should be noted that 6 additional extra switching events still exist when the relative magnitude of the three-phase reference voltages is changed in the adjacent switching periods.

Mode 2: $u_{max} \leq 0.5$

When $u_{max} \leq 0.5$, as illustrated above, there are three possible clamping modes, which are MIN_O, MID_O and MAX_O. As shown in Fig. 8.43, if one of the three clamping modes is applied, the switching state [O] is essential to generate the three phase voltages. And hence, the extra switching event can be eliminated if the switching sequence of each phase is started and ended with switching state [O]. To make the switching sequence of one phase started and ended with switching state [O], the concave VDIPC can be used if $u_{mo} > 0$ and the convex VDIPC can be used if $u_{mo} \leq 0$.

Based on the above analyses, the carrier types to eliminate the extra switching event in NTI are summarized in Table 8.13.

8.2 Fault-Tolerant Control Technology

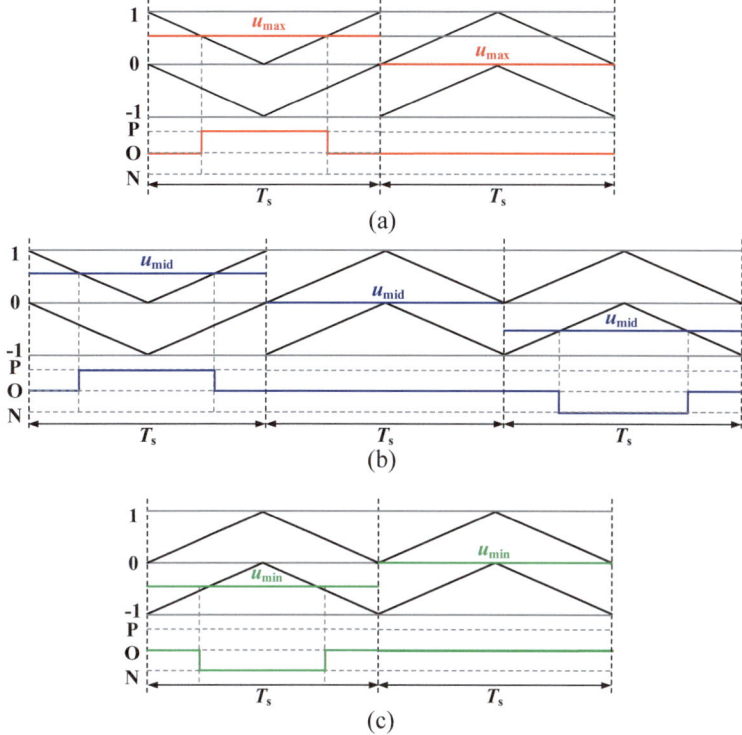

Fig. 8.43 When $u_{max} \leq 0.5$, the improved switching sequence of the phase corresponding to **a** u_{max}, **b** u_{mid}, **c** u_{min}

Table 8.13 Carrier types to eliminate the extra switching event in NTI

Conditions			Carrier types
$u_{max} > 0.5$	$u_{mo} = u_{max}$		Convex VDIPC
	$u_{mo} = u_{mid}$	$u_{mid} > 0$	Concave VDIPC
		$u_{mid} \leq 0$	Convex VDIPC
	$u_{mo} = u_{min}$		Concave VDIPC
$u_{max} \leq 0.5$	u_{mo}	$u_{mo} > 0$	Concave VDIPC
		$u_{mo} \leq 0$	Convex VDIPC

Part 2: Fault-Tolerant Control in ATI

The fault-tolerant control in ATI is demonstrated by dividing faults into two conditions: fault occurs in HB switches and fault occurs in NP switches.

Condition 1: Fault Occurs in HB Switches

When OCF occurs at HB switches, the HB switches fault-tolerant control (HBS-TC) is proposed. S_{a1} fault is given as a representative example. According to the analysis in the above section, the time interval B of faulty phase A belongs to ATI. The current will be distorted in time interval B of phase A, since the switching state [P] of the faulty phase will be changed into switching state [O]. Therefore, in order to eliminate the current distortion, the switching state [P] should be avoided in time interval B, which can be realized by adding a proper time offset. Besides, to reduce the switching loss within a switching period, the time offset corresponding to the clamping mode should be used. Then, the modulation wave of the faulty phase A is expressed as

$$u_{ao} = u_{a,ref,off} + u_{off} \tag{8.68}$$

To avoid current distortion in ATI, the time offset corresponding to the clamping modes should satisfy the following restriction.

$$u_{off} \leq -u_{a,ref,off} \text{ if } i_a \leq 0 \tag{8.69}$$

Accordingly, the clamping mode is classified into two types: the influenced clamping mode (ICM) which means the u_{off} of the corresponding clamping mode does not satisfy the above limitation and not influenced clamping mode (NICM) which means the u_{off} of the corresponding clamping mode satisfies the above limitation. To balance the NPV, the NICM which has the maximum i_{np} should be applied if $\Delta U_{np} < 0$. However, if $\Delta U_{np} \geq 0$, the NICM which has the minimum i_{np} should be used. Noted that the clamping modes of MAX_P and MIN_N are avoided when $u_{max} \leq 0.5$, which is similar with that in NTI.

To assist the understanding of the proposed HBS-TC, the modulation wave of phase A is given in Fig. 8.44. After the time offset u_{off} corresponding to the NICM is applied, u_{ao} is no larger than zero in the abnormal time interval B of phase A. Under this condition, the voltage of phase A will be generated by switching state [O] and [N], which means the impossible switching state [P] is avoided in ATI. Thus, the current distortion is eliminated.

Using the same operating principle, the appropriate clamping mode can be selected if OCF occurs at S_{a4}. The method to eliminate the extra switching event in NTI can be directly applied in ATI when OCF occurs at HB switches.

Condition 2: Fault Occurs in NP Switches

The NP switches fault-tolerant control (NPS-TC) is proposed in this section. For explanation of the proposed NPS-TC, S_{a2} OCF is given as an example. Under the faulty condition of S_{a2}, the ATI is comprised of two time intervals: time interval B and C of phase A, where the switching state [O] are utilized. In order to maintain the normal operation, the two-level modulation strategy for phase A is conducted in ATI. Figure 8.45 shows the implementation of the proposed two-level modulation strategy. Without breaking the principle of voltage-second balance, the modulation wave of the faulty phase is compared with a single carrier to avoid the use of the impossible switching state [O]. Similar with the modulation strategy for three-level converters, the carriers used in two-level modulation strategy are divided into two

8.2 Fault-Tolerant Control Technology

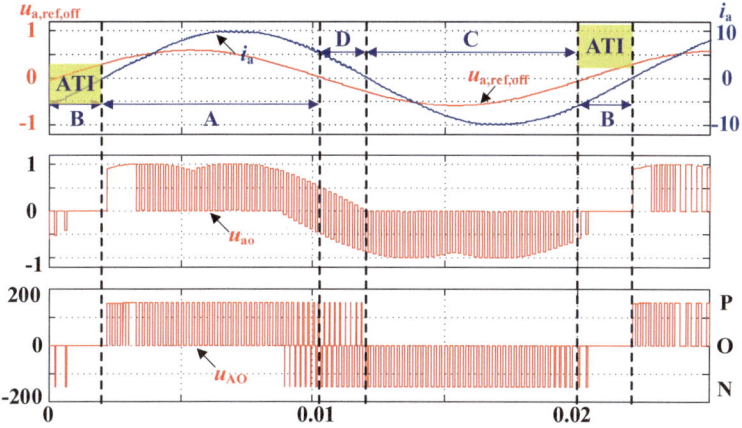

Fig. 8.44 The modulation wave of the faulty phase when the HBS-TC is applied

types: the concave carrier which is shown in Fig. 8.45a and the convex carrier which is shown in Fig. 8.45b.

When S_{a2} OCF occurs, the possible clamping modes in ATI are shown in Table 8.14. Besides, the average NP current in a switching period is also given. In ATI, the faulty phase A makes no contribution to the NP current, since the switching state [O] is avoided. Note that, when $u_{max} \leq 0.5$, the NP current corresponding to the clamping mode MAX_P is different from the NP current corresponding to the clamping mode MIN_O, which is different from that in NTI. Hence, the clamping mode MAX_P is used to balance the NPV in ATI when $u_{max} \leq 0.5$. Similarly, when $u_{max} \leq 0.5$, the clamping mode MIN_N is also used in ATI.

To balance the NPV, for instance $\Delta U_{dc} > 0$, the clamping mode which has the minimum i_{np} among the possible clamping modes under each condition should be used. Whereas, if $\Delta U_{np} \leq 0$, the clamping mode which has the maximum i_{np} among

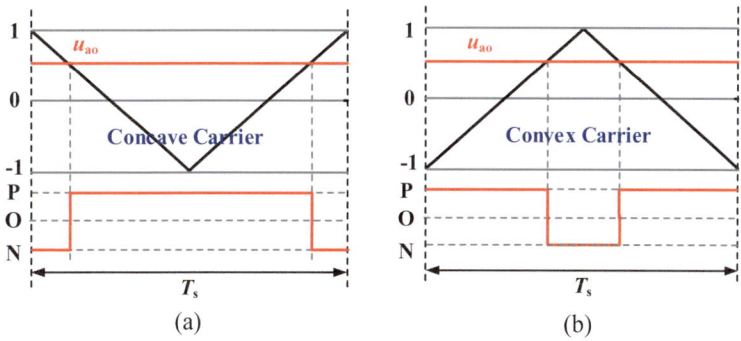

Fig. 8.45 Implementation of the two-level modulation strategy in the faulty phase A during ATI **a** concave carrier is used, **b** with convex carrier is used

Table 8.14 The possible clamping mode and revised NP current when OCF occurs at S_{a2} in ATI

Conditions		Possible clamping mode	NP current
$u_{max} > 0.5$	$u_{ao} = u_{max}$	MAX_P, MID_O, MIN_N	If $u_{ao} = u_{max}$ $i_{np} = [1 - \text{sgn}(u_{mid})u_{mid}]i_{mid}$ $+[1 - \text{sgn}(u_{min})u_{min}]i_{min}$
	$u_{ao} = u_{mid}$	MAX_P, MIN_N	
$u_{max} \leq 0.5$	$u_{ao} = u_{min}$	MAX_P, MID_O, MIN_N	If $u_{a,ref} = u_{mid}$ $i_{np} = [1 - \text{sgn}(u_{max})u_{max}]i_{max}$ $+[1 - \text{sgn}(u_{min})u_{min}]i_{min}$
	$u_{ao} = u_{max}$	MAX_P, MID_O, MIN_O, MIN_N	
	$u_{ao} = u_{mid}$	MAX_P, MAX_O, MIN_O, MIN_N	If $u_{ao} = u_{min}$ $i_{np} = [1 - \text{sgn}(u_{max})u_{max}]i_{max}$ $+[1 - \text{sgn}(u_{mid})u_{mid}]i_{mid}$
	$u_{ao} = u_{min}$	MAX_P, MAX_O, MID_O, MIN_N	

the possible clamping modes should be utilized. For a better understanding of the proposed NPS-TC, the modulation wave of the faulty phase is given in Fig. 8.46. To avoid current distortion, the two-level modulation strategy is adopted in the faulty phase during ATI, then the voltage of the faulty phase is generated by switching states [P] and [N]. Besides, a proper time offset corresponding to the possible clamping mode listed is added in the modulation wave to balance the NPV and reduce the switching event within a switching period.

The extra switching event in ATI when OCF occurs at NP switches, is also considered. To assist the understanding of the proposed method, S_{a2} fault is given as an example. The proposed method is discussed according to the value of u_{max}.

Mode 1: $u_{max} > 0.5$

When $u_{max} > 0.5$ and $u_{ao} = u_{max}$, there are three possible clamping modes MAX_P, MID_O and MIN_N. If the clamping mode MAX_P is used, the voltage of the faulty

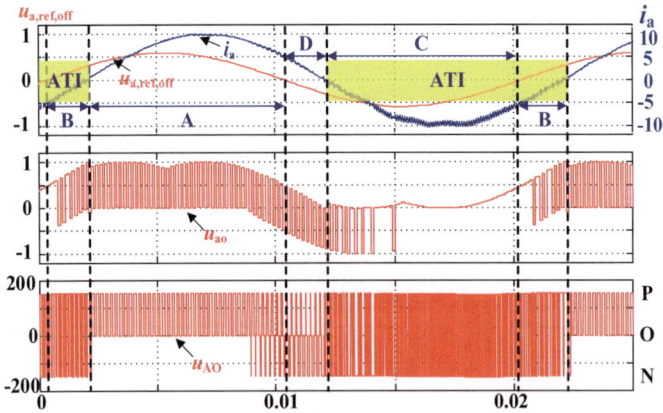

Fig. 8.46 The modulation wave of the faulty phase when the NPC-TC is applied

8.2 Fault-Tolerant Control Technology

phase will be generated by switching state [P]. However, if the other two clamping modes are used, the voltage of the faulty phase will be made by switching states [N] and [P], since the two-level modulation strategy is used. Therefore, under this condition, if the switching sequence of the faulty phase is started and ended with switching state [P], which is shown in Fig. 8.47a, the extra switching event will be eliminated. To realize this switching sequence, the convex carrier should be used in the faulty phase.

Under the condition of $u_{max} > 0.5$ and $u_{ao} = u_{mid}$, there are two possible clamping modes MAX_P and MIN_N. The voltage of the faulty phase will be generated by switching states [N] and [P] when one of the possible clamping modes is utilized. Under this condition, the concave carrier should be used for the faulty phase to eliminate the extra switching event, which is shown in Fig. 8.47b.

In the case of $u_{max} > 0.5$ and $u_{ao} = u_{min}$, three clamping modes MAX_P, MID_O and MIN_N are possible. As shown in Fig. 8.47c, if the concave carrier is used for the faulty phase, which makes the switching sequence be started and ended with switching state [N], the extra switching event in the faulty phase will be eliminated.

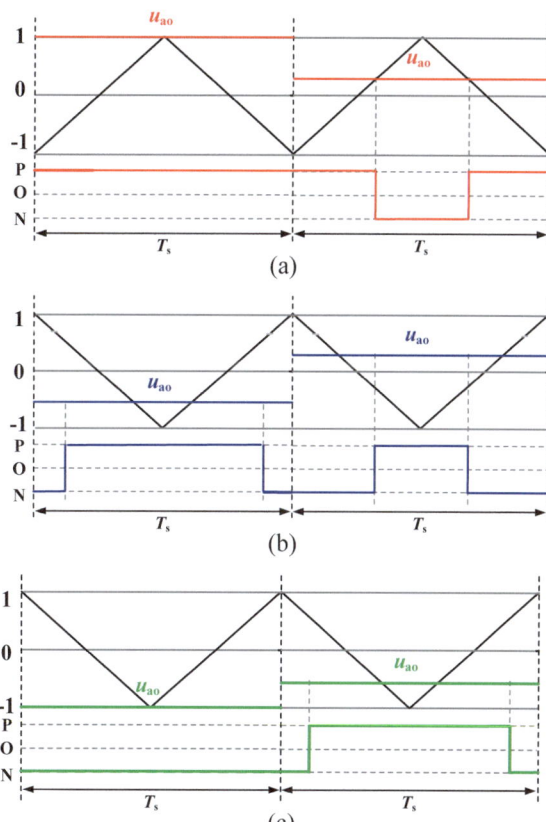

Fig. 8.47 The improved switching sequence of the faulty phase a when $u_{max} > 0.5$ under the condition of **a** $u_{ao} = u_{max}$, **b** $u_{ao} = u_{mid}$, **c** $u_{ao} = u_{min}$

As for the two normal phases, the extra switching event can be eliminated using the method proposed in NTI.

Mode 2: $u_{max} \leq 0.5$

Under the condition of $u_{max} \leq 0.5$ and $u_{ao} = u_{max}$, four possible clamping modes, MAX_P, MID_O, MIN_O and MIN_N, can be used to balance the NPV. If the clamping mode MAX_P is used, the voltage of the faulty phase A will be generated by switching state [P]. However, if the other three possible clamping modes are applied, the switching states [N] and [P] will be used to generate the faulty phase voltage. And thus, the extra switching event of the faulty phase is reduced when a convex carrier is used, since the switching sequence is started and ended with the switching state [P]. Under the same condition, as for the phase corresponding to u_{mid}, the switching state [O] is used to synthesize the phase voltage when one of the clamping modes is applied. As a result, the extra switching event of the phase corresponding to u_{mid} will be eliminated when the switching sequence is started and ended with switching state [O]. For the phase corresponding to u_{min}, if MIN_N is used, the phase voltage will be generated by switching state [N]. However, if the other three possible clamping modes are used, the switching state [O] is used to generate the phase voltage. To reduce the extra switching event as much as possible, the switching sequence of the phase corresponding to u_{min} should be started and ended with switching state [O]. Consequently, based on the above analysis, to reduce the extra switching event of the two normal phases, the method proposed in NTI can be applied.

Using a similar operating principle, the carrier types to reduce the extra switching event under the conditions of $u_{ao} = u_{mid}$ and $u_{ao} = u_{min}$ can be obtained.

In summary, the carrier types for the faulty phase to eliminate the extra switching event is shown in Table 8.15. However, the extra switching event of the two normal phases can be reduced using the method proposed in NTI.

Due to the asymmetry of the modulation strategy when OCF occurs at NP switches, the NPV variation caused by the time offset in a switching period will always be larger than zero during a fundamental period in the pale blue region of Fig. 8.48. And it will lead to a dc unbalance on NPV. In order to avoid this, the fault-tolerant control proposed in ATI should also be used in NTI.

In order to further reduce the NPV ac unbalance, both in the NTI and ATI, a dc bias $u_{dc,bias}$ is introduced to the difference of the two capacitors' voltages. Then, the difference of the two capacitors' voltages is rewritten as

$$\Delta U_{np} = U_p - U_n + u_{dc,bias} \tag{8.70}$$

Table 8.15 Carrier types to eliminate the extra switching event for the faulty phase in ATI

Conditions	Faulty phase A
$u_{ao} = u_{max}$	Convex carrier
$u_{ao} = u_{mid}$	Concave carrier
$u_{ao} = u_{min}$	Concave carrier

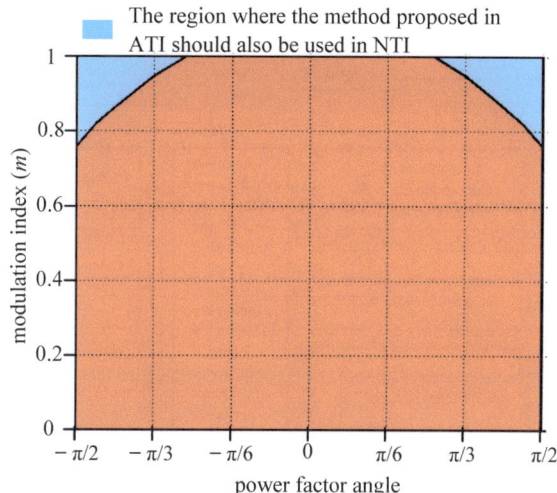

Fig. 8.48 The region where the fault-tolerant control proposed in ATI should also be applied in NTI

However, $u_{dc,bias}$ is different from each other under different m and pf. Therefore, an online estimation scheme is proposed to obtain a more general value of $u_{dc,bias}$. This scheme is realized by the following formula

$$u_{dc,bias} = \begin{cases} U_{p,max} - U_{p,min} & \text{if } U_{p,max} + U_{p,min} > U_{dc} \\ U_{p,min} - U_{p,max} & \text{if } U_{p,max} + U_{p,min} \leq U_{dc} \end{cases} \quad (8.71)$$

where the $U_{p,max}$ and $U_{p,min}$ are the maximum and minimum values of U_p in the last fundamental period of the ac voltage, respectively.

Figure 8.49 shows the block diagram of the proposed fault-tolerant control strategy when OCF occurs at switch S_{a1} or S_{a2}. The fault-tolerant control algorithm when OCF occurs to other switches can be easily obtained using the similar operating principle.

Step 3: Performance and Comparison of Fault-Tolerant Controls Based on Discontinuous CB-PWM

Performance and comparison are given to illustrate the effectiveness and the advantages of the proposed method in fault-tolerant control, NPV oscillation suppression, and switching losses reduction. The main performance includes the following five cases: AR with different modulation index and power factor, the capability of fault-tolerant control, the control capability of NPV on dc unbalance and ac unbalance, effectiveness of switching losses reduction, and average current THD.

Case 1: AR with Different Modulation Index and Power Factor

When OCF occurs at S_{a1}, according to the combinations of pf and m, the AR of the HBS-TC is shown in the shaded area of Fig. 8.50. The phase difference is caused by the pf and filter impedance. When only the pf is considered, the proposed HBS-TC is feasible over the entire factor range when $m \leq 0.5$. By increasing m from 0.5, the

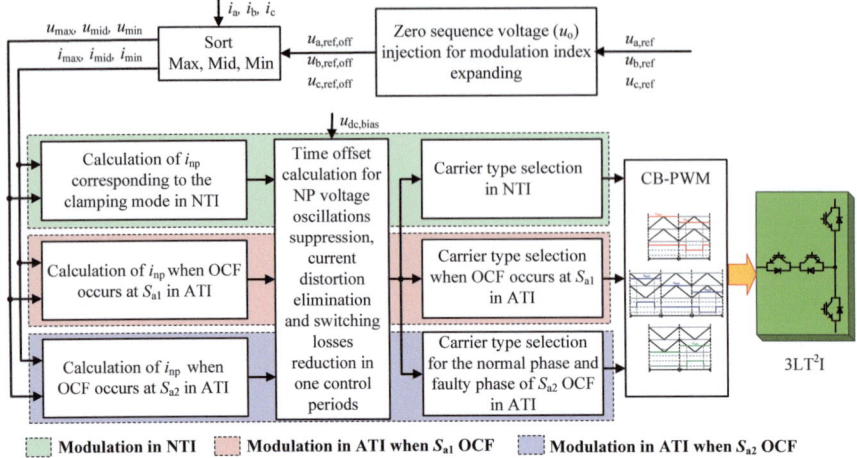

Fig. 8.49 Block diagram of the proposed fault-tolerant control with NPV balance and loss reduction

AR of the HBS-TC decreases. While in some applications where a large current is required, the filter impedance cannot be neglected. The AR is decreased when the $3LT^2I$ operates with lagging pf, while increased with leading pf operation. If the rectifier does not operate within the applicable range, we should properly regulate the m and pf to avoid current distortion when OCF occurs.

Case 2: Capability of Fault-Tolerant Control

Figure 8.51a shows the waveforms of the input current, phase voltage (u_{AO}), dc-link voltage and NPV when OCF occurs at S_{a1}, where $m = 0.55$, $pf = 0.8$. When OCF occurs at S_{a1}, the three-phase current is distorted in ATI (time interval B) due to the impossible current path. In addition, the NPV dc unbalance appears. After 0.4 s, the HBS-TC is applied, the impossible switching state [P] is avoided in ATI, and thus, the

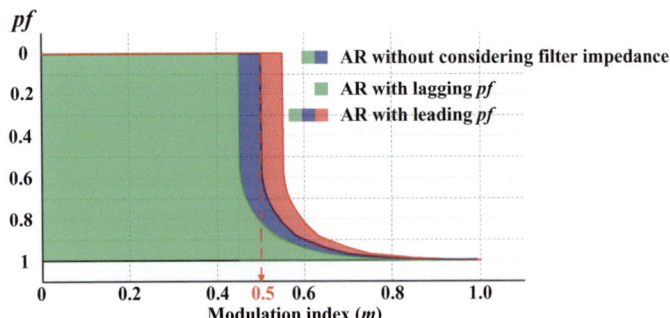

Fig. 8.50 AR corresponding to modulation index and power factor when S_{a1} OCF using the HBS-TC method considering the effect of the filter impedance

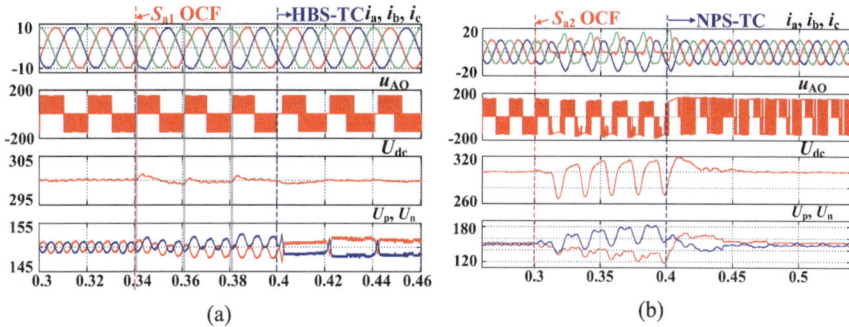

Fig. 8.51 The waveforms under different fault-tolerant control methods **a** when S_{a1} OCF occurs ($m = 0.55, pf = 0.8$), **b** when S_{a2} OCF occurs ($m = 0.87, pf = 0.8$)

distortion of the three-phase current is eliminated. Besides, the NPV dc unbalance disappears, due to the use of proper time offset u_{off} and the dc bias $u_{dc,bias}$.

Figure 8.51b shows the waveforms when OCF occurs at S_{a2} in the case of $m = 0.87$ and $pf = 0.8$. The three-phase current is distorted and the dc-link voltage has a ripple which is about 40 V. When the proposed NPS-TC is applied after 0.4 s, since the two-level modulation strategy is used, the impossible switching state [O] is avoided to generate the pole voltage of the faulty phase during the ATI. As a result, the distortion of the input current is eliminated, and dc-link voltage ripple as well as the NPV are well controlled.

Case 3: Control Capability of NPV on dc Unbalance and ac Unbalance

Figure 8.52 shows the NPV waveforms with the conventional CVA-TC and the HBS-TC when OCF occurs at S_{a1}. Figure 8.53 shows the NPV waveforms when 3LSO-TC, OP2LS-TC and NPS-TC, are implemented under S_{a2} faulty condition. In this test, the NPV dc deviation is set to 30 V. At 0.3 s, the fault-tolerant controls are applied. As shown in Fig. 8.52, when OCF occurs at S_{a1}, compared with CVA-TC, the HBS-TC has an extremely short dynamic response in NPV dc unbalance control under the two cases: $u_{max} > 0.5$ and $u_{max} \leq 0.5$, due to the proposed NPV balance algorithm. As it can be seen in Fig. 8.53, if OCF occurs at S_{a2}, compared with the other fault-tolerant controls, a relatively short NPV dc unbalance elimination process is achieved by the proposed NPS-TC under both conditions, since the NPV balance is seldom considered in the two fault-tolerant controls.

Figure 8.54 shows the fault-tolerant controls' performance in NPV ac unbalance suppression when OCF occurs at S_{a1} and S_{a2}, respectively. Compared with the CVA-TC, the magnitude of the NPV oscillations is reduced to half of that after the HBS-TC is applied. In OP2LS-TC, the two-level modulation strategy is used in the faulty phase. In the 3LSO-TC, the three-level modulation method is maintained in sectors as much as possible. However, in both methods, the NPV ac unbalance is never considered. And hence, the proposed NPS-TC maintains the best performance in NPV ac unbalance suppression when OCF occurs at S_{a2}. After the NPS-TC is applied,

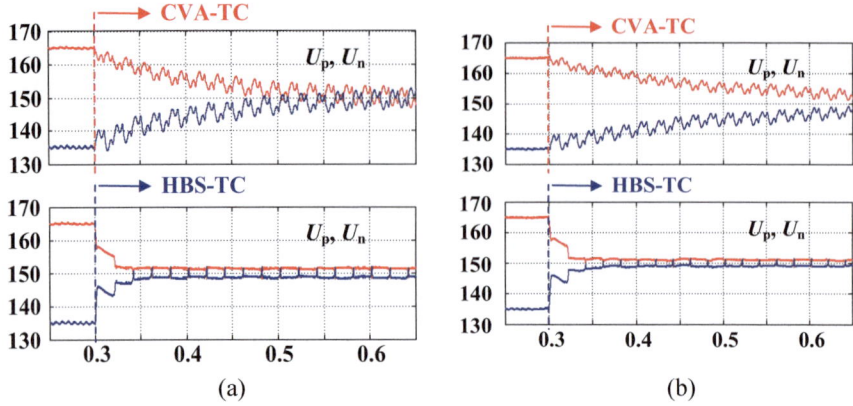

Fig. 8.52 Control capability of NPV on dc unbalance when OCF occurs at S_{a1} under different conditions **a** $m = 0.55$, $pf = 0.8$, **b** $m = 0.435$, $pf = 0.8$

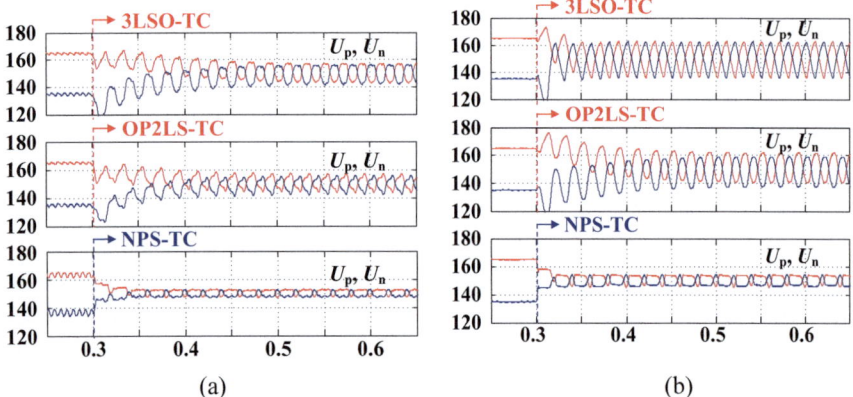

Fig. 8.53 Control capability of NPV on dc unbalance when OCF occurs at S_{a2} under different conditions **a** $m = 0.87$, $pf = 0.8$, **b** $m = 0.435$, $pf = 0.8$

the NPV ac unbalance is reduced to about 1/3 of that in the 3LSO-TC and 2/5 of that in the OP2LS-TC.

For further understanding, analysis and investigation on normalized amplitude of the NPV ac unbalance under various m and pf are carried out.

When OCF occurs at S_{a1}, normalized amplitudes of the NPV ac unbalance with different control methods are shown in Fig. 8.55. In which, the filter impedance is not considered and the magnitude of the NPV ac unbalance in the area where the HBS-TC is infeasible is set to zero. In the AR, the NPV ac unbalance in the proposed HBS-TC is reduced to at least a half of that in the CVA-TC method.

Figure 8.56 shows the comparison of the three fault-tolerant controls in terms of NPV ac unbalance under the condition of S_{a2} OCF. When the $3LT^2I$ operates with

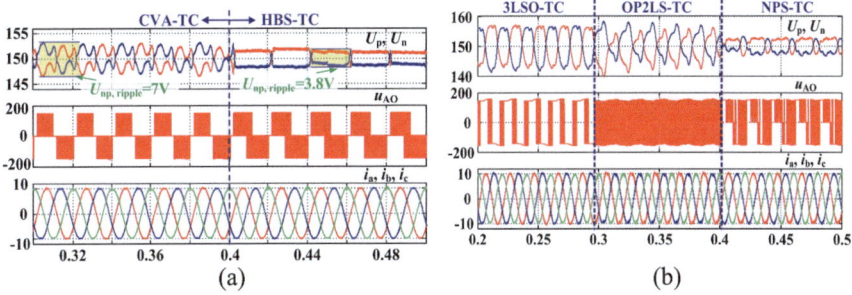

Fig. 8.54 Control capability of NPV on ac unbalance using different control methods **a** OCF occurs at S_{a1}, **b** OCF occurs at S_{a2}

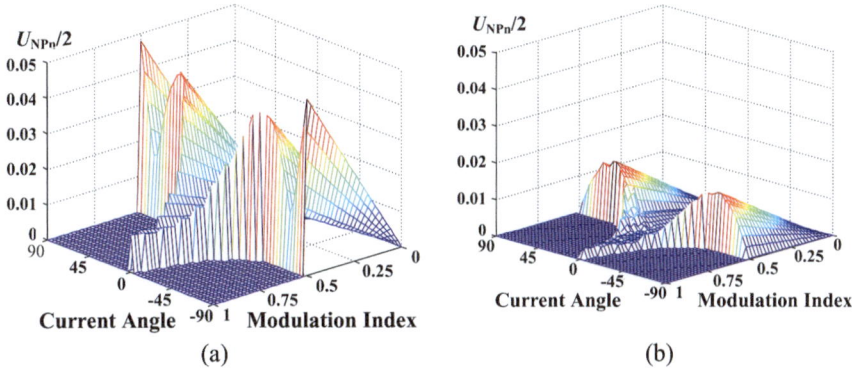

Fig. 8.55 Normalized amplitude of NPV ac unbalance using different control methods when OCF occurs at S_{a1} **a** CVA-TC, **b** HBS-TC

a higher modulation index, the largest NPV ac unbalance will be generated in the 3LSO-TC. On the contrary, if the $3LT^2I$ operates with a lower modulation index, the OP2LS-TC will maintain the largest NPV ac unbalance. Compared with the two methods, it can be clearly seen that the proposed NPS-TC has the smallest NPV ac unbalance.

Case 4: Effectiveness of Switching Losses Reduction

The total losses under different operating conditions are compared, which is carried out by PLECS simulation. When OCF occurs at S_{a1}, as shown in Fig. 8.57, under the two cases $u_{max} > 0.5$ and $u_{max} \leq 0.5$, the conduction losses below the red dotted line for the two methods are almost the same. Nevertheless, the switching loss is largely reduced in the proposed HBS-TC, due to the reduction of the switching event within and between the switching periods. Therefore, the loss reduction of the HBS-TC is confirmed.

Figure 8.58 shows the total loss of the fault-tolerant controls under different operating conditions when OCF occurs at S_{a2}. As it can be seen, compared with

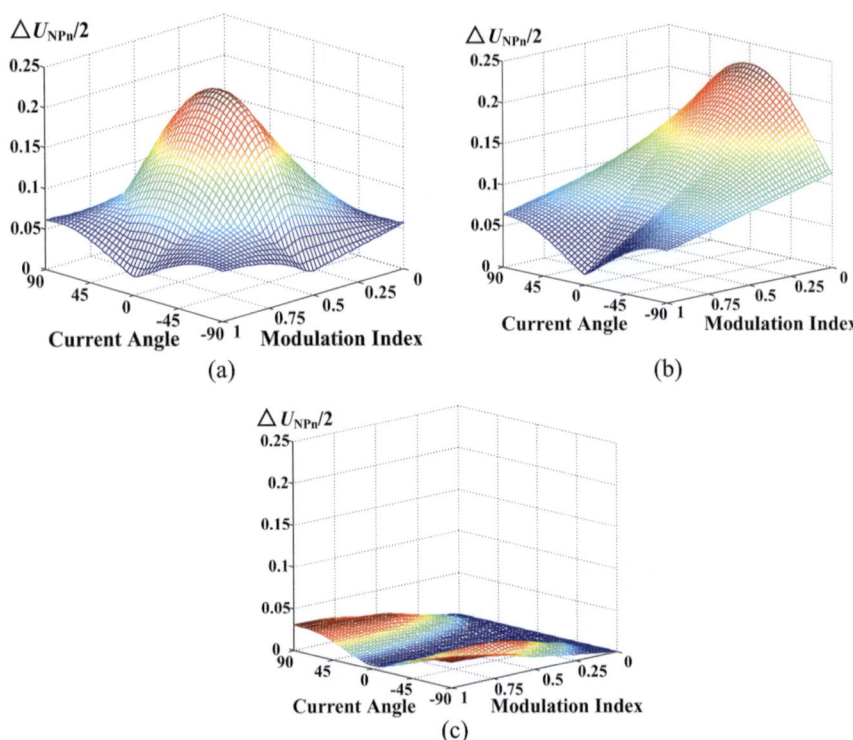

Fig. 8.56 Normalized amplitude of NPV ac unbalance using different control methods when OCF occurs at S_{a2} **a** 3LSO-TC, **b** OP2LS-TC, **c** NPS-TC

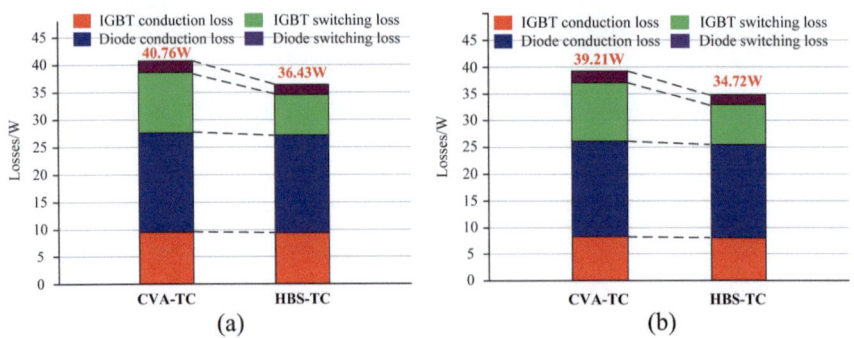

Fig. 8.57 The total loss comparison using different control methods when OCF occurs at S_{a1} under different modulation index **a** $m = 0.435$, $pf = 0.8$, **b** $m = 0.55$, $pf = 0.8$

8.2 Fault-Tolerant Control Technology

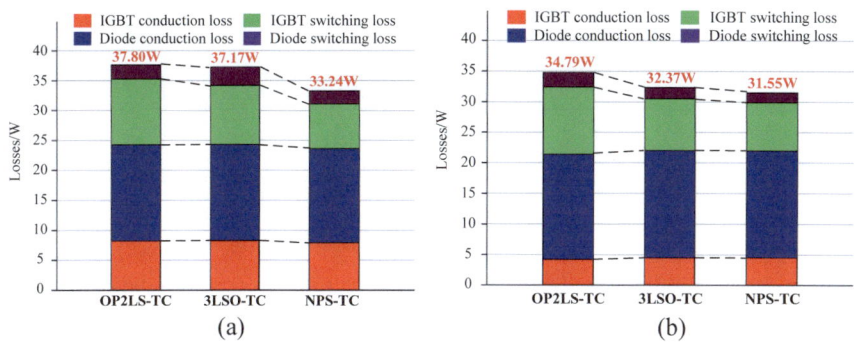

Fig. 8.58 The total loss comparison using different control methods when OCF occurs at S_{a2} under different modulation index **a** $m = 0.435, pf = 0.8$, **b** $m = 0.87, pf = 0.8$

the OP2LS-TC and 3LSO-TC, the switching losses of NPS-TC are the smallest among the three methods, and thus, the NPS-TC maintains the best performance in losses reduction.

Case 5: Average Current THD

The averaged input current THDs of the three phases under different combinations of m and pf to verify the performance of the proposed HBS-TC and NPS-TC are given. As shown in Table 8.16, when OCF occurs at S_{a1}, the proposed HBS-TC is advantageous in current THD compared with the method of CVA-TC. When OCF occurs at S_{a2}, the averaged current THD of the three phases is shown in Table 8.17. In OP2LS-TC, the two-level modulation strategy is used in the faulty phase. Besides, this method has a large NPV oscillation. Hence, the averaged current THD of the OP2LS-TC is the largest among the three fault-tolerant controls. In 3LSO-TC, the averaged current THD slightly decreases, because of the wide application range of the three-level switching method. The proposed NPS-TC maintains the best performance in current THD, due to the reduction of the NPV ac unbalance and applicable range of the two-level modulation method in the faulty phase.

In this section, a discontinuous CB-PWM fault-tolerant control strategy for $3LT^2I$ operated in charging mode considering NPV balance and loss reduction is proposed. This algorithm is demonstrated by dividing the fundamental period into two parts: NTI and ATI. In NTI, the NPV and loss reduction within a switching period are well controlled by the proposed time offset injection method. Besides, in ATI, the fault-tolerant control for half bridge switches OCF is achieved by introducing a restriction

Table 8.16 The averaged current THD using different fault-tolerant control methods in case of S_{a1} fault

Condition	CVA-TC (%)	HBS-TC (%)
$m = 0.435, pf = 0.8$	2.90	2.05
$m = 0.55, pf = 0.8$	2.59	1.95

Table 8.17 The averaged current THD using different fault-tolerant control methods in case of S_{a2} fault

Conditions	3LSO-TC (%)	OP2LS-TC (%)	NPS-TC (%)
$m = 0.435, pf = 0.8$	4.34	5.08	3.07
$m = 0.87, pf = 0.8$	3.41	3.97	3.18

of the injected time offset. While if OCF occurs at NP switches, the two-level modulation strategy is utilized in the faulty phase to avoid the impossible switching state [O]. Additionally, under this condition, the time offset is revised correspondingly to realize NP voltage control and loss reduction within a switching period. In addition, the extra switching events both in NTI and ATI are fully considered and reduced by employing an appropriate carrier type.

8.3 Summary

In the PV inverter system, the power semiconductor switch is one of the most fragile components due to the limitations of the material and the manufacturing process. Statistics indicate that about 34% of the power converter system's failures are caused by the power device and soldering failures.

To improve the reliability of the PV inverter, OCF diagnosis and software-based fault-tolerant control strategies are presented in this chapter. The proposed OCF diagnosis method is based on the instant voltage error of the line-to-line voltages and the dc-link voltage, which has the advantages of wide power factor application range and fast response. The software-based fault-tolerant control strategies are proposed from the perspective of PWM method. The essence of fault-tolerant control is to avoid using the current paths affected by the fault power semiconductor switches, so as to ensure the correct output current. For the $3LT^2I$ with specific power flowing directions, three fault-tolerant control methods are presented in this chapter. They are the methods based on SVPWM method, based on continuous CB-PWM method, and discontinuous CB-PWM method.

To facilitate the reader's understanding, the theoretical analyses and implementation process of each control technology are illustrated step by step by the author. Moreover, the comprehensive waveforms are given for intuitive understanding of control algorithms.

References

1. S. Yang, D. Xiang, A. Bryant, et al., Condition monitoring for device reliability in power electronic converters: a review. IEEE Trans. Power Electron. **25**(11), 2734–2752 (2010)
2. S. Xu, J. Zhang, J. Hang, Investigation of a fault-tolerant three-level T-type inverter system. IEEE Trans. Ind. Appl. **53**(5), 4613–4623 (2017)
3. J.S. Lee, K.B. Lee, An open-switch fault detection method and tolerance controls based on SVM in a grid-connected T-type rectifier with unity power factor. IEEE Trans. Ind. Electron. **61**(12), 7092–7104 (2014)
4. L.M.A. Caseiro, A.M.S. Mendes, Real-time IGBT open-circuit fault diagnosis in three-level neutral-point-clamped voltage-source rectifiers based on instant voltage error. IEEE Trans. Ind. Electron. **62**(3), 1669–1678 (2015)
5. J. Chen, C. Zhang, X. Xing, A. Chen, C. Du, An open-circuit fault diagnosis method for t-type three-level rectifiers, in *2019 IEEE Energy Conversion Congress and Exposition (ECCE)*, (2019), pp. 5502–5506
6. S. Xu, J. Zhang, J. Hang, Investigation of a fault-tolerant three-level T-type inverter system, in *2015 IEEE Energy Conversion Congress and Exposition (ECCE)*, (2015), pp. 1632–1638
7. U.M. Choi, K.B. Lee, F. Blaabjerg, Diagnosis and tolerant strategy of an open-switch fault for T-type three-level inverter systems. IEEE Trans. Ind. Appl. **50**(1), 495–508 (2014)
8. U.M. Choi, F. Blaabjerg, K.B. Lee, Reliability improvement of a T-type three-level inverter with fault-tolerant control strategy. IEEE Trans. Power Electron. **30**(5), 2660–2673 (2015)
9. J. Pou, R. Pindado, D. Boroyevich, P. Rodriguez, Evaluation of the low-frequency neutral-point voltage oscillations in the three-level inverter. IEEE Trans. Ind. Electron. **52**(6), 1582–1588 (2005)
10. J. Chen, C. Zhang, A. Chen, X. Xing, Fault-tolerant control strategies for t-type three-level inverters considering neutral-point voltage oscillations. IEEE Trans. Ind. Electron. **66**(4), 2837–2846 (2019)
11. J.S. Lee, U.M. Choi, K.B. Lee, Comparison of tolerance controls for open-switch fault in a grid-connected T-type rectifier. IEEE Trans. Power Electron. **30**(10), 5810–5820 (2015)
12. J.S. Lee, K.B. Lee, F. Blaabjerg, Open-switch fault detection method of a back-to-back converter using NPC topology for wind turbine systems. IEEE Trans. Ind. Appl. **51**(1), 325–335 (2015)
13. J.S. Lee, K.B. Lee, Open-circuit fault-tolerant control for outer switches of three-level rectifiers in wind turbine systems. IEEE Trans. Power Electron. **31**(5), 3806–3815 (2016)
14. J. Chen, C. Zhang, A. Chen, X. Xing, F. Gao, A carrier-based fault-tolerant control strategy for t-type rectifier with neutral-point voltage oscillations suppression. IEEE Trans. Power Electron. **34**(11), 10988–11001 (2019)
15. J.S. Lee, S. Yoo, K.B. Lee, Novel discontinuous PWM method of a three-level inverter for neutral-point voltage ripple reduction. IEEE Trans. Ind. Electron. **63**(6), 3344–3354 (2016)
16. U.M. Choi, H.H. Lee, K.B. Lee, Simple neutral-point voltage control for three-level inverters using a discontinuous pulse width modulation. IEEE Trans. Energy Convers. **28**(2), 434–443 (2013)
17. J. Chen, C. Zhang, X. Xing, A. Chen, A fault-tolerant control strategy for T-type three-level rectifier with neutral point voltage balance and loss reduction. IEEE Trans. Power Electron. **35**(7), 7492–7505 (2020)

Open Access This chapter is licensed under the terms of the Creative Commons Attribution 4.0 International License (http://creativecommons.org/licenses/by/4.0/), which permits use, sharing, adaptation, distribution and reproduction in any medium or format, as long as you give appropriate credit to the original author(s) and the source, provide a link to the Creative Commons license and indicate if changes were made.

The images or other third party material in this chapter are included in the chapter's Creative Commons license, unless indicated otherwise in a credit line to the material. If material is not included in the chapter's Creative Commons license and your intended use is not permitted by statutory regulation or exceeds the permitted use, you will need to obtain permission directly from the copyright holder.

MIX
Papier aus verantwortungsvollen Quellen
Paper from responsible sources
FSC® C105338

If you have any concerns about our products,
you can contact us on
ProductSafety@springernature.com

In case Publisher is established outside the EU,
the EU authorized representative is:
**Springer Nature Customer Service Center GmbH
Europaplatz 3, 69115 Heidelberg, Germany**

Printed by Libri Plureos GmbH
in Hamburg, Germany